Quantum Geometry, Matrix Theory, and Gravity

Building on mathematical structures familiar from quantum mechanics, this book provides an introduction to quantization in a broad context before developing a framework for quantum geometry in Matrix theory and string theory. Taking a physics-oriented approach to quantum geometry, that framework helps explain the physics of Yang–Mills-type matrix models, leading to a quantum theory of spacetime and matter. This novel framework is then applied to Matrix theory, which is defined through distinguished maximally supersymmetric matrix models related to string theory. A mechanism for gravity is discussed in depth, which emerges as a quantum effect on quantum spacetime within Matrix theory. Using explicit examples and exercises, readers will develop a physical intuition for the mathematical concepts and mechanisms. This book will benefit advanced students and researchers in theoretical and mathematical physics, and provides a useful resource for physicists and mathematicians interested in the geometrical aspects of quantization in a broader context.

Harold C. Steinacker is a senior scientist at the University of Vienna. He obtained the Ph.D. degree in physics at the University of California at Berkeley and has held research positions at several universities. He has published more than 100 research papers, contributing significantly to the understanding of quantum geometry and matrix models in fundamental physics.

T0338720

Quantum Geometry, Matrix Theory, and Gravity

Harold C. Steinacker

University of Vienna, Austria

CAMBRIDGE UNIVERSITY PRESS

Shaftesbury Road, Cambridge CB2 8EA, United Kingdom

One Liberty Plaza, 20th Floor, New York, NY 10006, USA

477 Williamstown Road, Port Melbourne, VIC 3207, Australia

314–321, 3rd Floor, Plot 3, Splendor Forum, Jasola District Centre,
New Delhi – 110025, India

103 Penang Road, #05–06/07, Visioncrest Commercial, Singapore 238467

Cambridge University Press is part of Cambridge University Press & Assessment,
a department of the University of Cambridge.

We share the University's mission to contribute to society through the pursuit of
education, learning and research at the highest international levels of excellence.

www.cambridge.org
Information on this title: www.cambridge.org/9781009440783

DOI: 10.1017/9781009440776

First published 2024

A catalogue record for this publication is available from the British Library

A Cataloging-in-Publication data record for this book is available from the Library of Congress

ISBN 978-1-009-44078-3 Hardback

To Fiona

Contents

Preface

This book explores physical and mathematical aspects of quantum geometry and matrix models in the context of fundamental physics. This is motivated by efforts to incorporate gravity into a comprehensive quantum theory, which is expected to entail some quantization of spacetime and geometry.

In the first part of this book, a mathematical framework for quantum geometry is developed, using the mathematical concepts of quantum mechanics to describe spacetime and geometry. Gauge theory on quantum spaces is defined through matrix models, which lead to dynamical quantum geometries naturally interpreted in terms of gravity. Consistency of the quantum theory then leads to Matrix theory, which is defined in terms of a distinguished maximally supersymmetric matrix model related to string theory. Matrix theory is considered as a possible foundation for a quantum theory of fundamental interactions including gravity. In the last and more exploratory part of the book, the mechanism for emergent gravity arising within Matrix theory is elaborated in some detail. This leads directly to gravity in $3 + 1$ dimensions, and provides an alternative to the more conventional approach to string theory.

The term "quantum geometry" is used here to indicate certain noncommutative mathematical structures replacing classical spacetime, analogous to the quantum mechanical description of phase space in terms of operators. In a mathematical context, this is often denoted as "noncommutative geometry," which is a priori independent of possible $\hbar$ corrections to the geometry arising from quantum theory. However, the present book is not an introduction to noncommutative geometry. Rather, it provides a systematic development of a specific physics-oriented approach to this topic, starting with the underlying mathematical structures and culminating in a specific model known as "Matrix theory." This physical theory is defined in terms of distinguished maximally supersymmetric matrix models known as Ishibashi–Kawai–Kitazawa–Tsuchiya (IKKT) and Banks–Fischer–Shenker–Susskind (BFSS) models, which are related to string theory. The main emphasis is on the IKKT model, which is best suited to exhibit a novel mechanism for gravity on quantum spacetime. The reason for this particular focus is simple: among all Yang–Mills-type matrix models, it is the unique one that is free from UV/IR pathologies upon quantization, leading to sufficiently local low-energy physics. The relation to other approaches is briefly discussed in the given context, and some supplementary topics and steps are delegated to exercises.

The book starts with a thorough discussion of the mathematical structures and tools to describe geometry through matrices, with emphasis on explicit examples known as fuzzy spaces. These are quantized symplectic spaces with finitely many degrees of freedom per unit volume. These results are then applied to physical models analogous to quantum field theory, leading to a discussion of noncommutative field theory as relevant in the context of matrix models. In particular, the novel nonlocal and stringy features are exhibited, which are often overlooked. Gauge theories are defined in terms of matrix models, which can alternatively be viewed as models for dynamical quantum geometry. Reconciling these points of view leads to a specific picture of emergent gravity, which as a classical theory differs significantly from general relativity. The appropriate geometrical structures are discussed in detail, including a particularly interesting type of covariant quantum spaces. In the last chapters, the quantization of these matrix models in terms of $\hbar$ is discussed. The explicit geometrical form of the one-loop effective action is obtained using the previously developed tools, exhibiting the Einstein–Hilbert action as a quantum effect. This suggests that gravity may be understood as a quantum effect on quantum spacetime in the framework of matrix models.

The book contains both well-known material as well as unpublished results, notably in Part III. Some of the more specialized or technical sections are marked with an asterisk and may be omitted at first reading. Parts of this book are suitable for lecture courses, notably Chapter 2 on quantization of symplectic manifolds, possibly supplemented with parts of Chapter 3 and some of the physics-related topics in the later chapters. This was tested in a two-semester course at the University of Vienna in 2021/2022, leading to numerous improvements due to attentive students, to whom I would like to express my gratitude.

I would also like to thank Emmanuele Battista, Jun Nishimura, Jurai Tekel, and Tung Tran who helped to improve the book in many ways, as well as Masanori Hanada for collaboration on Sections 5.5 and 13.3. Special thanks go to Vince Higgs from Cambridge University Press for his support during the publishing process. Finally, this book could not have been written without the help and inspiration of many colleagues and teachers during my attempts to understand fundamental physics, including Nima Arkani-Hamed, Chong-Sun Chu, Stefan Fredenhagen, Harald Grosse, Pei-Ming Ho, Hikaru Kawai, John Madore, Bogdan Morariu, Peter Schupp, Julius Wess, George Zoupanos, and Bruno Zumino, among many others.

The trouble with spacetime

Nature is governed by quantum mechanics. This was the most profound insight of twentieth-century physics, which has been established by overwhelming experimental evidence, but it was also driven by theoretical inconsistencies of classical physics. Such inconsistencies arise e.g. in the description of atoms coupled to electromagnetic radiation, which is classically unstable. This problem is cured in quantum mechanics because phase space is quantized. Indeed, all phenomena accessible to terrestrial experiments appear to be consistently described by the standard model of elementary particle physics, which is a renormalizable quantum field theory.

However, our efforts to incorporate gravity into a comprehensive quantum theory are plagued by profound difficulties. The currently accepted description of gravity is provided by general relativity (GR), which is a classical theory describing spacetime as a manifold, equipped with a dynamical metric with Lorentzian signature. This framework provides the stage for matter and fields, and hence for all known physics. General relativity also suffers from inconsistencies, such as the singularities in the center of black holes. However, GR does not seem to allow a straightforward quantization.

A simple way to see the problem is as follows. The standard model of elementary particles and interactions is governed by dimensionless coupling constants. This means that the coupling constants are the same at any distances (up to quantum corrections) and remain small. In more technical terms, these theories are called renormalizable, which means that they can be defined consistently as quantum theories at any length scale. However, gravity is governed by the dimensionful Newton constant $G_N \sim L_{Pl}^2$, where $L_{Pl} \sim 10^{-33}$ cm is the Planck length. This means that gravity becomes strongly coupled at short distances, where the quantum structure of matter and fields is significant. Since spacetime is determined by matter through the Einstein equations, that quantum structure must be strongly imprinted to spacetime at short scales. Indeed by naively applying the Einstein equations,[1] the quantum fluctuations of any field confined in a volume L_{Pl}^3 would entail strong curvature fluctuations with associated Schwarzschild radius $r_S \sim L_{Pl}$, so that the concept of a classical spacetime becomes meaningless at that scale.

In view of these issues, it seems unreasonable to insist that spacetime remains classical at all scales. It is more plausible that spacetime, matter, and fields should have a unified quantum description in a fundamental theory, possibly based on different degrees of freedom. This leads to the idea of quantizing spacetime and geometry, which goes back to the early days of quantum field theory[2] and is a recurring theme in various approaches to quantum gravity including string theory. However, the proper formulation of such a thoroughly "quantum" theory of gravity is far from obvious. In view of the aforementioned difficulties, it seems pointless trying to adapt some of the standard formulations of GR into a generalized framework. Rather, there should be a special, simple, and thoroughly well-defined definition of such an underlying theory, even if it looks unusual to the traditional eye.

Such a situation is not new in physics: the theory of strong interactions has an effective low-energy description via chiral perturbation theory, which is not renormalizable, and does not make sense as a quantum theory beyond a certain scale. The correct fundamental description is given by quantum chromodynamics, which is formulated in terms of totally different degrees of freedoms (called gluons), and makes sense as a quantum theory at any scale. It would be a bad idea to try to quantize the low-energy effective theory beyond its low-energy regime of applicability. A more mundane example to illustrate the point is given by the Navier–Stokes equations, which provide a perfectly adequate description of classical hydrodynamics above the molecular scale. We know that hydrodynamics is part of a comprehensive quantum theory, which is well understood, but it would be utter nonsense to try to quantize the Navier–Stokes equations directly beyond the molecular scale, even with the most sophisticated mathematical methods. Since the Planck scale plays

---

[1] There are many variants of this argument, which is more or less folklore. For a more mathematically refined discussion, see [62].

[2] Perhaps the first specific such proposal was published by Snyder [172] in 1947.

a fundamental role in gravity, it is plausible that GR provides its effective description only up to L_{Pl}, while the fundamental formulation of the theory may take a very different form.

These considerations suggest that spacetime should be described as *a dynamical physical system with intrinsic quantum structure*, treated at the same footing as the fields that live on it. Gravity should then arise through a universal metric, which governs the emergent physics. We will see that this idea can be realized through simple matrix models, where spacetime arises as solution, and physical fields arise as fluctuations of the spacetime structure. In other words, spacetime along with physical fields will *emerge* from the basic matrix degrees of freedom. If successful, this would take the idea of unification in physics one step further, as the new description is simpler than the previous one(s). There is in fact one preferred model within the framework of matrix models, known as IKKT or IIB model, which turns out to be closely related to (super)string theory.

String theory is a profound approach to reconcile gravity with quantum mechanics, which tackles the problem at its roots by giving up the concept of localized particles and fields in classical spacetime. However, there is no clear-cut comprehensive definition, and its standard world-sheet formulation leads to a number of issues. Consistency of the quantum theory requires target space to be 10-dimensional, in obvious clash with observation. This problem is typically addressed by postulating that 6 of these 10 (or rather $9 + 1$) spacetime dimensions should be curled up, or "compactified," leading effectively to a $3 + 1$ dimensional spacetime. This is a fruitful idea, which led to a wealth of new insights into the resulting quantum field theories. However, no convincing mechanism has been found which could select a realistic spacetime among the zillions of different possible compactifications. The resulting vast collection of possible worlds is dubbed the "landscape."

Avoiding this landscape problem is one motivation for the approach in this book. The matrix model(s) under consideration are closely related to string theory and thus inherit much of its magic, yet they lead to a different mechanism for $3 + 1$ dimensional spacetime and emergent (quantum) gravity, without requiring target space compactification and thereby avoiding the landscape problem. The foundations of that theory are exhibited in this book, but its detailed physical properties and its physical viability remain to be understood.

Quantum geometry and Matrix theory

This book provides an introduction to a specific approach toward a quantum theory of spacetime, matter, and gravity based on matrix models. The framework of matrix models is extremely simple, yet it provides all ingredients for a fundamental physical theory. This leads to a nonstandard formulation of field theory and gravity where the required mathematical structures emerge naturally, rather than being imposed by hand. The aim of this book is to provide the appropriate tools to understand the physical significance of these models.

All fundamental interactions are described by some type of gauge theory. Gauge theories provide the only known consistent description of quantum fields with spin. Spin 1 gauge fields include photons, gluons, and W- and Z-bosons and are described by Yang–Mills (or Maxwell) gauge fields $A_\mu(x)$. The requirement of unitarity and the absence of ghosts (i.e. negative-norm states) in quantum theory entails certain constraints for these fields, and a consistent description is achieved through gauge theories, which allow to consistently get rid of unphysical components. The same applies to gravitons in GR, which is governed by a different type of gauge theory.

There have been many attempts to formulate gauge theories on various quantum spaces, which typically encounter all sorts of issues. Rather than attempting to discuss and compare different approaches, we will focus on one approach based on matrix models. There are many reasons for this choice, but the first and foremost is simplicity; after all, a fundamental theory ought to be simple. Despite their simplicity, the models are rich enough to accommodate the basic structures required in physics. Most interestingly, they provide a novel mechanism for (quantum) gravity, which may overcome the problems in the traditional approaches.

One important message of this book is that matrix models preserve much of the power and magic of string theory. In fact, the preferred models under consideration were proposed as a constructive way to define string theory, or at least some sector of it. On the other hand, they also provide a novel mechanism for gravity on certain $3 + 1$ dimensional quantized "brane" solutions or backgrounds, which play the role of spacetime. No compactification of target space is required, thereby avoiding the landscape problem. These backgrounds are $3 + 1$ dimensional quantum geometries, and the model is expected to provide an intrinsic mechanism to choose a preferred one. Although that selection mechanism is not yet understood, matrix models certainly provide a clear-cut framework and definition, and they can even be simulated on a computer.

We will see that within the class of matrix models under consideration, there is one model that is special and preferred, and the alluded mechanism only works in this unique model. This model is known as the IKKT or IIB matrix model [102], which was first proposed in the context of string theory. We will understand the reasons for its unique standing on independent grounds, and develop a framework to understand the resulting physics. This model and its resulting physics will be denoted as *Matrix theory* in this book, even though that name is often associated with a related model of matrix quantum mechanics known as the BFSS model [21, 57]. These two models are in fact closely related and should be considered as siblings, although the proposed mechanism for emergent spacetime and gravity is better understood within the IKKT model. In any case, these models are sufficiently remarkable and rich to warrant serious efforts to understand their physics.

Part I

Mathematical background

1 Differentiable manifolds

We recall some basic concepts of differentiable manifolds, calculus, and classical geometry. This is not intended as an introduction, but to collect some important facts and to establish the notation. Chapter 2 will provide a lightning review of some basic concepts of Lie groups, including (co)adjoint orbits, which will play an important role later. However, most of this book is accessible without sophisticated mathematics, and readers familiar with the basic concepts can skip this section.

Let $\mathcal{M}$ be an n-dimensional *differentiable manifold* defined in terms of local coordinate charts and their transition functions. *Vector fields* V on $\mathcal{M}$ are best viewed as derivations acting on the algebra of smooth functions $\mathcal{C}(\mathcal{M})$, i.e.

$$V[fg] = fV[g] + gV[f], \qquad f,g \in \mathcal{C}(\mathcal{M}). \tag{1.0.1}$$

In terms of local coordinates x^μ, they can be written as $V = V^\mu \frac{\partial}{\partial x^\mu} \equiv V^\mu \partial_\mu$. Vector fields can also be viewed as sections of the *tangent bundle* $T\mathcal{M}$, which is dual to the *cotangent bundle* $T^*\mathcal{M}$, whose sections are one-forms $\alpha = \alpha_\mu dx^\mu \in \Omega^1(\mathcal{M})$.

Differential forms

The vector space of differential forms or k-forms $\Omega^k(\mathcal{M})$ on $\mathcal{M}$ consists of elements of the form

$$\omega = \frac{1}{k!}\omega_{\mu_1...\mu_k}dx^{\mu_1} \wedge ... \wedge dx^{\mu_k} \qquad \in \Omega^k(\mathcal{M}). \tag{1.0.2}$$

Here $\omega_{\mu_1...\mu_k}(x)$ is totally antisymmetric, and $\wedge$ denotes the antisymmetric wedge product

$$dx^\mu \wedge dx^\nu = -dx^\nu \wedge dx^\mu, \tag{1.0.3}$$

which is sometimes suppressed. The Einstein sum convention will be used throughout. This wedge product defines an algebra structure on the space of all differential forms, denoted by $\Omega^*(\mathcal{M})$. The *exterior derivative*

$$d \colon \Omega^n(\mathcal{M}) \to \Omega^{n+1}(\mathcal{M}) \tag{1.0.4}$$

is defined by

$$df = (\partial_\mu f)\,dx^\mu, \tag{1.0.5}$$

and the graded Leibniz rule,

$$d(\alpha \wedge \beta) = (d\alpha) \wedge \beta + (-1)^p \alpha \wedge d\beta, \tag{1.0.6}$$

where α is a p-form and β is an arbitrary differential form, and imposing that $ddf = 0$ for $f \in \mathcal{C}(\mathcal{M})$. Then d satisfies more generally

$$d \circ d = 0. \tag{1.0.7}$$

In local coordinates,

$$d\omega = \frac{1}{k!} d\omega_{\mu_1 \ldots \mu_k} \wedge dx^{\mu_1} \wedge \ldots \wedge dx^{\mu_k} \tag{1.0.8}$$

for $\omega \in \Omega^k(\mathcal{M})$ as in (1.0.2).

The *interior product* or contraction of a vector field $V = V^\mu \partial_\mu$ with a one-form $\alpha = \alpha_\mu dx^\mu$ is defined through the dual evaluation

$$i_V \alpha = \langle V, \alpha \rangle = V^\mu \alpha_\mu \qquad \in \mathcal{C}(\mathcal{M}), \tag{1.0.9}$$

which is extended to k-forms through

$$i_V(\alpha \wedge \beta) = (i_V \alpha) \wedge \beta + (-1)^p \alpha \wedge i_V \beta. \tag{1.0.10}$$

Here α is a p-form and β is an arbitrary differential form. In local coordinates, this takes the form

$$i_V \Big(\frac{1}{k!} \alpha_{\mu_1 \ldots \mu_k} dx^{\mu_1} \wedge \ldots \wedge dx^{\mu_k} \Big) = \frac{1}{(k-1)!} V^\mu \alpha_{\mu \ldots \mu_k} dx^{\mu_2} \wedge \ldots \wedge dx^{\mu_k}. \tag{1.0.11}$$

Push-forward and pullback maps

Any smooth map

$$\phi: \quad \mathcal{M} \to \mathcal{N} \tag{1.0.12}$$

between two manifolds $\mathcal{M}$ and $\mathcal{N}$ defines by differentiation a *tangential map* or push-forward

$$d\phi: \quad T_x \mathcal{M} \to T_{\phi(x)} \mathcal{N}, \qquad (d\phi)(V)[f] = V[\phi^* f], \tag{1.0.13}$$

where $(\phi^* f)(y) = f(\phi(y))$ is the pullback of the function f from $\mathcal{N}$ to $\mathcal{M}$. Note that this map is a priori defined only point-wise. If $\phi: \mathcal{M} \to \mathcal{N}$ is a *diffeomorphism*, i.e. a bijective smooth map whose inverse is also smooth, then the push-forward defines a map from vector fields on $\mathcal{M}$ to vector fields on $\mathcal{N}$. If ϕ is not injective, then the push-forward of a vector field is not defined, since vectors on different points in $\mathcal{M}$ can be mapped to the same point in $\mathcal{N}$. This observation will play an important role in the higher-spin theories discussed in Section 5.

By duality, this push-forward map defines a *pullback map* for one-forms

$$\phi^*: \quad T^*_{\phi(x)} \mathcal{N} \to T^*_x \mathcal{M}, \tag{1.0.14}$$

through $\langle V, \phi^* \alpha \rangle = \langle (d\phi)(V), \alpha \rangle$, which extends to a map

$$\phi^*: \quad \Omega^*(\mathcal{N}) \to \Omega^*(\mathcal{M}). \tag{1.0.15}$$

In local coordinates x^μ on $\mathcal{N}$ and y^ν on $\mathcal{M}$, these maps reduce to the familiar covariant transformation laws for vectors and covectors. For example,

$$\phi^*(dx^\mu) = \frac{\partial x^\mu}{\partial y^\nu} dy^\nu,$$

$$(d\phi)\left(\frac{\partial}{\partial y^\mu}\right) = \frac{\partial x^\nu}{\partial y^\mu} \frac{\partial}{\partial x^\nu}, \qquad (1.0.16)$$

where $x^\mu(y) = \phi^* x^\mu = x^\mu(\phi(y))$ is understood.

Lie derivative

The *Lie derivative* $\mathcal{L}_V$ along a vector field V on $\mathcal{M}$ generalizes the action of V on functions $f \in \Omega^0(\mathcal{M})$ to an action on any forms $\omega \in \Omega^*(\mathcal{M})$. This is again a derivation, which satisfies

$$\mathcal{L}_V f = V[f] = i_V df$$
$$\mathcal{L}_V(\alpha \wedge \beta) = (\mathcal{L}_V \alpha) \wedge \beta + \alpha \wedge \mathcal{L}_V \beta$$
$$\mathcal{L}_V(d\omega) = d(\mathcal{L}_V \omega), \qquad \omega \in \Omega^*(\mathcal{M}). \qquad (1.0.17)$$

Cartan's magic formula then states that

$$\mathcal{L}_V \omega = (d i_V + i_V d)\omega. \qquad (1.0.18)$$

This formalism is particularly useful in the context of symplectic manifolds, which will play a central role in this book.

The Lie derivative $\mathcal{L}_V$ can be extended to act also on vector fields and general tensor fields, but this requires a different perspective. The idea is that any vector field V on $\mathcal{M}$ defines a flow

$$\phi: \quad \mathbb{R} \times \mathcal{M} \to \mathcal{M}$$
$$(t, x) \mapsto \phi_t(x) \qquad (1.0.19)$$

through

$$\frac{d}{dt}\phi_t(x) = V(x) \quad \in T_x\mathcal{M}. \qquad (1.0.20)$$

In other words, ϕ_t is the diffeomorphism that realizes the integral flow along the vector field V. This can be used to drag any tensorial fields along the flow via the differential map

$$d\phi_t: \quad T_x\mathcal{M} \to T_{\phi_t(x)}\mathcal{M} \qquad (1.0.21)$$

and similar for $T^*\mathcal{M}$; note that flows are always invertible. Then the Lie derivative is simply the derivative of any tensor fields along this flow, where vectors at different points are subtracted after transporting them to the same point along the flow. This leads to the following explicit formulas:

$$\mathcal{L}_V W = [V, W],$$
$$\mathcal{L}_V(X \otimes Y) = \mathcal{L}_V X \otimes Y + X \otimes \mathcal{L}_V Y, \qquad (1.0.22)$$

where $[V, W]$ is the Lie bracket or commutator of the vector fields V, W on $\mathcal{M}$, and X, Y are any tensor fields. In local coordinates x^μ on $\mathcal{M}$, this takes the form

$$(\mathcal{L}_V W)^\mu = V^\rho \partial_\rho W^\mu - W^\rho \partial_\rho V^\mu. \tag{1.0.23}$$

It is important to keep in mind the difference between the Lie derivative $\mathcal{L}_V$ and a connection ∇_V. By definition, a connection (such as the Levi–Civita connection defined in terms of a metric) satisfies $\nabla_{fV} = f\nabla_V$ for any $f \in C(\mathcal{M})$, and therefore defines a tensor. In contrast, Lie derivations are not tensorial, since $\mathcal{L}_{fV} \neq f\mathcal{L}_V$ in general.

Bundles

In physics, a manifold $\mathcal{M}$ typically carries some extra structure, such as matter fields that live in some vector space over each point of $\mathcal{M}$ or gauge fields that allow to consistently differentiate these. Such structures are typically described by the notion of a bundle over $\mathcal{M}$. The general definition is as follows.

Definition 1.1 A *fiber bundle* is defined in terms of a base manifold $\mathcal{M}$, a total (or bundle) space $\mathcal{B}$ which is also a manifold, and a map $\Pi\colon \mathcal{B} \to \mathcal{M}$ such that the local structure of $\mathcal{B}$ is that of a product manifold,

$$\mathcal{B} \overset{\text{loc}}{\cong} \mathcal{M} \times \mathcal{F}. \tag{1.0.24}$$

Here $\mathcal{F}$ is called the fiber. All maps are understood to be smooth.

A *section* of a bundle is a map

$$\sigma\colon \quad \mathcal{M} \to \mathcal{B} \qquad \text{such that} \qquad \Pi \circ \sigma = \mathrm{id}_{\mathcal{M}}. \tag{1.0.25}$$

For example, vector fields on $\mathcal{M}$ can be viewed as sections of the tangent bundle $T\mathcal{M}$, and the space of such sections is denoted as $\Gamma(\mathcal{M})$.

A fiber bundle is called a *vector bundle* if $\mathcal{F}$ is a vector space, and it is called a *principal bundle* if $\mathcal{F}$ is a Lie group (cf. Section 2). Basic examples of vector bundles are the tangent bundle $T\mathcal{M}$ and the cotangent bundle $T^*\mathcal{M}$. In field theory, matter fields are typically described by sections of some vector bundle. Simple examples of principal bundles are $U(1)$ bundles, where the fiber is given by $S^1 \cong U(1)$. We will also encounter other types of bundles such as sphere bundles, where the fiber is a two-sphere $\mathcal{F} \cong S^2$.

1.1 Symplectic manifolds and Poisson structures

To motivate the notion of a symplectic manifold, consider a classical system whose configuration space is an n-dimensional manifold $\mathcal{N}$ with coordinates q^i, $i = 1, \ldots, n$, which in the simplest case is $\mathbb{R}^n$. In Hamiltonian mechanics, it is useful to consider the phase space

$$\mathcal{M} = T^*\mathcal{N} \tag{1.1.1}$$

associated with $\mathcal{N}$. This is by definition the cotangent bundle over $\mathcal{N}$, which consists of all one-forms

$$\alpha = \sum_i \alpha_i(q) dq^i \tag{1.1.2}$$

on $\mathcal{N}$. The cotangent bundle $T^*\mathcal{N}$ captures not only the kinematical information as explained in Section 1.2, but it contains an extra structure, which does not exist on the tangent bundle $T\mathcal{N}$. To see this, we recall the canonical isomorphism $(V^*)^* \cong V$, where V is any vector space and V^* is its dual. Applying this isomorphism to the case of $V = T_q\mathcal{N}$ at some given point $q \in \mathcal{N}$, we obtain canonical maps

$$p_i \colon \mathcal{M} \to \mathbb{R}, \qquad p_i(dq^j) = \delta_i^j. \tag{1.1.3}$$

Thus, p_i recovers the coefficient functions of a one-form α (1.1.2) expanded in the basis dq^i. In this sense, the p_i are dual to the one-forms dq^i, and they are called canonical momenta. Together, the (q^i, p_j) form the so-called canonical coordinates on $\mathcal{M} = T^*\mathcal{N}$. This means that the p_i are canonically associated with q^i, i.e. there is no need for any extra structure such as a metric.

In particular, there is a canonical one-form on $\mathcal{M} = T^*\mathcal{N}$

$$\theta = \sum_i p_i dq^i = \sum_i \tilde{p}_i d\tilde{q}^i, \tag{1.1.4}$$

which[1] has the same form in any coordinates $\tilde{q}_i$ on $\mathcal{N}$. The exterior derivative of θ defines a canonical two-form ω on $\mathcal{M}$, which is automatically closed,

$$\omega = d\theta = dp_i \wedge dq^i \in \Omega^2(\mathcal{M}), \qquad d\omega = 0. \tag{1.1.5}$$

Recall that the Einstein sum convention is understood, so that $dp_i \wedge dq^i \equiv \sum_i dp_i \wedge dq^i$. Clearly, ω is also nondegenerate, so that ω is a *symplectic form* on $\mathcal{M}$, i.e. a closed nondegenerate two-form. The general definition is as follows.

Definition 1.2 A *symplectic manifold* is a manifold $\mathcal{M}$ equipped with a closed nondegenerate two-form ω.

Note that all symplectic manifolds have even dimension. For example, $\mathcal{M} = T^*\mathcal{N}$ is naturally a symplectic manifold, which is always noncompact; its relation with Hamiltonian mechanics will be recalled in Section 1.2. In this book, we will also encounter other types of symplectic manifolds, including compact symplectic manifolds.

The magic feature of symplectic forms is that they naturally define a Poisson bracket. To see this, consider local coordinates x^a on $\mathcal{M}$, and write ω as

$$\omega = \frac{1}{2}\omega_{ab} dx^a \wedge dx^b. \tag{1.1.6}$$

Recall that ω is closed $d\omega = 0$ if and only if

$$\partial_c \omega_{ab} + \partial_a \omega_{bc} + \partial_b \omega_{ca} = 0 \qquad \text{with} \qquad \omega_{ab} + \omega_{ba} = 0. \tag{1.1.7}$$

---

[1] In more abstract language, the defining property of θ is that $\alpha^*(\theta) = \alpha$ for any one-form $\alpha = p_i dq^i$ on $\mathcal{N}$. Here α^* is the pullback map associated with α, viewed as section $\alpha \colon \mathcal{N} \to T^*\mathcal{N} = \mathcal{M}$ of the cotangent bundle.

Since ω_{ab} is nondegenerate, we can consider the inverse tensor field θ^{ab}

$$\theta^{ab}\omega_{bc} = \delta^a_c, \tag{1.1.8}$$

which is also antisymmetric. This defines a bivector field $\theta^{ab}\partial_a \otimes \partial_b$ on $\mathcal{M}$, or equivalently a bracket on the algebra $\mathcal{A} = \mathcal{C}^\infty(\mathcal{M})$ of functions on $\mathcal{M}$ via

$$\{f,g\} := \theta^{ab}\partial_a f \partial_b g. \tag{1.1.9}$$

It is easy to see that this bracket satisfies the following properties:

$$
\begin{array}{lll}
\{f,g\} = -\{g,f\} & \text{antisymmetry} & \\
\{f, g + \lambda h\} = \{f,g\} + \lambda\{f,h\} & \text{linearity} & \\
\{f, gh\} = g\{f,h\} + h\{f,g\} & \text{Leibniz (product) rule} & \\
\{f,\{g,h\}\} + \{g,\{h,f\}\} + \{h,\{f,g\}\} = 0 & \text{Jacobi identity} & (1.1.10)
\end{array}
$$

for any $f,g,h \in \mathcal{A}$, and $\lambda \in \mathcal{R}$, which constitutes the definition of a Poisson bracket; the Jacobi identity follows from (1.1.7). In terms of local coordinates, the Jacobi identity takes the form

$$\theta^{ad}\partial_d\theta^{bc} + \theta^{bd}\partial_d\theta^{ca} + \theta^{cd}\partial_d\theta^{ab} = 0, \tag{1.1.11}$$

which is a consequence of (1.1.7).

This is a simple but profound result since the Jacobi relation is a nonlinear partial differential equation (PDE) for θ^{ab}. The framework of symplectic forms allows to recast this nonlinear PDE as a linear PDE (1.1.7) for ω_{ab}, which is much easier to handle. Hence, symplectic forms allow to understand and classify nondegenerate Poisson brackets. In particular, the canonical variables q^i and p_j on $\mathcal{M} = T^*\mathcal{N}$ satisfy the following Poisson brackets:

$$
\begin{aligned}
\{q^i, p_j\} &= \delta^i_j \\
\{q^i, q^j\} &= 0 = \{p_i, p_j\}.
\end{aligned}
\tag{1.1.12}
$$

The following important theorem states that this "canonical" form of the Poisson brackets can always be achieved locally for every symplectic manifold.

Theorem 1.3 (Darboux)   *For any point x in a symplectic manifold $\mathcal{M}$, there is an open neighborhood $\mathcal{U} \ni x$ and local coordinates $q^i, p_j : \mathcal{U} \to \mathbb{R}$ such that*

$$\omega = \sum_i dp_i \wedge dq^i. \tag{1.1.13}$$

In these local coordinates, the Poisson brackets take the form (1.1.12).

We introduce some further concepts and notation. A Poisson manifold is a manifold $\mathcal{M}$ carrying a Poisson structure, i.e. a bracket satisfying the relations (1.1.10). A Poisson structure can be conveniently encoded as a bivector field $\theta^{ab}\partial_a f \partial_b g$, where

$$\theta^{ab} = \{x^a, x^b\} \tag{1.1.14}$$

in local coordinates x^a. The θ^{ab} will be denoted as Poisson tensor.

If the tensor field θ^{ab} is nondegenerate, we can invert it and obtain a symplectic structure. Then Darboux theorem implies that, locally, the Poisson tensor takes the standard form

$$\theta^{ab} = \begin{pmatrix} 0 & \mathbb{1}_n \\ -\mathbb{1}_n & 0 \end{pmatrix} \tag{1.1.15}$$

in canonical coordinates $x^a = (p_i, q^i)$. If the Poisson tensor is degenerate and hence not invertible, one can still use it to define a *symplectic foliation* into *symplectic leaves*. The general statement is that any (regular, finite-dimensional) Poisson manifold is the disjoint union of its symplectic leaves.[2] For a more detailed discussion, see e.g. [137].

More generally, a (commutative) Poisson algebra $\mathcal{A}$ is an algebra together with a bracket

$$\{.,.\}: \quad \mathcal{A} \times \mathcal{A} \to \mathcal{A} \tag{1.1.16}$$

such that the relations (1.1.10) are satisfied. The most important case is where $\mathcal{A} = \mathcal{C}(\mathcal{M})$ is the algebra of (smooth, typically) functions on some manifold $\mathcal{M}$. Then the Poisson bracket can be identified with a bivector field $\theta^{ab} \partial_a f \partial_b g$.

Another useful result related to Darboux's theorem is Moser's lemma, which is a statement about local deformations of a symplectic structure. The statement is the following.

Lemma 1.4 (Moser) *Let ω_t be a family of symplectic forms depending smoothly on $t \in \mathbb{R}$. Then for any point $p \in \mathcal{M}$, there exists a local neighborhood $\mathcal{U}$ of p and a family of maps $g_t: \mathcal{U} \to \mathcal{U}$ with $g_0 = \mathrm{id}$ and $g_t^* \omega_t = \omega_0$.*

This means that any smooth deformation of a symplectic form can be absorbed locally by some diffeomorphism, which is the essence of Darboux's theorem. The idea of the proof is to consider the infinitesimal version of this relation, given by

$$0 = \frac{d}{dt}(g_t^* \omega_t) = g_t^* \left(\mathcal{L}_{V_t} \omega_t + \frac{d}{dt} \omega_t \right), \tag{1.1.17}$$

where the vector field V_t generates g_t. Since ω_t is closed, this reduces to $0 = i_{V_t}\omega_t + \mu_t$ using Cartan's magic formula (1.0.18), where μ_t is defined locally by $\frac{d}{dt}\omega_t = d\mu_t$ via Poincare's lemma. This can be solved for V_t as ω_t is nondegenerate, which upon integration yields g_t.

1.1.1 Hamiltonian vector fields

For any Poisson manifold, one can define an important class of special vector fields as follows.

Definition 1.5 For any $f \in \mathcal{C}(\mathcal{M})$, the *Hamiltonian vector field* V_f is defined by

$$V_f[g] := \{f, g\} \qquad \forall\, g \in \mathcal{C}(\mathcal{M}). \tag{1.1.18}$$

---

[2] This is proved using Frobenius' theorem, using the fact that Hamiltonian vector fields are always in involution.

We use here the fact that vector fields are naturally identified with derivations of the algebra $\mathcal{C}(\mathcal{M})$ of smooth functions on a manifold. If the Poisson structure is nondegenerate, then (1.1.18) is equivalent to the well-known relation

$$i_{V_f}\omega = df, \tag{1.1.19}$$

where $i_V\omega$ is the contraction of the vector field V with the two-form ω. Indeed writing the Hamiltonian vector field $V_f = (V_f)^a\partial_a$ in local coordinates, this gives

$$i_{V_f}\omega = (V_f)^a\omega_{ab}\,dx^b \overset{!}{=} df = \partial_b f dx^b. \tag{1.1.20}$$

Hence,

$$(V_f)^a = (\omega^{-1})^{ba}\partial_b f = \theta^{ba}\partial_b f = \{f, x^a\} \tag{1.1.21}$$

using (1.1.8), consistent with (1.1.18). The relation between symplectic forms and Poisson brackets can now be stated in a coordinate-free form as follows:

$$\{f, g\} = \omega(V_g, V_f), \tag{1.1.22}$$

which in local coordinates reduces to

$$\theta^{ab}\partial_a f\partial_b g = \omega_{ab}\theta^{ca}\theta^{db}\partial_c f\partial_d g. \tag{1.1.23}$$

In particular, Cartan's magic formula $\mathcal{L}_V\omega = (i_V d + d i_V)\omega$ gives immediately Liouville's theorem.

Theorem 1.6 (Liouville)

$$\mathcal{L}_{V_f}\omega = 0 = \mathcal{L}_{V_f}\omega^{\wedge n} \tag{1.1.24}$$

for any Hamiltonian vector field V_f on a symplectic manifold $\mathcal{M}$ of dimension $2n$.

The second statement implies that any symplectic manifold is equipped with a natural volume form

$$\Omega := \frac{1}{n!}\omega^{\wedge n} = \rho_M(x)d^{2n}x, \tag{1.1.25}$$

which is preserved by Hamiltonian vector fields; we will often write $\omega^n \equiv \omega^{\wedge n}$. Here

$$\rho_M(x) := \mathrm{pf}(\omega) = \frac{1}{2^n n!}\varepsilon^{a_1 b_1 \ldots a_n b_n}\omega_{a_1 b_1}\cdots\omega_{a_n b_n} = \sqrt{\det\omega_{ab}} \tag{1.1.26}$$

is the symplectic density on $\mathcal{M}$, which is given by the Pfaffian $\mathrm{pf}(\omega)$ of the antisymmetric matrix ω_{ab}. The relation with the determinant can be seen by writing ω_{ab} in block-diagonal form, noting that both the Pfaffian and the determinant factorize.

The first statement of (1.1.24) also deserves some discussion: it means that the symplectic structure is invariant under Hamiltonian vector fields:

$$\mathcal{L}_{V_f}\omega = 0, \qquad \mathcal{L}_{V_f}\theta^{ab} = 0. \tag{1.1.27}$$

Therefore, the flow defined by any Hamiltonian vector field is a *symplectomorphism*, i.e. a diffeomorphism that leaves ω invariant.

We conclude this brief introduction with some important observations. First, the Jacobi identity implies the following useful identity for any Poisson tensor:

$$0 = \partial_a(\rho_M \theta^{ab}), \tag{1.1.28}$$

where ρ_M is the symplectic volume density (1.1.26). To see this, consider

$$\begin{aligned}
\partial_a \theta^{ab} &= -\theta^{aa'} \partial_a \theta^{-1}_{a'b'} \theta^{b'b} = \theta^{aa'} \theta^{b'b} (\partial_{a'} \theta^{-1}_{b'a} + \partial_{b'} \theta^{-1}_{aa'}) \\
&= -\theta^{aa'} \partial_{a'} \theta^{b'b} \theta^{-1}_{b'a} - \theta^{b'b} \partial_{b'} \theta^{aa'} \theta^{-1}_{aa'} \\
&= -\partial_a \theta^{ab} - 2\theta^{ab} \rho_M^{-1} \partial_a \rho_M,
\end{aligned} \tag{1.1.29}$$

noting that $2\rho_M^{-1} \partial_b \rho_M = \partial_b \theta^{aa'} \theta^{-1}_{aa'}$ and using the fact that $\omega_{ab} = \theta^{-1}_{ab}$ is closed in the second step. Then (1.1.28) follows. As an application of this formula, we note that all Hamiltonian vector fields satisfy the following divergence constraint:

$$\partial_a(\rho_M V_f^a) = -\partial_a(\rho_M \theta^{ab} \partial_b f) = 0, \tag{1.1.30}$$

which expresses the fact that V_f preserves $\Omega = \rho_M d^{2n}x$. Another way to see this is by computing (1.1.24) directly:

$$\begin{aligned}
0 = \mathcal{L}_V \Omega = d(i_V \Omega) &= \frac{1}{(2n-1)!} d(\rho_M V^b \varepsilon_{ba_2 \dots a_{2n}} dx^{a_2} \dots dx^{a_{2n}}) \\
&= \frac{1}{(2n-1)!} \varepsilon_{ba_2 \dots a_{2n}} \partial_c(\rho_M V^b) dx^c dx^{a_2} \dots dx^{a_{2n}} \\
&= \partial_b(\rho_M V^b)\Omega
\end{aligned} \tag{1.1.31}$$

since $\varepsilon_{ba_2 \dots a_{2n}} \varepsilon^{ca_2 \dots a_{2n}} = (2n-1)! \delta_b^c$.

Finally, we note the important identity

$$[V_f, V_g] = V_{\{f,g\}}, \tag{1.1.32}$$

where the left-hand side is the Lie bracket of the Hamiltonian vector fields associated with f and g. This relation follows from the Jacobi identity, upon acting on some test function h:

$$[V_f, V_g][h] = \{f, \{g, h\}\} - \{g, \{f, h\}\} = \{\{f, g\}, h\} = V_{\{f,g\}}[h]. \tag{1.1.33}$$

It means that the Hamiltonian vector fields satisfy the same Lie algebra as the Poisson brackets of their generators. In more abstract terms, the map $f \to V_f$ is a Lie algebra homomorphism from the (Poisson) Lie algebra on the space of functions $\mathcal{C}(\mathcal{M})$ on $\mathcal{M}$ to the Lie algebra of vector fields on $\mathcal{M}$. These relations are cornerstones of Hamiltonian mechanics, and they will play an important role in the following.

Exercise 1.1.1   Verify the Jacobi identity (1.1.11) explicitly using $d\omega = 0$ (1.1.7).

1.2 The relation with Hamiltonian mechanics

We have seen that the symplectic structure of the cotangent bundle $T^*\mathcal{N}$ is canonical, i.e. independent of any extra structure such as a Lagrangian. It is nevertheless worthwhile to recall the relation with Lagrangian and Hamiltonian mechanics. Assume that some physical system is described through some Lagrangian $\mathcal{L}(q^i, v^j)$, which is a function on $T\mathcal{N}$. Here q^i are local coordinates on $\mathcal{N}$, and $v^j = \dot{q}^j$ are the associated tangent space coordinates. We assume that $\mathcal{L}$ is convex in the velocities v^j, which is tantamount to stability. Then the (fiber-wise) map $\phi \colon T_q\mathcal{N} \to T_q^*\mathcal{N}$ defined through

$$\phi(v)[w] = \frac{d}{ds}\Big|_0 \mathcal{L}(q, v + sw) \tag{1.2.1}$$

is invertible. Explicitly, v^i is expressed in terms of p_j via

$$p_i = \frac{\partial \mathcal{L}}{\partial v^i}, \tag{1.2.2}$$

where p_i is the canonical momentum (1.1.3). Then the Legendre transformation maps the function $\mathcal{L}$ on $T\mathcal{N}$ to the function H on $T^*\mathcal{N}$ via

$$\mathcal{C}(T\mathcal{N}) \to \mathcal{C}(T^*\mathcal{N})$$
$$\mathcal{L}(q^i, v^j) \mapsto H(q^i, p_j) := p_i\phi_*(v^i) - \phi_*\big(\mathcal{L}(q^i, v^i)\big). \tag{1.2.3}$$

Here ϕ_* indicates the identification via ϕ by substituting (1.2.2). The Legendre transformation is involutive, i.e. it is its own inverse, and maps convex functions to convex functions. The Euler–Lagrange equations can then be recast in Hamiltonian form,

$$\dot{f} = \{f, H\} = -V_H[f], \tag{1.2.4}$$

for any function f on $T^*\mathcal{N}$, with canonical Poisson structure given by (1.1.5). This formulation leads to powerful tools such as canonical transformations (which by definition preserve the Poisson or symplectic structure) in Hamiltonian mechanics, and it is essential to understand the relation with quantum mechanics.

Lie groups and coadjoint orbits

We have seen that the cotangent bundle $T^*\mathcal{M}$ of any manifold $\mathcal{M}$ is naturally a symplectic manifold. There is another important class of symplectic manifolds, which arises in the context of Lie groups: (co)adjoint orbits. They provide interesting examples of compact symplectic manifolds, which will play an important role in the following.

To set the stage, we briefly collect the main concepts from the theory of Lie groups as needed in this context. This is not meant as a fully self-contained introduction, and some familiarity with the basics of Lie groups and algebras will be assumed. For a more detailed introduction we refer to the literature, e.g. [78, 137]. The qualitative features can be understood from the basic example of the symplectic sphere underlying fuzzy S_N^2, which requires only some familiarity with $SU(2)$.

Definition 2.1 A **Lie group** G is a group that is a differentiable manifold, such that the multiplication map and the inverse map are smooth. Its **Lie algebra** $\mathfrak{g}$ is defined to be the vector space of left-invariant vector fields on G, and it can be identified with the tangent space of G at the identity element.

$$\mathfrak{g} = \{X \in \Gamma(G); \ g \cdot X = X \ \forall g \in G\} \cong T_e G \cong \mathbb{R}^n, \tag{2.0.1}$$

where $n = \dim(\mathfrak{g}) = \dim(G)$.

Here $\Gamma(G)$ denotes the space of vector fields on G and e is the identity element. The vector space $\mathfrak{g}$ acquires the structure of a Lie algebra through

$$[.,.]\colon \mathfrak{g} \times \mathfrak{g} \to \mathfrak{g}$$
$$(X, Y) \mapsto [X, Y], \tag{2.0.2}$$

which is defined to be the Lie bracket of (left-invariant) vector fields and satisfies the Jacobi identity. An explicit relation between a Lie group G and its Lie algebra $\mathfrak{g}$ is given by the exponential map as follows:

$$\exp\colon \mathfrak{g} \to G, \tag{2.0.3}$$

which is a local (but typically not global) diffeomorphism. In particular, each generator $X \in \mathfrak{g}$ in the Lie algebra defines a one-parameter subgroup of G denoted by $\exp(tX)$, which satisfies

$$\exp(sX)\exp(tX) = \exp((s+t)X), \qquad \frac{d}{dt}\exp(tX)\Big|_{t=0} = X. \tag{2.0.4}$$

Definition 2.2   A **representation** of a Lie group G is a Lie group homomorphism

$$\Pi : G \rightarrow GL(\mathcal{H}), \tag{2.0.5}$$

for some vector space $\mathcal{H}$, i.e. $\Pi(g_1 g_2) := \Pi(g_1) \circ \Pi(g_2)$. Here $GL(\mathcal{H})$ is the group of all invertible linear maps $\mathcal{H} \rightarrow \mathcal{H}$.

Given a representation Π, a dual representation acting on $\mathcal{H}^*$ is naturally defined in terms of $\Pi_*(g) = \Pi(g^{-1})^T$ acting on the dual basis.

A **representation** of a Lie algebra $\mathfrak{g}$ is a Lie algebra homomorphism

$$\pi : \mathfrak{g} \rightarrow \text{End}(\mathcal{H}) \tag{2.0.6}$$

for some vector space $\mathcal{H}$, i.e. $\pi([X, Y]) = [\pi(X), \pi(Y)]$. Here $\text{End}(\mathcal{H})$ is the space of all linear maps $\mathcal{H} \rightarrow \mathcal{H}$, equipped with a bracket given by the commutator $[X, Y] = XY - YX$.

A particularly important concept in our context is the group action on some manifold:

Definition 2.3   Let G be a Lie group and $\mathcal{M}$ some manifold. A **group action** of G on $\mathcal{M}$ is a map:

$$G \times \mathcal{M} \rightarrow \mathcal{M},$$
$$(g, x) \mapsto g \cdot x, \tag{2.0.7}$$

which is compatible with the group law, i.e.

$$(g_1 g_2) \cdot x = g_1 \cdot (g_2 \cdot x), \qquad e \cdot x = x \tag{2.0.8}$$

for all $g_1, g_2 \in G$ and $x \in \mathcal{M}$.

Given such a group action of G on $\mathcal{M}$, the one-parameter subgroup $\exp(tX)$ generated by some generator $X \in \mathfrak{g}$ defines a flow on $\mathcal{M}$, i.e. one-parameter subgroup of diffeomorphisms $\phi_t : \mathcal{M} \rightarrow \mathcal{M}$ via

$$\phi_t(p) = \exp(tX) \cdot p, \qquad p \in \mathcal{M}. \tag{2.0.9}$$

This flow is generated by vector fields V_X on $\mathcal{M}$,

$$\frac{d}{dt} \phi_t \Big|_{t=0} = V_X. \tag{2.0.10}$$

Then by differentiating the group law (2.0.8) it follows that the Lie brackets of the vector fields V_X satisfy the Lie algebra of $\mathfrak{g}$,

$$[V_X, V_Y] = V_{[X,Y]}. \tag{2.0.11}$$

Further important notions are as follows:

Definition 2.4   Given two group actions of G on $\mathcal{M}$ and $\mathcal{N}$, respectively, an **intertwiner** is a map

$$\Psi : \mathcal{M} \rightarrow \mathcal{N}, \tag{2.0.12}$$

which is compatible with the group actions, i.e. $\Psi(g \cdot m) = g \cdot \Psi(m)$ for all $g \in G$ and $m \in \mathcal{M}$.

(Co)adjoint action

An important example of a group action is the **adjoint action**, where G acts on itself via

$$G \times G \to G,$$
$$(g, h) \mapsto ghg^{-1}. \tag{2.0.13}$$

By differentiating the second factor at the unit element this leads to the adjoint action of G on its Lie algebra $\mathfrak{g}$, which can be obtained by differentiating $h = \exp(\varepsilon X)$ at $\varepsilon = 0$:

$$G \times \mathfrak{g} \to \mathfrak{g},$$
$$(g, X) \mapsto gXg^{-1}. \tag{2.0.14}$$

By dualizing one obtains similarly the **coadjoint action** of G on the dual $\mathfrak{g}^*$ of the Lie algebra

$$G \times \mathfrak{g}^* \to \mathfrak{g}^*,$$
$$(g, \alpha) \mapsto g\alpha g^{-1} \tag{2.0.15}$$

for $\alpha \in \mathfrak{g}^*$.

Each representation of G induces a representation of $\mathfrak{g}$ by differentiating at e, and conversely representations of $\mathfrak{g}$ can be "lifted" to representations of G provided some additional global conditions are satisfied (which is always the case if G is simply connected). In particular, since the map (2.0.14) is linear in $\mathfrak{g}$, it can be re-interpreted as adjoint representation of G

$$Ad: \ G \to GL(\mathfrak{g}), \qquad Ad_g(Y) = gYg^{-1}. \tag{2.0.16}$$

By differentiating, one obtains the adjoint representation of $\mathfrak{g}$

$$ad: \ \mathfrak{g} \to \mathrm{End}(\mathfrak{g}), \qquad ad_X(Y) = [X, Y]. \tag{2.0.17}$$

Similarly, (2.0.15) can be re-interpreted as coadjoint representation $G \to GL(\mathfrak{g}^*)$. For semi-simple Lie groups, the Killing form

$$\kappa(X, Y) = \mathrm{tr}(ad_X ad_Y) \tag{2.0.18}$$

provides a nondegenerate ad-invariant inner product on $\mathfrak{g}$. Then $\mathfrak{g}^*$ can be naturally identified with $\mathfrak{g}$, and we can largely restrict ourselves to the adjoint action, which is perhaps more intuitive.

Poisson action

Now consider a group action on a Poisson or symplectic manifold $\mathcal{M}$. Assume that the vector fields V_X are Hamiltonian,

$$V_X = \{H_X, .\} = V_{H_X}, \tag{2.0.19}$$

for some functions $H_X \in \mathcal{C}(\mathcal{M})$. Such an action of G is called *Hamiltonian* or *Poisson* action, and H_X are the generators of the symmetry. Then together with (1.1.32), we obtain

$$V_{\{H_X,H_Y\}} = [V_{H_X}, V_{H_Y}] = [V_X, V_Y] = V_{[X,Y]} = V_{H_{[X,Y]}}. \qquad (2.0.20)$$

This implies that

$$\boxed{\{H_X, H_Y\} - H_{[X,Y]} =: c_{X,Y}} \qquad (2.0.21)$$

is central under the Poisson bracket, i.e. $\{c_{X,Y}, .\} = 0$. In the symplectic case, this implies that $c_{X,Y} = const$. Hence, the generators H_X satisfy the Lie algebra relations of $\mathfrak{g}$ with respect to the Poisson brackets, possibly up to a central extension with central charges $c_{X,Y}$. This can be expressed by saying that the map

$$\mathfrak{g} \to \mathcal{C}(\mathcal{M})$$
$$X \mapsto H_X \qquad (2.0.22)$$

provides a *projective representation* of the Lie algebra $\mathfrak{g}$ on $\mathcal{C}(\mathcal{M})$ via Poisson brackets. This map can be considered as moment map[1] of the Hamiltonian action. The appearance of central charges is a typical quantum phenomenon, but, in fact, it arises already at the level of Poisson brackets. For semi-simple Lie algebras, the central charges can always be removed by a redefinition of the generators.

As an example, consider $\mathcal{M} = \mathbb{R}^{2n}$ with the canonical symplectic structure (1.1.15). The group $G \cong \mathbb{R}^{2n}$ of translations on $\mathcal{M}$ is an abelian Lie group that acts on $\mathcal{M}$ as

$$x \to T_a(x) = x + a, \qquad x \in \mathcal{M} \cong \mathbb{R}^{2n}, \qquad a \in G \cong \mathbb{R}^{2n}, \qquad (2.0.23)$$

so that $\mathcal{M}_{ab} = -\mathcal{M}_{ba}$ for $a, b \in \{1, \ldots, 2n\}$. The Lie algebra $\mathfrak{g} \cong \mathbb{R}^{2n}$ consists of translationally invariant vector fields whose brackets vanish. These vector fields are Hamiltonian, given by $\{q^i, .\} = \partial_{p_i}$ and $\{-p_j, .\} = \partial_{q^j}$ using the canonical variables (1.1.12). Thus, $H_{\partial_{p_i}} = q^i$ and $H_{\partial_{q_i}} = -p_i$. Then, (2.0.21) is verified by inspection, and a non-trivial central charge indeed arises for the mixed brackets:

$$\{H_{\partial_{p_i}}, H_{\partial_{p_j}}\} = \{q^i, q^j\} = 0, \qquad \{H_{\partial_{q^i}}, H_{\partial_{q^j}}\} = \{p_i, p_j\} = 0 \qquad \text{but}$$
$$\{H_{\partial_{q^i}}, H_{\partial_{p_j}}\} = -\{p_i, q^j\} = \delta^i_j. \qquad (2.0.24)$$

Hence the nonvanishing terms δ^i_j are recognized as central charges of the translation group $\mathbb{R}^{2n}$. Similarly, $SO(n)$ rotations acting within the $\mathbb{R}^n_q$ and $\mathbb{R}^n_p$ subspaces are generated by the functions

$$H_{\underline{a}} := \sum_{i,j} a_{ij} p_i q^j \qquad \in \mathcal{C}(\mathbb{R}^{2n}) \qquad (2.0.25)$$

specified by some (constant) antisymmetric matrix $a_{ij} = -a_{ji}$. Indeed,

$$\{H_{\underline{a}}, q^k\} = -a_{kj} q^j \qquad (2.0.26)$$

---

[1] This is usually stated in an equivalent dualized guise.

and similar for p_j. Furthermore, it is straightforward to verify (2.0.21), i.e.

$$\{H_{\underline{a}}, H_{\underline{b}}\} = \sum_k (a_{ik}b_{kj} - b_{ik}a_{kj})p_i q^j = H_{[a,b]} \tag{2.0.27}$$

for antisymmetric matrices $\underline{a}, \underline{b}$. Since $\mathfrak{so}(n)$ can be identified with antisymmetric $n \times n$ matrices, this means that the $H_{\underline{a}}$ indeed generate $\mathfrak{so}(n)$, without central charge.

As a last example, we note that the $SU(n)$ action on the complex coordinate functions $z^i = \frac{1}{\sqrt{2}}(q^i + ip_i)$ on $\mathbb{C}^n \cong \mathbb{R}^{2n}$ is also Hamiltonian. To see this, one first verifies the oscillator relations

$$\{z^i, \bar{z}_j\} = -i\delta^i_j, \qquad \{z^i, z^j\} = 0 = \{\bar{z}_i, \bar{z}_j\}, \tag{2.0.28}$$

where $\bar{z}_i = \frac{1}{\sqrt{2}}(q^i - ip_i)$. Then it is easy to see that the functions

$$H_{\underline{A}} = i \sum \bar{z}_i A^i_j z^j \tag{2.0.29}$$

for $A^i_j = -(A^j_i)^*$ generate the $\mathfrak{su}(n)$ action on z^i as

$$\{H_{\underline{A}}, z^k\} = -A^k_j z^j, \tag{2.0.30}$$

and the generators satisfy the Poisson bracket relations

$$\{H_{\underline{A}}, H_{\underline{B}}\} = \sum \bar{z}_i (A^i_k B^k_j - B^i_k A^k_j) z^j = H_{[A,B]}. \tag{2.0.31}$$

This is indeed the Lie algebra $\mathfrak{su}(n)$ without central charge.

The construction in these examples is known as Jordan–Schwinger or second-quantized representation of Lie algebras. It plays an important role in quantum field theory, and also in some standard examples of quantized geometries discussed in the following.

Exercise 2.01   Verify (2.0.27) and (2.0.31) in detail. As an application, show that the n-dimensional harmonic oscillator defined by the Hamilton function $H = \sum_i \frac{1}{2}(q_i^2 + p_i^2) = \sum_i \bar{z}_i z^i$ is invariant not only under $SO(n)$ but also under $SU(n)$, in the above sense.

2.1 $\mathbb{C}P^{n-1}$ as symplectic reduction of $\mathbb{C}^n$

The canonical Poisson brackets on $\mathbb{R}^{2n} \cong \mathbb{C}^n$ can be used to induce a Poisson structure on less trivial spaces. Consider for example the complex projective space $\mathbb{C}P^{n-1}$, which can be viewed as a quotient of the sphere $S^{2n-1} \subset \mathbb{R}^{2n} \cong \mathbb{C}^n$:

$$\mathbb{C}P^{n-1} \cong \mathbb{C}^n / \mathbb{C}^* \cong S^{2n-1} / U(1), \tag{2.1.1}$$

where $\mathbb{C}^* = \mathbb{C} \setminus \{0\}$. To define a Poisson structure – and, in fact, a symplectic structure – on $\mathbb{C}P^{n-1}$, we equip $\mathbb{C}^n$ with the canonical Poisson structure (2.0.28), corresponding to the symplectic form

$$\omega = dp_i \wedge dq^i = i d\bar{z}_i \wedge dz^i. \tag{2.1.2}$$

This is clearly invariant[2] under $U(n)$ acting on z^i. Now consider the constraint

$$C^2 := \sum_i \bar{z}_i z^i = R^2, \tag{2.1.3}$$

which defines a sphere $S^{2n-1} \subset \mathbb{R}^{2n}$ of radius R. Since this is an odd-dimensional manifold, it cannot be symplectic. Accordingly, the pull-back of the symplectic form ω to S^{2n-1} is degenerate. However, it does induce a proper symplectic form on the quotient $\mathbb{C}P^{n-1} \cong S^{2n-1}/U(1)$ as follows: S^{2n-1} can be viewed as a $U(1)$ bundle over $\mathbb{C}P^{n-1}$, with bundle projection

$$\begin{array}{c} S^{2n-1} \\ \pi \downarrow \\ \mathbb{C}P^{n-1}. \end{array} \tag{2.1.4}$$

The algebra of functions on S^{2n-1} decomposes accordingly into the direct sum

$$\mathcal{C}(S^{2n-1}) \cong \bigoplus_{k \in \mathbb{Z}} \mathcal{C}^{(k)}, \qquad \mathcal{C}^{(k)} = \{\phi(e^{i\varphi}z) = e^{ik\varphi}\phi(z)\} \tag{2.1.5}$$

corresponding to the k-th harmonics on S^1. The generator C^2 respects S^{2n+1} since $\{C^2, C^2\} = 0$, but it acts nontrivially on the harmonics, since the Poisson structure on $\mathbb{R}^{2n}$ is nondegenerate. Indeed C^2 measures the "winding number" k, in the sense

$$\{C^2, z^i\} = iz^i, \qquad \{C^2, .\}|_{\mathcal{C}^{(k)}} = ik\mathbb{1}. \tag{2.1.6}$$

This means that C^2 generates rotations of the $U(1)$ fiber over $\mathbb{C}P^{n-1}$. In particular, $\mathcal{C}^{(0)}$ are the functions that are invariant under $U(1)$, which can thus be identified with the space of functions on $\mathbb{C}P^{n-1}$:

$$\mathcal{C}(\mathbb{C}P^{n-1}) \cong \mathcal{C}^{(0)}. \tag{2.1.7}$$

Hence (polynomial) functions on $\mathbb{C}P^{n-1}$ are given by polynomials in $z^j, \bar{z}_j$ with equal numbers of z and $\bar{z}$ generators. This means that the Poisson brackets respect $\mathcal{C}^{(0)}$, leading to a Poisson structure on $\mathcal{C}^{(0)}$ that is easily seen to be nondegenerate i.e. symplectic. Moreover, that Poisson structure is clearly invariant under $SU(n)$.

The bottom line is that the Poisson structure on $\mathbb{R}^{2n}$ reduces to a $SU(n)$-invariant symplectic structure on $\mathbb{C}P^{n-1}$. We can identify the generators of $\mathfrak{su}(n)$ acting on $\mathbb{C}P^{n-1}$ in the sense of (2.0.19) as bilinears

$$H_a = \bar{z}_i (\lambda_a)^i_j z^j, \tag{2.1.8}$$

where the λ_a are (generalized) Gell–Mann matrices that satisfy the $\mathfrak{su}(n)$ Lie algebra

$$[\lambda_a, \lambda_b] = if_{abc}\lambda_c \tag{2.1.9}$$

(using physics conventions). It is easy to verify that then the H_a realizes this Lie algebra via the Poisson brackets:

$$\{H_a, H_b\} = f_{abc}H_c \tag{2.1.10}$$

---

[2] This implies that $U(n)$ is a subgroup of the symplectic group $SP(2n, \mathbb{R})$ of canonical transformations.

cf. (2.0.21). We can recognize the Jordan–Schwinger construction (2.0.31), which implements $\mathfrak{su}(n)$ as a Poisson action on $C(\mathbb{C}P^{n-1})$. Since the Poisson brackets are invariant under $SU(n)$, the corresponding symplectic form on $\mathbb{C}P^{n-1}$ coincides (up to trivial rescaling) with the Kirillov–Kostant–Souriau symplectic form, which can be defined for any coadjoint orbit on discussed in Section 2.2.

The present construction can be viewed as "symplectic reduction" of a canonical symplectic structure on some constraint surface. The need to quotient out some group action is prototypical for gauge theories in general. Moreover, the entire construction is manifestly compatible with the action of $SU(n)$: the z^i transform in the fundamental representation of $SU(n)$, the $\bar{z}_i$ transform in the anti-fundamental, and the H^a (2.1.8) transform in the adjoint. This induces an action of $SU(n)$ on S^{2n-1}, and on the quotient $\mathbb{C}P^{n-1}$. These two actions are clearly compatible in the sense of definition 2.5, which means that S^{2n-1} is an equivariant bundle over $\mathbb{C}P^{n-1}$.

In the case of $n = 2$, these considerations amount to the well-known statement that S^3 is a $U(1)$ bundle over $\mathbb{C}P^1 \cong S^2$, which is $SU(2)$- equivariant. In that case, (2.1.8) can be interpreted as Hopf map.

$$S^3 \to S^2$$
$$(z^1, z^2) \mapsto H^a := \bar{z}_i(\sigma^a)^i_j z^j, \tag{2.1.11}$$

where σ^a, $a = 1, 2, 3$ are the Pauli matrices. Their completeness relation

$$(\sigma^a)^i_j(\sigma_a)^k_l + \delta^i_j\delta^k_l = 2\delta^i_l\delta^k_j \tag{2.1.12}$$

implies

$$H_a H^a = (\bar{z}_i z^i)^2, \tag{2.1.13}$$

which means that the image of the Hopf map (2.1.11) is indeed S^2. Clearly the overall phase factor on $S^3 \subset \mathbb{C}^2$ drops out, so that the image can be identified with $\mathbb{C}P^1$:

$$\mathbb{C}P^1 \cong S^2. \tag{2.1.14}$$

In quantum mechanics, this is known as the Bloch sphere, which represents the pure states of a quantum mechanical system with Hilbert space $\mathcal{H} = \mathbb{C}^2$. In later sections, various generalizations of this construction and its quantization will play an important role, notably the generalized Hopf map $S^7 \to S^4$ in the context of the fuzzy four-sphere.

Exercise 2.1a)   Verify the Jordan–Schwinger relations (2.1.10).

Exercise 2.1b)   Verify the radial constraint (2.1.13), assuming $|z^1|^2 + |z^2|^2 = 1$.

2.2 (Co)adjoint orbits as symplectic manifolds

Given some element $X \in \mathfrak{g}$, we can define the **adjoint orbit** of G through X as the orbit of the group action (2.0.14) for fixed X. More explicitly, **adjoint orbits** of G are conjugacy classes of the form

$$\mathcal{O}[X] = \{gXg^{-1}; \ g \in G\} \subset \mathfrak{g}. \tag{2.2.1}$$

We can assume that X is in the Cartan subalgebra $\mathfrak{g}_0 \subset \mathfrak{g}$, i.e. traceless diagonal matrices in the case of $G = SU(N)$. Let $K_X = \{g \in G : gXg^{-1} = X\}$ be the stabilizer of X. Then, $\mathcal{O}[X]$ can be viewed as homogeneous space or coset

$$\mathcal{O}[X] \cong G/K_X. \tag{2.2.2}$$

Regular conjugacy classes are those with $K_X = T = \exp(\mathfrak{g}_0)$ being a maximal torus T of G, and they are isomorphic to G/T. In particular, their dimension is $\dim(\mathcal{O}[X]) = \dim(G) - \mathrm{rank}(G)$. *Degenerate conjugacy classes* have a larger stability group K_X, hence their dimension is smaller; for example, $\mathcal{O}[0] = \{0\}$ is just a point. These conjugacy classes are invariant under the adjoint action

$$g^{-1}\mathcal{O}[X]g = \mathcal{O}[X] \qquad \forall g \in G. \tag{2.2.3}$$

We can also consider the orbit of G through some element $t \in G$ in the adjoint action (2.0.13) on the group, which leads to a submanifold of G:

$$\mathcal{O}[t] = \{gtg^{-1}; \ g \in G\} \subset G. \tag{2.2.4}$$

Here one can again assume that t belongs to a maximal torus T of G. This is simply obtained from $\mathcal{O}[t]$ by the exponential map: if $t = \exp(X)$ then $\mathcal{O}[t] = \exp(\mathcal{O}[X])$, so that there is no need to consider this separately.

Similarly, the **coadjoint orbits** of G are conjugacy classes through $\alpha \in \mathfrak{g}^*$ of the form

$$\mathcal{O}[\alpha] = \{g\alpha g^{-1}; \ g \in G\} \subset \mathfrak{g}^*. \tag{2.2.5}$$

In the case of semi-simple Lie groups, they can be identified with adjoint orbits via the Killing form. Conceptually, the coadjoint orbits are preferred because they are symplectic manifolds, via a canonical symplectic form known as *Kirillov–Kostant–Souriau symplectic form*. Rather than dwelling on its abstract definition, it is better to provide a simple explicit description of the corresponding Poisson brackets. Consider a basis $X^a \in \mathfrak{g}$, $a = 1, \ldots, n$ of the Lie algebra, with structure constants

$$[X^a, X^b] = f^{ab}_{\ \ c} X^c \tag{2.2.6}$$

(in mathematics conventions without i). By definition, the X^a can be considered as coordinate functions on the dual $\mathfrak{g}^*$. We will denote these coordinate functions as x^a, in order to emphasize that the Lie algebraic structure plays no role here; the x^a simply generates the commutative algebra of functions on $\mathfrak{g}^*$. Since $\mathcal{O}[\alpha] \subset \mathfrak{g}^*$, these x^a can be considered as functions on the orbit, which specify the *embedding*

$$\boxed{x^a: \quad \mathcal{O}[\alpha] \hookrightarrow \mathfrak{g}^* \cong \mathbb{R}^n} \tag{2.2.7}$$

as a submanifold in "target space" $\mathbb{R}^n$. This embedding is specified by certain relations among the x^a. For example, the classical quadratic Casimir function

$$C^2 = x^a x^b \kappa_{ab} \tag{2.2.8}$$

(here κ_{ab} is the Killing metric) is invariant under the (co)adjoint action of G, and therefore constant on $\mathcal{O}[\alpha]$. For higher-rank Lie groups there are further relations on each $\mathcal{O}[\alpha]$, some of which will be illustrated in the following.

We can now define a Poisson bracket on the space of functions on $\mathfrak{g}^*$ or on the orbits $\mathcal{O}[\alpha]$ via

$$\{x^a, x^b\} := f^{ab}_{\ \ c} x^c, \tag{2.2.9}$$

where $f^{ab}_{\ \ c}$ are the structure constants. The Jacobi identity of this bracket follows from the Jacobi identity of the Lie algebra $\mathfrak{g}$. This simple observation underlies most of the quantum spaces under consideration in this book. In particular, the (coadjoint) action of $X^a \in \mathfrak{g}$ on the orbit $\mathcal{O}[\alpha]$ is Hamiltonian, given by (2.2.9). However, this Poisson bracket is degenerate; for example, the quadratic Casimir C^2 is central w.r.t. the Poisson bracket. This is equivalent to the statement that C^2 is a Casimir for the Lie algebra $\mathfrak{g}$,

$$\{C^2, x^a\} = 2\kappa_{cb}x^c\{x^b, x^a\} = 2f_c^{\ a}_{\ d}x^c x^d = 0, \tag{2.2.10}$$

where $f_c^{\ a}_{\ d} = \kappa_{cb}f^{ba}_{\ \ d}$, using the fact that f_{cad} is totally antisymmetric for semi-simple Lie groups. The same is true for all higher Casimirs of higher-rank Lie groups. The symplectic leaves of this Poisson structure are precisely the coadjoint orbits $\mathcal{O}[\alpha]$, which are homogeneous spaces embedded in $\mathfrak{g}^*$, and thereby symplectic manifolds. Indeed, the generic dimension of $\mathcal{O}[\alpha]$ is $\dim(G) - \mathrm{rank}(G)$. However, their dimensions is reduced in degenerate cases. This will play an important role later, and we will study their structure in in some examples. In particular, we will learn how to quantize the embedding functions x^a (2.2.7) for "quantizable" orbits $\mathcal{O}[\alpha]$ in terms of irreducible representations of $\mathfrak{g}$, and the degenerate orbits will correspond to "short" representations of the underlying Lie group.

For the sake of completeness, we quote the abstract, coordinate-free definition of the Kirillov–Kostant–Souriau symplectic form on the coadjoint orbit $\mathcal{O}[\alpha]$:

$$\omega_\alpha(V_X, V_Y) = -\alpha([X, Y]). \tag{2.2.11}$$

Here V_X is the vector field on $\mathfrak{g}^*$ generated by the coadjoint action of $X, Y \in \mathfrak{g}$ on $\mathfrak{g}^*$. Using an orthonormal basis of $\mathfrak{g}^* \cong \mathfrak{g}$ and recalling (1.1.22), this reduces to (2.2.9). One can show that ω_α is closed and nondegenerate on $\mathcal{O}[\alpha]$; for more details, see e.g. [119]. However, the explicit form (2.2.9) of the Poisson brackets is more transparent and intuitive, and will be used throughout this book.

To see that (2.2.11) is equivalent to (2.2.9), consider a compact semisimple Lie algebra $\mathfrak{g}$ with orthonormal basis X^a w.r.t. the Killing form. The basis X^a can also be viewed as Cartesian coordinates on $\mathfrak{g}^*$, which we denote by x^a as above, to emphasize that they generate a commutative algebra of functions. The adjoint action is generated by the vector fields $V_{(a)} := [X^a, .]$ on $\mathfrak{g}$, which acts on X^b as $V_{(a)}[X^b] = [X^a, X^b] = f^{ab}_{\ \ c}X^c$. Hence in the coordinates x^a, the coadjoint action is generated by the vector fields

$$V_{(a)} = f^{ab}_{\ \ c}x^c \partial_b = \{x^a, .\}, \tag{2.2.12}$$

where the Poisson bracket on the rhs is the one defined in (2.2.9). This means that the $V_{(a)}$ are Hamiltonian vector fields generated by x^a. Since the coadjoint orbit $\mathcal{O}[\alpha]$ is generated by these vector fields acting on α, it is precisely a symplectic leaf of the Poisson bracket $\{.,.\}$ (2.2.9). Then (1.1.22) states that

$$\omega(V_{(a)}, V_{(b)}) = -\{x^a, x^b\} = -f^{ab}_{\ \ c} x^c$$
$$= -\alpha([X^a, X^b]), \qquad (2.2.13)$$

where the second line is evaluated at $\alpha = x^c X^*_c$ of the orbit $\mathcal{O}[\alpha]$, where X^*_c is dual to X^c. This is nothing but the rhs of (2.2.11).

For matrix groups such as $SU(n)$ or $SO(n)$, this symplectic structure can be written down more explicitly as follows:

$$\omega(V_X, V_Y) = -\text{tr}(A[X, Y]) \qquad (2.2.14)$$

on the orbit through $A \in \mathcal{O}$, where $X, Y \in \mathfrak{g}$ generate the vector fields $V_X = [X, A]$. Moreover for many simple coadjoint orbits,[3] the symplectic form can be written down explicitly as

$$\omega = f_{abc} x^a dx^b \wedge dx^c \qquad (2.2.15)$$

up to normalization, as $d\omega = f_{abc} dx^a \wedge dx^b \wedge dx^c$ vanishes on $\mathcal{O}$.

To summarize, (co)adjoint orbits are a particular class of homogeneous spaces G/K, which are naturally equipped with a symplectic structure. However, not every homogeneous space is a coadjoint orbit. For example, S^4 is not a coadjoint orbit of $SO(5)$, and it is known that S^4 does not admit any symplectic structure. We conclude this discussion with a few important examples of coadjoint orbits.

S^2 as coadjoint orbit of $SO(3)$

The basic example is the sphere S^2, which can be considered as a (co)adjoint orbit of $SO(3)$ or $SU(2)$. We recall that the Lie algebra $\mathfrak{su}(2)$ of $SO(3)$ is spanned by three generators J^a, $a = 1, 2, 3$, with

$$[J^a, J^b] = i\varepsilon^{abc} J^c \qquad (2.2.16)$$

(using physics conventions with i). To be specific, we choose the generator of J^3, and consider the (co)adjoint orbit $\mathcal{O}[J^3]$. The stabilizer of J^3 is clearly $K = e^{icJ_3} \cong U(1)$, so that

$$\mathcal{O}[J^3] = \{O J^3 O^{-1}; O \in SO(3)\} = SO(3)/U(1) \cong S^2. \qquad (2.2.17)$$

This is very intuitive: considering J^3 as unit vector pointing at the north pole, its orbit under $SO(3)$ spans the sphere, while the rotations around the north pole act trivially on J_3. The Poisson structure is also easy to understand. Consider the embedding $S^2 \subset \mathbb{R}^3 \cong \mathfrak{so}(3)$ with Cartesian coordinates x^i on $\mathfrak{so}(3) \cong \mathbb{R}^3$. They specify the embedding

$$x^a : \quad S^2 \hookrightarrow \mathbb{R}^3, \qquad (2.2.18)$$

---

[3] including $\mathbb{C}P^n$, and more generally for coadjoint orbits that are also symmetric spaces i.e. where $\mathfrak{g} = \mathfrak{t} \oplus \mathfrak{r}$ satisfies the structure $[\mathfrak{r}, \mathfrak{r}] = \mathfrak{r}$, $[\mathfrak{t}, \mathfrak{r}] = \mathfrak{t}$, $[\mathfrak{t}, \mathfrak{t}] = \mathfrak{r}$, where $\mathfrak{r}$ is the Lie algebra of the stabilizer group.

in target space $\mathbb{R}^3 \cong \mathfrak{so}(3)$. We can use x^1, x^2 as local coordinates on the upper hemisphere, whose Poisson bracket is then

$$\{x^1, x^2\} = x^3 = \sqrt{R^2 - (x^1)^2 - (x^2)^2} = R\Big(1 + O\big(\frac{(x^i)^2}{R^2}\big)\Big), \qquad (2.2.19)$$

near the north pole. Hence, the Poisson structure is nondegenerate, and approximates that of $\mathbb{R}^2_\theta$ near the north pole (or any other reference point). It is an easy exercise to verify that the underlying symplectic structure on S^2 is given by

$$\omega = \frac{1}{2}\epsilon_{abc}x^a dx^b \wedge dx^c, \qquad (2.2.20)$$

which is clearly the unique $SO(3)$-invariant two-form on S^2 up to normalization. This is the classical structure underlying the fuzzy two-sphere discussed in Section 3.3.1. The two-sphere $S^2 \cong \mathbb{C}P^1$ can also be viewed as the simplest example of complex projective spaces $\mathbb{C}P^n$, all of which are coadjoint orbits as discussed in the following paragraphs.

$\mathbb{C}P^{n-1}$ as coadjoint orbit of $SU(n)$

More generally, all complex projective spaces $\mathbb{C}P^{n-1}$ for $n \geq 2$ can be viewed as a (co)adjoint orbits. To see this, we identify the Lie algebra $\mathfrak{su}(n)$ with traceless Hermitian $n \times n$ matrices, and consider the adjoint $SU(n)$ orbit $\mathcal{O}[X]$ through

$$X = \mathrm{diag}(n-1, -1, \ldots, -1) = -\mathbb{1} + nP_{|\psi\rangle} \qquad \in \mathfrak{su}(n). \qquad (2.2.21)$$

Here,

$$P_{|\psi\rangle} = |\psi\rangle\langle\psi| \qquad (2.2.22)$$

is the rank 1 projector on

$$|\psi\rangle = \begin{pmatrix} 1 \\ 0 \\ \vdots \\ 0 \end{pmatrix} \in \mathbb{C}^n. \qquad (2.2.23)$$

Noting that

$$UXU^{-1} = -\mathbb{1} + nP_{U|\psi\rangle} \qquad (2.2.24)$$

for $U \in U(n)$, we see that the orbit

$$\begin{aligned}
\mathcal{O}[X] = \{UXU^{-1}, \; U \in U(n)\} &\cong \mathbb{C}P^{n-1} = \mathbb{C}^n/_\sim \\
-\mathbb{1} + nP_{|\psi\rangle} &\leftrightarrow |\psi\rangle/_\sim
\end{aligned} \qquad (2.2.25)$$

is indeed in one-to-one correspondence to $\mathbb{C}P^{n-1}$, where $\sim$ indicates equivalence modulo multiplication with $\mathbb{C} \setminus \{0\}$. Both characterizations lead to

$$\mathbb{C}P^{n-1} = SU(n)/K, \qquad K = S(U(n-1) \times U(1)) \qquad (2.2.26)$$

where K is the stabilizer of X. However, the present construction as coadjoint orbit provides an embedding in target space $\mathfrak{su}(n) \cong \mathbb{R}^D$ with $D = n^2 - 1$, as well as a symplectic structure on $\mathbb{C}P^{n-1}$, which is suitable for quantization. To make this explicit, let λ^a, $a = 1, \ldots, D$ be an orthonormal basis of $\mathfrak{su}(n)$, with Lie brackets $[\lambda^a, \lambda^b] = if^{abc}\lambda^c$ (2.1.9). In the case of $\mathfrak{su}(3)$, the λ^a can be identified with the Gell-Mann matrices. We can choose this basis such that one of the generators coincides with the element X (2.2.21) which generates $\mathbb{C}P^{n-1} \cong \mathcal{O}[X]$. In the case of $\mathfrak{su}(3)$ this is the generator λ^8 of the Gell-Mann matrices (up to normalization).

Following the general scheme, we view the λ^a as coordinate functions on $\mathfrak{su}(n)^* \cong \mathbb{R}^D$, which we denote as x^a to avoid confusion. These generate the algebra of functions on $\mathbb{R}^D$, equipped with the Poisson bracket

$$\{x^a, x^b\} = f^{abc}x^c. \tag{2.2.27}$$

This is precisely the Poisson structure obtained in (2.1.10), which is now understood in a more general setting. Again, the x^a are viewed as embedding functions of the orbit

$$x^a: \quad \mathcal{O}[X] \cong \mathbb{C}P^n \hookrightarrow \mathbb{R}^D, \tag{2.2.28}$$

as such they satisfy a number of constraints. One of them is again the quadratic Casimir constraints $C^2 = \text{const}$ where:

$$C^2[\mathfrak{su}(n)] = x^a x^b \delta_{ab}, \tag{2.2.29}$$

but there are further constraints, which encode the $2n$-dimensional nature of $\mathbb{C}P^n$.

To exhibit the remaining constraints, it is useful to introduce the following matrix-valued function on $\mathbb{R}^D$ as follows:

$$\Xi := x^a \lambda_a, \tag{2.2.30}$$

which maps each point of $\mathcal{O}(X) \subset \mathbb{R}^D$ to itself. Here λ_a are viewed as $n \times n$ matrices given by the fundamental representation of $\mathfrak{su}(n)$, so that Ξ is a matrix-valued function on $\mathcal{O}[X]$. By $SU(n)$ invariance, it suffices to consider the point where $\Xi = X$ has the diagonal form (2.2.21). Then $\Xi + \mathbb{1}$ is a projector (up to normalization), which satisfies the characteristic relation

$$(\Xi + \mathbb{1})(\Xi - (n-1)\mathbb{1}) = 0. \tag{2.2.31}$$

Plugging in $\Xi = x^a \lambda_a$, this amounts to the following relations:

$$x^a x_a = 1, \qquad d_{abc} x^a x^b = x_c, \tag{2.2.32}$$

where d_{abc} is the totally invariant tensor of $\mathfrak{su}(n)$.

The identification (2.2.25) provides the link between the definition of $\mathbb{C}P^{n-1}$ as set of rays in $\mathbb{C}^n$ with the description as highly degenerate coadjoint orbit of $SU(n)$, with maximal stabilizer group $SU(n-1) \times U(1)$. There are other classes of coadjoint orbits $\mathcal{O}[X]$ with larger dimensions and smaller stabilizer group. The regular coadjoint orbits $X \in \mathfrak{su}(n)$ have minimal stabilizer group $U(1)^{n-1}$, leading to a space with maximal dimension $n^2 - n$. For large n, there are many different types of coadjoint orbits in between these extremes, all of which are symplectic (and in fact Kähler) manifolds that have a natural quantization as discussed in Section 3.3.3.

The special case of $\mathbb{C}P^3$ as coadjoint orbit of $SU(4)$ (or rather a non-compact version thereof) will play a central role to describe gravity on cosmological spacetime in the matrix model. We will also exploit the Lie algebra identity $\mathfrak{su}(4) \cong \mathfrak{so}(6)$, which means that $SU(4)$ is the universal covering group of $SO(6)$. Various types of coadjoint orbits of $SU(3)$ will appear in Section 4.6, including $\mathbb{C}P^2$ but also other types of orbits. Quantized versions of these turn out to be solutions in the matrix model describing fuzzy extra dimensions, and lead to structures reminiscent of particle physics.

Coadjoint orbits of $SO(n)$

Finally we discuss some coadjoint orbits of the orthogonal groups $SO(n)$, because some of them will play an important role in later chapters. We start with the Lie algebra of $\mathfrak{so}(n)$, which can be written as

$$[\mathcal{M}_{ab}, \mathcal{M}_{cd}] = i\left(\delta_{ac}\mathcal{M}_{bd} - \delta_{ad}\mathcal{M}_{bc} - \delta_{bc}\mathcal{M}_{ad} + \delta_{bd}\mathcal{M}_{ac}\right) \tag{2.2.33}$$

for $a, b, c, d = 1, \ldots, n$, in physics conventions with hermitian generators λ_a and an explicit i. These generators satisfy the antisymmetry constraints

$$\mathcal{M}_{ab} = -\mathcal{M}_{ba}. \tag{2.2.34}$$

Again we can consider the $\mathcal{M}_{ab}$ as coordinate functions on $\mathfrak{so}(n)^* \cong \mathbb{R}^D$ for $D = n(n-1)/2$, which we will denote as m_{ab} to avoid confusion. They are of course also antisymmetric in ab, and generate the (commutative) algebra of functions on $\mathbb{R}^D$. Viewed as functions on $\mathcal{O}[X]$ they define the embedding

$$m_{ab}: \quad \mathcal{O}[X] \hookrightarrow \mathbb{R}^D. \tag{2.2.35}$$

According to (2.2.9), we can introduce the canonical Poisson brackets

$$\{m_{ab}, m_{cd}\} = \delta_{ac}m_{bd} - \delta_{ad}m_{bc} - \delta_{bc}m_{ad} + \delta_{bd}m_{ac}, \tag{2.2.36}$$

where the explicit i is dropped. For $n = 3$ this reduces to (2.2.19) via $m_{ab} = \epsilon_{abc}x_c$. They satisfy the constraint

$$C^2[\mathfrak{so}(n)] = m_{ab}m_{a'b'}\delta^{aa'}\,\delta^{bb'} = \text{const} \tag{2.2.37}$$

and others in general, which depend on the nature of the orbit under consideration.

One example that will play an important role later are coadjoint orbits of $SO(6)$. Due to the relation $\mathfrak{so}(6) \cong \mathfrak{su}(4)$, they can be considered as coadjoint orbits of $SU(4)$, which includes in particular the maximally degenerate orbit $\mathbb{C}P^3$. This equivalence provides a natural action of $SO(5) \subset SO(6)$ on $\mathbb{C}P^3$. Non-compact orbits e.g. of $SO(2, n-2)$ are obtained by replacing δ_{ab} with the appropriate invariant tensor η_{ab}.

Equivariant bundles from coadjoint orbits

We have seen that the dimension of a (co)adjoint orbit $\mathcal{O}[X]$ depends on the generating element $X \in \mathfrak{g}$. For example, $\mathbb{C}P^n$ was recognized in (2.2.25) as coadjoint orbit through a maximally degenerate $X \in \mathfrak{su}(n)$. A natural question is what happens if we slightly perturb such a maximally degenerate case as $\tilde{X} = X + Y$, with $Y \ll X$ in some sense. Then the orbit $\mathcal{O}[\tilde{X}]$ is certainly "close" to $\mathcal{O}[X]$, but its dimension will be larger. It should therefore not be surprising that $\mathcal{O}[\tilde{X}]$ is naturally a bundle over $\mathcal{O}[X]$. But this bundle has a special feature: both the bundle and its base manifold carry a G action, and these actions are compatible with the bundle projection. This situation is captured by the following concept, which will play an important role in later chapters:

Definition 2.5   An **equivariant bundle** is a bundle $\Pi \colon \mathcal{B} \to \mathcal{M}$ with total space $\mathcal{B}$, base manifold $\mathcal{M}$ and bundle projection Π, where both $\mathcal{B}$ and $\mathcal{M}$ carry an action of a Lie group G that is compatible with the bundle projection Π. This means that

$$\Pi(g \cdot b) = g \cdot \Pi(b) \tag{2.2.38}$$

for all $b \in \mathcal{B}$ and $g \in G$, where the appropriate group actions on $\mathcal{B}$ and $\mathcal{M}$ are understood, and both denoted with $\cdot$.

Exercise 2.2.1   Verify (2.2.20)

Toward noncommutative geometry

The basic idea of noncommutative geometry is to get rid of the notion of a spacetime continuum, and to replace classical manifolds with an analogous quantum structure. Some first steps in this direction can be taken within the framework of classical geometry, by reformulating the description of geometry in a mathematical language that does not use the notion of a point. The idea is to describe a space $\mathcal{M}$ not as a set of points, but in terms of the *algebras of continuous functions* $\mathcal{A} = \mathcal{C}_0(\mathcal{M})$ on $\mathcal{M}$. This algebra is clearly commutative. The (commutative) Gelfand–Naimark theorem states that the converse is also true: every unital commutative C^*-algebra $\mathcal{A}$ is equivalent to an algebra of continuous functions $\mathcal{C}_0(\mathcal{M})$ on some compact space $\mathcal{M}$. This means that compact spaces can be characterized in terms of their algebras of functions. This algebraic point of view – suitably amended with extra structure encoding differential structure and a metric – allows a generalization to noncommutative algebras, which are then interpreted as functions on a quantum space. Similar ideas have been contemplated for a long time by many authors including John von Neumann who coined the term "pointless geometry," as well as Heisenberg, Schrödinger, Snyder, Connes, and others.

The prime physical realization of a quantum space is quantum mechanics, where classical phase space is replaced by a quantized phase space. All observations are specified in terms of observables, which are quantized functions on phase space, rather than points. These functions form an algebra, namely the algebra of operators $\mathrm{End}(\mathcal{H})$ on a Hilbert space. The basic idea of *noncommutative geometry* is to add a metric structure into such a framework. One way to makes this precise was developed by Alain Connes, who axiomatized the above ideas in terms of spectral triples $(\mathcal{A}, \mathcal{H}, \slashed{D})$. However, mathematical axioms should be regarded with caution in physics, and the appropriate mathematical description of physics can only be identified in retrospect. We will therefore not start with mathematical axioms, but extract the appropriate mathematical structures from a specific approach based on matrix models, which is motivated by fundamental physics. This will lead to a framework for quantum geometry that is well adapted to noncommutative field theory, gauge theory, and string theory. The basic mathematical structure is based on quantized symplectic spaces, amended with extra geometrical structure arising from their realization in matrix models. This framework will be denoted as *quantum geometry* in this book, which should be viewed as a specific version of noncommutative geometry.

Part II

Quantum spaces and geometry

3 Quantization of symplectic manifolds

The conceptual relation between quantum mechanics and classical mechanics was first exhibited by Dirac. Given the formulation of analytical mechanics on phase space, the crucial insight is to relate the Poisson brackets $\{.,.\}$ of functions on phase space with commutators $-\frac{i}{\hbar}[.,.]$ of observables acting on a Hilbert space $\mathcal{H}$. This procedure is known as quantization, and its generalization will play a central role in the following. The basic idea is to map the space of functions $\mathcal{C}(\mathcal{M})$ on a symplectic manifold to the space of linear maps or endomorphisms $\text{End}(\mathcal{H})$ on some Hilbert space, preserving as much structure as possible. This allows to obtain a (semi-) classical picture of some quantum system. While in general there is no unique way to define such a map, there are preferred and sometimes unique quantization maps in the presence of sufficient symmetry.

Armed with the required mathematical structures and tools, we can properly address the problem of quantization. We start by recalling the correspondence between classical Hamiltonian mechanics and quantum mechanics, summarized in Table 3.1. Classical mechanics can be formulated in terms of the algebra of (smooth, complex-valued) functions $\mathcal{A} = \mathcal{C}(\mathcal{M})$ on phase space $(\mathcal{M}, \{.,.\})$, equipped with a Poisson bracket. Observables are given by real functions $f = f^* \in \mathcal{A}$, while states are characterized by some positive density $\rho \in \mathcal{A}$ with $\int \rho = 1$. The central idea of quantum mechanics is to replace the commutative algebra $\mathcal{A}$ by the noncommutative algebra $\mathcal{A}_\hbar = \text{End}(\mathcal{H})$ of operators on a Hilbert space $\mathcal{H}$, and the Poisson bracket by the commutator. Observables are then self-adjoint operators $F = F^\dagger$, while states are given by some positive density matrix $\hat{\rho} \in \mathcal{A}_\hbar$ with $\text{Tr}\hat{\rho} = 1$. The dynamics is defined through $\dot{f} = \{f, H\}$ for some Hamiltonian function $H \in \mathcal{A}$, or $\dot{F} = -\frac{i}{\hbar}[F, H]$ in terms of a Hamiltonian $H \in \mathcal{A}_\hbar$, respectively. The invariant functionals given by the integral and trace allow to define an inner product and a Hilbert space structure on (subsets of) $\mathcal{C}(\mathcal{M})$ and $\text{End}(\mathcal{H})$, as discussed in the following.

The correspondence principle then stipulates that the classical theory should be recovered in a suitable semi-classical limit or regime. In particular, there should be a correspondence between classical functions on phase space and hermitian operators on $\mathcal{H}$. This amounts to some map,

$$\mathcal{Q}: \quad \mathcal{C}(\mathcal{M}) \to \text{End}(\mathcal{H}) \tag{3.0.1}$$

called *quantization map*, which should be bijective at least in some appropriate physical regime. Following Dirac, one would like to require that $\mathcal{Q}$ maps Poisson brackets to commutators.

Table 3.1 Correspondence between classical Hamiltonian mechanics and quantum mechanics.

Structure	Classical mechanics	Quantum mechanics
Algebra of functions	$\mathcal{A} = \mathcal{C}(\mathcal{M})$	$\mathcal{A}_\hbar = \text{End}(\mathcal{H})$
Observables	$f = f^* \in \mathcal{A}$	$F = F^\dagger \in \mathcal{A}_\hbar$
States	$0 \leq \rho \in \mathcal{A}$	$0 \leq \hat{\rho} \in \mathcal{A}_\hbar$
Dynamics	$\dot{f} = \{f, H\}$	$\dot{F} = -\frac{i}{\hbar}[F, H]$
Invariant functional	$\frac{1}{(2\pi)^n} \int_{\mathcal{M}} \Omega f$	$\text{Tr}(F)$

However, it turns out that this is too much to ask for. We can choose canonical coordinates (q^i, p_j) on phase space, and try to replace the canonical Poisson brackets (1.1.12) with the canonical commutation relations

$$[Q^i, P_j] = i\hbar \delta^i_j \mathbb{1}$$
$$[Q^i, Q^j] = 0 = [P^i, P^j]. \tag{3.0.2}$$

Here Q^i, P_j are self-adjoint operators acting irreducibly on a Hilbert space $\mathcal{H}$. This suggests defining $\mathcal{Q}(q^i) = Q^i$ and $\mathcal{Q}(p_i) = P_i$, but it remains to define a general correspondence between functions $\mathcal{C}(\mathbb{R}^{2n})$ on phase space and operators in $\text{End}(\mathcal{H})$. Due to the ordering ambiguities on the operator side, it is not evident how to define such a map. In fact, there is no way to define such a map such that all Poisson brackets are mapped to commutators, as shown in a theorem of Groenewold and van Howe. This theorem states that there is no irreducible[1] Lie algebra homomorphism

$$\mathcal{Q}: \quad \text{Pol}(\mathbb{R}^{2n}) \rightarrow \text{End}(\mathcal{H}) \tag{3.0.3}$$

with $\mathcal{Q}(1) = \mathbb{1}$, where $\text{Pol}(\mathbb{R}^{2n}) \subset \mathcal{C}(\mathbb{R}^{2n})$ denotes polynomials in q^i, p_j viewed as Lie algebra defined by the canonical Poisson brackets (1.1.12). It can be proved by assuming that such a map $\mathcal{Q}$ exists, which leads to a contradiction for monomials of sufficiently high order [79].

This result means that there is no unique concept of quantization. That should hardly be surprising, since quantum mechanics is more fundamental and different from classical physics. We will therefore adopt a weaker concept of quantization, which is not unique but nevertheless useful. Moreover, we will see that preferred quantization maps $\mathcal{Q}$ do exist for spaces $\mathcal{M}$ with sufficient symmetries.

With the above motivation, a quantization of a symplectic manifold $\mathcal{M}$ will be defined in terms of a linear quantization map $\mathcal{Q}$, which associates to every function $f \in \mathcal{C}(\mathcal{M})$ an operator $\mathcal{Q}(f) = \hat{f} \in \text{End}(\mathcal{H})$, such that the Poisson brackets on $\mathcal{M}$ are recovered from

---

[1] Irreducibility is essential for the statement. Reducible homomorphisms (3.0.3) do exist in terms of operators on $\mathcal{M}$, denoted as pre-quantization in the context of geometric quantization. However, they act on a Hilbert space that is far "too big" and pathological from a physics point of view, containing far too many modes.

the commutators in $\mathrm{End}(\mathcal{H})$ in the semi-classical limit or regime. Clearly $\mathcal{Q}$ cannot respect the multiplication i.e. it cannot be an algebra map, but it should respect the algebra in the semi-classical limit. More precisely, the following conditions will be required:

Definition 3.1   Let $\mathcal{M}$ be a symplectic manifold, with Poisson tensor $\theta^{\mu\nu} = \theta\bar{\theta}^{\mu\nu}$ scaled by $\theta \in \mathbb{R}$. A quantization map is a linear map

$$\mathcal{Q}: \quad \mathcal{C}(\mathcal{M}) \to \mathrm{End}(\mathcal{H}) \tag{3.0.4}$$

satisfying the axioms

1.

$$\mathcal{Q}(1) = \mathbb{1} \tag{3.0.5}$$

2.

$$\mathcal{Q}(f)^{\dagger} = \mathcal{Q}(f^{*}) \quad \forall f \in \mathcal{C}(\mathcal{M}) \tag{3.0.6}$$

3.

$$\lim_{\theta \to 0} \left\| \mathcal{Q}(fg) - \mathcal{Q}(f)\mathcal{Q}(g) \right\| = 0 \qquad \forall f, g \in \mathcal{C}(\mathcal{M}),$$

$$\lim_{\theta \to 0} \frac{1}{\theta} \left\| [\mathcal{Q}(f), \mathcal{Q}(g)] - i\mathcal{Q}(\{f, g\}) \right\| = 0 \qquad \forall f, g \in \mathcal{C}(\mathcal{M}) \tag{3.0.7}$$

4. *(asymptotic) isometry*

$$\langle \mathcal{Q}(f), \mathcal{Q}(g) \rangle_{\mathrm{End}(\mathcal{H})} \to \langle f, g \rangle_{L^2(\mathcal{M})} \qquad \text{as } \theta \to 0 \tag{3.0.8}$$

5. *irreducibility*: If $F \in \mathrm{End}(\mathcal{H})$ commutes with the image of $\mathcal{Q}$, then $F \propto \mathbb{1}$.

In items 3. and 4., the functions are assumed to be normalized $\|f\| = \|g\| = 1$, and the norm and inner product will be explained in the following. We will argue that the isometry property is typically a consequence of the other requirements. For compact $\mathcal{M}$ the Hilbert space $\mathcal{H}$ should be finite-dimensional and depends on θ, which is typically discrete. Note that the above "semi-classical" limit $\theta \to 0$ is not uniform in $f, g \in \mathcal{C}(\mathcal{M})$, but holds only for (sufficiently nice) fixed f, g.

The precise meaning of the semi-classical limit is debatable in general, and various refinements or variations of these requirements can be considered, cf. [211]. We will not dwell on these subtleties, because the concept of such a limit is not satisfactory in physics anyway: Planck's constant $\hbar$ and similarly the scale of noncommutativity θ are dimensionful scales that cannot be tuned. Hence, in physics, the "semi-classical limit" should be interpreted more properly as **semi-classical regime**, where the physical wavelengths under consideration are large compared to the scale of noncommutativity set by $\theta^{\mu\nu}$.

Integration, trace, and inner product

To understand the inner products in (3.0.8), we recall that symplectic manifolds are equipped with a symplectic volume form[2] Ω (1.1.25), which allows to define an integral over $\mathcal{M}$ as

$$\int_{\mathcal{M}} \Omega f, \qquad f \in C(\mathcal{M}). \tag{3.0.9}$$

This integral is invariant under Hamiltonian vector fields:

$$\int_{\mathcal{M}} \Omega \mathcal{L}_{V_g} f = 0 = \int_{\mathcal{M}} \Omega \{g, f\}, \tag{3.0.10}$$

which follows from the invariance of the symplectic form $\mathcal{L}_V \Omega = 0 = \mathcal{L}_V \omega$ (1.1.24). The analogous concept on the quantum side is clearly the trace, which satisfies the invariance property

$$\mathrm{Tr}([F, G]) = 0. \tag{3.0.11}$$

It is therefore natural to expect or require that a quantization map $\mathcal{Q}$ should relate the integral with the trace,

$$\mathrm{Tr}\mathcal{Q}(f) = c \int_{\mathcal{M}} \Omega f \qquad \text{at least for } \theta \to 0, \tag{3.0.12}$$

up to some normalization constant c. To understand this constant, consider some compact $\mathcal{M}$ with finite symplectic volume. Then (3.0.12) reduces for $f = 1$ to

$$\dim(\mathcal{H}) = \mathrm{Tr}\mathbb{1} = c \int_{\mathcal{M}} \Omega \tag{3.0.13}$$

using $\mathcal{Q}(1) = \mathbb{1}$ (3.0.5). In particular, $\mathcal{H}$ must be finite-dimensional for compact $\mathcal{M}$. Since the lhs is an integer, this amounts to a quantization condition for the symplectic volume. We will see that the normalization is typically given by the "Bohr–Sommerfeld" quantization $c = \frac{1}{(2\pi)^n}$ where $\dim \mathcal{M} = 2n$,

$$\mathrm{Tr}\mathcal{Q}(f) = \int_{\mathcal{M}} \frac{\Omega}{(2\pi)^n} f \qquad \text{at least for } \theta \to 0. \tag{3.0.14}$$

This condition will be understood in Section 4.2.1 as a quantization condition for the field strength of a connection on some $U(1)$ bundle.

The above quantization of symplectic volume will play a fundamental role in the following, since it entails that a physical theory on such a space has only finitely many degrees of freedom per volume. Quantum spaces with that property will often be denoted as

---

[2] We will often suppress Ω in later chapters.

fuzzy spaces. This is the distinctive feature of the quantum geometry under consideration here, as opposed to other notions of noncommutative geometry[3].

The compatibility of the integral and the trace has important implications. The integral allows to define an inner product on the algebra of functions on $\mathcal{M}$ via

$$\langle f, g \rangle := \int_{\mathcal{M}} \frac{\Omega}{(2\pi)^n} f^* g, \qquad f, g \in C(\mathcal{M}). \tag{3.0.15}$$

For example, given a probability density $\rho \geq 0$ on $\mathcal{M}$ with $\frac{1}{(2\pi)^n} \int \Omega \rho = 1$, the expectation value of some classical observable $f \in C(\mathcal{M})$ is given by

$$E_\rho(f) = \int \frac{\Omega}{(2\pi)^n} \rho f = \langle \rho, f \rangle. \tag{3.0.16}$$

Similarly, there is a natural inner product on the quantum observables known as Hilbert–Schmidt inner product, which is defined in terms of the trace on $\mathcal{H}$ as follows:

$$\langle F, G \rangle := \mathrm{Tr}(F^\dagger G), \qquad F, G \in \mathrm{End}(\mathcal{H}). \tag{3.0.17}$$

The space of observables $\mathrm{End}(\mathcal{H})$ thereby becomes a Hilbert space,[4] which is the analog of the space $L^2(\mathcal{M})$ of square-integrable functions on $\mathcal{M}$. This is the basis for the probability interpretation in quantum mechanics: the expectation value of some quantum observable $F \in \mathrm{End}(\mathcal{H})$ in a system described by the density matrix $\hat{\rho} \geq 0$ is given by

$$E_\rho(F) = \mathrm{Tr}\hat{\rho}F = \langle \hat{\rho}, F \rangle. \tag{3.0.18}$$

The correspondence principle thus suggests that the quantization map $\mathcal{Q}$ should also respect this inner product as in (3.0.8),

$$\langle \mathcal{Q}(f), \mathcal{Q}(g) \rangle_{\mathrm{End}(\mathcal{H})} = \langle f, g \rangle_{L^2(\mathcal{M})} \qquad \text{at least for } \theta \to 0. \tag{3.0.19}$$

In other words, $\mathcal{Q}$ should be an isometry, at least in the semi-classical regime. As explained earlier, these inner products are invariant under transformations generated by Hamiltonian vector fields $V_f = \{f, .\}$ and their quantum counterparts $[F, .]$ acting on $C(\mathcal{M})$ and $\mathrm{End}(\mathcal{H})$, respectively. Therefore, the norm in (3.0.7) is naturally taken to be the Hilbert–Schmidt norm $\|f\|^2 = \langle f, f \rangle$, or a suitable rescaling thereof. We will see that for many homogeneous spaces, there is typically a unique quantization map $\mathcal{Q}$ that is an exact isometry and respects the group invariance.

If $\mathcal{M}$ is a Poisson manifold but not symplectic, then analogous statements should be required for the symplectic leaves. The quantization is then no longer irreducible, and we will not discuss this case any further.

---

[3] The mathematical literature on noncommutative geometry typically considers spaces with infinitely many degrees of freedom per unit volume, such as the so-called noncommutative torus T_θ^2. In contrast, the fuzzy torus T_N^2 discussed in Section 3.4 meets this requirement, and so does the Moyal–Weyl quantum plane $\mathbb{R}_\theta^{2n}$.

[4] More precisely, the space of normalizable operators known as Hilbert–Schmidt operators forms a Hilbert space.

Quantum mechanics versus quantum geometry

Even though the above structures originate from quantum mechanics, we will apply them in a very different physical context. The algebra $\mathrm{End}(\mathcal{H})$ and its inner product structure will be interpreted as quantized algebra of functions $\mathcal{C}(\mathcal{M})$ on space(time) or toy models thereof, which sets the stage for physics in the framework of matrix models. In that context, the states in the Hilbert space $|\psi\rangle \in \mathcal{H}$ will have no direct physical meaning, and probability interpretation of measurements of observables plays no role. Nevertheless, certain (coherent) states will be very useful as a tool to understand these quantum geometries.

This different context also leads to different classes of relevant functions or observables. In quantum mechanics, the dynamics is governed by a Hamiltonian of type $H \sim \frac{1}{2m}P^2 + V(X)$. This implies that physically relevant low-energy states are restricted to $P \approx 0$, while they can be completely nonlocal in X, leading to strongly nonclassical behavior. In the present context, fields will be functions on $\mathcal{M}$, and the low-energy sector will correspond to slowly varying functions in all directions of $\mathcal{M}$. This is in some sense closer to classical field theory on $\mathcal{M}$. In fact, much of this book will be concerned with the analog of classical physics or field theory on quantum spaces. Quantum physics on quantum space(time) will be defined in a second step in terms of a (matrix) path integral on quantum spaces, as discussed in Sections 6.3, 7.2, and Chapter 11.

Quantization versus de-quantization

The concept of a quantization map $\mathcal{Q}$ is central to the understanding of quantum geometry, and we will gain some intuition through a variety of examples. Throughout this book, we will freely switch via $\mathcal{Q}$ between the classical and quantum point of view. The classical side provides the intuitive understanding of the quantum setting.

Nevertheless, the above formulation is not fully satisfactory. It requires a whole family of quantizations for each θ, while the quantization parameter in physics (i.e. $\hbar$ in the case of quantum mechanics) is a fixed number, which cannot be changed. One is thus faced with the opposite problem of *de-quantization*, i.e. extracting the *semi-classical (Poisson) limit* from a given quantum system. Similarly in quantum geometry, we are typically given some quantum system, and aim to extract a classical picture. Thus, we cannot assume a continuous family of quantization maps $\mathcal{Q}$ for any θ. Therefore, the classical picture of a given quantum system can only be approximate, and applies only in some *scale regime*. In quantum mechanics, the scale that separates the classical from the quantum regime is set by the area quantum $\hbar$ in phase space.

An important lesson is that quantum mechanics should not be viewed as a quantization of classical mechanics, but as the fundamental starting point. In the same vein, quantum geometry should not be viewed as a quantization of classical geometry. Rather, we will consider quantum geometry as more fundamental, and try to understand if and how classical geometry (and physics, hopefully) emerges as an effective description of quantum geometry in some regime.

Therefore, we should address the following problem: Given some "quantum" or matrix geometry (given in terms of a set of matrices $X^a \in \mathrm{End}(\mathcal{H})$), how can we extract an underlying semi-classical Poisson manifold? Clearly the answer can only be approximate, and requires extra structure to identify some scale, which separates the semi-classical regime from the quantum regime. Such a structure will be provided in the context of matrix models through the metric δ_{ab} on the target space $\mathbb{R}^d$. As discussed in Section 4, this will indeed allow to extract an effective semi-classical (Riemann–Poisson) geometry from suitable matrices X^a.

3.1 The Moyal–Weyl quantum plane $\mathbb{R}^{2n}_\theta$

3.1.1 Canonical Moyal–Weyl quantum plane or quantum mechanical phase space

The prototype of a quantum space is provided by quantum mechanical phase space. This is defined in terms of hermitian (more precisely selfadjoint) operators Q^i, P_j, which provide an irreducible representation of the canonical commutation relations (3.0.2) on a Hilbert space $\mathcal{H}$

$$[Q^i, P_j] = i\theta\delta^i_j\mathbb{1}, \qquad [Q^i, Q^j] = 0 = [P_i, P_j] \tag{3.1.1}$$

for $i, j = 1, \ldots, n$, where θ is interpreted as $\hbar$ in Quantum Mechanics. To introduce a more systematic notation, we rename these generators as follows:

$$X^a = \begin{pmatrix} Q^i \\ P_j \end{pmatrix}, \qquad a = 1, \ldots, 2n. \tag{3.1.2}$$

Introducing the antisymmetric matrix

$$\theta^{ab} = \theta \begin{pmatrix} 0 & \mathbb{1}_n \\ -\mathbb{1}_n & 0 \end{pmatrix} \tag{3.1.3}$$

the canonical commutation relations take the following more systematic form

$$[X^a, X^b] = i\theta^{ab}\,\mathbb{1}, \qquad a, b = 1, \ldots, 2n. \tag{3.1.4}$$

Representation and some identities

According to the Stone–von Neumann theorem, we can assume that $\mathcal{H} \cong L^2(\mathbb{R}^n)$ is the standard Hilbert space of square-integrable functions on $\mathbb{R}^n$, and

$$Q^i\phi(q) = q^i\phi(q)$$
$$P_j\phi(q) = -i\theta\,\partial_j\phi(q). \tag{3.1.5}$$

$\mathcal{H}$ is spanned by the improper basis of plane waves $|e_p\rangle = e^{ip_iq^i}$, which satisfy

$$P_j|e_p\rangle = \theta p_j|e_p\rangle$$
$$e^{ib^jP_j}|e_p\rangle = e^{i\theta b^jp_j}|e_p\rangle$$

$$e^{ia_jQ^j}|e_p\rangle = |e_{p+a}\rangle$$
$$\langle e_k|e_p\rangle = (2\pi)^n\delta^n(p-k). \tag{3.1.6}$$

This allows to write a completeness relation and a trace formula as follows:

$$\mathbb{1} = \int \frac{d^np}{(2\pi)^n}\,|e_p\rangle\langle e_p|,$$

$$\mathrm{Tr}A = \int \frac{d^np}{(2\pi)^n}\,\langle e_p|A|e_p\rangle. \tag{3.1.7}$$

It is natural to consider the quantum plane waves

$$V_k = e^{ik_aX^a} = e^{i(a_jQ^j+b^jP_j)} = e^{\frac{i}{2}\theta a_jb^j}e^{ia_jQ^j}e^{ib^jP_j}, \tag{3.1.8}$$

where $k_a = (a^i, b_j)$, which satisfy the following trace formula:

$$\begin{aligned}
\mathrm{Tr}_{\mathcal{H}}(e^{ik_aX^a}) &= \int \frac{d^np}{(2\pi)^n}\,\langle e_p|e^{i(a_jQ^j+b^jP_j)}|e_p\rangle\\
&= \int \frac{d^np}{(2\pi)^n}\,e^{\frac{i}{2}\theta a_jb^j}e^{i\theta b^jp_j}\langle e_p|e^{ia_jQ^j}|e_p\rangle\\
&= \int \frac{d^np}{(2\pi)^n}\,e^{i\theta b^jp_j}\langle e_p|e_{p+a}\rangle\\
&= \int d^np\,e^{i\theta b^jp_j}\delta^n(a) = \left(\frac{2\pi}{\theta}\right)^n\delta^n(b)\delta^n(a)\\
&= \left(\frac{2\pi}{\theta}\right)^n\delta^{2n}(k). \tag{3.1.9}
\end{aligned}$$

The same formula also holds on a general Moyal–Weyl quantum plane as discussed in the following. Even though these formulas are somewhat formal for the sake of readability, they can easily be made rigorous.

Exercise 3.1.1   Show the following identity

$$e^{tA}Be^{-tA} = e^{t[A,\cdot]}B = \sum_{n=0}^\infty \frac{t^n}{n!}[A,[A,\dots[A,B]\dots]] \tag{3.1.10}$$

for any operators $A, B \in \mathrm{End}(\mathcal{H})$.

Exercise 3.1.2   (Baker–Campbell–Hausdorff formula)
Show that for operators A, B which satisfy $[A, [A, B]] = 0 = [B, [A, B]]$, the following identity holds:

$$e^{A+B} = e^{-\frac{1}{2}[A,B]}e^Ae^B. \tag{3.1.11}$$

<u>Hint:</u> Observe that $e^{tA}Be^{-tA} = B + t[A, B]$, and use this to establish that the function $F(t) := e^{tA}e^{tB}$ satisfies

$$\frac{d}{dt}F(t) = (A+B+t[A,B])F(t),$$

with solution $F(t) = e^{t(A+B)+\frac{t^2}{2}[A,B]}$. The Baker–Campbell–Hausdorff formula follows for $t = 1$.

3.1.2 General Moyal–Weyl quantum plane $\mathbb{R}^{2n}_\theta$

Now consider the general Moyal–Weyl quantum plane denoted as $\mathbb{R}^{2n}_\theta$, which is defined by

$$[X^a, X^b] = i\theta^{ab} \mathbb{1} \tag{3.1.12}$$

for a general nondegenerate antisymmetric matrix $\theta^{ab} = -\theta^{ba}$. The X^a are viewed as quantizations of the Cartesian coordinate functions x^a on $\mathbb{R}^{2n}$ satisfying the Poisson brackets

$$\{x^a, x^b\} = \theta^{ab}, \tag{3.1.13}$$

which correspond to the symplectic form

$$\omega = \frac{1}{2}\theta^{-1}_{ab} dx^a dx^b. \tag{3.1.14}$$

It is well known that every antisymmetric matrix $\theta^{ab} = -\theta^{ba}$ can be brought to a block-diagonal standard form by an orthogonal transformation $O \in O(2n)$:

$$(O^T \theta O)^{ab} = \begin{pmatrix} \begin{pmatrix} 0 & \theta_1 \\ -\theta_1 & 0 \end{pmatrix} & & \\ & \ddots & \\ & & \begin{pmatrix} 0 & \theta_n \\ -\theta_n & 0 \end{pmatrix} \end{pmatrix} =: \bar{\theta}^{ab}. \tag{3.1.15}$$

Using a suitable rescaling of generators, it follows that for any Moyal–Weyl quantum plane, the generator X^a can be expressed as a linear combination of canonical Q^i and P_j generators,

$$X^a = \begin{pmatrix} A & B \end{pmatrix} \begin{pmatrix} Q \\ P \end{pmatrix} = A^a_i Q^i + B^{aj} P_j. \tag{3.1.16}$$

This in turn implies that the commutation relations (3.1.4) are invariant under symplectic transformations $SP(n)$ acting on the canonical variables Q^i, P_j. It follows that the trace formula (3.1.9) also holds in the general case,

$$\mathrm{Tr}_{\mathcal{H}}(e^{ik_a X^a}) = \left(\frac{2\pi}{\theta}\right)^n \delta^{2n}(k), \tag{3.1.17}$$

where the **noncommutativity scale** is defined by

$$L^2_{\mathrm{NC}} := \theta := |\theta^{ab}|^{1/2n} = (\theta_1 \ldots \theta_n)^{1/n}. \tag{3.1.18}$$

The Moyal–Weyl quantum plane is not just a particularly nice and simple quantum space. Since any symplectic space can be locally brought into canonical form due to Darboux's theorem, the Moyal–Weyl quantum plane can be viewed as a local approximation to a generic nondegenerate quantum space. This idea will be developed further in the general discussion of matrix geometries in Section 4.

3.1.3 Weyl quantization

Consider the general Moyal–Weyl quantum plane $\mathbb{R}^{2n}_\theta$ in more detail. This space is characterized by the existence of a translation group $\mathbb{R}^{2n}$ which acts on the space of functions $\text{End}(\mathcal{H})$. To describe this group action, consider the noncommutative plane waves on $\mathbb{R}^{2n}_\theta$

$$V_k = e^{ik_a X^a}. \tag{3.1.19}$$

Using the Baker–Campbell–Hausdorff formula, the commutation relations (3.1.4) imply

$$V_k^{-1} X^a V_k = X^a + k_b \theta^{ba} \mathbb{1},$$
$$V_k V_p = e^{-ik_a \theta^{ab} p_b} V_p V_k = e^{-\frac{i}{2} k_a \theta^{ab} p_b} V_{k+p}. \tag{3.1.20}$$

This takes a more familiar form in terms of the translation or displacement operators

$$U_y := e^{i\tilde{y}_a X^a} = e^{iy^b \theta_{ba}^{-1} X^a}, \tag{3.1.21}$$

where

$$\tilde{y}_a := y^b \theta_{ba}^{-1}. \tag{3.1.22}$$

Then (3.1.20) becomes

$$\boxed{\begin{aligned} U_y^{-1} X^a U_y &= X^a + y^a \mathbb{1}, \\ U_y U_x &= e^{iy^a \theta_{ab}^{-1} x^b} U_x U_y = e^{\frac{i}{2} y^a \theta_{ab}^{-1} x^b} U_{x+y}. \end{aligned}} \tag{3.1.23}$$

Therefore, the operators U_k provide a unitary representation of the translation group $\mathbb{R}^{2n}$ on $\text{End}(\mathcal{H})$,

$$\mathbb{R}^{2n} \times \text{End}(\mathcal{H}) \to \text{End}(\mathcal{H})$$
$$(y, \Phi) \mapsto T_y(\Phi) := U_y^{-1} \Phi U_y \tag{3.1.24}$$

and a projective representation on $\mathcal{H}$,

$$\mathbb{R}^{2n} \times \mathcal{H} \to \mathcal{H}$$
$$(y, |\psi\rangle) \mapsto U_y |\psi\rangle. \tag{3.1.25}$$

"Projective" indicates the extra phase factor in (3.1.23), and the commutation relations (3.1.4) can accordingly be viewed as a central extension of the abelian Lie algebra of translations. Since the representations are unitary, they decompose into the direct sum (or the direct integral) of irreducible representations. These irreps are one-dimensional, since the translation group $\mathbb{R}^{2n}$ is abelian. For $\text{End}(\mathcal{H})$, they are given by the noncommutative plane waves V_k (3.1.19),

$$U_y^{-1} V_k U_y = e^{ik_a y^a} V_k \tag{3.1.26}$$

using (3.1.23). Hence, the V_k are eigenvectors of T_y with eigenvalue $e^{ik_a y^a}$, and in this sense they correspond to the classical plane waves on $\mathbb{R}^{2n}$,

$$v_k(x) = e^{ik_a x^a}, \tag{3.1.27}$$

which are also eigenvectors of the translation operator

$$T_y: \quad x^a \to x^a + y^a \tag{3.1.28}$$

with the same eigenvalue $e^{ik_a y^a}$. Since the sets of (NC and commutative) plane waves are complete, there is a unique quantization map defined through

$$\begin{aligned} \mathcal{Q}: \quad & \mathcal{C}(\mathbb{R}^{2n}) \to \mathrm{End}(\mathcal{H}) \\ & v_k = e^{ikx} \mapsto e^{ikX} = V_k, \end{aligned} \tag{3.1.29}$$

which intertwines the action of the translation group and is isometric. The definition can be stated more rigorously in terms of square-integrable wave-packets:

$$\mathcal{Q}: \quad L^2(\mathbb{R}^{2n}) \to \mathrm{End}(\mathcal{H})$$

$$\phi = \int d^{2n}k\, \hat\phi(k) e^{ikx} \mapsto \mathcal{Q}(\phi) = \int d^{2n}k\, \hat\phi(k) e^{ikX}, \tag{3.1.30}$$

where $\hat\phi$ denotes the Fourier transform of ϕ. This map is called **Weyl quantization**, which will sometimes be indicated by a subscript $\mathcal{Q} \equiv \mathcal{Q}_W$. It is an *isometry*

$$\langle \mathcal{Q}(\phi), \mathcal{Q}(\psi) \rangle = \langle \phi, \psi \rangle_{L^2} \tag{3.1.31}$$

with respect to the natural inner products on $L^2(\mathbb{R}^{2n})$ and $\mathrm{End}(\mathcal{H})$

$$\langle \phi, \psi \rangle = \int\limits_{\mathbb{R}^{2n}} \frac{\Omega}{(2\pi)^n} \phi(x)^* \psi(x), \qquad \langle \Phi, \Psi \rangle = \mathrm{Tr}(\Phi^\dagger \Psi), \tag{3.1.32}$$

where

$$\Omega = \frac{1}{n!}\omega^n = \sqrt{|\theta_{ab}^{-1}|}\, d^{2n}x = L_{\mathrm{NC}}^{-2n}\, d^{2n}x \tag{3.1.33}$$

is the symplectic volume form. More precisely,

$$\mathcal{Q}_W: \quad L^2(\mathbb{R}^{2n}) \to \mathrm{HS}(\mathcal{H}) \subset \mathrm{End}(\mathcal{H}) \tag{3.1.34}$$

is an isometry from square-integrable functions to Hilbert–Schmidt operators $\mathrm{HS}(\mathcal{H})$, which is the Hilbert space of operators with respect to the norm (3.1.32). Explicitly, the inner product of plane waves is obtained from (3.1.17) as

$$\langle v_k, v_p \rangle = \left(\frac{2\pi}{L_{\mathrm{NC}}^2}\right)^n \delta^{2n}(k - p) = \langle V_k, V_p \rangle. \tag{3.1.35}$$

These properties determine the map $\mathcal{Q}$ uniquely. Of course $\mathcal{Q}$ should be viewed as a map of vector spaces, since it does not respect the algebra structures on both sides. It is easy to verify

$$\mathcal{Q}(f^*) = \mathcal{Q}(f)^\dagger. \tag{3.1.36}$$

Moreover, the quantization condition (3.0.7) holds in the following sense: for given k, p, we have

$$[V_k, V_p] = -2i \sin\left(\frac{1}{2}k_a\theta^{ab}p_b\right)V_{k+p}$$

$$\{v_k, v_p\} = -k_a\theta^{ab}p_b\, v_{k+p}. \tag{3.1.37}$$

Keeping k, p fixed, this clearly implies[5]

$$\lim_{\theta\to 0}\left(\mathcal{Q}(v_k)\mathcal{Q}(v_p) - \mathcal{Q}(v_k v_p)\right) = 0,$$

$$\lim_{\theta\to 0}\frac{1}{\theta}\left([\mathcal{Q}(v_k), \mathcal{Q}(v_p)] - i\mathcal{Q}(\{v_k, v_p\})\right) = 0. \tag{3.1.38}$$

Finally for $k = 0$ we obtain $\mathcal{Q}(1) = \mathbb{1}$. Therefore, Weyl quantization satisfies all the requirements for quantization maps in (3.0.7), and provides a correspondence between classical functions on $\mathbb{R}^{2n}$ and operators in $\mathrm{End}(\mathcal{H})$. Moreover, the following property follows from (3.1.17):

$$\int \frac{\Omega}{(2\pi)^n}\phi(x) = \mathrm{Tr}(\mathcal{Q}(\phi)). \tag{3.1.39}$$

This encodes the Bohr–Sommerfeld quantization: each "unit cell" in quantized phase space has volume $2\pi\theta$, where θ is replaced by $\hbar$ in the context of quantum mechanics.

The particular properties of Weyl quantization are based on the fact that $\mathbb{R}^{2n}$ is a homogeneous space. We will obtain analogous quantization maps for more general coadjoint orbits associated to some Lie group G, whose symplectic form is invariant under G. In particular, while both sides of

$$\mathrm{Vol}_\omega(\mathcal{M}) = (2\pi)^n\mathrm{Tr}\mathbb{1} = (2\pi)^n\dim\mathcal{H} \tag{3.1.40}$$

diverge on $\mathcal{M} = \mathbb{R}^{2n}$, this will become a meaningful and important relation for compact $\mathcal{M}$. Hence the symplectic volume is quantized in multiples of $(2\pi)^n$, which is the essence of the Bohr–Sommerfeld quantization condition in phase space.

Although $\mathcal{Q}$ was defined only for square-integrable functions, it extends naturally to more general functions including polynomials. By differentiating $\mathcal{Q}(e^{i\varepsilon kx}) = e^{i\varepsilon kX}$ (3.1.29) n times with respect to ε at $\varepsilon = 0$, we obtain

$$\mathcal{Q}\left((k_a x^a)^n\right) = (k_a X^a)^n. \tag{3.1.41}$$

Expanding both sides in polynomials of k_a, it follows that Weyl quantization always maps polynomials in x^a to totally symmetrized polynomials in X^a:

$$\mathcal{Q}(x^a) = X^a,$$

$$\mathcal{Q}(x^a x^b) = \frac{1}{2}(X^a X^b + X^b X^a),$$

$$\mathcal{Q}(x^{a_1}\ldots x^{a_n}) = \left(X^{a_1}\ldots X^{a_n}\right)_{\mathrm{symm}}. \tag{3.1.42}$$

Here the appropriate factor $\frac{1}{n!}$ is understood in the symmetrization [6].

---

[5] Note that the limit $\theta \to 0$ is not uniform, i.e. the correspondence holds only for fixed k, p.

[6] This argument can be made more rigorous by adding a Gaussian damping term.

There exist other quantization maps besides Weyl quantization, such as normal ordering, or coherent state quantization. These have slightly different properties, and are useful in certain contexts. Coherent state quantization turns out to be particularly useful, which will be discussed in Section 3.1.5.

Application: Quantum–classical correspondence in statistical physics

Classical statistical physics is based on the canonical partition function, which is defined by

$$Z_{\text{class}} = \int_{\mathcal{P}} \Omega \, e^{-\beta H}. \tag{3.1.43}$$

Here H is the classical Hamilton function on the (high-dimensional) phase space $\mathcal{P}$ of dimension $2N$, typically describing some many-particle system, and $\beta = kT$. In the corresponding quantum system, the canonical partition function is defined as

$$Z_{\text{quant}} = \text{Tr} e^{-\beta \hat{H}}, \tag{3.1.44}$$

where $\hat{H} = \mathcal{Q}(H)$ is the Hamiltonian. Using the correspondence (3.1.39) and approximating $\mathcal{Q}(e^{-\beta H}) \approx e^{-\beta \hat{H}}$ for sufficiently small β, this is approximated by

$$Z_{\text{quant}} \approx \int_{\mathcal{P}} \frac{\Omega}{(2\pi)^N} \, e^{-\beta H} = \frac{1}{(2\pi)^N} Z_{\text{class}}, \tag{3.1.45}$$

where the trace is replaced by an integral over phase space. This establishes the correspondence between classical and quantum statistical mechanics valid at high temperatures[7].

Let us illustrate this in the example of a noninteracting quantum gas. Consider an ideal gas of N noninteracting particles with mass m in a volume $V \subset \mathbb{R}^3$ at temperature T. The corresponding classical phase space is given by $\mathcal{P} = V^N \times \mathbb{R}^{3N}$ with Hamilton function

$$H(q^i, p_j) = \sum \frac{\vec{p}_i^2}{2m}. \tag{3.1.46}$$

The classical canonical partition function is given by

$$Z_{\text{class}} = \int_{\mathcal{P}} \Omega \, e^{-\beta H} = V^N \int d^{3N} p_i \, e^{-\beta \sum \frac{\vec{p}_i^2}{2m}} = V^N \Big(\frac{2\pi m}{\beta}\Big)^{3N/2} \tag{3.1.47}$$

assuming $\omega = \sum_i dp_i dq^i$. This gives for the free energy

$$\beta F = -\ln Z = -N\Big(\ln V + \frac{3}{2} \ln\Big(\frac{2\pi m}{\beta}\Big)\Big) \tag{3.1.48}$$

and

$$U = \frac{d}{d\beta} F = \frac{3}{2} NkT. \tag{3.1.49}$$

---

[7] Of course there are many reasons why the quantum case is preferred. For example, Z_{quant} is dimensionless as it should due to the presence of $\hbar$, while Z_{class} is not a priori. Also, positivity of the entropy is manifest in the quantum case but not in the classical case.

In quantum mechanics, the Hamilton operator is

$$\hat{H}(Q^i, P_j) = Q(H(q^i, p_j)) = \sum \frac{\vec{P}_i^2}{2m}, \tag{3.1.50}$$

acting on the Hilbert space $\mathcal{H} = L^2(V^N) = L^2(V)^{\otimes N}$ of square-integrable functions on V^N in the case of nonidentical particles. We assume for simplicity that $V = L^3$ with periodic boundary conditions, i.e. we work on a torus T^{3N}. Then a basis of energy eigenstates is given by

$$|\vec{p}_i\rangle = \mathcal{N} \, e^{\frac{i}{\hbar} p_i q^i}, \qquad p_i = \frac{2\pi \hbar n_i}{L}, \quad n_i \in \mathbb{Z}$$

$$\vec{P}_i |\vec{p}_i\rangle = -i\hbar \frac{\partial}{\partial q^i} (\mathcal{N} \, e^{\frac{i}{\hbar} p_i q^i}) = \vec{p}_i |\vec{p}_i\rangle \tag{3.1.51}$$

for some normalization $\mathcal{N}$. The partition function for nonidentical particles is then

$$Z_{\text{quant}} = \text{Tr} \, e^{-\beta \hat{H}} = \sum_{p_i} \langle p_i | e^{-\beta \hat{H}} | p_i \rangle. \tag{3.1.52}$$

The sum $\sum_{p_i}$ can be approximated by a Riemann integral:

$$
\begin{aligned}
Z_{\text{quant}} &\approx \frac{V^N}{(2\pi \hbar)^{3N}} \int dp_i e^{-\beta \sum \frac{\vec{p}_i^2}{2m}} \\
&= \frac{1}{(2\pi \hbar)^{3N}} \int_{V^N} dq^i \int dp_i e^{-\beta \sum \frac{\vec{p}_i^2}{2m}} \\
&= \frac{1}{(2\pi)^{3N}} \int_{\mathcal{P}} \Omega \, e^{-\beta H(q^i, p_j)} = \frac{1}{(2\pi \hbar)^{3N}} Z_{\text{class}}.
\end{aligned} \tag{3.1.53}
$$

The approximation is justified as long as the function $e^{-\beta H}$ is approximately constant on the sampling size, i.e. the thermal wavelength $\lambda_T = \sqrt{\frac{2\pi\beta\hbar^2}{m}}$ is much smaller than the inter-particle spacing L/N.

For identical particles, the correct Hilbert space is given by the totally symmetrized wave-functions $\mathcal{H} = L^2(V)^{\otimes_S N}$. This leads to a factor $\frac{1}{N!}$ in the partition function compared with (3.1.53), which is essential to obtain the correct extensive property (notably $\ln V \to \ln \frac{V}{N}$ in (3.1.48)) of the thermodynamic quantities.

Mathematical refinements and Wigner map

Weyl quantization can be cast into a closed form using the operator-valued distribution

$$\Delta(x) = \int \frac{d^{2n}k}{(2\pi)^{2n}} \, e^{-ik_a(x^a - X^a)}. \tag{3.1.54}$$

Then the Weyl quantization map can be written as

$$\mathcal{Q}(\phi) = \int d^{2n}x \, \phi(x) \Delta(x). \tag{3.1.55}$$

Using the relation (3.1.17), the inverse of Weyl quantization can be written down as follows:

$$Q^{-1}: \quad \mathrm{End}(\mathcal{H}) \to \mathcal{C}(\mathcal{M})$$
$$\Phi \mapsto \phi(x) = (2\pi\theta)^n \, \mathrm{Tr}(\Phi\Delta(x)) \tag{3.1.56}$$

for $\Phi = \int d^{2n}k \, \hat{\phi}(k)e^{ikX}$. In the context of quantum mechanics this is known as **Wigner map**, which maps an observable Φ to the classical (Wigner) function $\phi(x)$ on phase space.

The Wigner map can be made more explicit in the position space representation. For simplicity we restrict ourselves to the canonical case, where $x^a = (q^j, p_j)$. Consider the improper position eigenstates $Q^j|q\rangle = q^j|q\rangle$ with $\langle q|r\rangle = \delta(q-r)$. Then the Wigner map takes the form[8]

$$\phi(x^a) = \phi(q^j, p_j) = 2^n \int d^n r \, e^{2ipr/\theta} \langle q-r|\Phi|q+r\rangle. \tag{3.1.57}$$

To see this, we use the Baker–Campbell–Hausdorff formula

$$e^{i(a_j Q^j + b^j P_j)} = e^{-i\theta a_j b^j/2} e^{ib^j P_j} e^{ia_j Q^j} \tag{3.1.58}$$

and compute

$$\mathrm{Tr}_{\mathcal{H}}(\Phi e^{ik_a X^a}) = \int d^n q \, \langle q|\Phi e^{i(a_j Q^j + b^j P_j)}|q\rangle = \int d^n q \, e^{ia_j q^j} e^{-i\theta a_j b^j/2} \langle q|\Phi e^{ib^j P_j}|q\rangle$$

$$= \int d^n q \, e^{ia_j(q^j - \theta b^j/2)} \langle q|\Phi|q - \theta b\rangle$$

$$= \int d^n q \, e^{ia_j q^j} \langle q + \frac{\theta b}{2}|\Phi|q - \frac{\theta b}{2}\rangle \tag{3.1.59}$$

for $k^a = (a_j, b^j)$, using $e^{ib^j P_j}|q\rangle = |q - \theta b\rangle$. Therefore, (3.1.56) gives

$$\phi(x) = (2\pi\theta)^n \, \mathrm{Tr}(\Phi\Delta(x)) = \frac{\theta^n}{(2\pi)^n} \int d^{2n}k \, e^{-ik_a x^a} \mathrm{Tr}(\Phi e^{ik_a X^a})$$

$$= \frac{\theta^n}{(2\pi)^n} \int d^{2n}k \, e^{ik_a x^a} \mathrm{Tr}(\Phi e^{-ik_a X^a})$$

$$= \frac{\theta^n}{(2\pi)^n} \int d^{2n}k \, e^{ik_a x^a} \int d^n q \, e^{-ia_j q^j} \langle q - \frac{\theta b}{2}|\Phi|q + \frac{\theta b}{2}\rangle \tag{3.1.60}$$

and we obtain

$$\phi(q,p) = \frac{\theta^n}{(2\pi)^n} \int d^n a \, d^n b \, e^{i(a_j q^j + b^j p_j)} \int d^n \tilde{q} \, e^{-ia_j \tilde{q}^j} \langle \tilde{q} - \frac{\theta b}{2}|\Phi|\tilde{q} + \frac{\theta b}{2}\rangle$$

$$= \theta^n \int d^n b \, e^{ib^j p_j} \langle q - \frac{\theta b}{2}|\Phi|q + \frac{\theta b}{2}\rangle$$

$$= 2^n \int d^n r \, e^{2i\theta^{-1} r^j p_j} \langle q-r|\Phi|q+r\rangle.$$

The isometry property (3.1.31) implies immediately

$$\mathrm{Tr}(\Phi^\dagger \Psi) = \int \frac{d^n p \, d^n q}{(2\pi\theta)^n} \, \phi^*(q,p)\psi(q,p). \tag{3.1.61}$$

[8] Notice that here $|q\rangle$ are not coherent states but position eigenstates.

In particular, we can apply this to the density matrix $\rho = |\psi\rangle\langle\psi|$ corresponding to a pure state. The corresponding Wigner function is

$$\phi(q^j, p_j) = 2^n \int d^n r \, e^{2ipr/\theta} \psi(q-r)^* \psi(q+r) \tag{3.1.62}$$

in terms of the position space wave function $\psi(q)$. The Wigner function can be viewed as a quasi-probability density in phase space that captures all properties of the quantum mechanical state, but is in general not positive.

From the point of view of quantum geometry, the Wigner map provides a classical avatar $\phi(q^j, p_j)$ of some quantum "function" $\Phi \in \text{End}(\mathcal{H})$. In the context of matrix models, we will always work with (the analog of) such functions on phase space, and the Wigner map (3.1.56) – or some analog formula in terms of coherent states – will take center stage.

3.1.4 Derivatives and inner derivations

The above discussion of translations allows to identify the generators of translations on $\mathbb{R}^{2n}_\theta$ as derivative operators. In complete analogy to the commutative case, we can write the finite translation operators as exponentials of their generators as

$$T_y(\Phi) = \exp(y^a \partial_a)\Phi = U_y^{-1} \Phi U_y. \tag{3.1.63}$$

Together with (3.1.21) this gives

$$\boxed{\partial_a \Phi = -i\theta_{ab}^{-1}[X^b, \Phi],} \tag{3.1.64}$$

and the reader should verify the relation

$$\partial_a X^b = \delta_a^b. \tag{3.1.65}$$

This innocent-looking formula conveys a very important message: derivatives on quantum spaces arise as inner derivations, i.e. as commutators with appropriate generators. This observation is at the very heart of noncommutative gauge theory on quantum spaces, and leads to their formulation through matrix models, which is fundamentally different and simpler than the traditional formulation in terms of connections and fiber bundles. We also note that the derivations on $\mathbb{R}^{2n}_\theta$ are mutually commuting:

$$\partial_a \partial_b = \partial_b \partial_a. \tag{3.1.66}$$

These considerations can be extended to higher-order differential operators. With hindsight, we introduce a metric structure on $\mathbb{R}^{2n}_\theta$ in terms of the following **Matrix Laplacian**

$$\Box := \delta_{ab}[X^a, [X^b, .]] = -\gamma^{ab}\partial_a\partial_b : \quad \text{End}(\mathcal{H}) \to \text{End}(\mathcal{H}). \tag{3.1.67}$$

Here

$$\gamma^{ab} = \theta^{aa'}\theta^{bb'}\delta_{a'b'} \tag{3.1.68}$$

is the **effective metric** on $\mathbb{R}^{2n}_\theta$. The origin of this Laplacian and the role of δ_{ab} will become clear in the context of matrix models.

3.1.5 Coherent states on the Moyal–Weyl quantum plane

So far, we have focused on the observables or operators in $\text{End}(\mathcal{H})$ and their relation to the functions in $\mathcal{C}(\mathcal{M})$. Now we will consider the Hilbert space $\mathcal{H}$ itself, and single out preferred states in $\mathcal{H}$ denoted as coherent states, which have special localization and transformation properties. Their generalizations will play a central role in the subsequent developments. In particular, we will discuss an alternative to Weyl quantization denoted as coherent state quantization, which can be generalized to more general quantum spaces.

Coherent states on the Moyal–Weyl quantum plane are states $|x\rangle \in \mathcal{H}$ that are optimally localized at some point $x \in \mathbb{R}^{2n}$. Since the translation group $\mathbb{R}^{2n}$ acts on $\mathcal{H}$, it suffices to consider the state $|0\rangle$ localized at the origin; all others are then obtained as

$$|x\rangle := U_x|0\rangle, \qquad x \in \mathbb{R}^{2n}. \tag{3.1.69}$$

"Optimally localized" means by definition that the expectation value

$$\langle 0|H|0\rangle \quad \text{is minimal}, \qquad H = \frac{1}{2}X^a X^b \delta_{ab} \tag{3.1.70}$$

and similarly for the states $|x\rangle$ centered at x. Note that the metric δ_{ab} on $\mathbb{R}^{2n}$ enters here, which will be discussed in more detail later. This minimization condition holds if and only if $|0\rangle$ is the ground state of H, i.e.

$$H|0\rangle = \lambda|0\rangle, \tag{3.1.71}$$

where λ the minimal eigenvalue of H. Since H is nothing but the quantum mechanical harmonic oscillator in the canonical case $X^a = (X^i, P_j)$, the coherent states are ground states of the harmonic oscillator, shifted in phase space. The general case will be discussed in detail later.

These coherent states have a number of important properties. The most important property is the completeness relation

$$\mathbb{1}_{\mathcal{H}} = \int_{\mathbb{R}^{2n}} \frac{\Omega}{(2\pi)^n} |x\rangle\langle x| \tag{3.1.72}$$

that is equivalent to the trace identity

$$\text{Tr}(\Phi) = \int_{\mathbb{R}^{2n}} \frac{\Omega}{(2\pi)^n} \langle x|\Phi|x\rangle. \tag{3.1.73}$$

The normalization will be verified in (3.1.112), and Ω is the symplectic volume form (3.1.33). Apart from the normalization, this relation follows from group theory: the relations (3.1.69) and (3.1.23) imply

$$U_y|x\rangle = e^{i\varphi}|x+y\rangle \tag{3.1.74}$$

for some phase φ, which will be determined in the following. Since the integral measure is translation invariant, it follows that

$$U_y\left(\int \Omega|x\rangle\langle x|\right)U_y^{-1} = \int \Omega|x\rangle\langle x| \tag{3.1.75}$$

for any $y \in \mathbb{R}^{2n}$. This means that the operator (3.1.72) is a trivial representation of the translation group. Since $\mathcal{H}$ is an irreducible representation of the CCR, it must be proportional to $\mathbb{1}$.

Given these coherent states, we can define the following *coherent state quantization* map:

$$\mathcal{Q}: \quad C(\mathbb{R}^{2n}) \to \text{End}(\mathcal{H})$$

$$\phi(x) \mapsto \int \frac{\Omega}{(2\pi)^n} \, \phi(x) \, |x\rangle\langle x| \qquad (3.1.76)$$

that associates[9] to every classical function on $\mathbb{R}^{2n}$ an operator or observable in $\text{End}(\mathcal{H})$. By construction, this map satisfies

$$\mathcal{Q}(\phi^*) = \mathcal{Q}(\phi)^\dagger \qquad (3.1.77)$$

and the completeness relation implies

$$\mathcal{Q}(1) = \mathbb{1}. \qquad (3.1.78)$$

$\mathcal{Q}$ is also an intertwiner of the translation group, since

$$\int \Omega\phi(x-y) \, |x\rangle\langle x| = \int \Omega\phi(x) \, |x+y\rangle\langle x+y| = U_y\left(\int \Omega\phi(x) \, |x\rangle\langle x|\right) U_y^{-1}. \qquad (3.1.79)$$

This might suggest that $\mathcal{Q}$ coincides with Weyl quantization map (3.1.29), but this is not quite true. The intertwiner property implies only that plane waves are mapped to plane waves

$$\mathcal{Q}(v_k) = c_k V_k \qquad \text{i.e.} \qquad \int \frac{\Omega}{(2\pi)^n} e^{ik_a x^a} \, |x\rangle\langle x| = c_k e^{ik_a X^a} \qquad (3.1.80)$$

up to some normalization constant c_k, which will be computed in (4.6.34). We will see that $c_k \to 1$ for $k \to 0$, so that the isometry property is recovered if the wavelengths are much larger than L_{NC}.

Coherent state quantization is very useful for many reasons. It is very transparent, and is naturally defined also for functions that are not square-integrable, e.g. $\mathcal{Q}(1) = \mathbb{1}$, $\mathcal{Q}(x^a) \propto X^a$, etc. Moreover, it generalizes to a much larger class of quantum spaces, as discussed in the following paragraph.

In particular, coherent states allow to define a "de-quantization" map as follows:

$$\text{End}(\mathcal{H}) \to C(\mathbb{R}^{2n})$$

$$\Phi \mapsto \langle x|\Phi|x\rangle =: \phi(x). \qquad (3.1.81)$$

The function $\phi(x)$ is known as the "symbol" of the operator Φ. This map is again an intertwiner of the translation group, and it is therefore the inverse of $\mathcal{Q}$ up to normalization,

---

[9] To avoid confusion, one should use different symbols $\mathcal{Q}_W$ and $\mathcal{Q}_c$ for the Weyl and coherent state quantization. The choice is usually clear from the context, and we often drop the subscript.

which depends on the wavenumber k of the mode. More precisely, for modes with wavenumber bounded by some cutoff Λ the following holds[10]

$$\lim_{\theta \to 0} \langle x | \mathcal{Q}(\phi) | x \rangle \to \phi(x), \qquad \lim_{\theta \to 0} \mathcal{Q}(\langle x | \Phi | x \rangle) \to \Phi. \tag{3.1.82}$$

This can be established using the inner products of different coherent states, which requires explicit formulas for the coherent states given below.

Exercise 3.1.3   Show that

$$Y^a := \int \Omega x^a \, |x\rangle \langle x| \overset{!}{\propto} X^a \tag{3.1.83}$$

up to normalization (Hint: work out the transformation property of Y^a under translations, and obtain the commutators $[X^a, Y^b]$).

Coherent states in the canonical case

Consider first the canonical case $X^a = (Q^i, P_j)$ familiar from Quantum Mechanics, for $i, j = 1, \ldots, n$. Introducing the creation and annihilation operators

$$Z^i := \frac{1}{\sqrt{2\theta}}(Q^i + iP_i),$$

$$Z^\dagger_i := \frac{1}{\sqrt{2\theta}}(Q^i - iP_i), \tag{3.1.84}$$

the CCR take the form

$$[Z^i, Z^\dagger_j] = \delta^j_i \mathbb{1}, \qquad [Z^i, Z^j] = 0 = [Z^\dagger_i, Z^\dagger_j], \tag{3.1.85}$$

and the distance square operator can be written as follows:

$$H = \frac{1}{2} \delta_{ab} X^a X^b = \frac{1}{2} \sum_i ((Q^i)^2 + P_i^2) = \sum_i \theta \left(Z^\dagger_i Z^i + \frac{1}{2} \mathbb{1} \right). \tag{3.1.86}$$

As familiar from Quantum Mechanics, this implies that the ground state satisfies

$$Z^i | 0 \rangle = 0. \tag{3.1.87}$$

Then the coherent state localized at $x \in \mathbb{R}^{2n}$ defined by $|x\rangle := U_x |0\rangle$ (3.1.69) satisfies

$$Z^i | x \rangle = z^i | x \rangle \tag{3.1.88}$$

in terms of the complex coordinates

$$z^i = \frac{1}{\sqrt{2\theta}}(q^i + ip_i), \qquad \bar{z}_i = \frac{1}{\sqrt{2\theta}}(q^i - ip_i), \tag{3.1.89}$$

where $x^a = (q^i, p_i)$. To make this more explicit, we rewrite the shift operators U_x (3.1.21) using

$$z^i Z^\dagger_i - \bar{z}_i Z^i = \frac{i}{\theta}(p_i Q^i - q^i P_i) = i \tilde{x}_a X^a, \tag{3.1.90}$$

---

[10] Note that the convergence is not uniform in Λ.

where $\tilde{x}_a = x^b \theta_{ba}^{-1} = \frac{1}{\theta}(p_i, -q^j)$. Then

$$U_x = e^{i\tilde{x}_a X^a} = e^{z^i Z_i^\dagger - \bar{z}_i Z^i} = e^{-\frac{1}{4}|x|_g^2} e^{z^i Z_i^\dagger} e^{-\bar{z}_i Z^i}, \tag{3.1.91}$$

where the **quantum metric** is defined by

$$g_{ab} = \frac{1}{L_{\mathrm{NC}}^2} \delta_{ab}, \qquad |x|_g^2 = \frac{1}{L_{\mathrm{NC}}^2}(x^a x^b \delta_{ab}) = 2\bar{z}^i z_i. \tag{3.1.92}$$

Here $L_{\mathrm{NC}}^2 = \theta$ denotes the **noncommutativity scale** (3.1.18). The inner product of different coherent states is now obtained using (3.1.91) as

$$\langle 0|x \rangle = \langle 0|U_x|0 \rangle = e^{-\frac{1}{4}|x|_g^2} \langle 0|e^{z^i Z_i^\dagger} e^{-z_i Z^i}|0 \rangle = e^{-\frac{1}{4}|x|_g^2} \tag{3.1.93}$$

and more generally using (3.1.20)

$$\begin{aligned}
\langle x|y \rangle &= \langle 0|e^{-i\tilde{x}_a X^a} e^{i\tilde{y}_a X^a}|0 \rangle \\
&= e^{\frac{i}{2}\tilde{x}_a \theta^{ab} \tilde{y}_b} \langle 0|e^{i(\tilde{y}_a - \tilde{x}_a)X^a}|0 \rangle \\
&= e^{\frac{i}{2}\tilde{x}_a \theta^{ab} \tilde{y}_b} \langle 0|y_a - x_a \rangle \\
&= e^{\frac{i}{2}\tilde{x}_a \theta^{ab} \tilde{y}_b} e^{-\frac{1}{4}|x-y|_g^2}.
\end{aligned} \tag{3.1.94}$$

Coherent states on a general Moyal–Weyl plane

Now consider a generic Moyal–Weyl quantum plane $\mathbb{R}_\theta^{2n}$. We are looking for the ground state of

$$H = \frac{1}{2}\delta_{ab} X^a X^b. \tag{3.1.95}$$

After an orthogonal transformation $O \in SO(2n)$, we can assume that $\theta^{\mu\nu}$ has the standard block-diagonal form (3.1.15); note that H is invariant under this rotation. Then the X^a can be related to canonical variables $\bar{X}^a = (Q^i, P_j)$ via

$$\begin{aligned}
X^a &= \frac{1}{\sqrt{\theta}}\left(\sqrt{\theta_i}Q^i, \sqrt{\theta_i}P_i\right) = D_b^a \bar{X}^b \\
\bar{X}^a &= (Q^i, P_j),
\end{aligned} \tag{3.1.96}$$

where

$$D = \mathrm{diag}\left(\sqrt{\frac{\theta_i}{\theta}}\mathbb{1}_2\right) \tag{3.1.97}$$

is block-diagonal in the basis (3.1.15), and

$$\theta = (\det \theta^{\mu\nu})^{1/2n} = L_{\mathrm{NC}}^2 \tag{3.1.98}$$

is the noncommutativity scale (3.1.18). Note that $\det D = 1$. Then

$$H = \frac{1}{2}\delta_{ab} X^a X^b = \sum_i \theta_i \left(Z_i^\dagger Z^i + \frac{1}{2}\mathbb{1}\right) = \frac{1}{2}\sum_i \frac{\theta_i}{\theta}\left((Q^i)^2 + P_i^2\right) \tag{3.1.99}$$

in terms of (canonical) creation-and annihilation operators (3.1.84), i.e.

$$Z^i = \frac{1}{\sqrt{2\theta}}(Q^i + iP_i), \qquad Z_i^\dagger = \frac{1}{\sqrt{2\theta}}(Q^i - iP_i). \qquad (3.1.100)$$

The ground state of H is therefore the same as in the canonical case, and it is again annihilated by

$$Z^i|0\rangle = 0. \qquad (3.1.101)$$

The coherent state localized at $x \in \mathbb{R}^{2n}$ defined by $|x\rangle := U_x|0\rangle$ satisfies

$$Z^i|x\rangle = z^i|x\rangle. \qquad (3.1.102)$$

The translation operators are the same as in (3.1.21), and can therefore be written as

$$U_x = e^{i\tilde{x}_a X^a} = e^{ix^c \theta_{ca}^{-1} X^a} = e^{i\bar{x}^d D_d^c \theta_{ca}^{-1} D_b^a \bar{X}^b} = e^{i\bar{x}^d \bar{\theta}_{ca}^{-1} \bar{X}^b} = \bar{U}_{\bar{x}} \qquad (3.1.103)$$

where

$$x^a = D_b^a \bar{x}^b \qquad (3.1.104)$$

and $\bar{U}_{\bar{x}}$ is the shift operator in the canonical case. Therefore,

$$\langle 0|x\rangle = \langle 0|U_x|0\rangle = \langle 0|\bar{U}_{\bar{x}}|0\rangle = e^{-\frac{1}{4}|\bar{x}|_{\bar{g}}^2} = e^{-\frac{1}{4}|x|_g^2} \qquad (3.1.105)$$

and more generally using (3.1.20)

$$\langle x|y\rangle = \langle 0|e^{-i\tilde{x}_a X^a} e^{i\tilde{y}_a X^a}|0\rangle = e^{\frac{i}{2}\tilde{x}_a \theta^{ab}\tilde{y}_b} e^{-\frac{1}{4}|x-y|_g^2}. \qquad (3.1.106)$$

Hence the decay width is characterized by the distance square

$$|x|_g^2 = g_{ab}x^a x^b = \bar{x}^a \bar{x}^b \bar{g}_{ab} = |\bar{x}|_{\bar{g}}^2 \qquad (3.1.107)$$

in terms of the **quantum metric** g, which in the coordinates x^a and $\bar{x}^a$ is given by

$$g_{ab} = (D^{-1})_a^{a'} D^{-1}{}_b^{b'} \bar{g}_{a'b'}, \qquad \bar{g}_{ab} = \frac{1}{L_{\mathrm{NC}}^2}\delta_{ab}. \qquad (3.1.108)$$

The quantum metric enjoys the following compatibility relation[11] with the symplectic structure

$$\theta_{ca}^{-1}\theta_{db}^{-1}g^{cd} = g_{ab}, \qquad \tilde{k}_a \tilde{k}_b g^{ab} = k^a k^b g_{ab} \qquad (3.1.109)$$

where $\tilde{k}_b = k^a \theta_{ab}^{-1}$ (3.1.21). Therefore, the quantum metric characterizes the "decay width" of the coherent states, and we can define an associated **coherence scale** parameter as

$$L_{\mathrm{coh}}^2 := (\det g_{\mu\nu})^{-1/2n} = (\prod \theta^i)^{1/n}. \qquad (3.1.110)$$

This scale coincides with L_{NC}^2 on $\mathbb{R}_\theta^{2n}$, but these structures will be defined more generally in Sections 4.2.1. The above scale parameters are also related to the *uncertainty scale* Δ^2 of the X^a through

$$\Delta^2 := \langle 0|\sum_i X_a^2|0\rangle = \sum_i \theta_i \gtrsim nL_{\mathrm{NC}}^2, \qquad (3.1.111)$$

---

[11] This means that $(\mathbb{R}^4, \theta_{ab}^{-1}, g_{ab})$ is a Kähler manifold, as discussed more generally in Section 4.5.

using (3.1.99). Therefore, the lowest eigenvalue of H encodes the noncommutativity or uncertainty scale.

The normalization in the completeness relation (3.1.72) can now be justified by considering

$$1 = \langle 0|0 \rangle = \frac{1}{(2\pi\theta)^n} \int\limits_{\mathbb{R}^{2n}} d^{2n}x \, \langle 0|x \rangle \langle x|0 \rangle = \frac{1}{(2\pi\theta)^n} \int\limits_{\mathbb{R}^{2n}} d^{2n}x \, e^{-\frac{1}{2}|x|_g^2} \qquad (3.1.112)$$

which is verified by a Gaussian integral. Finally, we can obtain the action of translation operators on the coherent states as follows:

$$U_y|x\rangle = U_y U_x |0\rangle = e^{\frac{i}{2}y^a \theta_{ab}^{-1} x^b} U_{x+y}|0\rangle = e^{\frac{i}{2}y^a \theta_{ab}^{-1} x^b} |x+y\rangle$$
$$\langle y|x\rangle = e^{-\frac{i}{2}y^a \theta_{ab}^{-1} x^b} \langle 0|x-y\rangle = e^{-\frac{i}{2}y^a \theta_{ab}^{-1} x^b} e^{-\frac{1}{4}|x-y|_g^2} \qquad (3.1.113)$$

consistent with (3.1.106). This implies that the coherent states can be characterized as ground states of the shifted harmonic oscillator:

$$H_x|x\rangle = \lambda|x\rangle, \qquad \lambda = \frac{1}{2}\sum_i \theta_i$$

$$H_x = \frac{1}{2}\sum_a (X^a - x^a \mathbb{1})^2 \qquad (3.1.114)$$

for any $x \in \mathbb{R}^{2n}$, and λ is the minimal eigenvalue of H_x. That property will be the starting point of the general considerations in Section 4.

It should also be mentioned that the coherent state quantization map $\mathcal{Q}$ respects not only the translation group, but also the action of the symplectic group $SP(n)$ which leaves θ^{ab} invariant. In canonical variables, this includes not only the $SO(n)$ rotations acting on the Q^i and P_j, but also the $SU(n)$ transformations acting on Z^i, as well as squeezing transformations. This group leaves the commutation relation invariant, and acts unitarily on $\mathcal{H}$ and $\text{End}(\mathcal{H})$. All this is well-known from Quantum Mechanics. Furthermore, we will see later that more general unitary transformations on $\mathcal{H}$ can be interpreted as quantized symplectomorphisms.

Wigner functions of coherent state density matrices

It is interesting to compute the Wigner function of the density matrix $\rho_y = |y\rangle\langle y|$, which should be viewed as optimally localized function on quantum phase space centered at y. Using

$$\text{Tr}(e^{ik_a X^a} |y\rangle\langle y|) = \langle y|e^{ik_a X^a}|y\rangle = e^{ik_a y^a} e^{-\frac{1}{4}k_a k_b g^{ab}} \qquad (3.1.115)$$

one obtains its Wigner function using (3.1.56) and (3.1.113)

$$\phi_y(x) = \mathcal{Q}_{(W)}^{-1}(\rho_y) = \frac{(2\pi\theta)^n}{(2\pi)^{2n}} \int d^{2n}k \, e^{-ik_a(x^a-y^a)} e^{-\frac{1}{4}k_a k_b g^{ab}} = 2^n \, e^{-|x-y|_g^2}. \qquad (3.1.116)$$

This is a Gaussian in $\mathbb{R}^{2n}$ localized at y with characteristic size L_{NC}, properly normalized $\langle \phi_y, \phi_y \rangle = 1$ w.r.t. the inner product (3.1.32) as it must. It should not be confused with the

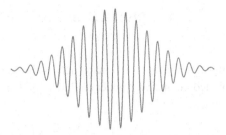

Figure 3.1 Sketch of the plane wave packet (3.1.118) corresponding to a string mode $|y\rangle\langle z|$.

position space representation $\langle q|x\rangle$ of a coherent state, which is a Gaussian wave packet in $\mathbb{R}^n$.

For later use, we also compute the Wigner function of the off-diagonal "string mode" $|y\rangle\langle z|$. Using

$$\mathrm{Tr}(e^{ik_a X^a}|y\rangle\langle z|) = \langle z|e^{ik_a X^a}|y\rangle = e^{-\frac{i}{2}z\theta^{-1}y}e^{\frac{i}{2}k_a(y^a+z^a)}e^{-\frac{1}{4}|(y-z)+k\theta|_g^2} \tag{3.1.117}$$

we obtain

$$\phi_{y;z}(x) = \mathcal{Q}^{-1}_{(W)}(|y\rangle\langle z|) = \frac{\theta^n}{(2\pi)^n}\int d^{2n}k\, e^{-ik_a x^a}\langle z|e^{ik_a X^a}|y\rangle$$
$$= 2^n e^{-\frac{i}{2}y\theta^{-1}z}e^{i(y-z)\theta^{-1}x}e^{-|x-(y+z)/2|_g^2}, \tag{3.1.118}$$

which is a Gaussian "plane wave packet" centered at $\frac{1}{2}(y+z)$ with wavenumber $\tilde{k}_a = (y-z)^b\theta^{-1}_{ba}$ and characteristic size L_{NC}, as indicated in Figure 3.1. This observation will be very useful to understand quantum field theory and gravity on quantum spaces, as discussed in Sections 6.2 and 11.1.

Exercise 3.1.4 Verify (3.1.118).

Holomorphic properties

We have seen that for any point $x \in \mathbb{R}^{2n}$, there is a coherent state $|x\rangle \in \mathcal{H}$. This defines a map

$$\mathbb{R}^{2n} \to \mathcal{H}, \qquad x \mapsto |x\rangle. \tag{3.1.119}$$

We have also seen that the canonical commutation relations take a particularly simple form in terms of the complexified generators Z^i and $Z^\dagger_i$ (3.1.84), and similarly for the Poisson brackets. This suggests to consider the Moyal–Weyl plane as quantized complex space $\mathbb{R}^{2n}_\theta \cong \mathbb{C}^n_\theta$. Remarkably, it turns out that the above map $x \mapsto |x\rangle$ then becomes *holomorphic*, provided the normalization of the coherent states is modified appropriately. This will be discussed and understood in a broader context in Section 4.6.2.

3.2 Star products and deformation quantization

Given a bijective quantization map such as the Weyl quantization (3.1.30), one can pull back the noncommutative product on $\mathrm{End}(\mathcal{H})$ to the classical space of functions, and define the following star product

$$\star: \quad \mathcal{C}(\mathcal{M}) \times \mathcal{C}(\mathcal{M}) \to \mathcal{C}(\mathcal{M})$$

$$(f, g) \quad \mapsto f \star g := \mathcal{Q}^{-1}\big(\mathcal{Q}(f)\mathcal{Q}(g)\big). \tag{3.2.1}$$

This defines a new noncommutative but associative product, which depends on the quantization parameter θ. It allows a systematic expansion of the product in powers of θ. A noncommutative theory can then be formulated in a way which looks similar to the classical case, which may provide some sense of familiarity. However, this sense is deceptive and we will de-emphasize this point of view, even though it is widely used in the literature. In the present framework, it will be very important that quantum spaces admit only finitely many degrees of freedom in any given volume. That feature is completely hidden in the star product point of view. For the same reason, the star product is seldom used in quantum mechanics, and the underlying quantum geometry is better described in terms of operators on Hilbert spaces.

The definition (3.2.1) of a star product works for any quantized symplectic space, given some quantization map $\mathcal{Q}$. For the Moyal–Weyl quantum plane $\mathbb{R}^{2n}_\theta$, the star product can be written down explicitly: for plane waves, the formula (3.1.20) gives

$$e^{ik_a x^a} \star e^{il_a x^a} = e^{-\frac{i}{2} k_a \theta^{ab} l_b} e^{i(k_a + l_a) x^a}. \tag{3.2.2}$$

This can be written in the explicit form

$$(f \star g)(x) = f(x) \exp\Big(\frac{i}{2} \overleftarrow{\partial}_a \theta^{ab} \overrightarrow{\partial}_b\Big) g(x), \tag{3.2.3}$$

which extends formally to more general functions $f, g \in \mathcal{C}(\mathbb{R}^{2n})$. A more concise integral formula for the Moyal–Weyl star product is obtained as follows:

$$\begin{aligned}
\mathcal{Q}(f)\mathcal{Q}(g) &= \int d^{2n}k \hat{f}(k) e^{ikX} \int d^{2n}p \hat{g}(p) e^{ipX} \\
&= \int d^{2n}k \int d^{2n}p \hat{f}(k) \hat{g}(p) e^{-\frac{i}{2} k_a \theta^{ab} p_b} e^{i(k+p)X} \\
&= \int d^{2n}k \int d^{2n}p \hat{f}(k) \hat{g}(p-k) e^{-\frac{i}{2} k_a \theta^{ab} p_b} e^{ipX}
\end{aligned} \tag{3.2.4}$$

so that

$$\begin{aligned}
(f \star g)(x) &= \int d^{2n}k \, d^{2n}p \hat{f}(k) \hat{g}(p-k) e^{-\frac{i}{2} k_a \theta^{ab} p_b} e^{ip_a x^a} \\
&= \int d^{2n}k \, d^{2n}p \int \frac{d^{2n}z}{(2\pi)^{2n}} \frac{d^{2n}y}{(2\pi)^{2n}} f(z) e^{-ikz} g(y) e^{-i(p-k)y} e^{-\frac{i}{2} k_a \theta^{ab} p_b} e^{ip_a x^a} \\
&= \int d^{2n}p \int \frac{d^{2n}z}{(2\pi)^{2n}} d^{2n}y \, \delta(z^a - y^a + \frac{1}{2}\theta^{ab} p_b) f(z) g(y) e^{-ipy} e^{ip_a x^a}
\end{aligned}$$

$$= \int \frac{d^{2n}p}{(2\pi)^{2n}} \int d^{2n}z f(z)g(z^a + \frac{1}{2}\theta^{ab}p_b)e^{ip(x-z)}$$

$$= \int \frac{d^{2n}p}{(2\pi)^{2n}} \int d^{2n}z f(z+x)g(x+\frac{1}{2}\theta \cdot p)e^{-ipz}. \tag{3.2.5}$$

Here $(\theta \cdot p)^a = \theta^{ab}p_b$, and $z \to z+x$, $p \to p - 2\theta^{-1}\cdot z$ was used in the last step. For small θ^{ab} and slowly varying f and g, this approximates the usual local product of functions.

It is instructive to elaborate the star product for low orders in θ. The leading order is obtained from (3.2.3) as

$$f \star g = fg + \frac{i\theta^{ab}}{2}(\partial_a f)(\partial_b g) + \mathcal{O}(\theta^2)$$

$$[f,g]_\star = f \star g - g \star f = i\{f,g\} + \mathcal{O}(\theta^2), \tag{3.2.6}$$

so that the Poisson bracket is reproduced to leading order in the quantization parameter θ. For monomials, we obtain e.g.

$$x^a \star x^b = x^a x^b + \frac{i}{2}\theta^{ab}$$

$$x^a \star x^b - x^b \star x^a = i\theta^{ab}. \tag{3.2.7}$$

Star products can be used to extend the concept of quantization to cases where no operator quantization is available. This is the idea of **deformation quantization**, which sometimes allows to prove mathematical result that are not easily available for operator quantization. We shall briefly discuss this concept, although it will not be used in the following.

A formal star product can be defined as follows: First, one replaces the space of functions $\mathcal{A} = \mathcal{C}^\infty(\mathcal{M})$ by the formal power series $\mathcal{A}[[\theta]] = \oplus_n\{f_n(x)\theta^n, f_n \in \mathcal{A}\}$ in some quantization or deformation parameter θ. One then proceeds as follows:

Definition 3.2 Let $(\mathcal{M}, \{.,.\})$ be a Poisson manifold. A (formal) star-product on $\mathcal{A} = \mathcal{C}^\infty(\mathcal{M})$ is a bilinear associative map

$$\star: \quad \mathcal{A}[[\theta]] \times \mathcal{A}[[\theta]] \to \mathcal{A}[[\theta]] \tag{3.2.8}$$

given by

$$f \star g = \sum_{n=0}^\infty \theta^n C_n(f,g) \tag{3.2.9}$$

in terms of bilinear maps $C_n(f,g)\colon \mathcal{A} \times \mathcal{A} \to \mathcal{A}$, such that

1. $\star$ is associative
2. $C_0(f,g) = gf$
3. $C_1(f,g) - C_1(g,f) = \{f,g\}$
4. $1 \star f = f \star 1 = f$.

These conditions imply that the quantization axioms (3.0.7) hold formally, without requiring or assuming a Hilbert space representation. This amounts to a technical simplification, which allows to prove theorems on the existence and classification of star products, cf. [56]. For example, an important result of Kontsevich states that all Poisson manifolds can be quantized through a formal star product. No analogous result is established in the framework of operator quantization, except for special cases such as Kähler manifolds.

The star product obtained from Weyl quantization can clearly be re-formulated such that the above definition is satisfied. If the power series in θ converge, one can hope to recover standard quantum mechanics in some sense [26].

In general, star products and quantization maps $\mathcal{Q}$ are far from unique. One merit of the deformation approach is that one can classify the inequivalent star products. Two star products $\star, \star'$ are said to be equivalent if there is a map $\mathcal{I}$ such that

$$f \star' g = \mathcal{I}^{-1}(\mathcal{I}(f) \star \mathcal{I}(g)); \tag{3.2.10}$$

here $\mathcal{I}$ is a linear map given by

$$
\begin{aligned}
\mathcal{I}: \quad & \mathcal{A}[[\theta]] \; \rightarrow \; \mathcal{A}[[\theta]] \\
& \mathcal{I} = \mathbb{1} + \sum_{n>0} \theta^n \mathcal{I}_n.
\end{aligned}
\tag{3.2.11}
$$

Then the corresponding quantization maps are related as $\mathcal{Q}' = \mathcal{Q} \circ \mathcal{I}$. For example, instead of Weyl quantization we can define a modified quantization map denoted as normal-ordered quantization, where a polynomial $f(q^i, p_j)$ is quantized by ordering all P_i in front of all Q^j:

$$
\begin{aligned}
\mathcal{Q}_N: \quad & \mathcal{A} \rightarrow \mathrm{End}(\mathcal{H}) \\
& q^i \mapsto Q^i \\
& p_i \mapsto P_i \\
q^{i_1} \dots q^{i_r} p_{j_1} \dots p_{j_s} & \mapsto Q^{i_1} \dots Q^{i_r} P_{j_1} \dots P_{j_s}.
\end{aligned}
\tag{3.2.12}
$$

Then the quantization maps $\mathcal{Q}_W$ and $\mathcal{Q}_N$ defined by Weyl ordering and normal ordering are related by $\mathcal{I}$:

$$
\begin{aligned}
\mathcal{I}(q^{i_1} \dots q^{i_r} p_{j_1} \dots p_{j_s}) &= \mathcal{Q}_N^{-1}(\mathcal{Q}_W(q^{i_1} \dots q^{i_r} p_{j_1} \dots p_{j_s})) \\
&= \mathcal{Q}_N^{-1}(\mathrm{Sym}[Q^{i_1} \dots Q^{i_r} P_{j_1} \dots P_{j_s}]), \\
\mathcal{I}(q^1 p_1) &= \frac{1}{2} \mathcal{Q}_N^{-1}(Q^1 P_1 + P_1 Q^1) = q^1 p_1 - \frac{i}{2}\theta.
\end{aligned}
\tag{3.2.13}
$$

One can show [89, 90] that if the Poisson structure is nondegenerate (corresponding to some symplectic form) and the second de Rham cohomology of $\mathcal{M}$ vanishes $H^2_{dR}(\mathcal{M}) = 0$, then any two star products are equivalent in the above sense. In particular, on $\mathbb{R}^{2n}$ all such quantizations are equivalent up to such ("generalized ordering") maps $\mathcal{I}$.

Exercise 3.2.1   Compute the Moyal–Weyl star product $f \star f$ for the Gaussian function $f(x) = e^{-|x-y|^2_g}$ where g_{ab} is the quantum metric (3.1.108). Compare the result with (3.1.116).

3.3 Compact quantum spaces from coadjoint orbits

3.3.1 The fuzzy sphere S^2_N

The basic example of a compact quantum geometry is the fuzzy sphere S^2_N. This is a quantization or matrix approximation of the usual sphere S^2 which fully respects its symmetry, with a cutoff in angular momentum. It was first formulated explicitly by Jens Hoppe [101] and developed further by John Madore [131]. It provides a canonical prototype for compact quantum spaces.

The classical sphere

Consider the two-sphere S^2 as a submanifold embedded in $\mathbb{R}^3$ via

$$x^a : \quad S^2 \hookrightarrow \mathbb{R}^3, \tag{3.3.1}$$

where x^a, $a = 1, 2, 3$ are Cartesian coordinates on target space[12] $\mathbb{R}^3$, and the embedding is defined by the relation

$$x^a x^b \delta_{ab} = 1. \tag{3.3.2}$$

Here the target space $(\mathbb{R}^3, \delta)$ is equipped with the standard Euclidean metric, and we choose the radius of the sphere to be 1 for simplicity. The embedding is compatible with the standard group action of $SO(3)$ on $\mathbb{R}^3$, which induces a group action

$$SO(3) \times S^2 \to S^2$$
$$(g, x) \mapsto gx. \tag{3.3.3}$$

S^2 can be equipped with a $SO(3)$-invariant symplectic structure which is unique up to normalization, given by

$$\omega_N = \frac{N}{4} \varepsilon_{abc} x^a dx^b dx^c, \qquad \int_{S^2} \omega_N = 2\pi N \tag{3.3.4}$$

for any $N \in \mathbb{N}$, cf. (2.2.20). The normalization is chosen such that the symplectic volume is $2\pi N$. This symplectic form corresponds via (1.1.8) to the Poisson brackets

$$\{x^a, x^b\} = \frac{2}{N} \varepsilon^{abc} x^c, \tag{3.3.5}$$

which is also manifestly $SO(3)$ invariant. To see the relation with ω_N and to gain some intuition, it is useful to pick some point on S^2 – the north pole $\xi = (0, 0, 1)$, say – and use the two tangential Cartesian coordinate functions x^μ, $\mu = 1, 2$ on the embedding space $\mathbb{R}^3$ as local coordinates. We shall denote such coordinates as *local embedding coordinates* henceforth. Then, the symplectic form takes the form

$$\omega_N|_\xi = \frac{N}{4} \varepsilon_{3\mu\nu} dx^\mu dx^\nu = \frac{1}{2} (\omega_N)_{\mu\nu} dx^\mu dx^\nu \tag{3.3.6}$$

---

[12] In stringy terminology, S^2 plays the role of a brane in target space.

at the north pole ξ with

$$(\omega_N)_{\mu\nu} = \frac{N}{2}\epsilon_{\mu\nu}, \tag{3.3.7}$$

or $\omega_N = \frac{N}{2}\sin\vartheta\, d\vartheta\, d\varphi$ in spherical coordinates. This has the canonical form (3.1.15), hence (3.3.5) follows at ξ, which extends to all of S^2 by $SO(3)$ invariance. The normalization $\int \omega_N = 2\pi N$ follows since the area of S^2 is 4π.

Now consider the algebra $C(S^2)$ of smooth functions on S^2, equipped with the inner product

$$\langle \phi, \psi \rangle := \int\limits_{S^2} \frac{\omega_N}{2\pi}\, \phi^*\psi \tag{3.3.8}$$

(cf. (3.1.32)), which can be completed as a Hilbert space $L^2(S^2)$. For most purposes it is enough to consider polynomial functions, which are dense in $C(S^2)$. These polynomials are generated by the Cartesian coordinate functions x^a of $\mathbb{R}^3$ modulo the relation $x^a x^b \delta_{ab} = 1$. The $SO(3)$ group action on S^2 induces an action on the algebra of functions[13] on S^2

$$SO(3) \times C(S^2) \to C(S^2)$$

$$(g, \phi(x)) \mapsto (g \cdot \phi)(x) = \phi(g^{-1}x). \tag{3.3.9}$$

The g^{-1} on the rhs is needed to ensure the laws of a group action, notably $(g_1 g_2) \cdot \phi = g_1 \cdot (g_2 \cdot \phi)$. The above inner product is clearly invariant under $SO(3)$.

In particular, we can view the x^a via (3.3.1) as functions on S^2 that transform as a vector under $SO(3)$, i.e. $g \cdot x^a = \pi(g)^a_b x^b$. More generally, it is well-known that the totally symmetric traceless homogeneous polynomials of degree l in the x^a constitute an irreducible spin l representation of $SO(3)$, which are nothing but the spherical harmonics $Y^l_m(x)$:

$$Y^l_m(x) = Y^{(m)}_{a_1 \dots a_l} x^{a_1} \dots x^{a_l}, \tag{3.3.10}$$

where the $Y^{(m)}_{a_1 \dots a_l}$ are totally symmetric and traceless, $Y^{(m)}_{a_1 \dots a_l} \delta^{a_i a_j} = 0$. For example, for $l = 2$ there are five polynomials of the form $\frac{1}{2}(x^a x^b + x^b x^a) - \frac{1}{3}\delta^{ab}(x^c x_c)$, which in spherical coordinates reduce to the standard expressions in terms of the associated Legendre polynomials. These $Y^l_m(x)$ with $|m| \leq l$ form a basis of the space of all polynomials on S^2, which is dense in $C(S^2)$. We can therefore decompose the vector space $C(S^2)$ into irreps

$$C(S^2) = \bigoplus_{l=0}^{\infty} (l), \qquad (l) = \{Y^l_m(x),\ m = -l, \dots, l\}. \tag{3.3.11}$$

Here (l) denotes the spin l irrep of $SO(3)$, which is realized here by the spherical harmonics $Y^l_m(x)$. This decomposition of the algebra of functions into irreps of the symmetry group is the key to define and understand the fuzzy sphere.

---

[13] Note that x denotes some point in $\mathbb{R}^3$ or S^2 here, while x^a denotes a (coordinate) function on $\mathbb{R}^3$ or S^2.

The fuzzy sphere

The fuzzy sphere S_N^2 is defined in terms of three $N \times N$ hermitian matrices $X^a \in \text{End}(\mathcal{H})$ for $a = 1, 2, 3$ and $\mathcal{H} = \mathbb{C}^N$, which satisfy the relations

$$[X^a, X^b] = \frac{i}{\sqrt{C_N^2}} \varepsilon^{abc} X^c, \qquad \sum_{a=1}^{3} X^a X^a = \mathbb{1}, \qquad C_N^2 = \frac{1}{4}(N^2 - 1). \qquad (3.3.12)$$

By inspection, the X^a are nothing but rescaled angular momentum generators

$$X^a := \frac{1}{\sqrt{C_N^2}} J_{(N)}^a, \qquad [J_{(N)}^a, J_{(N)}^b] = i\varepsilon^{abc} J_{(N)}^c \qquad (3.3.13)$$

acting on the N-dimensional $\mathfrak{su}(2)$ irrep $\mathcal{H} = \mathbb{C}^N$, and C_N^2 is the value of the quadratic Casimir. Irreducibility implies that the X^a generate the full matrix algebra $\text{End}(\mathcal{H})$. We claim that $\text{End}(\mathcal{H})$ can be naturally interpreted as quantized algebra of functions on the symplectic space (S^2, ω_N). This will be justified by constructing a suitable isometric quantization map $\mathcal{Q}$, in complete analogy to Weyl quantization. $\text{End}(\mathcal{H})$ is again equipped with the structure of a Hilbert space with inner product

$$\langle \Phi, \Psi \rangle := \text{Tr}_{\mathcal{H}}(\Phi^\dagger \Psi), \qquad (3.3.14)$$

and provides a unitary representation of $SO(3)$ generated by

$$\mathcal{J}^a \Phi = [J_{(N)}^a, \Phi]. \qquad (3.3.15)$$

We can then interpret the matrices X^a as quantized embedding functions in the Euclidean target space $\mathbb{R}^3$,

$$X^a \sim x^a : \quad S^2 \hookrightarrow \mathbb{R}^3. \qquad (3.3.16)$$

To justify this, we need to construct a quantization map $\mathcal{Q}$ according to the general scheme discussed in Section 3. This is achieved by decomposing $\text{End}(\mathcal{H})$ into irreps of the adjoint action of $SO(3)$:

$$\text{End}(\mathcal{H}) \cong \left(\frac{N-1}{2}\right) \otimes \left(\frac{N-1}{2}\right)^* = (0) \oplus (1) \oplus \ldots \oplus (N-1)$$
$$= \{\hat{Y}_0^0\} \oplus \ldots \oplus \{\hat{Y}_m^{N-1}\}, \qquad (3.3.17)$$

where (l) denotes the spin l irrep of $\mathfrak{su}(2)$ or $\mathfrak{so}(3)$. This decomposition corresponds precisely to the decomposition (3.3.11) of polynomial functions on S^2, up to the cutoff N. It is therefore natural to **define** the fuzzy spherical harmonics $\hat{Y}_m^l$ as the irreps (l) on the rhs of (3.3.17). As in the classical case, they are given by totally symmetric traceless polynomials in X^a of degree l

$$\hat{Y}_m^l \propto Y_{a_1 \ldots a_l}^{(m)} X^{a_1} \ldots X^{a_l}. \qquad (3.3.18)$$

This follows from the fact that the tensors $Y^{(m)}_{a_1 \ldots a_l}$ define $\mathfrak{su}(2)$ irreps. We can now write down the quantization map

$$
\begin{aligned}
\mathcal{Q}: \quad \mathcal{C}(S^2) &\rightarrow \quad \mathrm{End}(\mathcal{H}) \\
Y^l_m &\mapsto \quad
\begin{cases}
\hat{Y}^l_m, & l < N \\
0, & l \geq N.
\end{cases}
\end{aligned}
\tag{3.3.19}
$$

$\mathcal{Q}$ is by definition an $SO(3)$ *intertwiner*, i.e. it respects the $SO(3)$ action, and the normalization[14] is fixed by requiring that $\mathcal{Q}$ is an *isometry* with respect to the inner products (3.3.8) and (3.3.14). Moreover, the following compatibility between trace and integral holds

$$
\mathrm{Tr}(\mathcal{Q}(f)) = \int\limits_{S^2} \frac{\omega_N}{2\pi} f
\tag{3.3.20}
$$

cf. (3.0.14). Up to normalization this is a consequence of the intertwiner property, noting that the only harmonic with nonvanishing trace or integral is the constant functions $\mathbb{1}$ or 1. The normalization follows for $f = 1$ using (3.3.4).

An explicit realization of the fuzzy spherical harmonics in terms of the Wigner 3j-symbols (which are suitably normalized Clebsch–Gordon coefficients) is as follows:

$$
\hat{Y}^l_m = (-1)^m \sqrt{2l+1} \sum_{r,s}
\begin{pmatrix} l & \alpha & \alpha \\ m & r & -s \end{pmatrix}
|r\rangle \langle s|
\tag{3.3.21}
$$

for $0 \leq l \leq 2\alpha$, where

$$
\alpha = (N-1)/2.
\tag{3.3.22}
$$

Here $|r\rangle$, $r = -\alpha, \ldots, \alpha$ is the weight basis of $\mathcal{H}$. Then the reality property

$$
\hat{Y}^{l\dagger}_m = (-1)^m \hat{Y}^l_{-m}
\tag{3.3.23}
$$

holds, and the orthogonality relation of the 3j symbols

$$
(2l+1) \sum_{rs}
\begin{pmatrix} l & \alpha & \alpha \\ m & r & -s \end{pmatrix}
\begin{pmatrix} l' & \alpha & \alpha \\ m' & r & -s \end{pmatrix}
= \delta^{ll'} \delta_{mm'}
\tag{3.3.24}
$$

implies the orthogonality relations and the normalization of the fuzzy harmonics

$$
\mathrm{Tr}\!\left(\hat{Y}^{l\dagger}_m \hat{Y}^{l'}_{m'}\right) = \delta^{ll'} \delta_{mm'}.
\tag{3.3.25}
$$

Moreover, the product of functions on S^2_N in this basis is given in terms of the $6J$ symbols of $su(2)$ as follows:

$$
\hat{Y}^I_i \hat{Y}^J_j = \sqrt{\frac{N}{4\pi}} \sum_{K,k} (-1)^{2\alpha+I+J+K+k} \sqrt{(2I+1)(2J+1)(2K+1)}
$$

$$
\cdot
\begin{pmatrix} I & J & K \\ i & j & -k \end{pmatrix}
\begin{Bmatrix} I & J & K \\ \alpha & \alpha & \alpha \end{Bmatrix}
\hat{Y}^K_k,
\tag{3.3.26}
$$

---

[14] The normalization factors are different from the classical ones, and we refrain from working them out explicitly.

where the sum is over $0 \leq K \leq 2\alpha, -K \leq k \leq K$. This is most transparent upon representing the matrices $(\hat{Y}^l_m)^i_j$ by a triple vertex with strands corresponding to irreps labeled by $(\alpha i), (\alpha j)$, and (lm). Useful approximation formulas exist for the asymptotic behavior of the $6J$ symbols, but they are highly oscillatory for large quantum numbers. A much more convenient representation of functions with non-oscillatory behavior will be discussed in Section 6.2.

In particular, the above definition yields

$$\mathcal{Q}(\mathbb{1}) = \mathbb{1}, \qquad \mathcal{Q}(x^a) = X^a. \tag{3.3.27}$$

The latter can be seen noting that both x^a and X^a transform as spin 1 irreps, and are normalized such that

$$\langle x_a, x^a \rangle = \int \frac{\omega_N}{2\pi} x_a x^a = N = \mathrm{Tr}(X_a X^a) = \langle X_a, X^a \rangle \tag{3.3.28}$$

(sum over a is understood).

The commutation relations imply that the generators approximately commute for large N. Therefore, the space of polynomials in X^a up to some degree l modulo the relations (3.3.12) reduces for $N \to \infty$ to the space of polynomials in x^a up to some degree l modulo the relations $x_a x^a = 1$ and $[x^a, x^b] = 0$. Since the inner products on both spaces are uniquely specified by the radial constraint as in (3.3.28), it follows that $\mathcal{Q}(x^{a_1} \ldots x^{a_l}) \to X^{a_1} \ldots X^{a_l}$ as $N \to \infty$, hence

$$\mathcal{Q}(fg) \to \mathcal{Q}(f)\mathcal{Q}(g) \qquad \text{as} \quad N \to \infty \tag{3.3.29}$$

for any polynomials f, g with fixed degree. Furthermore, we observe that

$$\{x^a, x^b\} = \frac{2}{N}\varepsilon^{abc} x^c, \qquad [X^a, X^b] = \frac{2i}{\sqrt{N^2-1}}\varepsilon^{abc} X^c. \tag{3.3.30}$$

Since $\frac{2}{\sqrt{N^2-1}} \sim \frac{2}{N}$ as $N \to \infty$, the Poisson structure is properly reproduced for large N. In local embedding coordinates x^μ around the north pole, the Poisson tensor is given by

$$\theta^{\mu\nu} = L^2_{\mathrm{NC}}\epsilon^{\mu\nu}, \tag{3.3.31}$$

where we identify the scale of noncommutativity by

$$L^2_{\mathrm{NC}} = \frac{2}{N} =: \theta. \tag{3.3.32}$$

In particularly, we can locally approximate S^2_N by the quantum plane $\mathbb{R}^2_\theta$. More precisely,

$$\mathcal{Q}(i\{x^a, x^b\}) - [X^a, X^b] = i\Big(\frac{2}{N} - \frac{2}{\sqrt{N^2-1}}\Big)\varepsilon^{abc} X^c = O\Big(\frac{1}{N^3}\Big)\varepsilon^{abc} X^c. \tag{3.3.33}$$

This implies that the quantization condition (3.0.7) is satisfied for the generators X^a. Using the Leibniz rule, this extends immediately to any given polynomial $f, g \in \mathcal{C}(S^2)$, and (3.0.8) is a consequence of (3.0.7) together with (3.3.20).

To make sense of $\lim_{\theta \to 0}$ in (3.0.7), we need to define a norm on $\mathrm{End}(\mathcal{H}_N)$. The inner products (3.3.28) define such a norm via $\|f\|^2 = \langle f, f \rangle$. Since $\mathrm{End}(\mathcal{H}_N)$ depends on

N, the normalization also depends on N. For compact quantum spaces, it is natural to re-scale these norms such that $\|1\|$ is independent of N, e.g.

$$\|1\| = 1 = \|\mathbb{1}\|. \tag{3.3.34}$$

With this convention we have $\|x^a\| = \frac{1}{3} = \|X^a\|$. Then (3.0.7) holds in the sense that

$$\lim_{N \to \infty} \frac{1}{1/N} \left\| \mathcal{Q}(i\{f,g\}) - [\mathcal{Q}(f), \mathcal{Q}(g)] \right\| = 0 \tag{3.3.35}$$

for fixed normalized polynomials f, g in x. Note that this relation does not hold uniformly in $\mathcal{C}(S^2)$, i.e. the convergence rate depends on the functions. A similar remark applies to (3.3.29), where the difference $\|\mathcal{Q}(fg) - \mathcal{Q}(f)\mathcal{Q}(g)\|$ is negligible only if the angular momenta l are bounded by $l^2 \ll N$. This bound is best understood in terms of coherent states discussed in Sections 4.6.1 and 6.2. Analogous observations apply to the more general compact quantum spaces discussed in the following.

The present map $\mathcal{Q}$ is clearly irreducible, and therefore satisfies all requirements of a quantization map stated in definition 3.1. However, it is important to note that $\mathcal{Q}$ is not injective, as it annihilates all modes with angular momentum $l \geq N$. Therefore, the fuzzy sphere S_N^2 is a quantization of the classical sphere (S^2, ω_N), with intrinsic UV cutoff at

$$l_{UV} := N. \tag{3.3.36}$$

The correspondence is summarized and refined in Table 3.1; an analogous dictionary applies to all quantum spaces to be considered here.

It is interesting to observe that the symplectic structure or the quantization parameter $\theta \sim \frac{1}{N}$ is discrete, in contrast to the Moyal–Weyl quantum plane. Such a quantization phenomenon is characteristic for compact spaces. This is reflected in (3.3.20), which for $f = 1$ amounts to

$$N = \dim(\mathcal{H}) = \mathrm{Tr}\mathbb{1} = \int_{S^2} \frac{\omega_N}{2\pi}. \tag{3.3.37}$$

The origin of this quantization condition will be discussed in Section 4.2.1. Hence, S_N^2 comprises N quantum cells with length scale

$$L_{\mathrm{NC}} = \sqrt{2/N}, \tag{3.3.38}$$

which are probed by modes with angular momentum $l \leq \sqrt{N}$. This scale separates the space of operators or NC "functions" $\mathrm{End}(\mathcal{H})$ into two distinct regimes: the

semi-classical or almost-local regime: $\quad 0 \leq l \leq \sqrt{N} =: l_{\mathrm{NC}}, \tag{3.3.39}$

which can be interpreted as semi-classical functions, and the

deep quantum regime: $\quad l_{\mathrm{NC}} = \sqrt{N} \leq l \leq N = l_{UV}, \tag{3.3.40}$

which consists of modes which are highly noncommutative. The latter are best understood in terms of coherent states and string modes, as discussed in Section 6.2. A simple way to see this is by recalling that the $\hat{Y}^l_m$ are polynomials of order l in the generators X_a, whose commutator is of order

$$[X^{l_1}, X^{l_2}] = \frac{1}{N} O(l_1 l_2) \tag{3.3.41}$$

due to the derivation property of the commutator. An analogous separation of scales applies to all quantum space, including the Moyal–Weyl quantum plane $\mathbb{R}^{2n}_\theta$. Obviously the vast majority of modes is in the deep quantum regime, where the semi-classical picture is no longer appropriate. In that regime, the correspondence with classical functions given by $\mathcal{Q}$ is rather misleading, and should be used with great caution. A more appropriate picture of the deep quantum regime in terms of string modes will be discussed in Section 6.2.

Oscillator construction of $S^2_N \cong \mathbb{C}P^1_N$ and Hopf map

There is an illuminating and very useful construction of the fuzzy sphere in terms of a doublet of bosonic creation- and annihilation operators, which satisfy the canonical oscillator algebra

$$[Z^\alpha, Z^\dagger_\beta] = \delta^\alpha_\beta, \qquad [Z^\alpha, Z^\beta] = 0 = [Z^\dagger_\alpha, Z^\dagger_\beta], \qquad \alpha, \beta = 1, 2. \tag{3.3.42}$$

It is easy to verify that the following generators defined in terms of the Pauli matrices σ^a

$$J^a := \frac{1}{2} Z^\dagger_\alpha (\sigma^a)^\alpha_\beta Z^\beta \tag{3.3.43}$$

satisfy the $\mathfrak{su}(2)$ commutation relations

$$[J^a, J^b] = i\varepsilon^{abc} J^c. \tag{3.3.44}$$

This is the Jordan–Schwinger realization of $\mathfrak{su}(2)$ (2.0.29), which is familiar in quantum field theory as second-quantized symmetry. Now consider the Fock space spanned by all creation operators

$$|n_1, n_2\rangle = \frac{1}{\sqrt{n_1! n_2!}} (Z^\dagger_1)^{n_1} (Z^\dagger_2)^{n_2} |0\rangle \tag{3.3.45}$$

acting on the vacuum $|0\rangle$, which satisfies $Z_\alpha |0\rangle = 0$. Clearly all J^a commute with the number operator

$$\hat{N} = Z^\dagger_\alpha Z^\alpha, \tag{3.3.46}$$

and satisfy

$$J^a J_a = \frac{\hat{N}(\hat{N} + 2)}{4}. \tag{3.3.47}$$

Therefore the J^a define an irreducible representation of $\mathfrak{su}(2)$ on the N- particle subspace

$$\mathcal{H}_N := \{|n_1, n_2\rangle, \ n_1 + n_2 = N\} \cong \mathbb{C}^{N+1} \tag{3.3.48}$$

of the Fock space for any $N \geq 1$. It is easy to see (e.g. by computing the quadratic Casimir $C^2 = J_a J^a$) that $\mathcal{H}_N$ forms an $N + 1$ dimensional irreducible representation of $\mathfrak{su}(2)$. The generators of the fuzzy sphere with the normalization (3.3.12) are given by a rescaling as in (3.3.13), and the functions on S^2_N commutes with $\hat{N}$:

$$[\hat{N}, \text{End}(\mathcal{H}_N)] = 0. \tag{3.3.49}$$

A more conceptual understanding is obtained by comparing this construction with the discussion of $\mathbb{C}P^{n-1}$ in Section 2.1, for $n = 2$. We will recognize the $Z^\alpha, Z^\dagger_\alpha$ as quantizations of the complex coordinate functions on the canonical plane $\mathbb{C}^2 \cong \mathbb{R}^4_\theta$, and the constraint $\hat{N} = N$ as quantization of the classical constraint (2.1.3). Therefore, the fuzzy sphere can be viewed as quantized complex projective space $S^2_{N+1} \cong \mathbb{C}P^1_N$, and the oscillator construction is a quantum version of the symplectic reduction. This oscillator construction will generalize to $\mathbb{C}P^n_N$ and related spaces.

We can make this more transparent in the semi-classical limit. Then the above construction of J^a reduces to the Hopf map (up to normalization)

$$S^3 \mapsto S^2$$
$$z^\alpha \mapsto z^\dagger_\alpha (\sigma^a)^\alpha_{\ \beta} z^\beta \tag{3.3.50}$$

observing that $|z^1|^2 + |z^1|^2 = 1$ defines the three-sphere S^3. Since the image of this map is independent of the overall phase of z^α, we conclude that S^3 is a $U(1)$ bundle over S^2. Factoring out this phase provides the well-known identification $\mathbb{C}P^1 \cong S^2$. Moreover, this map is an intertwiner of the actions of $SU(2)$ and $SO(3)$ on S^3 and S^2, respectively, and the local stabilizer $U(1)$ of any point on S^2 acts nontrivially on the $U(1)$ fiber. This means that S^3 is an $SU(2)$-equivariant bundle, cf. Definition 2.5. Generalizations of this map will play an important role in Section 5.

As an extra bonus from the above oscillator construction, we recognize the "off-diagonal operators"

$$\mathcal{C}^{(k)} := \{A : \mathcal{H}_{N+k} \to \mathcal{H}_N\} \tag{3.3.51}$$

(cf. (2.1.5)) as quantized functions on a $U(1)$ bundle over S^2_N with winding number k, or equivalently as sections of a quantized $U(1)$ bundle over S^2 with $U(1)$ charge (and Chern number)

$$[\hat{N}, .]\big|_{\mathcal{C}^{(k)}} = k. \tag{3.3.52}$$

The quantization of the charge k is directly related to the quantization of the radius $\hat{N}$ in $\mathbb{C}^2$. We also observe that $\mathcal{C}^{(k)}$ is a left module of $\text{End}(\mathcal{H}_{N+k})$ and a right module of $\text{End}(\mathcal{H}_N)$. This interpretation of rectangular matrices or rather bi-modules as sections on nontrivial bundles generalizes to other quantum spaces.

Further structures and remarks

Finally, we note that the (round) metric on S^2 is encoded in the $SU(2)$-invariant "matrix Laplacian"

$$\square = [X^a, [X^b, .]]\delta_{ab}: \quad \text{End}(\mathcal{H}) \to \text{End}(\mathcal{H}), \tag{3.3.53}$$

which is nothing but the rescaled quadratic $SU(2)$ Casimir on S^2. It is easy to see that its spectrum coincides (up to a scale) with the spectrum of the classical Laplace operator on S^2 up to the cutoff,

$$\text{spec}(\Box) = \frac{1}{C_N^2}\{l(l+1), \quad l = 0, 1, 2, \ldots, N-1\}. \tag{3.3.54}$$

The corresponding eigenfunctions are given by the fuzzy spherical harmonics $\hat{Y}_m^l$. This means that the "spectral geometry" of the fuzzy sphere coincides with that of the classical sphere. Finally, S_N^2 is a solution of (7.0.10) with

$$\Box X^a = \frac{2}{C_N^2} X^a. \tag{3.3.55}$$

An interesting application of S_N^2 within the standard framework of theoretical physics is quantized phase space for the rotational dynamics of a rigid body. Its classical phase space for fixed angular momentum $\vec{L}^2$ carries the same Poisson structure in different guise, cf. [137]. A more exotic application is as lowest Landau levels for a magnetic monopole, cf. Section 4.8. We will consider S_N^2 mainly as a prototype for a quantum space, which plays the role of a configuration space for a noncommutative quantum field theory or gauge theory. This will be elaborated in Section 6. S_N^2 also arises as solution of certain matrix models, as discussed in Section 7. Then the matrix Laplace operator (3.3.53) arises naturally in the kinetic term, and governs the fluctuations around the background.

Flat Moyal–Weyl limit $S_N^2 \to \mathbb{R}_\theta^2$

If the fuzzy sphere is blown up around a given point, it can serve as a local approximation of the Moyal–Weyl quantum plane $\mathbb{R}_\theta^2$. To obtain this flat limit, it is convenient to introduce an alternative representation of the fuzzy sphere in terms of a stereographic projection. Consider the generators

$$Y_+ = 2RX_+(R-X_3)^{-1}, \quad Y_- = 2R(R-X_3)^{-1}X_-, \tag{3.3.56}$$

where $X_\pm = X_1 \pm iX_2$ are generators of the rescaled S_N^2 with radius R. The $Y_\pm \sim y_\pm$ are interpreted as quantized coordinate functions of the stereographic projection from the north pole, such that $y = 0$ corresponds to the south pole. Now we take the large N and large R limit, such that

$$N \to \infty, \quad R^2 = N\theta/2 \to \infty, \qquad \text{keeping } \theta \text{ fixed}. \tag{3.3.57}$$

Then,

$$[Y_+, Y_-] = -\frac{8}{N}R^3(R-X_3)^{-1} \quad + O(\frac{1}{N^2}). \tag{3.3.58}$$

Since $Y_+Y_- = 4R^2(R+X_3)(R-X_3)^{-1} + O(N^{-1/2})$, we can approximate the entire Y-plane with a small neighborhood of the south pole of the sphere as $N \to \infty$. The commutation relation of the Y generators takes the form

$$[Y_+, Y_-] = -2\theta \quad + O(\frac{1}{N^2}), \tag{3.3.59}$$

so that we recover the commutation relations of $\mathbb{R}^2_\theta$

$$[Y_1, Y_2] = -i\theta \qquad (3.3.60)$$

with $Y_\pm = Y_1 \pm iY_2$.

Exercise 3.3.1   Verify that $\mathcal{J}^a$ acts on $\text{End}(\mathcal{H})$ as a self-adjoint operator, i.e.

$$\langle \Phi, \mathcal{J}^a \Psi \rangle = \langle \mathcal{J}^a \Phi, \Psi \rangle \qquad (3.3.61)$$

and satisfies the $\mathfrak{su}(2)$ commutation relations

$$[\mathcal{J}^a, \mathcal{J}^b] = i\varepsilon^{abc} \mathcal{J}^c. \qquad (3.3.62)$$

3.3.2 Fuzzy $\mathbb{C}P^{n-1}_N$

The above oscillator construction of the fuzzy sphere $S^2_N \cong \mathbb{C}P^1_N$ generalizes naturally to higher-dimensional fuzzy complex projective spaces $\mathbb{C}P^n_N$. This is achieved by quantizing the construction of classical $\mathbb{C}P^{n-1}$ in Section 2.1. We start with the complex quantum plane $\mathbb{C}^n_\theta$ (3.1.85) generated by Z^i and $Z^\dagger_j$ with $[Z^i, Z^\dagger_j] = \delta^i_j$. Define the number operator

$$\hat{N} := Z^\dagger_i Z^i \qquad (3.3.63)$$

(sum over i understood) which satisfies

$$[\hat{N}, Z^i] = -Z^i. \qquad (3.3.64)$$

Consider the associated Fock space spanned by all creation operators

$$\mathcal{F} := \big\langle (Z^\dagger_1)^{k_1} \dots (Z^\dagger_n)^{k_n} |0\rangle \big\rangle_{\mathbb{C}} \qquad (3.3.65)$$

acting on the Fock vacuum, which satisfies $Z_i|0\rangle = 0$. The number operator $\hat{N}$ clearly respects the N- particle subspace

$$\mathcal{H}_N := \big\langle (Z^\dagger_1)^{k_1} \dots (Z^\dagger_n)^{k_n} |0\rangle; \;\; k_1 + \dots + k_n = N \big\rangle_{\mathbb{C}} \;\; \subset \mathcal{F} \qquad (3.3.66)$$

of the Fock space, with $\hat{N}|_{\mathcal{H}_N} = N$. Then the space of (quantized) functions on fuzzy $\mathbb{C}P^{n-1}_N$ is defined as

$$\mathcal{C}(\mathbb{C}P^{n-1}_N) := \text{End}(\mathcal{H}_N). \qquad (3.3.67)$$

We can establish the relation of $\mathcal{C}(\mathbb{C}P^{n-1})$ and $\text{End}(\mathcal{H}_N)$ by observing that both are generated by polynomials $P(Z^i, Z^\dagger_j)$ with the same number of creation- and annihilation operators, which respect $\mathcal{H}_N$:

$$\text{End}(\mathcal{H}_N) \cong \Big(\bigoplus_{k=0}^{N} \{Z^\dagger_{j_1} \dots Z^\dagger_{j_k} Z^{i_1} \dots Z^{i_k}\} \Big) \Big/_{Z^\dagger Z = N} \qquad (3.3.68)$$

$$\mathcal{C}(\mathbb{C}P^{n-1}) \cong \Big(\bigoplus_{k=0}^{\infty} \{\bar{z}_{j_1} \dots \bar{z}_{j_k} z^{i_1} \dots z^{i_k}\} \Big) \Big/_{\bar{z}z = N}; \qquad (3.3.69)$$

recall that only such "balanced" polynomials are invariant under $U(1)$, and therefore define functions on $\mathbb{C}P^{n-1}$ (2.1.7). Due to the simple commutation relations, there is clearly a one-to-one correspondence (e.g. via the Weyl quantization map for $\mathbb{C}^n_\theta$, or via normal ordering) between these classical and quantized polynomials modulo the constraint $\hat{N} = N$. This equivalence holds only up to the maximal degree N, since $Z^{i_1} \ldots Z^{i_{N+1}} = 0$ on $\mathcal{H}_N \subset \mathcal{F}$. This amounts to a UV cutoff on $\mathcal{C}(\mathbb{C}P^{n-1})$.

Another description of fuzzy $\mathbb{C}P^{n-1}_N$ as a quantized submanifold of $\mathfrak{su}(n) \cong \mathbb{R}^{n^2-1}$ (more specifically as a quantized coadjoint orbit of $SU(n)$ as discussed in Section 3.3.3) in terms of $SU(n)$ group theory works as follows. Consider the generators

$$T_a = Z_i^\dagger (\lambda_a)^i_j Z^j, \tag{3.3.70}$$

where λ_a are $\mathfrak{su}(n)$ generators that satisfy (2.1.9). Similarly as (2.1.10), it is straightforward to verify that the T_a realize the $\mathfrak{su}(n)$ Lie algebra

$$[T_a, T_b] = if_{abc}T_c. \tag{3.3.71}$$

The T^a clearly respect $\mathcal{H}_N$, since

$$[\hat{N}, T^a] = 0. \tag{3.3.72}$$

It is not hard to see that

$$\mathcal{H}_N \cong (N, 0, \ldots, 0) \tag{3.3.73}$$

is an irreducible representation of $\mathfrak{su}(n)$ with highest weight $(N, 0, \ldots, 0)$ in Dynkin label notation, which equivalent to the totally symmetric tensor product of the fundamental representation denoted with a Young diagram $\square\square \ldots \square$, and to the Hilbert space (3.3.66). Comparing this construction with the discussion of $\mathbb{C}P^{n-1}$ in Section 2.1, we recognize it as quantized complex projective space $\mathbb{C}P^{n-1}_N$, and the above construction is a quantum version of the symplectic reduction. The $T^a = (T^a)^\dagger$ are quantizations of the functions t^a (2.1.8), so that

$$\frac{1}{r_N}T^a \sim t^a : \quad \mathbb{C}P^{n-1} \hookrightarrow \mathfrak{su}(n) \cong \mathbb{R}^{n^2-1} \tag{3.3.74}$$

can be viewed as a quantized embedding of $\mathbb{C}P^{n-1}$ into flat target space $\mathbb{R}^{n^2-1}$. Here

$$T^a T_a = r_N^2 \mathbb{1}, \qquad r_N^2 = 2(n-1)(\frac{1}{n}N^2 + N). \tag{3.3.75}$$

They generate the algebra $\mathrm{End}(\mathcal{H}_N)$, which is a quantization of the algebra of functions $\mathcal{C}(\mathbb{C}P^{n-1}_N)$. Similarly as for the fuzzy sphere, one can define a quantization map

$$\mathcal{Q}: \quad \mathcal{C}(\mathbb{C}P^{n-1}) \to \mathrm{End}(\mathcal{H}_N)$$

$$Y_\Lambda \mapsto \begin{cases} \hat{Y}_\Lambda, & |\Lambda| < N \\ 0, & |\Lambda| \geq N \end{cases} \tag{3.3.76}$$

noting that both sides carry a natural action of $SU(n)$, and therefore decompose into harmonics i.e. irreducible representations labeled by Λ. More explicitly, one can show using $\mathfrak{su}(n)$ group theory (cf. Section 3.3.3) that the decomposition into harmonics of both sides is identical up to some cutoff:

$$\text{End}(\mathcal{H}_N) \cong \mathcal{H}_N \otimes \mathcal{H}_N^* \cong (N, 0, \ldots, 0) \otimes (0, \ldots, 0, N) \cong \bigoplus_{k=0}^{N} (k, 0, \ldots, 0, k)$$

$$\mathcal{C}(\mathbb{C}P^{n-1}) \cong \bigoplus_{k=0}^{\infty} (k, 0, \ldots, 0, k). \tag{3.3.77}$$

The map $\mathcal{Q}$ can therefore be defined as an intertwiner of $SU(n)$, which respects the inner products on $\text{End}(\mathcal{H}_N)$ and $\mathcal{C}(\mathbb{C}P^{n-1})$ up to the cutoff. Comparing the commutation relations (3.3.71) with the Poisson bracket (2.1.10) and adjusting for the different normalization

$$T^a T_a = r_N^2 \sim 2\frac{n-1}{n} N^2 \qquad \text{while} \qquad t^a t_a = 2\frac{n-1}{n}, \tag{3.3.78}$$

we see that $\mathcal{Q}$ satisfies all requirements of a quantization map, so that $\mathbb{C}P_N^{n-1}$ is a quantization of $(\mathbb{C}P^{n-1}, N\omega)$, where ω is the canonical Kirillov–Kostant symplectic form (2.2.9), (2.2.11).

Since $\mathbb{C}P^{n-1}$ is a $2(n-1)$ dimensional manifold, the $n^2 - 1$ generators T^a must satisfy some further constraints. These are best understood by combining the generators in the following $n \times n$ matrix

$$\Xi := \frac{n}{N} T^a \lambda_a \tag{3.3.79}$$

generalizing (2.2.30). By relating $T^a \lambda_a$ to the quadratic Casimir on $\mathcal{H}_N \otimes \mathbb{C}^n$, it is not hard to see that Ξ satisfies the characteristic relation

$$\left(\Xi - (n-1) \right)\left(\Xi + 1 + \frac{n}{N} \right) = 0, \tag{3.3.80}$$

which generalizes (2.2.31). This leads to the explicit constraints

$$T^a T_a = r_N^2 \mathbb{1}, \qquad d_{abc} T^a T^b = D_N T_c \tag{3.3.81}$$

for some constant D_N, which generalize (2.2.32). More details can be found e.g. in [41]. This description of $\mathbb{C}P_N^{n-1}$ in terms of the Lie algebra generators T^a provides a direct link with the description as a quantized coadjoint orbit discussed in Section 2.1. Both constructions are completely equivalent, but the present oscillator construction provides a useful basis for explicit computations.

Exercise 3.3.2   Verify (3.3.75) for T^a given by (3.3.70) and obtain r_N^2 using the $\mathfrak{su}(n)$ Fierz identity

$$(\lambda_a)_j^i (\lambda^a)_l^k = 2\big(\delta_l^i \delta_j^k - \frac{1}{n} \delta_j^i \delta_l^k \big) \tag{3.3.82}$$

for the normalization $\text{tr}(\lambda_a \lambda_b) = 2\delta_{ab}$.

Exercise 3.3.3   Consider the three-dimensional quantum mechanical harmonic oscillator

$$H = \frac{1}{2} \sum_i \big((Q^i)^2 + (P_i)^2 \big). \tag{3.3.83}$$

Show that $SU(3)$ acting linearly on the Z^i (3.1.84) is a symmetry of H which respects the commutation relations. Conclude that the energy eigenspaces are representations of $SU(3)$.

*Challenge: using $SU(3)$ group theory, show that the energy eigenstates are in fact irreducible representations of $SU(3)$, given by totally symmetrized tensor products of the fundamental representation.

3.3.3 *Quantized coadjoint orbits

The above examples of fuzzy S_N^2 and fuzzy $\mathbb{C}P_N^n$ are special cases of a more general construction, which works for any (co)adjoint orbit of a (semi-)simple Lie group. Even though this is technically somewhat involved and not essential for most of the remaining chapters, the construction is quite illuminating, and deserves to be discussed in some detail.

We have seen in Section 2.2 that coadjoint orbits are symplectic manifolds. It is therefore natural to expect that they can be quantized. It turns out that for compact (and at least for some noncompact) semi-simple Lie groups G they can indeed be quantized explicitly, provided some quantization conditions hold. The quantization is given in terms of matrix algebras $\mathrm{End}(\mathcal{H}_\Lambda)$, where $\mathcal{H}_\Lambda$ are unitary irreducible representations of G with highest weight Λ. We will discuss this result for the case of compact G in the following, which will be generalized later to some specific non-compact G in Section 5.2. A more elementary approach based on coherent states and quantum matrix geometry will be presented in Section 4.6.3.

The space of harmonics on the classical orbit $\mathcal{O}[\Lambda]$

Let $\Lambda \in \mathfrak{g}_0^*$ be the (unique) dominant weight in the coadjoint orbit $\mathcal{O}[\alpha] = \mathcal{O}[\Lambda]$. All information about this space can be obtained from the harmonic analysis, i.e. by decomposing functions on $\mathcal{O}[\Lambda]$ into harmonics under the adjoint action of G. This is particularly useful here, because quantized spaces are described in terms of their algebra of functions. The decomposition of the space of functions $\mathcal{C}(\mathcal{O}[\Lambda])$ into harmonics can be calculated explicitly using 2.2.2, and it must be preserved after quantization, at least up to some cutoff. Otherwise, the quantization would not be admissible. Let K be the stabilizer group of Λ, which we can assume to be integral. Then the coadjoint orbit is isomorphic to the homogeneous space or coset (2.2.2)

$$\mathcal{O}[\Lambda] \cong G/K. \tag{3.3.84}$$

As explained in Appendix E, the harmonics on $\mathcal{O}[\Lambda]$ are given by

$$\mathcal{C}(\mathcal{O}[\Lambda]) \cong \bigoplus_{\lambda \in P^+} \mathrm{mult}_{\lambda+}^{(K)} \, \mathcal{H}_\lambda. \tag{3.3.85}$$

Here λ runs over all dominant integral weights P^+, $\mathcal{H}_\lambda$ is the corresponding highest-weight G-module, and $\mathrm{mult}_{\lambda+}^{(K)}$ is the dimension of the subspace of $\mathcal{H}_{\lambda+}$ that is invariant under K.

For example, $\mathbb{C}P^{n-1} \cong SU(n)/K$, where the stabilizer of $\Lambda = (N, 0, \ldots, 0)$ is given by $K = SU(n-1) \times U(1)$, and the space of functions decomposes as (E.0.6)

$$C(\mathbb{C}P^{n-1}) \cong \bigoplus_{k=0}^{\infty} (k, 0, \ldots, 0, k). \qquad (3.3.86)$$

The space of harmonics on fuzzy $\mathcal{O}[\Lambda]$

Now we claim that the quantization of the above coadjoint orbit is given in terms of the operator algebra on a suitable (series of) irreducible representations of G or $\mathfrak{g}$, i.e. by the quantized algebra of functions

$$\mathrm{End}(\mathcal{H}_\Lambda) \cong \mathcal{H}_\Lambda \otimes \mathcal{H}_\Lambda^* \cong \oplus_\mu N_{\Lambda\Lambda^+}^\mu \, \mathcal{H}_\mu. \qquad (3.3.87)$$

Here $\mathcal{H}_\Lambda$ is the irreducible representation of $\mathfrak{g}$ with highest weight $\Lambda = (n_1, \ldots, n_k)$ labeled by Dynkin labels n_j. The decomposition of $\mathrm{End}(\mathcal{H}_\Lambda)$ into harmonics is given by the last form, where $N_{\lambda\lambda^+}^\mu$ are the usual fusion rules of highest weight representations of $\mathfrak{g}$. Here λ^+ is the conjugate weight to λ, so that $\mathcal{H}_\lambda^* \cong \mathcal{H}_{\lambda^+}$.

To establish a quantization map that respects G, we must match the space of harmonics in $\mathrm{End}(\mathcal{H}_\Lambda)$ with those in $C(\mathcal{O}[\Lambda])$ up to some cutoff and show that the map is an isomorphism below the cutoff. To obtain a large semi-classical regime, we should assume that Λ is large, possibly rescaling $\Lambda \to n\Lambda$ for large $n \in \mathbb{N}$; note that this does not change the geometry and the mode decomposition of $\mathcal{O}[\Lambda]$. As discussed in Appendix E, one can show that if μ is small enough (smaller than all *nonzero* Dynkin labels of λ, roughly speaking), then

$$N_{\lambda\lambda^+}^\mu = \mathrm{mult}_{\mu^+}^{(K_\lambda)}, \qquad (3.3.88)$$

where $K_\lambda \subset G$ is the stabilizer group of λ. Note that the mode structure for small μ does not depend on the particular λ, only on its stabilizer K_λ. Comparing this with the decomposition 3.3.85 of $C(\mathcal{O}[\Lambda])$, we see that indeed

$$\mathrm{End}(\mathcal{H}_\Lambda) \cong C(\mathcal{O}[\Lambda]) \qquad (3.3.89)$$

up to some cutoff in μ. This allows to define a quantization map

$$\mathcal{Q}: \quad C(\mathcal{O}[\Lambda]) \to \mathrm{End}(\mathcal{H}_\Lambda) \qquad (3.3.90)$$

as an isomorphism of the irreps, preserving the norms on both sides at least approximately. An explicit realization of such a map in terms of coherent states will be discussed in Section 4.6.3, which provides a geometric understanding of (3.3.89).

In particular, the brane is degenerate if Λ is invariant under a non-minimal subgroup K. These degenerate branes have smaller dimensions than the regular ones. An example for such degenerate quantized branes is fuzzy $\mathbb{C}P_N^{n-1}$, where

$$\mathrm{End}(\mathcal{H}_\Lambda) \cong (N, 0, \ldots, 0) \otimes (0, \ldots, 0, N) \cong \bigoplus_{k=0}^{N} (k, 0, \ldots, 0, k), \qquad (3.3.91)$$

which is clearly a truncation of $\mathcal{C}(\mathbb{C}P^{n-1})$ (3.3.86). The degeneracy of the underlying quantized coadjoint orbit is reflected in the characteristic equation (3.3.80) and (2.2.31) for Ξ in the fuzzy and classical case, respectively. These equations state that Ξ has only two different eigenvalues, rather than n for generic orbits.

3.4 The fuzzy torus T^2_N

The fuzzy torus is a simple example of a compact quantum space that is not the quantization of a coadjoint orbit.

The classical (Clifford) torus

Consider first the commutative torus T^2. As a two-dimensional manifold it can be identified with $T^2 \cong \mathbb{R}^2/_\sim$, with coordinates φ, ψ and periodic identifications $\varphi \sim \varphi + 2\pi, \psi \sim \psi + 2\pi$. Similar as for fuzzy S^2, we will not quantize the abstract manifold but an embedding of it in target space. Rather than considering a doughnut embedded in $\mathbb{R}^3$, it is more convenient to use the embedding as a flat submanifold of $\mathbb{R}^4$

$$x^a: \quad T^2 \hookrightarrow \mathbb{R}^4, \qquad a = 1, \ldots, 4 \tag{3.4.1}$$

$$\tag{3.4.2}$$

where $x^a(\varphi, \psi)$ are functions on T^2 defined by

$$u := x^1 + ix^2 = e^{i\varphi},$$
$$v := x^3 + ix^4 = e^{i\psi} \tag{3.4.3}$$

so that $|u|^2 = (x^1)^2 + (x^2)^2 = 1$ and $|v|^2 = (x^3)^2 + (x^4)^2 = 1$. They satisfy the reality conditions

$$u^* = u^{-1}, \qquad v^* = v^{-1}. \tag{3.4.4}$$

This "Clifford torus" inherits an induced metric that is flat, and admits a manifest $U(1) \times U(1)$ symmetry of translations

$$(U(1) \times U(1)) \times T^2 \to T^2$$
$$((e^{i\alpha}, e^{i\beta}), (u, v)) \mapsto (e^{i\alpha}u, e^{i\beta}v). \tag{3.4.5}$$

The symmetry group acts again on the algebra of functions $\mathcal{C}(T^2)$ analogous to (3.3.9). This infinite-dimensional vector spaces decomposes into a direct sum of irreducible representations, which are now one-dimensional and given by plane waves:

$$\mathcal{C}(T^2) = \bigoplus_{n,m=-\infty}^{\infty} (n, m), \qquad (n, m) := u^n v^m = e^{i(n\varphi + m\psi)}. \tag{3.4.6}$$

The torus is equipped with a $U(1) \times U(1)$-invariant symplectic structure given by

$$\omega = \frac{1}{2\pi} d\varphi d\psi, \qquad \int_{T^2} \frac{\omega}{2\pi} = 1, \qquad (3.4.7)$$

where the normalization is fixed by setting the symplectic volume to 2π. This symplectic form corresponds to the Poisson brackets

$$\{\psi, \varphi\} = 2\pi, \qquad \{u, v\} = 2\pi \, uv. \qquad (3.4.8)$$

The fuzzy torus

The fuzzy torus T_N^2 is a quantization of the Clifford torus, which is defined in terms of clock and shift operators U, V given by

$$U = \begin{pmatrix} 0 & 1 & 0 & \dots & 0 \\ 0 & 0 & 1 & \dots & 0 \\ & & \ddots & & \\ 0 & & \dots & 0 & 1 \\ 1 & 0 & \dots & & 0 \end{pmatrix}, \quad V = \begin{pmatrix} 1 & & & & \\ & e^{2\pi i \frac{1}{N}} & & & \\ & & e^{2\pi i \frac{2}{N}} & & \\ & & & \ddots & \\ & & & & e^{2\pi i \frac{N-1}{N}} \end{pmatrix} \qquad (3.4.9)$$

acting on $\mathcal{H} = \mathbb{C}^N$. These matrices provide an irreducible representation[15] of the relations

$$UV = qVU, \qquad q^N = 1, \quad U^N = V^N = 1 \qquad (3.4.10)$$

with $q = e^{2\pi i/N}$, and are unitary

$$U^\dagger = U^{-1}, \qquad V^\dagger = V^{-1}. \qquad (3.4.11)$$

They generate the algebra $\mathcal{A} \cong \text{End}(\mathcal{H})$, which can be viewed as quantization of the function algebra $\mathcal{C}(T^2)$ on the symplectic space (T^2, ω_N). To see this, we will define again a quantization map that respects a discrete $\mathbb{Z}_N \times \mathbb{Z}_N$ symmetry, which is the quantum analog of the $U(1) \times U(1)$ symmetry (3.4.5). It is defined as

$$\mathbb{Z}_N \times \mathcal{A} \to \mathcal{A} \qquad (3.4.12)$$

$$(\omega^k, \phi) \mapsto U^k \phi U^{-k} \qquad (3.4.13)$$

and the second $\mathbb{Z}_N$ is defined similarly by conjugation with V. Under this group action, the algebra of functions decomposes into harmonics i.e. irreps

$$\mathcal{A} = \text{End}(\mathcal{H}) = \bigoplus_{n,m=0}^{N-1} U^n V^m. \qquad (3.4.14)$$

This suggests to define the following quantization map:

$$\mathcal{Q}: \quad \mathcal{C}(T^2) \to \text{End}(\mathcal{H}) \qquad (3.4.15)$$

---

[15] Choosing other roots of unity for q leads to multiple coverings of the Clifford torus, cf. [169].

$$e^{in\varphi}e^{im\psi} \mapsto \begin{cases} q^{-nm/2} \, U^n V^m, & |n|, |m| < N/2 \\ 0, & \text{otherwise} \end{cases}$$

which is compatible with the $\mathbb{Z}_N \times \mathbb{Z}_N$ symmetry and satisfies $\mathcal{Q}(f^*) = \mathcal{Q}(f)^\dagger$. The underlying Poisson structure on T^2 is identified using $[U, V] = (q - 1)VU$ as

$$\{u, v\} = \frac{2\pi}{N} \, uv, \tag{3.4.16}$$

consistent with (3.4.8) for

$$\omega_N = N\omega = \frac{N}{2\pi} \, d\varphi d\psi. \tag{3.4.17}$$

Comparing with $\theta^{\mu\nu} = \theta \epsilon^{\mu\nu}$, we can again identify a scale of noncommutativity via

$$L_{\text{NC}}^2 = \frac{2\pi}{N} =: \theta. \tag{3.4.18}$$

Due to the simple algebraic structure, it is easy to verify that $\mathcal{Q}$ satisfies the requirements of a quantization map. Therefore, T_N^2 is the quantization of (T^2, ω_N). Moreover, the following integral relation holds

$$\text{Tr}(\mathcal{Q}(\phi)) = \int_{T^2} \frac{\omega_N}{2\pi} \phi. \tag{3.4.19}$$

This is a consequence of group invariance, and the normalization follows for $\phi = 1$ using (3.4.7).

To define a metric structure, we consider T_N^2 as quantized Clifford torus embedded in target space $\mathbb{R}^4$ via the four hermitian matrices X^a defined by

$$X^1 + iX^2 := U, \qquad X^3 + iX^4 := V. \tag{3.4.20}$$

As a consequence of (3.4.10) and unitarity, these satisfy the relations

$$[X^1, X^2] = 0 = [X^3, X^4]$$
$$(X^1)^2 + (X^2)^2 = 1 = (X^3)^2 + (X^4)^2,$$
$$(X^1 + iX^2)(X^3 + iX^4) = q(X^3 + iX^4)(X^1 + iX^2). \tag{3.4.21}$$

The X^a can hence be viewed as quantized embedding maps

$$X^a \sim x^a : \quad T^2 \hookrightarrow \mathbb{R}^4, \tag{3.4.22}$$

which reduce to (3.4.2) in the semi-classical limit. This allows to consider the matrix Laplace operator (3.3.53), and to compute its spectrum:

$$\Box\phi = [X^a, [X^b, \phi]]\delta_{ab}$$

$$= [U, [U^\dagger, \phi]] + [V, [V^\dagger, \phi]] = 4\phi - U\phi U^\dagger - U^\dagger \phi U - V\phi V^\dagger - V^\dagger \phi V. \tag{3.4.23}$$

Explicitly, this gives

$$\Box(U^n V^m) = c([n]_q^2 + [m]_q^2)\, U^n V^m \sim \frac{4\pi^2}{N^2}(n^2 + m^2)\, U^n V^m,$$

$$c = -(q^{1/2} - q^{-1/2})^2 = 4\sin^2(\pi/N) \sim \frac{4\pi^2}{N^2} \tag{3.4.24}$$

where

$$[n]_q := \frac{q^{n/2} - q^{-n/2}}{q^{1/2} - q^{-1/2}} = \frac{\sin(n\pi/N)}{\sin(\pi/N)} \sim n \tag{3.4.25}$$

for $n, m \ll N$. Clearly the spectrum of the matrix Laplacian approximates that of the classical Laplacian, sufficiently far below the cutoff. In this sense, T_N^2 with the embedding defined via the above embedding (3.4.20) has indeed the geometry of the Clifford torus. It is interesting to note that due to the periodicity of $[n]_q = [n + 2N]_q$, the "momentum space" of T_N^2 can be considered as compactified.

The fuzzy torus T_N^2 should not be confused with the so-called noncommutative torus T_θ^2, which is often considered in the literature on noncommutative geometry. The latter is defined in terms of an infinite-dimensional algebra of operators which satisfy the relations

$$UV = qVU \qquad \text{for} \quad q = e^{i\vartheta} \tag{3.4.26}$$

without imposing $U^N = V^N = 1$; then q need not be a root on unity.[16] The main difference is that T_θ^2 has infinitely many degrees of freedom, which leads to significant complications. For that reason we will not use it in the following, except for a discussion of compactifications[17] for the IKKT model in Section 11.1.8.

In view of the simple realization of two-dimensional fuzzy spaces with genus 0 and 1, it would seem natural to look for two-dimensional fuzzy spaces with genus g. Although there is no obstacle in principle, no nice and explicit realization has been found so far. This is not a problem in the present context, as we will be mainly concerned with higher-dimensional quantum spaces.

Further comments and reading

Needless to say that we have discussed only a few representative examples of a large list of quantized spaces that have been considered in the literature. In particular, we have not discussed degenerate spaces that are quantizations of manifolds with degenerate Poisson structure. This includes well-known examples such as κ- Minkowski spaces [130] or odd-dimensional spaces. On such spaces, the effective metric arising in the matrix model framework is degenerate, and the propagation is restricted to the symplectic leaves.

[16] If q is a root of unity, then T_N^2 is a quotient of T_θ^2 as an algebra.

[17] Such a compactification corresponds to the conventional approach to string theory by compactifying some of the 10 dimensions, where each compact direction supports infinitely many degrees of freedom. This could be considered as part of the "string theory landscape," which is exactly what we want to avoid.

There is a vast literature on coherent states in the context of quantum mechanics or the Moyal–Weyl quantum plane. For a discussion in the context of quantum optics, see e.g. [164]. A mathematical discussion of quantization for general compact Kähler manifolds can be found in [37, 165].

Deformation quantization has also attracted considerable attention; in particular, fundamental mathematical results obtained by Kontsevich [124] found applications in the context of noncommutative field theory [109, 133]. For a discussion in the context of quantum mechanics, see e.g. [26]. However, this approach is not adequate in the framework of matrix models and for quantum mechanical systems with a finite-dimensional Hilbert space.

The quantization of coadjoint orbits was studied long time ago in a mathematical context, notably by Kirillov and Berezin; a nice and accessible discussion can be found e.g. in [95]. Specific examples such as the fuzzy sphere were introduced several times in a physical context, starting with Hoppe [101] and Madore [131]. There is a large body of literature on their application in noncommutative quantum field theory starting with [84]; this will be discussed in Chapter 6. Many remarkable further structures arise on (quantized) coadjoint orbits, including coherent states and a Kähler structure; these will be discussed in Section 4.6.

Quantum spaces and matrix geometry

So far, we have viewed quantum spaces as quantizations of some given symplectic space. This is analogous to viewing quantum mechanics as quantization of classical mechanics. Now we take the opposite and more appropriate point of view: starting with some given quantum space, we aim to extract a symplectic manifold with appropriate symplectic structure, which can be viewed as semi-classical limit or picture of the quantum space. This is analogous to finding a de-quantization map for a given quantum mechanical system, i.e. some sort of inverse map of $\mathcal{Q}$. To achieve this, we first need to establish the framework in which the quantum spaces will arise, given by matrix models. This will allow to define quasi-coherent states as ground states of certain parameter-dependent Hamiltonians, and to exploit ideas familiar from the context of geometric phases in quantum mechanics.

Formulation of the problem and a first glimpse of matrix models

The only geometrical structure encountered up to now is the Poisson structure or the symplectic structure of the manifold $\mathcal{M}$, which is encoded in the commutation relations of the quantum case. To do interesting physics and in particular gravity, we must include a metric structure in the framework of quantum geometry. Many different approaches toward this goal have been considered in the literature, ranging from attempts to define some sort of noncommutative tensor calculus, to the idea of spectral triples introduced by Connes. However, there is no obvious way to generalize the standard concepts of metric and curvature.

Facing this situation, we will adopt a radically different approach. Rather than trying to mimic classical concepts, we will start with the simplest possible model(s) which are intrinsically noncommutative, and observe that the desired structures **emerge** in some semi-classical regime. It turns out that matrix models provide such a starting point. They are extremely simple, defined through an action $S[X]$ for any collection of matrices X^a. More precisely, a matrix configuration in such a model is a collection of D hermitian matrices $X^a \in \text{End}(\mathcal{H})$ for $a = 1, \ldots, D$. Here $\mathcal{H}$ is a separable Hilbert space, which we will mostly assume to be finite-dimensional. We will say that a matrix configuration is **irreducible** if it does not separate into a direct sum of smaller configurations, i.e. if the only matrix that commutes with all of them is the identity matrix:

$$[X^a, A] = 0 \quad \forall a \quad \Rightarrow \quad A = a\mathbb{1}. \tag{4.0.1}$$

The matrix models of interest are defined in terms of an action with the structure

$$S[X] = \text{Tr}([X^a, X^b][X_a, X_b] + m^2 X^a X_a) \tag{4.0.2}$$

or similar. Here the $X^a \in \text{End}(\mathcal{H})$, $a = 1, \ldots, D$ are hermitian matrices that define some matrix configuration, and indices are contracted with δ^{ab} or η^{ab}. The concept of a quantization map as discussed before suggests to view the matrices X^a as quantized embedding functions

$$X^a = Q(x^a) \sim x^a : \quad \mathcal{M} \hookrightarrow \mathbb{R}^D, \qquad (4.0.3)$$

where x^a are the Cartesian coordinate functions on target space $\mathbb{R}^D$ pulled back to some manifold $\mathcal{M}$. In other words, the matrices X^a are viewed as quantizations of the functions $x^a \in \mathcal{C}(\mathcal{M})$. This is indicated by the symbol $\sim$, which means "semi-classical limit" or regime. The matrices X^a generate a noncommutative algebra $\mathcal{A}$, which is interpreted as quantized algebra of functions on $\mathcal{M}$.

In the semi-classical limit, $\mathcal{M}$ carries a Poisson structure $\{x^a, x^b\} \sim -i[X^a, X^b]$ or even a symplectic structure ω. The embedding map x^a (4.0.3) also induces a metric structure on $\mathcal{M}$, via pullback of the metric in target space $\mathbb{R}^D$. This picture is very close to the concept of D-branes in string theory, which are sub-manifolds in target space, possibly carrying a noncommutative structure. The relation with string theory will be discussed in more detail in Chapter 11, and we will use the name "quantized brane" in a loose sense to indicate the setup in (4.0.3). While this is realized in many standard examples such as the fuzzy sphere, it is not evident a priori if that picture is generic. The aim of this chapter is to demonstrate that this is indeed the case, and there is a systematic way to extract a map (4.0.3) from any given set of hermitian matrices X^a.

We therefore want to address the following general problem: given some matrix configuration consisting of hermitian matrices X^a, $a = 1, \ldots, D$, is there a symplectic manifold $\mathcal{M} \subset \mathbb{R}^D$ embedded[1] in $\mathbb{R}^D$ via some map $x^a : \mathcal{M} \hookrightarrow \mathbb{R}^D$ such that the X^a can be viewed as quantizations of classical embedding functions x^a as in (4.0.3)? And if yes, how can we determine this manifold $\mathcal{M}$?

At first sight, this may seem like a hopeless problem: how should one possibly associate a geometry to e.g. some set of three 2×2 matrices X^a, or to any set of large but finite matrices? On the other hand, there is a large class of examples including S_N^2, T_N^2 etc. where the answer is obviously yes. The problem is comparable to the problem of identifying an appropriate classical system to some given quantum system.

Remarkably, there is indeed a general way to associate geometric structure to any set of hermitian matrices. If these matrices are completely random, the resulting geometrical structure will not provide an adequate picture, and such matrices should be considered as "white noise" or perhaps as "deep quantum foam." However, if the matrices are in some sense "almost-simultaneously diagonalizable," then this leads to a reasonable classical geometry, which provides a good description in some IR regime. Moreover, there is a systematic way to extract this geometry. This is achieved through quasi-coherent states $|x\rangle \in \mathcal{H}$, which are optimally localized in a suitable sense. They are approximate common eigenstates of the X^a, localized at some point in target space

$$x^a \approx \langle x | X^a | x \rangle \quad \in \mathbb{R}^{9,1}. \qquad (4.0.4)$$

---

[1] Here "embedding" is understood in a loose sense; the map can be degenerate, as in Section 5.

Such "quasi-coherent" states provide a general conceptual definition of a quantum geometry, and a powerful tool to understand the relation with classical geometry. We will also discuss some quantitative measures for the quality of the semi-classical description.

These concepts are well adapted to matrix models such as (4.0.2), which are expected to select configurations admitting such a geometric interpretation. Indeed, since the action is the square of commutators, matrix configurations are preferred whose commutators are small. Therefore, the dominant contributions are almost-commutative matrices X^a, which can "almost" be simultaneously diagonalized. These are the configurations to which we can associate some semi-classical geometry in a meaningful way. For matrix models with Minkowski signature as in Chapter 11, this argument is a little less compelling but is still expected to apply, since energy rather than the action itself should be minimized.

4.1 Quasi-coherent states

Now we return to the general discussion. Given some matrix configuration X^a and a point $x \in \mathbb{R}^D$, we define the following "displacement" Hamiltonian

$$H_x := \frac{1}{2} \sum_a (X^a - x^a)^2, \tag{4.1.1}$$

which is positive definite. Let $\lambda(x) > 0$ be the lowest eigenvalue of H_x. A **quasi-coherent state** $|x\rangle$ at x is then defined as normalized vector $\langle x|x \rangle = 1$ in the eigenspace E_x of H_x with eigenvalue $\lambda(x)$,

$$H_x|x\rangle = \lambda(x)|x\rangle. \tag{4.1.2}$$

This is analogous to the adiabatic ground states considered in the context of geometric phases [32]. We will see that H_x arises naturally as part of the matrix Laplacian (4.7.1), and measures the energy of a string state stretching from a brane $\mathcal{M}$ to a point (brane) $x \in \mathbb{R}^D$. Therefore, the eigenvalue $\lambda(x)$ will be denoted as **displacement energy**. We assume for simplicity that the lowest eigenspace is nondegenerate, except possibly on a singular set denoted as $\mathcal{K}$. We denote the set of generic points as $\tilde{\mathbb{R}}^D = \mathbb{R}^D \setminus \mathcal{K}$.

Given these states $|x\rangle$, we can associate to any operator in $\Phi \in \mathrm{End}(\mathcal{H})$ the **symbol** in $\mathcal{C}(\tilde{\mathbb{R}}^D)$ through the map

$$\mathrm{End}(\mathcal{H}) \to \mathcal{C}(\tilde{\mathbb{R}}^D)$$
$$\Phi \mapsto \langle x|\Phi|x\rangle =: \phi(x). \tag{4.1.3}$$

Elements of $\mathrm{End}(\mathcal{H})$ will be indicated by upper-case letters and functions by lower-case letters. The map (4.1.3) should be viewed as a de-quantization map, associating classical functions to noncommutative "functions" (or rather observables) in $\mathrm{End}(\mathcal{H})$. In particular, the symbol of the matrices X^a provides a map

$$\mathbf{x^a}: \quad \tilde{\mathbb{R}}^D \to \mathbb{R}^D$$
$$x \mapsto \mathbf{x^a}(x) := \langle x|X^a|x\rangle. \tag{4.1.4}$$

Generically $\mathbf{x^a}(x) \neq x^a$, but $\mathbf{x^a}(x) \approx x^a$, and the deviation is measured by the **displacement**

$$d^2(x) := \sum_a (\mathbf{x^a}(x) - x^a)^2. \tag{4.1.5}$$

The quality of the matrix configuration (or of the underlying quantum space) is measured by the **dispersion** or uncertainty

$$\Delta^2(x) := \sum_a (\Delta X^a)^2$$

$$(\Delta X^a)^2 := \langle x|(X^a - \mathbf{x^a}(x))^2|x\rangle = \langle x|X^a X^a|x\rangle - \mathbf{x^a}(x)\mathbf{x^a}(x). \tag{4.1.6}$$

If $\Delta^2(x)$ is small, then the X^a can be interpreted as operators or observables that approximate the functions $\mathbf{x^a}$ on $\tilde{\mathbb{R}}^D$, and if $d^2(x)$ is also small then $X^a \approx \mathbf{x^a} \approx x^a$. Note that (4.1.2) implies

$$2\lambda(x) = \Delta^2(x) + d^2(x), \tag{4.1.7}$$

hence a small $\lambda(x)$ implies that both $\Delta^2(x)$ and $d^2(x)$ are bounded by $\lambda(x) > 0$. $d^2(x)$ will be understood in Section 4.2 as displacement of x from the underlying quantum space or brane $\mathcal{M}$. In particular, quasi-coherent states are *optimally localized* in the sense that the dispersion $\Delta^2(x)$ is minimal among all states $|\psi\rangle$ with the same expectation value $\langle \psi|X^a|\psi\rangle = \langle x|X^a|x\rangle = \mathbf{x^a}$. In view of (4.1.6), they are hence an optimal set of states where all matrices X^a are "almost-simultaneously diagonal." Indeed, Yang–Mills matrix models prefer matrix configurations that are "almost-commuting." Hence, the quasi-coherent states provide a natural basis to describe the dominant configurations in the matrix model.

However, quasi-coherent states provide much more than just an approximate classical picture. They allow to associate a concise mathematical structure to the matrix configuration, which will justify the geometric interpretation of a quantized brane in target space via (4.0.3) of X^a. We will now unfold this geometric structure step by step.

4.2 The abstract quantum space $\mathcal{M}$

In the previous section, we considered quasi-coherent states $|x\rangle$ associated to points $x \in \tilde{\mathbb{R}}^D$. However, these states often coincide for different x. A more appropriate concept of quantum geometry is obtained by considering the abstract variety $\mathcal{M} \subset \mathbb{C}P^{N-1}$ of quasi-coherent states. This is the key to a deeper understanding of the quantum geometry associated to matrix configurations. The abstract space $\mathcal{M}$ can be viewed as an avatar of the quantum space, and it is naturally embedded in $\mathbb{R}^D$, thus recovering the previous view.

Consider the union of the normalized quasi-coherent states for all $x \in \tilde{\mathbb{R}}^D$

$$\mathcal{B} := \bigcup_{x \in \tilde{\mathbb{R}}^D} U(1)|x\rangle \subset \mathcal{H} \cong \mathbb{C}^N \tag{4.2.1}$$

as a subset of $\mathcal{H}$. Since E_x is assumed to be one-dimensional, $\mathcal{B}$ can be viewed as a $U(1)$ bundle

$$\mathcal{B} \to \mathcal{M}, \qquad \mathcal{M} := \mathcal{B}/_{U(1)} \hookrightarrow \mathbb{C}P^{N-1} \tag{4.2.2}$$

over $\mathcal{M}$, which is thus defined to be the base of the bundle $\mathcal{B}$. We denote $\mathcal{M}$ as **abstract quantum space associated to** X^a. Note that the union in (4.2.1) need not be disjoint, which is essential for obtaining compact quantum spaces $\mathcal{M}$. Using this definition, we can now refine the symbol map (4.1.3) as follows:

$$\mathrm{End}(\mathcal{H}) \to \mathcal{C}(\mathcal{M})$$
$$\Phi \mapsto \langle x|\Phi|x\rangle =: \boldsymbol{\phi}(x) \tag{4.2.3}$$

mapping matrices or operators to functions on $\mathcal{M}$. We will see that this refined concept of a symbol typically provides a faithful picture of the space of matrices. This is not the case for the naive map (4.1.3), which is highly degenerate if $\mathcal{M}$ is compact.

If the lowest eigenspace E_x of H_x is nondegenerate, we can give another, equivalent definition of $\mathcal{M}$ as follows: consider the equivalence relation on $\tilde{\mathbb{R}}^D$

$$x \sim y \; \Leftrightarrow \; |x\rangle = |y\rangle \tag{4.2.4}$$

by identifying points $x \in \tilde{\mathbb{R}}^D$ with identical eigenspace E_x. Then,

$$\mathcal{M} \cong \tilde{\mathbb{R}}^D/_\sim \tag{4.2.5}$$

so that $\mathcal{M}$ can be viewed as a quotient of $\tilde{\mathbb{R}}^D$. However, this does not provide the extra structure on $\mathcal{M}$ inherited from $\mathbb{C}P^{N-1}$ via (4.2.2).

In particular, a matrix configuration will be denoted as **quantum manifold** if the associated quantum space $\mathcal{M} \subset \mathbb{C}P^{N-1}$ is a regular (real) submanifold of $\mathbb{C}P^{N-1}$. This is indeed expected generically (possibly up to singularities), since standard theorems imply the existence of (local) smooth maps

$$\mathbf{q} \colon U \subset \tilde{\mathbb{R}}^D \to \mathcal{M} \subset \mathbb{C}P^{N-1}$$
$$x \mapsto |x\rangle. \tag{4.2.6}$$

However, $\mathbf{q}$ need not be injective. In particular, $\mathcal{M}$ inherits the induced (subset) topology and metric from $\mathbb{C}P^{N-1}$, and it is "locally translation invariant" with generators inherited from the $SU(N)$ symmetry of $\mathbb{C}P^{N-1}$. The structure of the resulting space or manifold is discussed in more detail in Section 4.2.

Now consider the following natural *embedding map*[2] provided by the symbol of X^a:

$$\boxed{\begin{aligned} \mathbf{x^a} &: \mathcal{M} \to \mathbb{R}^D \\ |x\rangle &\mapsto \mathbf{x^a} := \langle x|X^a|x\rangle = x^a - \partial^a\lambda, \end{aligned}} \tag{4.2.7}$$

---

[2] This is the quotient of the previously defined function $\mathbf{x^a}$ (4.1.4) on $\tilde{\mathbb{R}}^D$ under the identification (4.2.5), and the map is constant on the fibers $\mathcal{N}_x$.

where the formula on the rhs will be established in (4.2.55). The image

$$\boxed{\tilde{\mathcal{M}} := \mathbf{x}(\mathcal{M}) \quad \subset \mathbb{R}^D} \tag{4.2.8}$$

defines some variety in target space $\mathbb{R}^D$. In this way, we can associate to the abstract space $\mathcal{M}$ a subset $\tilde{\mathcal{M}} \subset \mathbb{R}^D$, and $\mathcal{B}$ can be considered as a $U(1)$ bundle over $\tilde{\mathcal{M}}$. This structure defines the **embedded quantum space** or **brane** associated to the matrix configuration. The concept is very reminiscent of noncommutative branes in string theory, which is borne out in the context of Yang–Mills matrix models as discussed in later chapters. Since the embedding may be degenerate, the abstract quantum space is clearly a more fundamental concept.

*Degeneracy and regularity results on $\mathcal{M}$

To understand better the structure of $\mathcal{M}$, we recall the equivalence class $\sim$ defined in (4.2.4). Denote the equivalence class through a point $x \in \tilde{\mathbb{R}}^D$ with $\mathcal{N}_x$. Due to the identity

$$H_x = H_y + \frac{1}{2}(x^a x_a - y^a y_a)\mathbb{1} - (x^a - y^a)X_a, \tag{4.2.9}$$

$x \sim y$ implies that $|x\rangle$ is an eigenvector of $(x^a - y^a)X_a$,

$$(x^a - y^a)X_a|x\rangle \propto |x\rangle. \tag{4.2.10}$$

But this means that the *equivalence classes $\mathcal{N}_x$ are always (segments of) straight lines or higher-dimensional planes*. This provides a remarkable link between local and global properties of $\mathbf{q}$: *whenever $\mathbf{q}(x) = \mathbf{q}(y)$ for $x \neq y$, a linear kernel $T\mathcal{N}_x \ni (x - y)$ of $d\mathbf{q}|_x$ arises*. In particular if $\operatorname{rank} d\mathbf{q} = D$ i.e. $\mathbf{q}$ is an immersion, $\mathbf{q}$ must be injective globally, since otherwise $d\mathbf{q}$ has some nontrivial kernel. This implies that $\mathbf{q}$ can be extended to $\tilde{\mathbb{R}}^D$, and we obtain the following result:

Theorem 4.1 *If $\mathbf{q}$ (4.2.6) is an immersion, then $\mathbf{q} : \tilde{\mathbb{R}}^D \to \mathcal{M}$ is bijective, and $\mathcal{M}$ is a D-dimensional quantum manifold. Moreover, x^a provide global coordinates.*

This applies e.g. to the Moyal–Weyl quantum plane. On the other hand, there are many interesting examples (such as the fuzzy sphere) where the rank of $d\mathbf{q}$ is reduced. Assuming that $\operatorname{rank} d\mathbf{q} = m$ is constant on $\tilde{\mathbb{R}}^D$, then the fibration $\tilde{\mathbb{R}}^D/_\sim$ is locally trivial, and according to a standard theorem we can choose functions y^μ, $\mu = 1, \ldots, m$ on a neighborhood of $\xi \in U \subset \tilde{\mathbb{R}}^D$ such that the image $\mathbf{q}|_U \subset \mathcal{M} \subset \mathbb{C}P^{N-1}$ is a submanifold of $\mathbb{C}P^{N-1}$. Since the only possible degeneracies of $\mathbf{q}$ are the linear fibers $\mathcal{N}$, it follows that $\mathcal{M}$ is an m-dimensional quantum manifold. In particular, there are no self-intersections of $\mathcal{M}$, and $\tilde{\mathbb{R}}^D$ has the structure of a bundle over $\mathcal{M}$.

4.2.1 $U(1)$ connection, would-be symplectic form, and quantum metric on $\mathcal{M}$

From now on we focus on quantum manifolds, so that $\mathcal{M} \subset \mathbb{C}P^{N-1}$ is a sub-manifold. Then $\mathcal{M}$ inherits extra structure from $\mathbb{C}P^{N-1}$, which plays a key role in the present framework. As the complex geometry of $\mathbb{C}P^{N-1}$ is fundamental to quantum theory, this justifies denoting the present framework as "quantum geometry."

Consider again the $U(1)$ bundle[3] $\mathcal{B}$ (4.2.1) over $\mathcal{M}$. Let x be local coordinates on $\mathcal{M}$, and let $|x\rangle$ be some local section of $\mathcal{B}$, i.e. some smooth choice of quasi coherent states. One can define a natural *connection one-form* A on $\mathcal{M}$ via

$$P \circ d|x\rangle = |x\rangle iA, \qquad iA := \langle x|d|x\rangle \ \in \Omega^1(\mathcal{M}), \tag{4.2.11}$$

where $P = |x\rangle\langle x|$ is the projector on E_x. Here $A(x)$ is real because

$$(\langle x|d|x\rangle)^* = d(\langle x|)|x\rangle = -\langle x|d|x\rangle, \tag{4.2.12}$$

and it transforms like a $U(1)$ gauge field

$$|x\rangle \rightarrow e^{i\Lambda(x)}|x\rangle, \qquad A \rightarrow A + d\Lambda. \tag{4.2.13}$$

This can be used to define a *parallel transport* of coherent states $|x\rangle$ along a path γ in $\mathcal{M}$, by choosing the phase factors along γ such that A vanishes. This connection is analogous to a Berry connection [32] on the hermitian line bundle $\mathcal{B}$. To gain more insights, consider the inner product

$$\langle x|y\rangle =: e^{i\varphi(x,y)-D(x,y)}, \tag{4.2.14}$$

which defines a distance function $D(x,y)$ and a phase function $\varphi(x,y)$. These functions satisfy

$$D(x,y) = D(y,x) \geq 0, \qquad \varphi(x,y) = -\varphi(y,x). \tag{4.2.15}$$

The phase $\varphi(x,y)$ clearly depends on the particular section $|x\rangle$ of the bundle $\mathcal{B}$, while $D(x,y)$ is independent of the gauge. Differentiating (4.2.14) w.r.t. y gives

$$\langle x|d_y|y\rangle|_{y=x} = id_y\varphi(x,y)|_{y=x} - d_yD(x,y)|_{y=x} \tag{4.2.16}$$

and comparing with (4.2.11) we conclude

$$id_y\varphi(x,y)|_{y=x} = iA = iA_a dx^a$$
$$d_yD(x,y)|_{y=x} = 0. \tag{4.2.17}$$

Hence, the phase $\varphi(x,y)$ encodes the connection A. To make this more explicit, we consider local coordinates x^μ on $\mathcal{M}$ with $A = A_\mu dx^\mu$, and define the gauge-invariant hermitian $D \times D$ matrix $h_{\mu\nu}$ by

$$h_{\mu\nu} = \big((\partial_\mu + iA_\mu)\langle x|\big)(\partial_\nu - iA_\nu)|x\rangle = h^*_{\nu\mu}$$
$$=: \frac{1}{2}(i\omega_{\mu\nu} + g_{\mu\nu}). \tag{4.2.18}$$

This decomposes into the real symmetric and antisymmetric tensors $g_{\mu\nu}$ and $\omega_{\mu\nu}$. The symmetric part

---

[3] Much of these considerations generalize to the degenerate case, in terms of a $U(n)$ bundle.

$$g_{\mu\nu} = \big((\partial_\mu + iA_\mu)\langle x|\big)(\partial_\nu - iA_\nu)|x\rangle + (\mu \leftrightarrow \nu)$$
$$= (\partial_\mu \langle x|)\partial_\nu |x\rangle - A_\mu A_\nu + (\mu \leftrightarrow \nu) \tag{4.2.19}$$

(using (4.2.11)) is the pullback of the Riemannian metric[4] on $\mathcal{H}$ through the section $|x\rangle$. More precisely, $g_{\mu\nu}$ is the pullback of the Fubini–Study metric on $\mathbb{C}P^{N-1}$ to $\mathcal{M}$, and will be identified as quantum metric below. Hence $D(x,y)$ can be determined in principle by viewing $\mathcal{M}$ as subset of $\mathbb{C}P^{N-1}$ and using the formula

$$\cos^2(\gamma(x,y)) = e^{-2D(x,y)} \tag{4.2.20}$$

for $\mathbb{C}P^{N-1}$, where $\gamma(x,y)$ is the geodesic squared-distance between $|x\rangle$ and $|y\rangle$ in the Fubini–Study metric. On the other hand, the antisymmetric part of $h_{\mu\nu}$ encodes a two-form

$$i\omega_{\mu\nu} = i(\partial_\mu A_\nu - \partial_\nu A_\mu) = (\partial_\mu \langle x|)\partial_\nu |x\rangle - (\partial_\nu \langle x|)\partial_\mu |x\rangle$$
$$i\omega = \frac{i}{2}\omega_{\mu\nu}dx^\mu \wedge dx^\nu = d\langle x| \wedge d|x\rangle = d(\langle x|d|x\rangle) = idA, \tag{4.2.21}$$

which is the field strength of the $U(1)$ connection A and therefore closed,

$$\omega = dA, \qquad d\omega = 0. \tag{4.2.22}$$

We will see that ω typically – or at least in many interesting cases – defines a symplectic form on $\mathcal{M}$, and will be denoted in general as **would-be[5] symplectic form**. Using (4.2.17) and (4.2.19) and assuming local (approximate) translation invariance, the leading terms in a Taylor expansion of $D(x,y)$ and $\varphi(x,y)$ around $x = y$ are given explicitly by

$$\varphi(x,y) = A_\mu(y^\mu - x^\mu) + \frac{1}{2}\omega_{\mu\nu}x^\mu y^\nu + \dots$$
$$D(x,y) = \frac{1}{4}(x-y)^\mu (x-y)^\nu g_{\mu\nu} + \dots . \tag{4.2.23}$$

Hence the phase $\varphi(x,y)$ encodes the connection A, while the distance function $D(x,y)$ is gauge invariant and encodes the metric $g_{\mu\nu}$. Therefore, for small $x \approx y$, the inner product of coherent states is given by

$$\langle x|y\rangle \approx e^{-\frac{1}{4}|x-y|^2_g}\, e^{i\varphi(x,y)}, \tag{4.2.24}$$

where

$$|x-y|^2_g = (x-y)^\mu (x-y)^\nu g_{\mu\nu}. \tag{4.2.25}$$

In particular, we conclude that the quasi-coherent states are localized within a region of size

$$L^2_{\text{coh}} := (\det g_{\mu\nu})^{-1/2n} \tag{4.2.26}$$

denoted as **coherence scale**, cf. (3.1.110). Below this scale, the $|x\rangle$ are approximately constant. Therefore, $g_{\mu\nu}$ will be denoted as **quantum metric**, which characterizes the quantum regime of the geometry. This quantum metric should not be confused with the

---

[4] Here $\mathcal{H}$ is viewed as a Euclidean space with metric defined by the norm. Note that $g_{\mu\nu}$ is *not* related to the Euclidean metric δ_{ab} on target space $\mathbb{R}^D$.

[5] "would-be" because it might be degenerate.

effective metric on quantum spaces in Yang–Mills matrix models, which will be discussed in Section 9.1. We also define the **noncommutativity scale** as

$$L_{\mathrm{NC}}^2 := |\det \omega_{\mu\nu}|^{-1/2n} = |\det \theta^{\mu\nu}|^{1/2n}, \tag{4.2.27}$$

where $\theta^{\mu\nu}$ is the inverse of $\omega_{\mu\nu}$, restricted to the symplectic leaves if necessary.

To summarize, the embedding $\mathcal{M} \subset \mathbb{C}P^{N-1}$ (4.2.2) defines the quantum metric $g_{\mu\nu}$ and the would-be symplectic form $\omega_{\mu\nu}$ on $\mathcal{M}$ as pullbacks to $\mathcal{M}$ of the metric and the symplectic structure on $\mathbb{C}P^{N-1}$.

It is interesting to recall the Moyal–Weyl quantum plane $\mathbb{R}_\theta^{2n}$, where the inner product of coherent states was obtained explicitly in (3.1.94). This has the same from as (4.2.24), where the quantum metric and the symplectic form satisfy the additional relation (3.1.109)

$$\omega_{\rho\mu}\omega_{\sigma\nu}g^{\rho\sigma} = g_{\mu\nu}. \tag{4.2.28}$$

This condition can be rewritten as

$$J^2 = -\mathbb{1}, \quad J_\nu^\mu = \omega_{\nu\sigma}g^{\sigma\mu}, \tag{4.2.29}$$

which is known as *almost-Kähler* condition. It will be seen to hold in all the important spaces under consideration, which typically satisfy an even stronger Kähler condition discussed later. It is a necessary condition for a quantum space $\mathcal{M}$ to be locally approximated by some Moyal–Weyl quantum space. For generic quantum spaces, it is not evident why all eigenvalues of J^2 should be -1 (or 0). However, this does hold if locally

$$[X^a, X^b] \approx i\theta^{ab}\mathbb{1}, \tag{4.2.30}$$

so that the quantum space looks locally like $\mathbb{R}^{2n}_\theta$. It is hence plausible that the almost-Kähler condition holds to a good approximation for all "almost-commutative" spaces; this will be discussed in more detail in the next section.

Now consider the phase factors in more detail. As pointed out before, the phase of the quasi-coherent states $|x\rangle$ is a priori arbitrary. Parallel transporting $|x\rangle$ along a path γ is achieved by the phase factor

$$e^{i\int_\gamma A} \in U(1). \tag{4.2.31}$$

For a closed path γ, this phase factor defines the *holonomy*. If the closed path is contractible in $\mathcal{M}$ i.e. $\gamma = \partial\Omega$, then the corresponding phase change along γ is given in terms of the field strength via Stokes theorem

$$\oint_\gamma A = \int_\Omega dA. \tag{4.2.32}$$

If the connection is flat $dA = 0$, the phases $\varphi(x, y)$ can thus be gauged away completely, and the holonomy is trivial.[6] Typically the connection is not flat, but has field strength $dA = \omega$ given by the underlying symplectic form (4.2.22). This suggests the following natural phase convention: Fixing some (arbitrary) origin $0 \in \mathcal{M}$, we can fix the phases of $|x\rangle$ such that the phase $\varphi(x, y)$ in (4.2.14) is given by the symplectic area of the triangle formed by $(x, y, 0)$,

---

[6] Non-contractible curves may lead to nontrivial holonomies even if $dA = 0$, which are called monodromies.

with edges being geodesics with respect to g (for example). This convention is consistent, because it is compatible with (4.2.32).

Quantization and cohomology

The considerations in the previous paragraph lead to an important quantization condition, which results from the fact that ω is the field strength of a connection on a $U(1)$ bundle: the flux over every two-cycle S^2 in $\mathcal{M}$ is quantized,

$$\int_{S^2} \frac{\omega}{2\pi} = k, \qquad k \in \mathbb{Z}. \qquad (4.2.33)$$

This arises as consistency condition[7] on the $U(1)$ holonomy along a closed path γ on S^2. Indeed, γ can be viewed as boundary of two complementary segments of S^2. Then (4.2.32) states that the phase of the parallel transport of $|x\rangle$ along γ is given by the flux $\int \omega$ over either segment, but with opposite orientation. Consistency then implies that the sum of the fluxes is a multiple of 2π, which is (4.2.33). In more technical language, $c_1 = -\frac{1}{2\pi}\omega$ is the first Chern class of $\mathcal{B}$ viewed as line bundle, which is the pullback of the first Chern class of the tautological line bundle over $\mathbb{C}P^{N-1}$.

Consider (4.2.33) in more detail: we define

$$I_\xi^{(2)} := \int_{S_\xi^2} \frac{\omega}{2\pi}, \qquad (4.2.34)$$

where $\xi \in \mathbb{R}^D$ is any point, and $S_\xi^2 \subset \tilde{\mathbb{R}}^D = \mathbb{R}^D \setminus \mathcal{K}$ is a two-sphere around ξ. Since the integral is quantized, the value does not depend on the radius of S_ξ^2, and we can contract the sphere as long as we can avoid the singular set $\mathcal{K}$ where the lowest eigenspace of H_ξ is degenerate. Since $\tilde{\mathbb{R}}^D$ is open, $I_\xi \neq 0$ can be nonvanishing only for $\xi \in \mathcal{K}$, i.e. for noncontractible S^2. The equivalence classes of such spheres are denoted as homology two-cycles. Therefore every nontrivial homology two-cycle in $\tilde{\mathbb{R}}^D$ yields some integer $I_\xi^{(2)} \in \mathbb{Z}$. In more abstract language, the two-form $\frac{1}{2\pi}\omega$ generates the cohomology group $H^2(\mathcal{M}, \mathbb{Z})$.

One can generalize these considerations to higher-dimensional $\mathbb{C}P_\xi^n \subset \tilde{\mathbb{R}}^D$ centered at some point ξ, and define[8]

$$I_\xi^{(2n)} := \int_{\mathbb{C}P_\xi^n} \frac{\omega^n}{(2\pi)^n}. \qquad (4.2.35)$$

Now we use the fact that ω is the pullback to $\mathcal{M} \subset \mathbb{C}P^N$ or $\tilde{\mathbb{R}}^D$ of the symplectic form $\tilde{\omega}$ on $\mathbb{C}P^N$. Indeed, the cohomology of $\mathbb{C}P^N$ is given by $H^{2n}(\mathbb{C}P^N) = 1$ for any even $n \leq N$. This implies that the integral over $\tilde{\omega}^n$ over any non-contractible $\mathbb{C}P^n \subset \mathbb{C}P^N$ is given by some integer quantum number. Therefore, such a nontrivial homology $2n$-cycle in $\tilde{\mathbb{R}}^D$ yields some integer $I_\xi^{(2n)}$. The numbers characterize the topology of $\mathcal{M}$ or $\tilde{\mathbb{R}}^D$, and

---

[7] The argument is completely analogous to Dirac's derivation of the quantization of monopole charges.

[8] I am indebted to Siye Wu for clarifications on this topic.

they are associated to the singular set $\mathcal{K}$, which often consists only of isolated points ξ. This allows to associate nontrivial quantum numbers to any matrix configuration, which are insensitive to small perturbations.

Completeness relation

Now assume that ω is symplectic[9], and consider the (approximate) *completeness relation*

$$\mathcal{P} := \int_{\mathcal{M}} |x\rangle\langle x| \overset{!}{\approx} \mathbb{1} \tag{4.2.36}$$

for quasi-coherent states, which will play an important role in the following. The integral is defined for any compact $\mathcal{M}$ via the symplectic volume form

$$\int_{\mathcal{M}} \phi(x) := \int_{\mathcal{M}} \frac{\Omega}{(2\pi)^n} \phi(x), \qquad \Omega := \frac{1}{n!}\omega^{\wedge n} \tag{4.2.37}$$

assuming $\dim \mathcal{M} = 2n$. The completeness relation is equivalent to the trace identity

$$\mathrm{Tr}\Phi \approx \int_{\mathcal{M}} \langle x|\Phi|x\rangle \qquad \forall \Phi \in \mathrm{End}(\mathcal{H}). \tag{4.2.38}$$

For quantum spaces with sufficient symmetry such as $\mathbb{R}^{2n}_\theta$ or quantized coadjoint orbits, the completeness relation is exact and follows from Schur's lemma, noting that

$$[X^a, \mathcal{P}] = 0 \tag{4.2.39}$$

by group invariance. In the following, we would like to justify the approximate completeness relation for more generic quantum spaces.

Consider a quantum space $\mathcal{M}$ that is sufficiently close to classical, i.e. we assume that g and ω are approximately constant on the coherence scale L_{coh}. We also assume that

$$L_{\mathrm{coh}} \approx L_{\mathrm{NC}} \tag{4.2.40}$$

(which follows in particular from the almost-Kähler condition (4.2.28)), and that the quasi-coherent states span $\mathcal{H}$. Then consider

$$\langle x|\mathcal{P}|x\rangle = \int_{\mathcal{M}_y} \frac{\Omega_y}{(2\pi)^n} \langle x|y\rangle\langle y|x\rangle = \int_{\mathcal{M}_y} \frac{\Omega_y}{(2\pi)^n} |\langle x|y\rangle|^2 \approx 1 \tag{4.2.41}$$

for any x, noting that (4.2.24)

$$|\langle x|y\rangle|^2 \sim e^{-\frac{1}{2}|x-y|_g^2} \approx (2\pi)^n \delta(x,y) \tag{4.2.42}$$

is a good approximation to the delta function. Indeed if g and ω are approximately constant on $\mathcal{M}$, we can evaluate this integral as

$$\int_{\mathbb{R}^{2n}_y} \frac{\Omega_y}{(2\pi)^n} e^{-\frac{1}{2}|x-y|_g^2} = \frac{\sqrt{\det\omega_{\mu\nu}}}{\sqrt{\det g_{\mu\nu}}} \approx 1 \tag{4.2.43}$$

---

[9] This may require some reduction process as discussed in Section 4.7.1.

using (4.2.40), and recalling

$$\Omega = \frac{1}{n!}\omega^n = d^{2n}x\sqrt{\det\omega_{\mu\nu}}. \qquad (4.2.44)$$

Since this holds for any x and assuming that the quasi-coherent states are complete, it is highly plausible that $\mathcal{P} \approx \mathbb{1}$. We can justify this formally by picking $N = \dim\mathcal{H}$ linearly independent $|x_i\rangle$, and conclude that

$$\text{Tr}\mathcal{P} \approx \dim\mathcal{H}. \qquad (4.2.45)$$

Using a similar consideration, we obtain

$$\langle x|\mathcal{P}^2|x\rangle = \int\limits_{\mathcal{M}_y}\frac{\Omega_y}{(2\pi)^n}\int\limits_{\mathcal{M}_z}\frac{\Omega_z}{(2\pi)^n}\langle x|y\rangle\langle y|z\rangle\langle z|x\rangle \approx 1 \qquad (4.2.46)$$

and we can conclude that

$$\langle x|(\mathcal{P}^2 - \mathcal{P})|x\rangle \approx 0 \qquad (4.2.47)$$

for all x, and

$$\text{Tr}\mathcal{P}^2 \approx \text{Tr}\mathcal{P} \approx \dim\mathcal{H}. \qquad (4.2.48)$$

Since $\mathcal{P}$ is self-adjoint, these two relations imply

$$\mathcal{P} \approx \mathbb{1}. \qquad (4.2.49)$$

Yet another justification for the approximate completeness relation (4.2.36) will be given in Section 4.3. In particular, we obtain the Bohr–Sommerfeld relation

$$\dim\mathcal{H} \approx \int\limits_{\mathcal{M}}\frac{\Omega}{(2\pi)^n}, \qquad (4.2.50)$$

which is indeed quantized due to (4.2.35).

4.2.2 Differential structure of quasi-coherent states

Now assume that $|x\rangle$ is a local section of the quasi-coherent states, with

$$H_x|x\rangle = \lambda(x)|x\rangle. \qquad (4.2.51)$$

Using Cartesian coordinates x^a on target space $\mathbb{R}^D$, we observe

$$\partial_a H_x = -(X_a - x_a\mathbb{1}). \qquad (4.2.52)$$

Thus, differentiating (4.2.51) gives

$$(H_x - \lambda(x))\partial_a|x\rangle = -\partial_a(H_x - \lambda(x))|x\rangle = (X_a - x_a + \partial_a\lambda)|x\rangle. \qquad (4.2.53)$$

Since the lhs is orthogonal to $\langle x|$, it follows that

$$0 = \langle x|(X_a - x_a + \partial_a\lambda)|x\rangle \qquad (4.2.54)$$

so that the expectation value or symbol of the basic matrices X_a is given by

$$\mathbf{x_a} = \langle x|X_a|x\rangle = x_a - \partial_a\lambda.$$ (4.2.55)

To proceed, we introduce the "reduced resolvent"

$$(H_x - \lambda(x))^{'-1} := (\mathbb{1} - P_x)(H_x - \lambda(x))^{-1}(\mathbb{1} - P_x), \qquad P_x := |x\rangle\langle x| \qquad (4.2.56)$$

which is always regular and satisfies

$$(H_x - \lambda)(H_x - \lambda)^{'-1} = \mathbb{1} - P_x = (H_x - \lambda)^{'-1}(H_x - \lambda),$$
$$(H_x - \lambda)^{'-1}|x\rangle = 0. \qquad (4.2.57)$$

Then (4.2.53) gives upon decomposing $\mathcal{H} = P_x\mathcal{H} \oplus (\mathbb{1} - P_x)\mathcal{H}$

$$\partial_a|x\rangle = |x\rangle\langle x|\partial_a|x\rangle + (H_x - \lambda)^{'-1}(X_a - x_a + \partial_a\lambda)|x\rangle$$
$$(\partial_a - iA_a)|x\rangle = (H_x - \lambda)^{'-1}(X_a - x_a + \partial_a\lambda)|x\rangle$$
$$= (H_x - \lambda)^{'-1}X_a|x\rangle \qquad (4.2.58)$$

using (4.2.11) and observing $(H_x - \lambda)^{'-1}(x_a - \partial_a\lambda)|x\rangle = 0$ due to (4.2.51). This can be written as

$$(\partial_a - iA_a)|x\rangle = i\mathcal{X}_a|x\rangle \qquad (4.2.59)$$

for $\mathcal{X}_a = -i(H_x - \lambda)^{'-1}X_a$. Since $(H_x - \lambda)^{'-1}|x\rangle = 0$, this can be replaced by the hermitian generator

$$\mathcal{X}_a := -i[(H_x - \lambda)^{'-1}, X_a] = \mathcal{X}_a^\dagger.$$ (4.2.60)

Moreover, we note

$$\langle x|\mathcal{X}_a|x\rangle = 0. \qquad (4.2.61)$$

Hence, $\mathcal{X}_a$ generates the gauge-invariant tangential variations of $|x\rangle$, which take value in the orthogonal complement of $|x\rangle$. This will be the basis for defining the quantum tangent space in section 4.3.3. The local Section $|x\rangle$ over $\tilde{\mathbb{R}}^D$ can now be written as

$$|x\rangle = P\exp\left(i\int_\xi^x (\mathcal{X}_a + A_a)dx^a\right)|\xi\rangle \qquad (4.2.62)$$

near the reference point $\xi \in \tilde{\mathbb{R}}^D$. Here P indicates path ordering, which is just a formal way of writing the solution of (4.2.59). In a small local neighborhood, the $\mathcal{X}_a$ are approximately constant, and A_a can be gauged away. Then (4.2.62) can be written as

$$|x\rangle \approx e^{i(x-y)^a\mathcal{X}_a}|y\rangle, \qquad (4.2.63)$$

which means that the $\mathcal{X}_a$ generate the local translations on $\mathcal{M}$.

Exercise 4.2.1 Verify (4.2.23) using a second-order Taylor expansion around some reference point and assuming translational invariance.

4.3 The semiclassical structure of $\mathcal{M}$

4.3.1 Almost-local quantum spaces and Poisson tensor

Given the quasi-coherent states, it is natural to define a class $\mathrm{Loc}(\mathcal{H}) \subset \mathrm{End}(\mathcal{H})$ of **almost-local operators** that satisfy

$$\Phi|x\rangle \approx |x\rangle\langle x|\Phi|x\rangle = P_x\Phi|x\rangle = |x\rangle\phi(x) \qquad \forall x \in \tilde{\mathbb{R}}^D. \tag{4.3.1}$$

Here $\phi(x) = \langle x|\Phi|x\rangle$ is the symbol of Φ, and $P_x = |x\rangle\langle x|$ is the projector on the quasi-coherent state $|x\rangle$. The question is how to make the meaning of $\approx$ more precise: we should certainly require that $\Phi|x\rangle \approx |x\rangle\phi(x) \in \mathcal{H}$ for every x, but it is not obvious how to treat the dependence on x, and how to specify bounds. The guiding idea is that it should make sense to identify Φ with its symbol

$$\Phi \sim \phi(x) = \langle x|\Phi|x\rangle. \tag{4.3.2}$$

This identification will be indicated by $\sim$ from now on. A more concise definition will be given in the Section 4.4, by requiring that $\sim$ is an *approximate isometry* from $\mathrm{Loc}(\mathcal{H})$ to some class $\mathcal{C}_{\mathrm{IR}}(\mathcal{M})$ of "infrared" functions on the abstract quantum space $\mathcal{M}$ associated to the matrix configuration. With this in mind, we proceed to elaborate some consequences of (4.3.1) for fixed x without specifying bounds.

Since $(\mathbb{1} - P_x)$ is a projector, we can estimate

$$\langle x|\Phi^\dagger\Phi|x\rangle = \langle x|\Phi^\dagger P_x\Phi|x\rangle + \langle x|\Phi^\dagger(\mathbb{1} - P_x)\Phi|x\rangle \geq \langle x|\Phi^\dagger P_x\Phi|x\rangle = |\phi(x)|^2. \tag{4.3.3}$$

It follows that every hermitian almost-local operator $\Phi = \Phi^\dagger$ satisfies

$$\langle x|\Phi\Phi|x\rangle \approx \langle x|\Phi|x\rangle^2 = |\phi(x)|^2 \qquad \forall x \in \tilde{\mathbb{R}}^D, \tag{4.3.4}$$

i.e. the uncertainty of Φ is negligible,

$$\langle x|(\Phi - \langle x|\Phi|x\rangle)^2|x\rangle \approx 0 \qquad \forall x \in \tilde{\mathbb{R}}^D. \tag{4.3.5}$$

This means that $(\Phi - \phi(x))|x\rangle$ is approximately zero, which in turn implies (4.3.1). Therefore, almost-locality is essentially equivalent to (4.3.5), up to global issues and specific bounds. We also note that for two operators $\Phi, \Psi \in \mathrm{Loc}(\mathcal{H})$ the factorization properties

$$\Phi\Psi|x\rangle \approx \Phi|x\rangle\langle x|\Psi|x\rangle \approx |x\rangle\phi(x)\psi(x)$$
$$\langle x|\Phi\Psi|x\rangle \approx \phi(x)\psi(x) \tag{4.3.6}$$

follow formally. However, one should not jump to the conclusion that $\mathrm{Loc}(\mathcal{H})$ is an algebra, since the specific bounds may be violated by the product. For some given matrix configuration, $\mathrm{Loc}(\mathcal{H})$ may be empty or very small. This happens e.g. for fuzzy spaces with very small quantum numbers, and it is expected for generic, random matrix configuration. But even in these cases, the associated geometrical structures still provide useful insights.

For interesting quantum geometries, we expect that at least all the X^a are almost-local, and also polynomials $P_n(X)$ up to some maximal degree n due to (4.3.6). We will also see that $\text{Loc}(\mathcal{H})$ can often be characterized by some bound on the eigenvalue of the matrix Laplacian $\Box$ (4.7.1), and it is related to the uncertainty scale L_{NC} (4.2.27). However, it is important to keep in mind $\text{Loc}(\mathcal{H})$ can never be more than a small subset of $\text{End}(\mathcal{H})$. Most of $\text{End}(\mathcal{H})$ should be viewed as UV or "deep quantum" regime, which can be understood in terms of string modes without classical analog.

The crucial result established in the following is that a Poisson structure arises on $\tilde{\mathbb{R}}^D$ in the semi-classical or almost-local regime. This makes sense for **almost-local quantum spaces**, which are defined to be matrix configurations where all X^a as well as all $[X^a, X^b]$ are almost-local operators. We shall assume this in the present section. Then define the following real antisymmetric tensor field

$$\theta^{ab} := -i\langle x|[X^a, X^b]|x\rangle = -\theta^{ba} \tag{4.3.7}$$

on $\tilde{\mathbb{R}}^D$. We will see that for almost-local quantum spaces, this is essentially the inverse of the would-be symplectic structure $\omega_{\mu\nu}$ (4.2.21) on $\mathcal{M}$, and should therefore be interpreted as Poisson structure underlying the quantum space. The hard part is to establish the link between (4.3.7) and the symplectic form ω, which was obtained from the connection on the bundle $\mathcal{B}$.

4.3.2 Relating the algebraic and geometric structures

Since the derivatives of $|x\rangle$ are spanned by the $\mathcal{X}^a|x\rangle$, it should be possible to relate the $U(1)$ field strength $\omega_{\mu\nu}$ and the quantum metric $g_{\mu\nu}$ on $\mathcal{M}$ to algebraic properties for the translation generators $\mathcal{X}^a$. However, we must keep in mind that the dimension of the brane may be smaller than D. Generalizing (4.2.18) to target space $\mathbb{R}^D$, we define

$$h_{ab} = \langle x|\mathcal{X}_a\mathcal{X}_b|x\rangle = \frac{1}{2}(i\omega_{ab} + g_{ab}). \tag{4.3.8}$$

Then

$$i\omega_{ab} = h_{ab} - h_{ba} = \langle x|(\mathcal{X}_a\mathcal{X}_b - \mathcal{X}_b\mathcal{X}_a)|x\rangle \tag{4.3.9}$$

and

$$g_{ab} = h_{ab} + h_{ba} = \langle x|(\mathcal{X}_a\mathcal{X}_b + \mathcal{X}_b\mathcal{X}_a)|x\rangle. \tag{4.3.10}$$

These tensors reduce to $g_{\mu\nu}$ and $\omega_{\mu\nu}$ along $\mathcal{M}$, and vanish along the null directions $\mathcal{N}_x$ transversal to the brane. This provides a first link between the geometric and algebraic structures. To proceed, consider the following hermitian tensor

$$P_{ab}(x) := \langle x|X_a(H_x - \lambda)'^{-1}X_b|x\rangle = P_{ba}(x)^*$$
$$= i\langle x|X_a\mathcal{X}_b|x\rangle = -i\langle x|\mathcal{X}_aX_b|x\rangle. \tag{4.3.11}$$

Its symmetric part is obtained by taking derivatives of (4.2.55)

$$\partial_b \mathbf{x_a}(x) = \partial_b x_a - \partial_b\partial_a\lambda$$
$$= \partial_b\langle x|X_a|x\rangle = i\langle x|[X_a, \mathcal{X}_b]|x\rangle$$
$$= P_{ab} + P_{ba} \tag{4.3.12}$$

lowering indices with δ_{ab}. This will be recognized as projector on the embedded quantum space.

To obtain a relation with θ^{ab} (4.3.7), we need to assume that $\mathcal{M}$ is almost-local. Then consider

$$[X^a, X^b](X_b - x_b) + (X_b - x_b)[X^a, X^b] = 2[X^a, H_x]. \qquad (4.3.13)$$

Taking the expectation value, we obtain

$$\langle x|[X^a, X^b](X_b - x_b)|x\rangle + \langle x|(X_b - x_b)[X^a, X^b]|x\rangle = 2\langle x|[X^a, H_x]|x\rangle = 0. \qquad (4.3.14)$$

Since X^a is almost-local, this implies

$$0 \approx \langle x|[X^a, X^b]|x\rangle \langle x|(X_b - x_b)|x\rangle = -i\theta^{ab}\partial_b\lambda \qquad (4.3.15)$$

using (4.2.55), which indicates that $\lambda \approx$ **const on** $\mathcal{M}$. On the other hand, acting with $[X^a, H_x]$ on $|x\rangle$ and using (4.2.51) we obtain the identity

$$[X^a, H_x]|x\rangle = -(H_x - \lambda)X^a|x\rangle. \qquad (4.3.16)$$

Since X^a and $[X^a, X^b]$ approximately commute on almost-local quantum spaces, we obtain

$$\begin{aligned}
-2(H_x - \lambda)X^a|x\rangle &\approx 2(X_b - x_b)[X^a, X^b]|x\rangle \\
&\approx 2(X_b - x_b)|x\rangle \langle x|[X^a, X^b]|x\rangle = 2i(X_b - x_b)|x\rangle\theta^{ab} \\
&\approx 2i(X_b - x_b + \partial_b\lambda)|x\rangle\theta^{ab}
\end{aligned} \qquad (4.3.17)$$

using the factorization property and (4.3.15) in the last step. The first approximation is justified more explicitly if X^a is a solution of the **Yang–Mills equations** (9.3.1)

$$[X_b, [X^b, X^a]] = 0, \qquad (4.3.18)$$

which are equations of motion of Yang–Mills matrix models. Then (4.3.13) implies

$$2[X^a, H_x] = 2(X_b - x_b)[X^a, X^b] \qquad (4.3.19)$$

which leads to

$$\begin{aligned}
-2(H_x - \lambda)X^a|x\rangle &= 2[X^a, H_x]|x\rangle = 2(X_b - x_b)[X^a, X^b]|x\rangle \\
&\approx 2i(X_b - x_b + \partial_b\lambda)|x\rangle\theta^{ab}.
\end{aligned} \qquad (4.3.20)$$

The rhs is orthogonal to $\langle x|$ due to (4.2.55), and multiplying with $(H_x - \lambda)^{'-1}$ we conclude

$$\begin{aligned}
-(H_x - \lambda)^{'-1}(H_x - \lambda)X^a|x\rangle &= -(X^a - x^a + \partial^a\lambda)|x\rangle \\
&\approx i(H_x - \lambda)^{'-1}(X_b - x_b + \partial_b\lambda)|x\rangle\theta^{ab} \\
&= -\theta^{ab}\mathcal{X}_b|x\rangle = i\theta^{ab}(\partial_b - iA_b)|x\rangle.
\end{aligned} \qquad (4.3.21)$$

The second and last forms yields

$$(X^a - x^a + \partial^a\lambda)|x\rangle \approx -i\theta^{ab}(\partial_b - iA_b)|x\rangle \qquad (4.3.22)$$

and both sides are orthogonal to $|x\rangle$. By conjugating this gives

$$\langle x|(X^d - x^d + \partial^d\lambda) \approx i\theta^{dc}(\partial_c + iA_c)\langle x|. \qquad (4.3.23)$$

These relations are very useful. First, they imply the important relation

$$[X^a, |x\rangle\langle x|] \approx -i\theta^{ab}\partial_b(|x\rangle\langle x|), \qquad (4.3.24)$$

which can be understood by viewing $|x\rangle\langle x|$ as End$(\mathcal{H})$-valued function on $\mathcal{M}$. Furthermore, multiplying (4.3.22) with $(\partial_c + iA_c)\langle x|$ gives

$$\begin{aligned}
-i\theta^{ab}\big((\partial_c + iA_c)\langle x|\big)(\partial_b - iA_b)|x\rangle &\approx -i\langle x|\mathcal{X}_c(X^a - x^a + \partial^a\lambda)|x\rangle \\
&= -i\langle x|\mathcal{X}_c X^a|x\rangle = P_c{}^a,
\end{aligned} \qquad (4.3.25)$$

and similarly from (4.3.23)

$$i\theta^{ac}\big((\partial_c + iA_c)\langle x|\big)(\partial_b - iA_b)|x\rangle \approx i\langle x|X^a\mathcal{X}_b|x\rangle = P^a{}_b. \qquad (4.3.26)$$

Together with (4.3.8), this gives

$$\begin{aligned}
-i\theta^{ab}h_{cb} &\approx P_c{}^a \\
i\theta^{ab}h_{bc} &\approx P^a{}_c.
\end{aligned} \qquad (4.3.27)$$

Relation between θ^{ab} and ω_{ab}

Adding the last relations and using (4.3.12) and (4.2.18), we obtain the desired relation between θ^{ab} (4.3.7) and ω_{ab}

$$-\theta^{ab}\omega_{bc} \approx \partial_c\mathbf{x^a} \qquad (4.3.28)$$

in the semi-classical regime. To understand the rhs, recall that for almost-local quantum spaces $\tilde{\mathcal{M}}$ is an m-dimensional manifold, and $d^2(x) = |\mathbf{x^a} - x^a|^2$ can be assumed to be small. Then,

$$\partial_c\mathbf{x^a} \approx \partial_c x^a \qquad (4.3.29)$$

is the projector on the tangent space $T\tilde{\mathcal{M}} \subset \mathbb{R}^D$ of the embedded brane, which is spanned by $\partial_\mu\mathbf{x^a}$ for any local coordinates on $\mathcal{M}$. Therefore, (4.3.28) implies that both tensors θ^{ab} and ω_{bc} are tangential to $\tilde{\mathcal{M}}$, and inverse of each other on $\tilde{\mathcal{M}}$. This implies that the restriction (or rather the pullback) of ω to $\mathcal{M}$ is nondegenerate and defines a symplectic form $\omega = \frac{1}{2}\omega_{\mu\nu}dx^\mu dx^\nu$, corresponding to the Poisson structure

$$\{x^a, x^b\} := \theta^{ab} \qquad (4.3.30)$$

on $\mathcal{C}(\mathcal{M})$, extended (via push-forward) to $\tilde{\mathbb{R}}^D$. Together with (4.3.7) we obtain

$$[X^a, X^b]|x\rangle \approx i\theta^{ab}|x\rangle. \qquad (4.3.31)$$

We conclude that ω defines a symplectic form on $\mathcal{M}$, which corresponds to the Poisson structure defined by θ^{ac}. Moreover, we have thereby justified (4.2.31), i.e. **almost-local quantum spaces $\mathcal{M}$ can be locally approximated by some Moyal–Weyl quantum plane**

$\mathbb{R}_\theta^{2n}$. In particular, this implies that **the almost-Kähler condition** (4.2.28) holds at least approximately.

Furthermore, the inner product of (4.3.22) and (4.3.23) gives

$$(\Delta X^a)^2 = \langle x|(X^a - x^a + \partial^d \lambda)(X^a - x^a + \partial^a \lambda)|x\rangle = \theta^{ab}\theta^{ac}g_{bc} \qquad (4.3.32)$$

(no sum over a), where g_{bc} is the quantum metric (4.2.19). Hence, the uncertainty of X^a is characterized by the coherence length together with the noncommutativity scale L_{NC} (4.2.27). For the Moyal–Weyl quantum plane these scales coincide, cf. (3.1.18), and this can be expected to hold to a good approximation for more general almost-local quantum spaces.

Some further remarks are in order. First, these arguments imply that the embedding map (4.2.7) is locally nondegenerate; however, $\tilde{\mathcal{M}} \subset \mathbb{R}^D$ might still have self-intersections globally. For the abstract quantum space, such self-intersections can be excluded by regularity statements such as theorem 4.1. Furthermore, we have just seen that λ is (approximately) constant on almost-local quantum spaces $\mathcal{M}$. This means that $\tilde{\mathcal{M}}$ is the location in target space where λ assumes its approximate minimum, which provides a simple way to numerically measure and picture such branes [167, 166]. That method may be too crude for quantum spaces that are far from classical, where λ may not be sufficiently constant on $\mathcal{M}$. Then, the abstract quantum space provides a more refined picture and more information.

Completeness relation revisited

We have justified the (approximate) completeness relation (4.2.36)

$$\mathcal{P} := \int_{\mathcal{M}} |x\rangle\langle x| \overset{!}{\approx} \mathbb{1}$$

in Section 4.2.1 for almost-local and symplectic quantum spaces. Here we provide yet another argument or consistency check using (4.3.24):

$$[X^a, \mathcal{P}] \approx -i \int_{\mathcal{M}} \theta^{ab}\partial_b(|x\rangle\langle x|) = -i \int_{\mathcal{M}} \{x^a, |x\rangle\langle x|\} = 0 \qquad (4.3.33)$$

because the integration measure Ω (4.2.37) is invariant under Hamiltonian vector fields. It follows that $\mathcal{P}$ (approximately) commutes with all X^a, hence irreducibility implies

$$\int_{\mathcal{M}} |x\rangle\langle x| \propto \mathbb{1}. \qquad (4.3.34)$$

For quantum spaces with sufficient symmetry such as fuzzy $\mathbb{C}P^n_N$, this argument is easily made exact using Schur's Lemma.

4.3.3 Quantum tangent space

From now on, we will assume that $\mathcal{M}$ is a quantum manifold. Since $\mathcal{M} \subset \mathbb{C}P^{N-1}$ is a (sub)manifold, we can determine its tangent space, taking advantage of the geometric structures of $\mathbb{C}P^{N-1}$. Choose some point $\xi \in \mathcal{M}$. The results of Section 4.2.2 notably (4.2.59) imply that $T_\xi \mathcal{M}$ is spanned by the D vectors

$$(\partial_a - iA_a)|x\rangle = i\mathcal{X}_a|x\rangle \in T_\xi \mathbb{C}P^{N-1}; \tag{4.3.35}$$

note that $\langle x|(\partial_a - iA_a)|x\rangle = 0$, hence $i\mathcal{X}_a|x\rangle$ is indeed a tangent vector[10] of $\mathcal{M} \subset \mathbb{C}P^{N-1}$. We can make this more explicit by choosing *"normal embedding" coordinates* near ξ on $\mathcal{M}$, which are defined as follows: after a suitable $SO(D)$ rotation, we choose among the Cartesian coordinates x^a on $\mathbb{R}^D$ a subset $\{x^\mu, \ \mu = 1, \ldots, m\}$ such that the $\mathcal{X}_\mu|x\rangle$ are linearly independent. Then these can serve as local coordinates of $\mathcal{M}$ near ξ, and an explicit basis of the tangent vectors in $T_\xi \mathcal{M}$ is given by $i\mathcal{X}_\mu|x\rangle$ for $\mu = 1, \ldots, m$. This provides a natural definition of the **(real) quantum tangent space** of $\mathcal{M}$:

$$T_\xi \mathcal{M} = \left\langle i\mathcal{X}_\mu|x\rangle \right\rangle_\mathbb{R} = \left\langle i\mathcal{X}_a|x\rangle \right\rangle_\mathbb{R} \subset T_\xi \mathbb{C}P^{N-1} \tag{4.3.36}$$

with basis $i\mathcal{X}_\mu|x\rangle$, $\mu = 1, \ldots, m$. In particular, $\dim T_\xi \mathcal{M} = m = \dim \mathcal{M}$ as it should be. The last form is conceptually simple, but the explicit basis is useful in specific computations.

One can now repeat the considerations in Section 4.2.1, in terms of the local coordinates $x^\mu, \ \mu = 1, \ldots, m$ on $\mathcal{M}$. Thus, $\mathcal{M}$ is equipped with a $U(1)$ connection

$$iA = \langle x|d|x\rangle \tag{4.3.37}$$

and a closed two-form (4.2.22)

$$i\omega_\mathcal{M} = d\langle x|d|x\rangle = \frac{i}{2}\omega_{\mu\nu}dx^\mu \wedge dx^\mu = idA, \qquad d\omega_\mathcal{M} = 0 \tag{4.3.38}$$

as well as a quantum metric $g_{\mu\nu}$, which are simply the pullback of the symplectic structure and the Fubini–Study metric on $\mathbb{C}P^{N-1}$. These structures are intrinsic, and have nothing to do with target space $\mathbb{R}^D$. Given the basis $i\mathcal{X}_\mu|x\rangle$ of tangent vectors, we can evaluate the symplectic form and the quantum metric in local embedding coordinates as

$$i\omega_{\mu\nu} = \langle x|(\mathcal{X}_\mu\mathcal{X}_\nu - \mathcal{X}_\nu\mathcal{X}_\mu)|x\rangle$$
$$g_{\mu\nu} = \langle x|(\mathcal{X}_\mu\mathcal{X}_\nu + \mathcal{X}_\nu\mathcal{X}_\mu)|x\rangle. \tag{4.3.39}$$

It should be noted that the quantum tangent space $T_x\mathcal{M}$ of the abstract quantum space is a subspace of $\mathbb{C}P^{N-1}$, and has a priori nothing to do with the embedding in target space $\mathbb{R}^D$. This is indicated by the attribute "quantum." The embedding (4.2.7) in target space induces another metric on $\mathcal{M}$, which in turn is distinct from the effective metric discussed in Section 9.1.1.

It is tempting to conjecture that $\omega_\mathcal{M}$ is always nondegenerate for irreducible matrix configuration, and thus defines a symplectic form on $\mathcal{M}$. This is not true, as demonstrated by the minimal fuzzy torus T_2^2 or minimal fuzzy H_0^4 where $\omega_\mathcal{M}$ vanishes, cf. [189]. However, we will see in the next section that whenever there is a semi-classical regime,

---

[10] since A_μ can be gauged away at any given point, these are derivatives of sections of the respective $U(1)$ bundles over $\mathcal{M}$ and $\mathbb{C}P^{N-1}$, which can be taken as representatives of tangent vectors on $\mathcal{M}$ and $\mathbb{C}P^{N-1}$, respectively. Although the $\mathcal{X}_a$ depend implicitly on x, the result is independent of the point $x \in \mathcal{N}_x$ because $\mathcal{M}$ is a manifold.

$\omega_{\mathcal{M}}$ is indeed nondegenerate and hence a symplectic manifold. From now on, we will mostly drop the subscript from $\omega_{\mathcal{M}} = \omega$.

4.4 Quantization map, symbol, and Semiclassical regime

Given the quasi-coherent states and the symplectic structure on the manifold $\mathcal{M}$, we can define a **quantization map**

$$\mathcal{Q}: \quad \mathcal{C}(\mathcal{M}) \to \mathrm{End}(\mathcal{H})$$

$$\phi(x) \mapsto \int_{\mathcal{M}} \phi(x) |x\rangle \langle x|, \tag{4.4.1}$$

which associates to every classical function on $\mathcal{M}$ an operator or observable in $\mathrm{End}(\mathcal{H})$, recall that the integral is defined as $\int_{\mathcal{M}} \equiv \int_{\mathcal{M}} \frac{\Omega}{(2\pi)^n}$ (4.2.37). Both sides are equipped with a natural norm and inner product, given by

$$\langle \Phi, \Psi \rangle = \mathrm{Tr}(\Phi^{\dagger}\Psi) \quad \text{and} \quad \langle \phi, \psi \rangle = \int_{\mathcal{M}} \phi(x)^* \psi(x) \tag{4.4.2}$$

leading to the Hilbert–Schmidt norm $\|\Phi\|_{\mathrm{HS}}$ and the L^2 norm $\|\phi\|_2$, respectively. The map $\mathcal{Q}$ cannot be injective, since $\mathrm{End}(\mathcal{H})$ is finite-dimensional; the kernel is typically given by functions with high "energy." It is not evident in general if this map is surjective, which will be established in the following for the case of quantum Kähler manifolds. The trace is always related to the integral via

$$\mathrm{Tr}\mathcal{Q}(\phi) = \int_{\mathcal{M}} \phi(x). \tag{4.4.3}$$

For almost-local quantum spaces, we observe that (4.3.24) implies

$$[X^a, \mathcal{Q}(\phi)] \approx -i \int_{\mathcal{M}} \phi(x) \theta^{ab} \partial_b(|x\rangle\langle x|)$$

$$= -i \int_{\mathcal{M}} \phi(x)\{x^a, |x\rangle\langle x|\} = i \int_{\mathcal{M}} \{x^a, \phi(x)\}|x\rangle\langle x|$$

$$= i \int_{\mathcal{M}} \theta^{ab} \partial_b \phi(x)|x\rangle\langle x|$$

$$= \mathcal{Q}(i\theta^{ab} \partial_b \phi) \tag{4.4.4}$$

because the integration measure Ω (4.2.37) is invariant under Hamiltonian vector fields, and $|x\rangle\langle x|$ is viewed as $\mathrm{End}(\mathcal{H})$-valued function on $\mathcal{M}$. This implies the completeness relation (4.2.36) (up to normalization, cf. (4.3.33)) for irreducible configurations. Furthermore, the relation (4.3.31) states that the quantization map $\mathcal{Q}$ essentially maps commutators to Poisson brackets, at least in the almost-local case. We conclude that generic almost-local quantum spaces $\mathcal{M}$ can be understood as quantized symplectic spaces, and are described explicitly via the quasi-coherent states.

Now recall the abstract symbol map (4.2.3)

$$\text{End}(\mathcal{H}) \to \mathcal{C}(\mathcal{M})$$

$$\Phi \mapsto \langle x|\Phi|x\rangle =: \phi(x). \tag{4.4.5}$$

The symbol map can be viewed as de-quantization map, which makes sense for any quantum space in the present framework. We will show that $\mathcal{Q}$ is an approximate inverse of the symbol map (4.4.6) in a suitable sense.

Almost-local observables and IR correspondence

The concept of almost-local observables discussed in Section 4.3.1 can now be refined. We re-define $\text{Loc}(\mathcal{H}) \subset \text{End}(\mathcal{H})$ as a maximal (vector) space of operators such that the restricted symbol map

$$\text{Loc}(\mathcal{H}) \to \mathcal{C}_{\text{IR}}(\mathcal{M})$$

$$\Phi \mapsto \langle x|\Phi|x\rangle =: \phi(x) \tag{4.4.6}$$

is an "approximate isometry" with respect to the Hilbert–Schmidt norm on $\text{Loc}(\mathcal{H}) \subset \text{End}(\mathcal{H})$ and the L^2-norm on $\mathcal{C}_{\text{IR}}(\mathcal{M}) \subset L^2(\mathcal{M})$. We will then identify $\Phi \sim \phi$. Approximate isometry means that $|\|\phi\|_2 - 1| < \varepsilon$ whenever $\|\Phi\|_{\text{HS}} = 1$ for some given $0 < \varepsilon < \frac{1}{2}$, depending on the context. Then the polarization identity implies

$$\langle \Phi, \Psi \rangle_{\text{HS}} \approx \langle \phi, \psi \rangle_2, \tag{4.4.7}$$

hence an ON basis of $\text{Loc}(\mathcal{H})$ is mapped to a basis of $\mathcal{C}_{\text{IR}}(\mathcal{M})$ that is almost ON. This defines the **semi-classical regime**, which can be made more precise in some given context by specifying some ε. Accordingly, **almost-local quantum spaces** are (re)defined as matrix configurations, where all X^a and $[X^a, X^b]$ are in $\text{Loc}(\mathcal{H})$.

Of course, some given matrix configuration may be far from any semi-classical space, in which case $\text{Loc}(\mathcal{H})$ is trivial. We will see that for almost-local quantum space, $\text{Loc}(\mathcal{H})$ typically includes the almost-local operators in the sense of (4.3.1) up to some bound, and in particular polynomials in X^a up to some order. Moreover, $\mathcal{Q}$ is an approximate inverse of the symbol map (4.4.6) on $\text{Loc}(\mathcal{H})$,

$$\boxed{\begin{aligned} \text{Loc}(\mathcal{H}) &\cong \mathcal{C}_{\text{IR}}(\mathcal{M}) \\ \Phi &\to \phi(x) = \langle x|\Phi|x\rangle \\ \mathcal{Q}(\phi) &\leftarrow \phi. \end{aligned}} \tag{4.4.8}$$

The semi-classical regime should contain a sufficiently large class of functions and operators to characterize the geometry to a satisfactory precision. Thus, identifying $\text{Loc}(\mathcal{H})$ with $\mathcal{C}_{\text{IR}}(\mathcal{M})$, the usual semi-classical relation

$$\boxed{[\Phi, \Psi] \sim i\{\phi, \psi\}} \tag{4.4.9}$$

holds to a good approximation. In this sense, the semi-classical geometry is encoded in the matrix configuration X^a. The correspondence is summarized in Table 4.1, where the

Table 4.1 Correspondence between almost-local operators and infrared functions on $\mathcal{M}$ for almost-local quantum spaces.

almost-local operators $\text{Loc}(\mathcal{H}) \subset \text{End}(\mathcal{H})$	$\sim$	IR functions $\mathcal{C}_{\text{IR}}(\mathcal{M}) \subset L^2(\mathcal{M})$
Φ	$\sim$	$\phi(x) = \langle x|\Phi|x\rangle$
X^a	$\sim$	$\mathbf{x}^{\mathbf{a}}(x)$
$[.,.]$	$\sim$	$i\{.,.\}$
Tr	$\sim$	$\int_{\mathcal{M}}$
$\Box$	$\sim$	$e^{-\sigma}\Box_G$

metric structure is encoded in the Laplacian $\Box$ (9.2.55). This provides a suitable concept of quantization that does not require any limit $\theta \to 0$ as in Definition 3.1, and provides the starting point for the considerations on emergent geometry and gravity discussed in the later chapters.

Let us try to justify these claims. The first observation is that $\mathbb{1} \in \text{Loc}(\mathcal{H})$, because its symbol is the constant function $1_{\mathcal{M}}$, and the norm is preserved due to (4.2.50). The reverse direction amounts to the completeness relation (4.2.36), which was established for almost-local quantum spaces, and which we assume to hold to a sufficient precision. Now let Φ be an almost-local hermitian operator as defined in Section 4.3.1, with symbol ϕ. Then, the trace relation (4.2.38) gives

$$\|\Phi\|_{\text{HS}}^2 \approx \int_{\mathcal{M}} \langle x|\Phi\Phi|x\rangle \approx \int_{\mathcal{M}} \phi(x)^2 = \|\phi\|_2^2 \qquad (4.4.10)$$

using (4.3.1). Therefore, almost-local operators in the sense of (4.3.1) are indeed contained in $\text{Loc}(\mathcal{H})$, up to the specific bounds. Conversely, assume that $\|\Phi\|_{\text{HS}} \approx \|\phi\|_2$ for hermitian Φ. Then the completeness relation implies again (4.4.10), which in turn gives

$$\int_{\mathcal{M}} \langle x|(\Phi - \phi(x))(\Phi - \phi(x))|x\rangle \approx 0, \qquad (4.4.11)$$

which implies that $(\Phi - \phi(x))|x\rangle \approx 0 \; \forall x \in \mathcal{M}$. Hence, they are approximately local in the sense of (4.3.1). In particular they approximate commute due to (4.3.6),

$$\Phi\Psi \approx \Psi\Phi, \qquad \Phi, \Psi \in \text{Loc}(\mathcal{H}). \qquad (4.4.12)$$

Hence, the present definition of $\text{Loc}(\mathcal{H})$ is a refinement of the definitions in Section 4.3.1, turning the local statements into global ones[11].

These abstract statements can be understood more explicitly as follows. $\mathcal{C}_{\text{IR}}(\mathcal{M})$ is typically given by functions which are slowly varying on the length scale L_{coh},

---

[11] There is no contradiction with the fact that Weyl quantization is an exact isometry, because the symbol map (4.4.8) differs from the Wigner map (3.1.56). Hence, the coherent states are more physical, because they sense the breakdown of the semi-classical regime.

corresponding to the semi-classical or infrared regime. To see that $\mathcal{Q}$ is approximately inverse to the symbol map, we note that the completeness relation implies

$$|y\rangle \approx \int_{\mathcal{M}} |x\rangle \langle x|y\rangle. \qquad (4.4.13)$$

This means that

$$\langle x|y\rangle \approx \delta_y(x) \qquad (4.4.14)$$

for any $y \in \mathcal{M}$ w.r.t. the measure (4.2.37), consistent with $|\langle x|y\rangle| \sim e^{-\frac{1}{4}|x-y|_g^2}$ (4.2.14) (4.2.23). Then

$$\mathcal{Q}(\phi)|y\rangle = \int_{\mathcal{M}} \phi(x)|x\rangle \langle x|y\rangle \approx \phi(y)|y\rangle. \qquad (4.4.15)$$

for functions $\phi(x)$ which are slowly varying on L_{coh}. Therefore, $\mathcal{Q}(\phi)$ is almost-local and hence $\mathcal{Q}(\phi) \in \mathrm{Loc}(\mathcal{H})$ for slowly varying ϕ, and moreover $\mathcal{Q}$ is approximately the inverse of the symbol map on $\mathrm{Loc}(\mathcal{H})$, since (4.4.15) gives

$$\langle y|\mathcal{Q}(\phi)|y\rangle \approx \phi(y). \qquad (4.4.16)$$

For almost-local quantum spaces, $\mathrm{Loc}(\mathcal{H})$ contains in particular the basic matrices

$$X^a \approx \int_{\mathcal{M}} \mathbf{x}^{\mathbf{a}}|x\rangle \langle x|. \qquad (4.4.17)$$

The approximation is good as long as the classical function $\mathbf{x}^{\mathbf{a}}$ is approximately constant on L_{coh}. Identifying almost-local observables with functions, (4.3.31) gives the semi-classical relation

$$\boxed{[X^a, X^b] \sim i\{x^a, x^b\} = i\theta^{ab}.} \qquad (4.4.18)$$

We have seen that θ^{ab} is tangential to $\mathcal{M}$ and the inverse of the symplectic form ω on $\mathcal{M}$, hence $\{x^a, x^b\}$ are Poisson brackets on $\mathcal{M}$. This Poisson structure extends trivially (by push-forward) to $\tilde{\mathbb{R}}^D$, which for $D > \dim \mathcal{M}$ decomposes into symplectic leaves of ω_{ab} that are preserved by the Poisson structure. Functions that are constant on these leaves then have vanishing Poisson brackets, which leads to a degenerate effective metric.

In the UV or deep quantum regime, the semi-classical picture is no longer justified, and in fact it is very misleading. In particular, consider *string modes* that are defined as rank one operators built out of quasi-coherent states [105, 184]

$$\psi_{x,y} := |x\rangle \langle y| \qquad \in \mathrm{End}(\mathcal{H}). \qquad (4.4.19)$$

They are highly nonlocal for $x \neq y$, and should not be interpreted as function but rather as open strings (or dipoles) linking $|y\rangle$ to $|x\rangle$ on the embedded brane $\tilde{\mathcal{M}}$. These states provide a complete and more adequate picture of $\mathrm{End}(\mathcal{H})$, and are responsible for UV/IR mixing in noncommutative field theory, as discussed in Section 6.3. They exhibit the stringy nature of noncommutative field theory and Yang–Mills matrix models, which goes beyond conventional (quantum) field theory.

4.5 *Complex tangent space and quantum Kähler manifolds

Now we return to the exact analysis. For any quantum manifold $\mathcal{M}$, the embedding $\mathcal{M} \to \mathbb{C}P^{N-1}$ induces the tangential map

$$T_\xi \mathcal{M} \to T_\xi \mathbb{C}P^{N-1}. \tag{4.5.1}$$

Now we take into account that $\mathbb{C}P^{N-1}$ carries an intrinsic complex structure

$$\mathcal{J}: \quad T_\xi \mathbb{C}P^{N-1} \to T_\xi \mathbb{C}P^{N-1}, \qquad \mathcal{J}v = iv \tag{4.5.2}$$

for any $v \in T_\xi \mathbb{C}P^{N-1}$. Accordingly, $T\mathbb{C}P^{N-1} \cong T^{(1,0)}\mathbb{C}P^{N-1}$ can be viewed as holomorphic tangent bundle, thus bypassing an explicit complexification of its real tangent space. With this in mind, we define the **complex quantum tangent space** of $\mathcal{M}$ as

$$T_{\xi,\mathbb{C}}\mathcal{M} := \left\langle \mathcal{X}_a | x \right\rangle_{\mathbb{C}} \subset T_\xi \mathbb{C}P^{N-1} \cong T_{\xi,\mathbb{C}} \mathbb{C}P^{N-1}, \tag{4.5.3}$$

which carries the complex structure

$$\mathcal{J}\mathcal{X}_a|x\rangle := i\mathcal{X}_a|x\rangle \quad \in T_{\xi,\mathbb{C}}\mathcal{M}, \qquad \mathcal{J}^2 = -\mathbb{1}. \tag{4.5.4}$$

Again, this complex tangent space is not necessarily the complexification of the real one. Using the basis $i\mathcal{X}_\mu|x\rangle$, $\mu = 1, \ldots, m$ of the real tangent space $T_\xi\mathcal{M}$ in normal embedding coordinates, there may be relations of the form

$$(i\mathcal{X}_\mu - J_\mu^{\ \nu} \mathcal{X}_\nu)|x\rangle = 0 \quad \text{for} \quad J_\mu^{\ \nu} \in \mathbb{R}, \tag{4.5.5}$$

so that $T_{\xi,\mathbb{C}}\mathcal{M}$ has reduced dimension over $\mathbb{C}$. We will see that for quantum Kähler manifolds, the complex dimension is half of the same as the real one.

Quantum Kähler manifolds

Consider the maximally degenerate case where the complex dimension of $T_{\xi,\mathbb{C}}\mathcal{M}$ is given by $n = \frac{m}{2} \in \mathbb{N}$, where $m = \dim_\mathbb{R} \mathcal{M}$. Then $T_\xi\mathcal{M}$ is stable under the complex structure operator $\mathcal{J}$

$$T_{\xi,\mathbb{C}}\mathcal{M} = T_\xi\mathcal{M} \cong \mathbb{C}^n \tag{4.5.6}$$

so that $T_\xi\mathcal{M}$ should be viewed as holomorphic tangent space of $\mathcal{M}$. But this implies that $\mathcal{M} \subset \mathbb{C}P^{N-1}$ is a complex sub-manifold (i.e. defined by holomorphic equations), cf. [210]. Such quantum manifolds $\mathcal{M}$ will be called **quantum Kähler manifolds**, for reasons explained below. Indeed, all complex sub-manifolds of $\mathbb{C}P^{N-1}$ are known to be Kähler. Note that this is an intrinsic property of a quantum space $\mathcal{M}$, and no extra structure is introduced here.[12] We will see that this includes the well-known quantized or "fuzzy" spaces arising from quantized coadjoint orbits.[13]

---

[12] It is interesting to note that due to (4.2.28), H_x preserves the complex tangent space $T_{\xi,\mathbb{C}}\mathcal{M}$, at least in the semi-classical regime. However, (4.2.28) is still weaker than the Kähler condition.

[13] It is worth pointing out that $\mathbb{C}P^{N-1}$ is itself a quantum Kähler manifold, as minimal fuzzy $\mathbb{C}P_N^{N-1}$.

Consider the quantum Kähler case in more detail. We can introduce a local holomorphic parametrization of $\mathcal{M} \subset \mathbb{C}P^{N-1}$ near ξ in terms of $z^k \in \mathbb{C}^n$. Then any local (!) holomorphic section of the tautological line bundle over $\mathbb{C}P^{N-1}$ defines via pull back a local holomorphic section of the line bundle

$$\tilde{\mathcal{B}} := \bigcup_{x \in \tilde{\mathbb{R}}^D} E_x \to \mathcal{M} \hookrightarrow \mathbb{C}P^{N-1} \tag{4.5.7}$$

over $\mathcal{M}$, denoted by $\|z\rangle$. This $\|z\rangle$ can be viewed as holomorphic $\mathbb{C}^N$-valued function on $\mathcal{M}$, which satisfies

$$\frac{\partial}{\partial \bar{z}^k} \|z\rangle = 0, \qquad \|z\rangle \big|_\xi = |\xi\rangle, \tag{4.5.8}$$

where $\bar{z}^k$ denotes the complex conjugate of z^k. Hence, $\|z\rangle$ arises from $|x\rangle$ through a reparametrization and gauge transformation along with a nontrivial normalization[14] factor; this is indicated by the double line in $\|z\rangle$, and we will denote the $\|z\rangle$ as **coherent states**. Then the differential of the section

$$d\|z\rangle = dz^k \frac{\partial}{\partial z^k} \|z\rangle \qquad \in \Omega_z^{(1,0)}\mathcal{M} \tag{4.5.9}$$

is a $(1,0)$ one-form, where $\mathcal{J} = i$. As in (4.2.18), we can then define the hermitian metric

$$h(X,Y) = \left((d\|z\rangle)^\dagger \otimes d\|z\rangle\right)(X,Y) \qquad \in T^{(1,1)} \tag{4.5.10}$$

whose imaginary and real part define the symplectic form and the quantum metric via

$$\omega(X,Y) = -i(h(X,Y) - h(Y,X)^*) = -\omega(Y,X)$$
$$g(X,Y) = h(X,Y) + h(X,Y)^* = g(Y,X). \tag{4.5.11}$$

Since $h \in T^{(1,1)}$, they satisfy the compatibility condition

$$\omega(X, \mathcal{J}Y) = -i(h(X, \mathcal{J}Y) - h(\mathcal{J}Y, X)^*)$$
$$= -i(ih(X,Y) + ih(Y,X)^*)$$
$$= g(X,Y) \tag{4.5.12}$$

(recall that $\mathcal{J} = -i$ on anti-holomorphic $(0,1)$ forms). The Kähler two-form is then given by

$$i\omega := (d\|z\rangle)^\dagger \wedge d\|z\rangle = \omega_{\bar{k}l} d\bar{z}^k \wedge dz^l \qquad \in \Omega_z^{(1,1)}\mathcal{M}$$
$$i\omega_{\bar{k}l} = (d_k\|z\rangle)^\dagger d_l\|z\rangle \tag{4.5.13}$$

which is closed

$$id\omega = -(d\|z\rangle)^\dagger \wedge dd\|z\rangle + (dd\|z\rangle)^\dagger \wedge d\|z\rangle = 0 \tag{4.5.14}$$

using holomorphicity of $\|z\rangle$. The Kähler form then arises from a local Kähler potential,

$$\omega_{\bar{k}l} = -\frac{1}{2}\bar{\partial}_k \partial_l \rho \tag{4.5.15}$$

---

[14] $\|z\rangle$ cannot be normalized, since e.g. $\langle y\|z\rangle$ must be holomorphic in z. Apart from that, $\tilde{\mathcal{B}}$ is equivalent to $\mathcal{B}$.

given by the restriction of the (Fubini–Study) Kähler potential on $\mathbb{C}P^N$. This means that $\mathcal{M}$ is a Kähler manifold, and the name "quantum Kähler manifold" indicates its origin from the matrices X^a. The relation (4.5.12) implies the almost-Kähler condition (4.2.28), which characterize quantum spaces that can be locally approximated by some Moyal–Weyl quantum space. In fact, we have shown in Section 4.3.2 that almost-local quantum spaces can always be locally approximated by some Moyal–Weyl quantum plane, which implies that the Kähler condition holds at least approximately. The almost-Kähler condition will be used in Section 6.2.3 to derive local trace formulas that are important in deriving gravity, and all significant spaces examples considered is this book satisfy this condition.

This provides a rather satisfactory concept of quantum Kähler geometry, which arises naturally from the complex structure in the Hilbert space. There is no need to invoke any semi-classical or large N limit. However, not all quantum spaces are of this type, a counterexample being the minimal fuzzy torus T_2^2 as shown in [189].

4.5.1 Coherent states and quantization map for quantum Kähler manifolds

We can establish the following lemma, which is well known for standard coherent states:

Lemma 4.2 *Let $|x\rangle$ be the coherent states of a quantum Kähler manifold $\mathcal{M}$, and $\mathcal{H}_0 \subset \mathcal{H}$ their linear span. Assume $A \in \mathrm{End}(\mathcal{H}_0)$ satisfies $\langle x|A|x\rangle = 0$ for all $x \in \mathcal{M}$. Then $A = 0$.*

Proof Consider the function

$$A(\bar{y}, z) := \langle y\|A\|z\rangle, \tag{4.5.16}$$

where $\|z\rangle$, $\|y\rangle$ are local holomorphic sections of the coherent states in a neighborhood of $\xi \in \mathcal{M}$. Clearly this function is holomorphic in z and in $\bar{y}$. By assumption, the restriction of $A(\bar{y}, z)$ to the diagonal $A(\bar{z}, z) = \langle z\|A\|z\rangle$ vanishes identically. But then the standard properties of holomorphic functions imply (cf. [155]) that $A(\bar{y}, z) \equiv 0$ identically. This argument applies near any given point $\xi \in \mathcal{M}$, which implies that $A = 0$. □

Using this lemma, we can establish the diagonal realization of operators via coherent states:

Theorem 4.3 *Let $|x\rangle$ be the (normalized) coherent states of a quantum Kähler manifold $\mathcal{M}$, and $\mathcal{H}_0 \subset \mathcal{H}$ their linear span. Then all operators $A \in \mathrm{End}(\mathcal{H}_0)$ can be written as*

$$A = \int_{\mathcal{M}} A(x)\,|x\rangle\langle x| \tag{4.5.17}$$

for some suitable complex-valued function $A(x)$ on $\mathcal{M}$.

$A(x)$ is sometimes denoted as "upper symbol" or A, and $\langle x|A|x\rangle$ as lower symbol. Note that if the holomorphic coherent states $\|x\rangle$ are used instead of the normalized $|x\rangle$, then $A(x)$ might have some singularities.

Proof   Assume that the subspace in $\text{End}(\mathcal{H}_0)$ spanned by the rhs of (4.5.17) is smaller than $\text{End}(\mathcal{H}_0)$. Let $B \in \text{End}(\mathcal{H}_0)$ be in its orthogonal complement w.r.t. the Hilbert–Schmidt metric. Then

$$0 = Tr\,(AB) = \int_{\mathcal{M}} A(x)\langle x|B|x\rangle \qquad \forall A(x) \in \mathcal{C}(\mathcal{M}). \qquad (4.5.18)$$

But this implies $\langle x|B|x\rangle = 0 \; \forall x \in \mathcal{M}$, and then by Lemma 4.2 it follows that $B = 0$.

$\square$

The theorem states that the quantization map $\mathcal{Q}$ (4.4.1) is surjective for quantum Kähler manifolds. On the other hand, $\mathcal{Q}$ is typically not injective, and the kernel of $\mathcal{Q}$ is typically given by functions above some energy cutoff. Surjectivity of $\mathcal{Q}$ is rather remarkable in light of the string modes $|x\rangle\langle y|$ (4.4.19), which are highly nonlocal. Nevertheless, even such string modes, and more generally operators in the deep quantum regime, can be represented in the diagonal form (4.5.17). However, $A(x)$ is then rapidly oscillating, and this diagonal representation should be used with great caution. For such operators, a representation in terms of nonlocal string modes is more appropriate, as discussed in Section 6.2.1. This will not depend on holomorphicity properties of quantum Kähler spaces but only on a more generic completeness relation.

4.6 Examples of quantum (matrix) geometries

We have seen that the most important tool to understand matrix geometry is the quantization map $\mathcal{Q}$ (4.4.1) and its properties summarized in definition 3.1, as well as the symbol map (4.2.3). On almost-local quantum spaces, these two maps are approximate inverse to each other at low energies. They provide a correspondence between matrices or operators in $\text{End}(\mathcal{H})$ and functions on $\mathcal{M}$ as indicated in Table 4.1. Using this dictionary, we can freely switch between the matrix point of view and the semi-classical point of view, keeping in mind the restriction to low energies. This will be understood and used throughout this book. To develop some intuition and confidence, we will discuss a couple of basic examples for quantum spaces. These examples are special because they enjoy certain symmetries, which makes them analytically accessible. We will extract their structure using the results in this chapter, without resorting to representation theory as in Section 3.3. Moreover, due to Darboux's theorem, all symplectic spaces of a given dimension are locally isomorphic. Therefore, the following examples should be seen as representatives for quantum (matrix) geometries with a given dimension and topology.

4.6.1 The fuzzy sphere S_N^2 revisited

The basic example of a compact quantum space is the fuzzy sphere S_N^2, which was introduced in Section 3.3.1, and a quantization map $\mathcal{Q}$ was defined. Here we reconsider it from the point of view of matrix configurations, and discuss the associated (quasi-)

coherent states. S_N^2 can be viewed as a matrix configuration given by the three hermitian matrices

$$X^a := \frac{1}{C_N} J_{(N)}^a \qquad (4.6.1)$$

acting on the spin $j = \frac{N-1}{2}$ irrep $\mathcal{H} = \mathbb{C}^N$ of $\mathfrak{su}(2)$, where $C_N^2 = \frac{1}{4}(N^2 - 1)$ is the value of the quadratic Casimir. The configuration is clearly irreducible, since the X^a generate the full matrix algebra $\text{End}(\mathcal{H})$. The displacement Hamiltonian is

$$H_x = \frac{1}{2} \sum_{a=1}^{3} (X^a - x^a)^2 = \frac{1}{2}(1 + |x|^2)\mathbb{1} - \sum_{a=1}^{3} x^a X^a, \qquad (4.6.2)$$

where $|x|^2 = \sum_a x_a^2$. Using $SO(3)$ invariance, it suffices to consider the north pole $n := (0, 0, x^3)$, where

$$H_n = \frac{1}{2}(1 + |x|^2)\mathbb{1} - |x| X^3 \qquad (4.6.3)$$

assuming $x^3 > 0$ to be specific. The ground state of H_n is clearly given by the highest weight vector $|n\rangle$ of the $\mathfrak{su}(2)$ irrep $\mathcal{H}$, with maximal eigenvalue of $J_3|_{|n\rangle} = \frac{N-1}{2}$. For any other $x \in S^2$ with radius 1, we choose some $g_x \in SO(3)$ such that $x = g_x \cdot n$, and define

$$|x\rangle = g_x \cdot |n\rangle, \qquad g_x \in SU(2). \qquad (4.6.4)$$

This is clearly a quasi-coherent state at x, and all quasi-coherent states are of this form up to a phase. Hence, the bundle of quasi-coherent states is

$$\mathcal{B} = \bigcup_{g \in SU(2)} \{U(1) g \cdot |n\rangle\}, \qquad (4.6.5)$$

which is clearly an $SU(2)$- equivariant $U(1)$ bundle over the abstract quantum space

$$\mathcal{M} \cong SU(2)/_{U(1)} \cong S^2 \subset \mathbb{C}P^{N-1}, \qquad (4.6.6)$$

noting that $SU(2)|n\rangle \cong SU(2)/_{U(1)} \cong S^2$, where $U(1)$ is the stabilizer of $|n\rangle$. The expectation value is obtained as

$$\langle X^a \rangle \equiv \langle x | X^a | x \rangle =: \mathbf{x^a}, \qquad \mathbf{x^a x_a} = \frac{N-1}{N+1} =: r_N^2 \qquad (4.6.7)$$

where r_N^2 is the radius of the orbit of coherent states. The eigenvalue of H_x on $|x\rangle$ is easily found to be

$$\lambda(x) = \frac{1}{2}(1 + |x|^2) - |x| r_N. \qquad (4.6.8)$$

The minima of $\lambda(x)$ on $\mathcal{N}_x$ describe a sphere with radius $r_N = \sqrt{\frac{N-1}{N+1}} = 1 + \mathcal{O}(\frac{1}{N})$, which coincides precisely with the embedded quantum space (4.2.8)

$$\tilde{\mathcal{M}} = \{\langle x | X^a | x \rangle\} = \{x \in \mathbb{R}^3 : |x| = r_N\} \cong S^2 \qquad (4.6.9)$$

defined by the expectation value $\mathbf{x}^a$. Note that the quasi-coherent states are constant along the radial directions as observed below (4.2.10),

$$|x\rangle = |\alpha x\rangle \qquad \text{for} \quad \alpha > 0. \tag{4.6.10}$$

Accordingly, the would-be symplectic form ω_{ab} and the quantum metric g_{ab} vanish along the radial direction. The singular set consists of the point $\mathcal{K} = \{0\}$, where the Hamiltonian is $H_0 \propto \mathbb{1}$, so that all energy levels become degenerate and cross. Following $|x\rangle$ along the radial direction through the origin, it turns into the highest energy level. The inner product of these states can be shown to be[15]

$$|\langle x|y\rangle|^2 = \left(\frac{1 + x \cdot y}{2}\right)^{N-1}$$

$$\approx e^{-\frac{1}{2}|x-y|_g^2}, \qquad x \approx y$$

$$\langle x|y\rangle \approx e^{-\frac{1}{4}|x-y|_g^2} e^{i\varphi(x,y)} \tag{4.6.11}$$

(for $|x| = 1 = |y|$), which is exponentially suppressed by the "quantum distance" or metric $|x - y|_g$ defined by

$$|x - y|_g^2 := \frac{|x - y|_\|^2}{L_{\mathrm{NC}}^2}, \qquad \text{where} \quad |x - y|_\|^2 \equiv \sum_\mu (x^\mu - y^\mu)^2 \tag{4.6.12}$$

measures the tangential distance in target space, and the characteristic decay length L_{NC} is

$$L_{\mathrm{NC}}^2 = \frac{2}{N} = L_{\mathrm{coh}}^2 \tag{4.6.13}$$

cf. (3.3.38). This is consistent with the results for $\mathbb{R}_\theta^2$ in Section 3.1.5. It follows from the definition that coherent states are optimally localized, i.e. they minimize the uncertainty

$$\Delta^2 = \sum_a \langle (X^a)^2 \rangle - \langle X^a \rangle^2 = 1 - r_N^2 \approx L_{\mathrm{NC}}^2. \tag{4.6.14}$$

Indeed, if there would be some other state $|\psi\rangle$ with smaller uncertainty, the analog of (4.1.7) would lead to an expectation value $\langle H_x \rangle$ that is smaller than $1 - r_N^2$ for $x^a = \langle \psi | X^a | \psi \rangle$. Therefore, L_{NC} characterizes the minimal uncertainty on S_N^2. In particular, $\Phi(X)|x\rangle \approx \phi(x)|x\rangle$ (4.3.1) for all polynomials in X^a up to order $O(\sqrt{N})$, so that S_N^2 is an almost-local quantum space for large N.

The phase $\varphi(x,y)$ is a priori gauge dependent. Since the $SU(2)$ group action defines a parallel transport along the great circles, the would-be symplectic form $\omega = dA$ (4.2.21) is $SO(3)$-invariant. Using the explicit representation of $SU(2)$, it is not hard to see that it satisfies the quantization condition (4.2.33) with $n = N$, so that

$$\omega = \omega_N, \qquad \int \frac{\omega_N}{2\pi} = N \tag{4.6.15}$$

consistent with (3.3.4). The gauge can be chosen such that the phase $\varphi(x,y)$ is given by the symplectic area of the triangle formed by x, y and e.g. the north pole on S^2. Moreover,

---

[15] This can be seen by realizing the highest weight state in $\mathcal{H}$ as $|n\rangle = |\frac{1}{2}, \frac{1}{2}\rangle^{\otimes_S (N-1)}$.

the abstract quantum space $\mathcal{M} \cong S^2 \subset \mathbb{C}P^{N-1}$ is a quantum Kähler manifold, since the complex tangent space (4.5.3) is one-dimensional, spanned by

$$T_{n,\mathbb{C}}\mathcal{M} = \langle J^-|n\rangle\rangle_{\mathbb{C}} \tag{4.6.16}$$

(at $|n\rangle \in \mathcal{M}$). This follows from the fact that $|n\rangle$ is the highest weight state, so that

$$J^+|n\rangle = 0. \tag{4.6.17}$$

Therefore, the two tangent vectors $\mathcal{X}^1|n\rangle$, $\mathcal{X}^2|n\rangle \in T_n\mathcal{M}$ are related by i, while $\mathcal{X}^3|n\rangle$ vanishes at n. This reflects the well-known fact that the (quasi-)coherent states on S_N^2 form a Riemann sphere, i.e. they can be augmented as holomorphic coherent states $\|z\rangle$. This implies as usual the relations (4.2.28) between the quantum metric and the symplectic form.

For the **minimal fuzzy sphere** S_2^2 with $N = 2$, the generators reduce to the Pauli matrices $X^a = \sigma^a$, and the (quasi)coherent states form the well-known Bloch sphere $\mathcal{M} = S^2 \cong \mathbb{C}P^1$. This is still a quantum Kähler manifold even though the semi-classical regime is trivial and contains only the constant functions $\text{Loc}(\mathcal{H}) = \mathbb{C}\mathbb{1}$, since the coherence length is of the same order as the entire space $\mathcal{M}$.

Coherent state quantization

It is worthwhile to elaborate the coherent state quantization in detail. The starting point is the completeness relation of the coherent states $|x\rangle$ on S_N^2:

$$\mathbb{1}_{\mathcal{H}} = \int_{S^2} |x\rangle\langle x|, \tag{4.6.18}$$

where $\int_{S^2} \equiv \int \frac{\omega_N}{2\pi}$ (4.2.37). Up to normalization this follows easily from $SU(2)$ equivariance and Schur's lemma, and the normalization is obtained taking the trace. This motivates to define the coherent state quantization map (4.4.1)

$$\mathcal{Q}_c: \quad L^2(S^2) \to \text{End}(\mathcal{H})$$

$$\phi \mapsto \int \phi(x)|x\rangle\langle x| \tag{4.6.19}$$

and a symbol (or de-quantization) map

$$\text{End}(\mathcal{H}) \to L^2(S^2)$$

$$\Phi \mapsto \langle x|\Phi|x\rangle. \tag{4.6.20}$$

It follows immediately from the definitions that these maps are $SO(3)$ intertwiners, and moreover $\mathcal{Q}_c$ is surjective. This in turn implies that they map the classical and fuzzy spherical harmonics Y_m^l to $\hat{Y}_m^l$ and vice versa, up to normalization. Normalizing the modes as

$$\text{Tr}\big(\hat{Y}_m^{l\dagger}\hat{Y}_{m'}^{l'}\big) = \delta^{ll'}\delta_{mm'} = \int_{S^2} Y_m^{l*}(x)Y_{m'}^{l'}(x), \tag{4.6.21}$$

we define the normalization constants c_l by

$$\hat{Y}_m^l = c_l \int Y_m^l(x)|x\rangle\langle x|. \tag{4.6.22}$$

Then (4.6.21) gives

$$1 = \mathrm{tr}(\hat{Y}_m^{l\dagger}\hat{Y}_m^l) = c_l \int Y_m^{l*}(x)\langle x|\hat{Y}_m^l|x\rangle = \int Y_m^{l*}(x)Y_m^l(x) \tag{4.6.23}$$

so that

$$\langle x|\hat{Y}_m^l|x\rangle = \frac{1}{c_l}Y_m^l(x). \tag{4.6.24}$$

The c_l can be obtained as [190]

$$c_l = \sqrt{\frac{(N-l-1)!(N+l)!}{N((N-1)!)^2}} \approx \begin{cases} 1, & l < l_{\mathrm{NC}} \\ e^{l^2/N}, & l \gg l_{\mathrm{NC}}. \end{cases} \tag{4.6.25}$$

where $l_{\mathrm{NC}} := \sqrt{N}$ (3.3.39). We can check this for $l = 1$, where $\hat{Y}_0^1 = \sqrt{\frac{3}{N}}X^3$ and $Y_0^1 = \sqrt{\frac{3}{2\pi N}}x^3$ in the present normalization. Then,

$$X^a = c_1 \int_{S^2} |x\rangle\langle x|x^a, \qquad c_1 = \sqrt{\frac{N+1}{N-1}} \tag{4.6.26}$$

which can be verified using (4.6.9) by considering $\mathrm{Tr}(X_a X^a) = c_1 \int_{S^2}\langle x|X^a|x\rangle x_a$. This exhibits again the separation of the space of modes on S_N^2 into IR and UV regime, separated by l_{NC}. In the IR regime, the quantization map and the symbol map are approximately inverse maps:

$$\langle x|\mathcal{Q}_c(\phi)|x\rangle \approx \phi, \qquad \mathcal{Q}_c(\langle x|\Phi|x\rangle) \approx \Phi \qquad \text{in the IR regime.} \tag{4.6.27}$$

This is no longer true in the UV regime where the c_l blows up, and the symbol (4.6.24) is exponentially suppressed. The coherent state representation of a fuzzy UV mode is thus rapidly oscillating, and rather misleading. This is particularly obvious for fuzzy modes of the form $\Phi = |x\rangle\langle y|$, which are clearly nonlocal and have a highly oscillatory coherent state representation. These will be discussed in Section 6.2.

4.6.2 The Moyal–Weyl quantum plane revisited

The Moyal–Weyl quantum plane $\mathbb{R}_\theta^{2n}$ is an example of a non-compact quantum space, based on an infinite-dimensional (but separable) Hilbert space $\mathcal{H}$. It is defined in terms of operators $X^a \in \mathrm{End}(\mathcal{H})$ that satisfy

$$[X^a, X^b] = i\theta^{ab}\mathbb{1}, \qquad a, b = 1, \ldots, 2n \tag{4.6.28}$$

where $\theta^{\mu\nu}$ is a constant antisymmetric tensor. The displacement Hamiltonian is

$$H_x = \frac{1}{2}\delta_{ab}(X^a - x^a)(X^b - x^b) = U_x H_0 U_x^{-1},$$

$$H_0 = \frac{1}{2} \sum \delta_{ab} X^a X^b \tag{4.6.29}$$

where $U_x = \exp(i\tilde{x}_a X^a)$ is the translation operator (3.1.91). Hence, the quasi-coherent states were obtained in Section 3.1.5 as

$$|x\rangle = U_x |0\rangle, \tag{4.6.30}$$

with eigenvalue $\lambda = \frac{1}{2}(\sum_i \theta_i)$. Now consider the phase factor of their inner products in more detail. The phase factor for $\langle 0|x\rangle$ amounts to a particular gauge choice, but the phase factors for generic $\langle x|y\rangle$ are unavoidable. As explained in Section (4.2.1), a globally consistent choice of the phase in (4.2.14) is given by the symplectic area of the triangle formed by $(x, y, 0)$ in $\mathbb{R}^{2n}$,

$$\varphi(x, y) = -\frac{1}{2} x^a \theta_{ab}^{-1} y^b. \tag{4.6.31}$$

This is consistent with (3.1.113),

$$\langle x|y\rangle = e^{-\frac{i}{2} x^a \theta_{ab}^{-1} y^a} e^{-\frac{1}{4}|x-y|_g^2}. \tag{4.6.32}$$

The coherent state quantization map is given as

$$\mathcal{Q}: \quad \mathcal{C}(\mathbb{R}^{2n}) \to \mathrm{End}(\mathcal{H})$$

$$\phi(x) \mapsto \hat{\phi} := \int_{\mathbb{R}^{2n}} \phi(x) |x\rangle \langle x|. \tag{4.6.33}$$

Note that this is not quite equivalent to the Weyl quantization map (3.1.29): even though $\mathcal{Q}$ is an intertwiner for the translation group, it does not map the L^2 norm to the (Hilbert–Schmidt) norm on $\mathrm{End}(\mathcal{H})$. The symbol map $\Phi \to \phi(x) = \langle x|\Phi|x\rangle$ provides a reverse map that also respects translations, but reproduces the inverse of $\mathcal{Q}$ only in the semi-classical limit. In particular, the symbol of a plane wave is obtained using (3.1.113)

$$\langle x|e^{i\tilde{k}_a X^a}|x\rangle = e^{\frac{i}{2} k^a \theta_{ab}^{-1} x^b} \langle x|x+k\rangle = e^{\frac{i}{2} k^a \theta_{ab}^{-1} x^b} e^{-\frac{i}{2} x^a \theta_{ab}^{-1} k^b} e^{-\frac{1}{4}|k|_g^2}$$

$$= e^{i\tilde{k}x} e^{-\frac{1}{4}|k|_g^2}. \tag{4.6.34}$$

The plane wave $e^{i\tilde{k}x}$ is properly recovered, but the norm is respected only in the semi-classical regime, due to the factor $e^{-\frac{1}{4}|k|_g^2}$. The inner products of the coherent states $\langle x|y\rangle$ was elaborated in Section 3.1.5, in terms of the quantum metric (3.1.108)

$$g_{ab} = D_{ac}^{-1} D_{ac'}^{-1} \delta^{cc'}, \tag{4.6.35}$$

which satisfies the almost-Kähler compatibility relation (3.1.109). The latter can be reformulated by stating that

$$\mathcal{J}_b^a := g^{ac} \theta_{cb}^{-1} = -\theta^{ac} g_{cb} \tag{4.6.36}$$

defines an almost-complex structure, i.e.

$$\mathcal{J}^2 = -1. \tag{4.6.37}$$

For the Moyal–Weyl quantum plane this is in fact a complex structure, which suggests to define the holomorphic coherent states

$$\|z\rangle := e^{z^i \bar{Z}_i^\dagger} |0\rangle = e^{\frac{1}{4}|x|_g^2} |z\rangle \tag{4.6.38}$$

using (3.1.91). They obviously satisfy

$$\frac{\partial}{\partial \bar{z}_j} \|z\rangle = 0, \tag{4.6.39}$$

which implies that all Moyal–Weyl quantum planes are quantum Kähler manifolds. This can also be seen in terms of the complex tangent space, without introducing the re-scaled $\|z\rangle$:

$\mathbb{R}_\theta^2$ and complex tangent space

It is instructive to work out the complex tangent space explicitly. For simplicity we consider the two-dimensional case with generators $[X, Y] = i\mathbb{1}$, identifying $X_1 \equiv X$ and $X_2 \equiv Y$. Then, the translation operator is given by

$$\begin{aligned} U(z) &= \exp(i(yX - xY)) = \exp(zZ^\dagger - \bar{z}Z) \\ &= e^{-\frac{1}{2}|z|^2} \exp(zZ^\dagger) \exp(-\bar{z}Z) \\ &= e^{\frac{i}{2}xy} \exp(-ixY) \exp(iyX) = e^{-\frac{i}{2}xy} \exp(iyX) \exp(-ixY), \end{aligned} \tag{4.6.40}$$

where

$$Z = \frac{1}{\sqrt{2}}(X + iY), \qquad Z^\dagger = \frac{1}{\sqrt{2}}(X - iY)$$

$$z = \frac{1}{\sqrt{2}}(x + iy). \tag{4.6.41}$$

The coherent states $|z\rangle = U(z)|0\rangle$ satisfy (3.1.102)

$$(Z - z)|z\rangle = 0, \tag{4.6.42}$$

which implies

$$\langle z|(X + iY)|z\rangle = x + iy. \tag{4.6.43}$$

We can compute the partial derivatives of $|z\rangle$ explicitly using (4.6.40):

$$\partial_x |z\rangle = -i(Y - \frac{1}{2}y)|z\rangle$$

$$\partial_y |z\rangle = i(X - \frac{1}{2}x)|z\rangle. \tag{4.6.44}$$

The $U(1)$ connection is then found to be

$$iA_1 = \langle z|\partial_x|z\rangle = -i\langle z|(Y - \frac{1}{2}y)|z\rangle = -\frac{i}{2}y$$

$$iA_2 = \langle z|\partial_y|z\rangle = i\langle z|(X - \frac{1}{2}x)|z\rangle = \frac{i}{2}x \tag{4.6.45}$$

with nonvanishing field strength

$$F_{12} = \partial_1 A_2 - \partial_2 A_1 = 1 = \omega_{12}. \qquad (4.6.46)$$

This is indeed the inverse of the Poisson tensor $\theta^{12} = 1$ underlying the present background $\mathbb{R}^2_\theta$, consistent with the general result (4.3.28). The covariant derivatives (4.2.58) of the $|z\rangle$ are then

$$(\partial_x - iA_1)|z\rangle = -i(Y - y)|z\rangle = i\mathcal{X}_1|z\rangle$$
$$(\partial_y - iA_2)|z\rangle = i(X - x)|z\rangle = i\mathcal{X}_2|z\rangle. \qquad (4.6.47)$$

We can verify the quantum Kähler condition e.g. at the origin by recasting the constraint $(X + iY)|0\rangle = 0$ in the form

$$\mathcal{X}_2|0\rangle = X|0\rangle = -iY|0\rangle = i\mathcal{X}_1|0\rangle. \qquad (4.6.48)$$

This means that the complex tangent space $T_{0,\mathbb{C}}\mathcal{M} = T_0\mathcal{M}$ coincides with the real one, and this suffices to guarantee that $\mathcal{M}$ is Kähler. The holomorphic coherent states are then given explicitly by

$$\|z\rangle = e^{za^\dagger}|0\rangle = e^{za^\dagger}e^{-\bar{z}a}|0\rangle = e^{\frac{1}{2}|z|^2}|z\rangle. \qquad (4.6.49)$$

They cannot be normalized, since the map $z \mapsto \langle w\|z\rangle$ must be holomorphic and hence unbounded. Rather, $\|z\rangle$ should be viewed as holomorphic section of the line bundle $\tilde{\mathcal{B}}$.

Further perspectives

It is easy to see that $\mathbb{R}^{2n}_\theta$ is a solution of the massless model (4.0.2) with $\Box X^a = 0$. Adding fluctuations $X^a \to X^a + \mathcal{A}^a(X)$ to this background[16], one obtains noncommutative Yang–Mills theory on $\mathbb{R}^{2n}_\theta$ [11], where the covariant derivatives on $\mathbb{R}^{2n}_\theta$ arise via $[X^a + \mathcal{A}^a, .] = iD^a$. This will be discussed in more detail in Section 6.6.

In the context of the IKKT model as discussed in Section 11.1, this solution describes D-branes embedded in 10-dimensional target space. We will see that the interactions of these D-branes can be computed within the IKKT model, and the results are consistent with more traditional calculations within string theory. Moreover, the gauge theory that arises on the brane due to open strings ending on the brane (cf. Section 6.7) can be described as noncommutative gauge theory on $\mathbb{R}^{2n}_\theta$. These and other observations led to the conjecture that the IKKT model may provide a nonperturbative definition of IIB string theory, at least on some backgrounds.

From a physics perspective, one issue with such noncommutative gauge theories is that $\theta^{\mu\nu}$ explicitly breaks Lorentz or Euclidean invariance. This issue is resolved on *covariant quantum spaces* as discussed in Chapter 5, where $\theta^{\mu\nu}$ is replaced by a bundle of different $\{\theta^{\mu\nu}\}$ over the base space, which is effectively averaged. Then the geometry acquires a nontrivial internal structure, which restores invariance at least partially. However, the fluctuation modes then involve harmonics on this internal space, leading to a higher-spin gauge theory as discussed in Chapter 10.

---

[16] These fluctuating noncommutative coordinates are aptly called "covariant coordinates" in NC field theory.

The Moyal–Weyl quantum plane has several established applications in physics. The prime application is of course quantum mechanics, where $\mathbb{R}^{2n}_\theta$ is the quantized phase space of a point particle. $\mathbb{R}^{2n}_\theta$ also arises as effective configuration space of the lowest Landau level for a charged particle in a magnetic field, in the limit $B \to \infty$. This will be discussed briefly in Section 4.8.

Exercise 4.6.1   Verify the relations (4.6.44) and (4.6.47) for the tangential space of $\mathbb{R}^2_\theta$.

4.6.3 *Quantized coadjoint orbits revisited

Now consider a compact semi-simple Lie group G with Lie algebra $\mathfrak{g}$. For any irreducible representation $\mathcal{H}_\Lambda$ with highest weight $\Lambda = (n_1, \ldots, n_k)$ labeled by Dynkin labels n_j, consider the matrix configuration

$$X^a = T^a, \qquad a = 1, \ldots, D = \dim(\mathfrak{g}). \tag{4.6.50}$$

Here T^a are orthogonal generators of $\mathfrak{g} \cong \mathbb{R}^D$ acting on $\mathcal{H}_\Lambda$. The corresponding displacement Hamiltonian can be written as

$$H_x = C^2(\mathfrak{g}) + \frac{1}{2}x_a x^a - x_a T^a, \tag{4.6.51}$$

where $C^2(\mathfrak{g}) = \frac{1}{2}T^a T^g \delta_{ab} \propto \mathbb{1}$ is the quadratic Casimir, in an ON basis of $\mathfrak{g}$. Using G-invariance, we can assume that x is in the dual of the Cartan subalgebra $\mathfrak{g}_0$ with dominant weight, i.e. x is in the fundamental Weyl chamber. Then $x_a T^a \equiv H_x \in \mathfrak{g}_0$ satisfies $H_x|\lambda\rangle = (x, \lambda)_\kappa|\lambda\rangle$, where $(.,.)_\kappa$ is the Killing form and $|\lambda\rangle$ has weight λ. Therefore, H_x is already diagonal on the weight basis $|\lambda_k\rangle$ of the Hilbert space $\mathcal{H}_\Lambda$. These weights λ_k are integral and can be written as $\lambda_k = \Lambda - \sum k_i \alpha_i$ for $k_i \in \mathbb{N}$, where Λ is the highest weight and α_i are the simple roots. Any x in the fundamental Weyl chamber can be written as $x = \sum x_k \Lambda_k$ with $x_k \geq 0$. Then,

$$H_x|\lambda_k\rangle = (x, \lambda_k)_\kappa|\lambda_k\rangle = \left((x, \Lambda) - \sum k_i(x, \alpha_i)\right)|\lambda_k\rangle. \tag{4.6.52}$$

Since $(\Lambda_j, \alpha_i)_\kappa = \delta_{ij}$, the maximal eigenvalue is (x, Λ), which is achieved for the highest weight state $\lambda_k = \Lambda$. It may happen that this eigenspace is degenerate; this is the case if $\sum k_i(x, \alpha_i) = 0$, which means that x is on the wall of the fundamental Weyl chamber. These points belong to the singular set $\mathcal{K}$, which we exclude. The quasi-coherent state for all other dominant weights $x \in \tilde{\mathbb{R}}^D$ is hence given by the highest weight state $|\Lambda\rangle$

$$|x\rangle = |\Lambda\rangle. \tag{4.6.53}$$

This in turn means that the quasi-coherent states form the group orbit

$$\mathcal{M} = G \cdot |\Lambda\rangle / U(1) \cong G/K \cong \mathcal{O}[\Lambda] \tag{4.6.54}$$

of the highest weight state $|\Lambda\rangle$ with stabilizer K. This is a quantum Kähler manifold due to the highest weight property, and it is isomorphic to the coadjoint orbit $\mathcal{O}[\Lambda]$ through Λ due to (E.0.2). The quantum metric g_{ab} (4.2.19) and the symplectic form ω (4.2.21) are the canonical group-invariant structures on the Kähler manifold $\mathcal{M}$. For large Dynkin indices $n_j \geq n \gg 1$, the almost-local operators comprise all polynomials in X^a up to some order $O(\sqrt{n})$, so that $\mathcal{M}$ is an almost-local quantum space. In particular, we can define a quantization map $\mathcal{Q}$ via the general formula (4.4.1)

$$\mathcal{Q}: \quad \mathcal{C}(\mathcal{M}) \to \mathrm{End}(\mathcal{H})$$

$$\phi(x) \mapsto \int_{\mathcal{M}} \phi(x)\,|x\rangle\langle x|. \qquad (4.6.55)$$

By construction, $\mathcal{Q}$ is a G-intertwiner, although it is not an exact isometry.[17] This map has an approximate inverse given by the symbol map (4.2.3)

$$\mathrm{End}(\mathcal{H}) \to \mathcal{C}(\mathcal{M})$$

$$\Phi \mapsto \langle x|\Phi|x\rangle =: \boldsymbol{\phi}(x). \qquad (4.6.56)$$

To see that this is an approximate inverse, it suffices to take the symbol map of $\mathcal{Q}$:

$$\langle y|\mathcal{Q}(\phi)|y\rangle = \int_{\mathcal{M}} \phi(x)\,|\langle y|x\rangle|^2 \approx c\phi(y) \qquad (4.6.57)$$

since $|\langle y|x\rangle|^2 \approx c\delta(x, y)$ approximates the delta function for large representations with large Dynkin labels. We have thus established (3.3.89), and recovered the classical orbit from its quantization. In particular, the (quasi-) coherent states coincide with the standard coherent states introduced in [155].

If some of the Dynkin labels n_j are small, then $\mathcal{M}$ can be viewed as a bundle over some $\mathcal{M}_0$ defined by $n_j = 0$, with a fiber which is "very fuzzy." For an application of such a structure see e.g. Section 4.2 in [173].

This construction generalizes to highest weight (discrete series) unitary irreducible representation of non-compact semi-simple Lie groups. A particularly interesting example is given by the "short" series of unitary irreps of $SO(4, 2)$ known as doubletons, which lead to the fuzzy four-hyperboloid H_n^4, and to quantum spaces that can be viewed as cosmological spacetime. This will be discussed in Sections 5.2 and 5.4.

Minimal fuzzy $\mathbb{C}P_N^{N-1}$ as a quantum Kähler manifold

As an example we consider minimal fuzzy $\mathbb{C}P_N^{N-1}$, which is obtained using the general construction for $G = SU(N)$ and its fundamental representation $\mathcal{H} = (1, 0, \ldots, 0)$, so that $G/K \cong \mathbb{C}P^{N-1}$. This is the quantum Kähler manifold obtained from the matrix configuration

$$X^a = \lambda^a \quad \in \mathrm{End}(\mathcal{H}), \qquad \mathcal{H} = \mathbb{C}^N \qquad (4.6.58)$$

for $a = 1, \ldots, N^2 - 1$, where λ^a are a (Gell-Mann) ON basis of $\mathfrak{su}(N)$ in the fundamental representation. Then $\mathrm{End}(\mathcal{H}) \cong (0, \ldots, 0) \oplus (1, 0, \ldots, 0, 1)$ can be viewed as a minimal quantization of functions on $\mathbb{C}P^{N-1}$. The quantization map

$$\mathcal{Q}(\phi) = \int_{\mathbb{C}P^{N-1}} |x\rangle\langle x|\phi(x) \qquad (4.6.59)$$

---

[17] The normalization can be adjusted by hand if desired.

is then the partial inverse of the symbol map, apart from the constant function:

$$\mathcal{Q}(\langle x|\Phi|x\rangle) = c\Phi \qquad \text{if } \text{Tr}(\Phi) = 0 \tag{4.6.60}$$

for some $c > 0$. Near $|\Lambda\rangle = |1, 0 \ldots 0\rangle$, the quasi-coherent states $|x\rangle$ can be organized as holomorphic sections

$$\|z\rangle = \exp(z^k T_k^+)|\Lambda\rangle, \tag{4.6.61}$$

where the T_k^+, $k = 1, \ldots, N - 1$ are the rising operators of a Chevalley basis of $\mathfrak{su}(N)$. Hence, fuzzy $\mathbb{C}P_N^{N-1}$ is a quantum Kähler manifold that coincides with $\mathbb{C}P^{N-1}$, with Kähler form

$$\omega_{\bar{k}l} = \frac{\partial}{\partial \bar{z}^k} \langle z\| \frac{\partial}{\partial z^l} \|z\rangle. \tag{4.6.62}$$

The generalization to generic fuzzy $\mathbb{C}P_N^{n-1}$ should be obvious, where the highest weight state $|\Lambda\rangle = |N, 0 \ldots 0\rangle$ generates the orbit of (quasi-) coherent states.

The fuzzy torus T_N^2 revisited

The fuzzy torus can similarly be viewed as a matrix configuration defined in terms of the four hermitian matrices (3.4.20). In principle one could now compute the quasi-coherent states; however, these turn out to be difficult to describe analytically. An explicit computation for $N = 2$ reveals that the abstract quantum space $\mathcal{M} \cong S^1$, which is clearly not a Kähler manifold and not even symplectic [189]. Therefore, the minimal fuzzy torus T_2^2 is an example of a more exotic quantum space, which is of course not almost-local. Nevertheless for large N, the fuzzy torus T_N^2 is a well-behaved quantum space.

4.7 Further structures and properties

Matrix Laplacian and effective metric

A matrix configuration X^a for $a = 1, \ldots, D$ determines further structure. We have discussed in detail the displacement Hamiltonian H_x, which defines the quasi-coherent states. Furthermore, one can define a matrix analog of the Laplacian as follows:

$$\Box = \delta_{ab}[X^a, [X^b, .]]: \quad \text{End}(\mathcal{H}) \to \text{End}(\mathcal{H}). \tag{4.7.1}$$

We will see in Section 7 that this is not just some ad-hoc definition, but it arises naturally in the framework of matrix models, where it governs the kinematics and the propagation of fluctuations around the given matrix configuration.

Displacement Dirac operator and spinorial quantum space

The present framework has a natural extension to spinors and Dirac-type operators. For any matrix configuration X^a, $a = 1, \ldots, D$ we can consider the following *displacement Dirac operator* [29, 55, 116, 167]

$$\slashed{D}_x = \Gamma_a(X^a - x^a), \qquad x^a \in \mathbb{R}^D \tag{4.7.2}$$

acting on $\mathcal{H} \otimes \mathbb{C}^s$. Here Γ_a generate the Clifford algebra of $SO(D)$ on the irreducible representation $\mathbb{C}^s$. This $\slashed{D}_x$ arises as off-diagonal part of the matrix Dirac operator $\slashed{D} = \Gamma_a[\hat{X}^a, .]$ in Yang–Mills matrix models such as the IIB or IKKT model, for the matrix configuration extended by a point brane

$$\hat{X}^a = \begin{pmatrix} X^a & 0 \\ 0 & x^a \mathbb{1} \end{pmatrix}. \tag{4.7.3}$$

The off-diagonal sector describes a fermionic string stretched between the brane and the point x^a. Quite remarkably, numerical investigations suggest that the Dirac operator $\slashed{D}_x$ generally has a nice manifold of exact zero modes

$$\slashed{D}_x |x, s\rangle = 0 \tag{4.7.4}$$

with the appropriate dimension, so that there is no need to introduce the lowest eigenvalue function $\lambda(x)$. This can be justified rigorously for two-dimensional branes [29], as well as for quantum Käher manifolds. Heuristic arguments supporting this hypothesis can be given in more general cases, see [29, 55, 116, 167] for further work in this direction. This suggests to define a spinorial abstract quantum space

$$\mathcal{M}_\psi = \left\{ \bigcup_x U(1)|x, s\rangle \right\} / U(1) \quad \hookrightarrow \quad \mathbb{C}P^{sN-1} \tag{4.7.5}$$

as the set of exact zero modes of $\slashed{D}_x$, replacing $\mathcal{M}$. Upon tracing over the spinorial Hilbert space, this would lead to an exact completeness relation replacing (4.2.36). Such an approach would be in many ways nicer than the present one in terms of quasi-coherent states, and deserves more detailed study. However, the more elementary quasi-coherent states as defined in this book will suffice for our purpose.

4.7.1 Dimension, oxidation, and reduction

For generic quantum spaces without any intrinsic symmetries or special structures, the abstract quantum space $\mathcal{M}$ often has the full dimension of target space, even though it is "essentially" a lower-dimensional manifold. To understand this, consider some matrix configuration $\bar{X}^a \in \text{End}(\mathcal{H})$ with abstract quantum space $\bar{\mathcal{M}}$, such as a fuzzy sphere to be specific. Now consider some deformation of it given by the matrix configuration

$$X^a = \bar{X}^a + \mathcal{A}^a(\bar{X}), \tag{4.7.6}$$

where $\mathcal{A}^a(\bar{X}) \in \text{End}(\mathcal{H})$ is some function of the $\bar{X}^a$. As long as the $\mathcal{A}^a$ are sufficiently mild, this deformed matrix configuration should clearly be interpreted as a deformed embedding of the same underlying $\bar{\mathcal{M}}$ in target space $\mathbb{R}^D$

$$x^a = \bar{x}^a + \mathcal{A}^a(\bar{x}): \quad \bar{\mathcal{M}} \hookrightarrow \mathbb{R}^D. \tag{4.7.7}$$

However, the abstract quantum space $\mathcal{M}$ defined by X^a may have larger dimension than $\bar{\mathcal{M}}$; this can be interpreted as "oxidation," as $\bar{\mathcal{M}}$ grows some "thickness" in transverse directions. This is an undesirable and spurious effect, which is not easily recognized in terms of quasi-coherent states. It would be desirable to extract the underlying "reduced" symplectic of $\bar{\mathcal{M}}$, which is stable under deformations due to the rigidity of symplectic spaces.

For many theoretical considerations, notable in the context of emergent gravity as discussed in later chapters, the underlying irreducible $\bar{\mathcal{M}}$ is known explicitly and given by some quantized coadjoint orbit. Then the coherent states defined through $\bar{\mathcal{M}}$ can be used for further computations. If the underlying irreducible space $\bar{\mathcal{M}}$ is not known, various methods to remove the oxidation are conceivable, such as looking for minima of the lowest eigenvalue function $\lambda(x)$, or identifying a hierarchy of the eigenvalues of ω and g, cf. [70, 167]. Perhaps the best way is to work with the spinorial quantum space $\mathcal{M}_\psi$ (4.7.5). In any case, efficient tools and algorithms to numerically "measure" and determine the underlying quantum space corresponding to some generic matrix configuration would be desirable.

4.7.2 Band-diagonal matrices

We have seen that for almost-commutative matrix configurations, it is possible to extract some effective semi-classical geometry using quasi-coherent states. The characterization "almost-commutative" is closely related to another class of nice matrices, namely (almost) band-diagonal or almost-diagonal matrices. This means that there is a preferred basis $|i\rangle$ of the Hilbert space $\mathcal{H}$, such that all matrices X^a are dominated by elements at or near the diagonal. This property is easily seen to hold for all the standard quantum spaces discussed so far, such as $\mathbb{R}^{2n}_\theta$ and quantized coadjoint orbits. The basic idea is always to diagonalize the matrices as much as possible, which is essential to recover approximately some commutative space.

More specifically, we can choose one matrix, most naturally the time-like matrix X^0, and diagonalize it. If the eigenvalues are nondegenerate we can arrange them in increasing order, thus defining a preferred basis $|i\rangle$ of $\mathcal{H}$ where

$$X^0 = \mathrm{diag}(\xi_1, \xi_2, \ldots), \qquad \xi_1 < \xi_2 < \ldots. \tag{4.7.8}$$

In this basis, one expects that all other matrices X^a are not quite but "almost" diagonal,

$$X^a \sim \begin{pmatrix} * & * & 0 & 0 & 0 & 0 & \ldots \\ * & * & * & 0 & 0 & 0 & \ldots \\ 0 & * & * & * & 0 & 0 & \ldots \\ 0 & 0 & * & * & * & 0 & \ldots \\ 0 & 0 & 0 & * & * & * & \ldots \\ \vdots & & & & & \ddots \end{pmatrix}, \tag{4.7.9}$$

where the matrix elements $(X^a)^i_j$ strongly decay with $|i - j|$. Such a structure clearly amounts to the matrices being almost-commutative. This is indeed typically observed in numerical simulations of matrix models, which can be taken as an indication that there is an underlying geometric space $\mathcal{M}$ such that $X^a \sim x^a : \mathcal{M} \hookrightarrow \mathbb{R}^D$ as in (4.0.3).

Moreover, the resulting structure of smaller matrix blocks can be used to obtain a crude idea of the embedding [118, 147]. However, such a procedure does not properly capture the geometry. For example, the spectra of the 3 matrices defining a fuzzy sphere do not lie on a sphere but fill a three-dimensional volume. If we diagonalize X^3 and consider the block-matrices corresponding to some interval of the spectrum of X^3, then the eigenvalues of the block-matrices corresponding to X^1 and X^2 describe a square around the origin. This procedure also does not work for $\mathbb{R}^2_\theta$. The proper picture is provided by the approach based on quasi-coherent states, leading to the abstract quantum space $\mathcal{M}$.

4.8 Landau levels and quantum spaces

There is a simple physical system that realizes the mathematical structure of a quantum space. Consider the Lagrangian of a system of a charged particle in two dimensions, in the presence of a constant perpendicular magnetic field B:

$$\mathcal{L} = \frac{m}{2}(\dot{x}^\mu)^2 + \frac{e}{c}\dot{x}^\mu A_\mu. \tag{4.8.1}$$

The classical particle clearly describes a circular motion. Choosing the gauge $A_\mu = -\frac{B}{2}\epsilon_{\mu\nu}x^\nu$, this becomes

$$\mathcal{L} = \frac{m}{2}(\dot{x}^\mu)^2 - \frac{eB}{2c}\dot{x}^\mu \epsilon_{\mu\nu}x^\nu. \tag{4.8.2}$$

The canonical momentum is therefore

$$p_\mu = \frac{\partial \mathcal{L}}{\partial \dot{x}^\mu} = m\dot{x}_\mu + \frac{e}{c}A_\mu \tag{4.8.3}$$

so that the Hamiltonian is given by

$$H = \dot{x}^\mu p_\mu - \mathcal{L} = \frac{m}{2}(\dot{x}^\mu)^2 = \frac{1}{2m}(p_\mu - \frac{e}{c}A_\mu)^2. \tag{4.8.4}$$

In the limit $m \to 0$ or equivalently large B, the Lagrangian reduces to the second term,

$$\mathcal{L} \cong -\frac{eB}{2c}\dot{x}^\mu \epsilon_{\mu\nu}x^\nu \cong -\frac{eB}{c}\dot{x}^1 x^2 \tag{4.8.5}$$

up to a total derivative. Then the equations of motion $\dot{x}^\mu = 0$ state that the particle is basically frozen at some point. Since there is no term quadratic in $\dot{x}$, the effective phase space is then only two-dimensional rather than four-dimensional; for example, the canonical momentum $p_1 = -\frac{eB}{c}x^2$ is related to x^2. Therefore, the canonical pair $(x, p) \cong (x^1, p_1)$ generates the (reduced) phase space, leading to the canonical Poisson brackets $\{x, p\} = 1$ or equivalently [67]

$$\{x^\mu, x^\nu\} = \frac{c}{eB}\epsilon^{\mu\nu}. \tag{4.8.6}$$

Upon quantization, this becomes

$$[X^\mu, X^\nu] = i\frac{\hbar c}{eB}\epsilon^{\mu\nu}, \tag{4.8.7}$$

which incorporates the finite density of states available in the lowest Landau level. Another way to obtain (4.8.7) is to quantize the full problem for $m > 0$ and then consider the projection Π on the lowest Landau level; this is known as Peierls projection. One can then show that the operators

$$\tilde{X}^\mu = \Pi X^\mu \Pi \tag{4.8.8}$$

satisfy the commutation relations (4.8.7) in the limit of vanishing mass.

The bottom line is that charged particles in a strong magnetic field live effectively on a quantized space, which is nothing but the two-dimensional Moyal–Weyl quantum plane. This provides a realization of a quantum space in the context of condensed matter physics. At the second-quantized level, this leads to a noncommutative field theory on $\mathbb{R}^2_\theta$, where the effects of truncation to the lowest Landau level are incorporated by noncommutativity. Such a formalism can be used to describe the integer quantum Hall effect, see e.g. [27, 61, 171] using a different formalism of noncommutative geometry. It was also proposed that the fractional quantum Hall effect can be described by a noncommutative version of the Chern–Simons action [99, 193, 158]. Higher-dimensional analogs of the quantum Hall effect on compact spaces can be similarly described by quantized compact symplectic spaces, cf. [111].

Comments and further reading

The present framework of quantum geometry is largely motivated by matrix models in the context of noncommutative field theory, gauge theory and gravity. The most important tools are the quantization map $\mathcal{Q}$ and the symbol map (4.2.3), which can be defined quite generically using the quasi-coherent states. On almost-local quantum spaces, these two maps are approximate inverse to each other at low energies, and an underlying symplectic space $\mathcal{M}$ can be identified. This provides a correspondence between matrices (operators) in $\mathrm{End}(\mathcal{H})$ and functions on $\mathcal{M}$ as indicated in Table 4.1. Using this dictionary, we can freely switch between the matrix point of view and the semi-classical point of view, which will be used throughout this book.

Coherent states on quantized coadjoint orbits can be defined based on group theory, cf. [155]; they were applied in the context of fuzzy spaces by various authors, including [82]. The idea of quasi-coherent states is motivated from string theory, where the displacement Hamiltonian describes the energy of strings between the brane $\mathcal{M}$ and a point-brane at x [29, 167]. The present approach in terms of quasi-coherent states follows mostly [189], which in turn is based on previous work and ideas including notably [29, 103, 104, 167].

There are of course different approaches to quantum geometry, with other merits and motivations. This is not the place to give an overview of all these approaches. However, one should mention the spectral triple approach formulated by Alain Connes, which is a rigorous mathematical framework built on the idea of *spectral geometry*, rather than quantized Poisson manifolds. Its central concept is a *spectral triple* $(\mathcal{A}, \mathcal{H}, \slashed{D})$, where $\mathcal{A}$ is an algebra of operators on the Hilbert space $\mathcal{H}$, and the Dirac operator $\slashed{D}$ satisfies certain compatibility conditions; for an introduction see [207] and [22] for the relation to fuzzy spaces. The spectral action is then defined as trace over $\mathcal{D}$ with a suitable cutoff, which leads to an induced Einstein–Hilbert term, but it entails an unacceptably large cosmological constant term. That mechanism is known as induced gravity in the physics literature, which goes back to Sakharov [163]. A well-behaved variant of this mechanism will play a central role in Chapter 12, without pathological cosmological constant problem.

The spectral triple approach has also been used to describe the internal structure of the (classical) standard model of particle physics [45]. However, this does not address the issues that arise upon quantization. For efforts toward quantizing spectral triples see [23].

Other approaches to quantum geometry have been developed in the context of quantum groups, as introduced by Drinfel'd–Jimbo [65, 108]; Woronowicz [213]; and Faddeev, Reshetikhin, and Takhtadjan [69]. This led in particular to the idea of noncommutative differential calculi, which has been used in attempts to formulate physical theories [66, 212]. Again, these structural approaches do not address the profound issue of UV/IR mixing encountered in loop calculations of noncommutative field theory, and their success has been mostly restricted to two-dimensional models. We will therefore focus on the matrix model approach in the present book.

5 Covariant quantum spaces

The basic concept of quantum spaces under consideration is based on quantized symplectic spaces. We have argued that this concept is appropriate for generic "almost-commuting" matrix configurations. On the other hand, there is no indication of such a symplectic or Poisson structure on physical spacetime; moreover, any Poisson structure on $\mathbb{R}^{3,1}$ would necessarily break rotation invariance, which is a fundamental property of spacetime. It seems hard to reconcile these seemingly conflicting requirements.

However, there is a class of *covariant quantum spaces* that allows to reconcile the concept of a quantum space with a large symmetry group. By definition, these are matrix configurations X^a that are **vector operators**, i.e. they transform as

$$U_g^{-1} X^a U_g = \Lambda_b^a(g) X^b. \tag{5.0.1}$$

Here Λ_b^a is a representation of some geometric covariance group $G \subset SO(D)$ acting on the indices, and U_g is a representation of G on $\mathcal{H}$. The prototype of a covariant quantum space is of course the fuzzy sphere S_N^2. In this section, we will elaborate in detail some higher-dimensional examples, notably fuzzy S_N^4, fuzzy H_n^4, and a cosmological Friedmann–Lemaitre–Robertson–Walker (FLRW) quantum spacetime $\mathcal{M}^{3,1}$. Moreover, we will see in Section 7.6.2 that their covariance property makes these spaces strong candidates for dynamically stabilized backgrounds in the IKKT model.

The key geometric property of higher-dimensional covariant quantum spaces is that they are quantized bundles over the space(time) $\mathcal{M}$, with local structure

$$\mathcal{B} \overset{\text{loc}}{\cong} \mathcal{M} \times \mathcal{K}, \tag{5.0.2}$$

such that the entire bundle space $\mathcal{B}$ is a (quantized) symplectic manifold, which admits a symmetry $G \times \mathcal{B} \to \mathcal{B}$ compatible with the bundle projection. This induces an extended symmetry of the base manifold, while $\mathcal{M}$ is not required to be symplectic. In other words, they are quantized symplectic equivariant bundles $\mathcal{B}$ over $\mathcal{M}$. The symplectic structure on $\mathcal{B}$ averages out upon projection to $\mathcal{M}$. Due to the nontrivial action of G, the harmonics on $\mathcal{K}$ lead to higher-spin modes, in contrast to standard Kaluza–Klein compactifications.

5.1 The fuzzy four-sphere S_N^4

The fuzzy four-sphere is defined in terms of a certain matrix configuration given by five hermitian matrices X^a, $a = 1, \ldots, 5$ acting on a finite-dimensional Hilbert space $\mathcal{H}$, satisfying the constraint

$$X^a X^b \delta_{ab} = R^2 \, \mathbb{1} \tag{5.1.1}$$

for some $R \in \mathbb{R}$. These matrices should transform as vectors under $SO(5)$, in analogy to the fuzzy two-sphere where the matrices transform as vectors under $SO(3)$. Hence there should be generators $M^{ab} = -M^{ba}$ for $a, b \in \{1, \dots, 5\}$ that provide a suitable representation of $\mathfrak{so}(5)$ on $\mathcal{H}$, and satisfy

$$[M^{ab}, X^c] = i(\delta^{ac} X^b - \delta^{bc} X^a),$$
$$[M^{ab}, M^{cd}] = i(\delta^{ac} M^{bd} - \delta^{ad} M^{bc} - \delta^{bc} M^{ad} + \delta^{bd} M^{ac}). \tag{5.1.2}$$

Indices are raised and lowered with $g_{ab} = \delta_{ab}$ here. The relations (5.1.1) and (5.1.2) constitute a *covariant quantum four-sphere*, and the commutator of the X^a will be denoted by

$$[X^a, X^b] := i\Theta^{ab}. \tag{5.1.3}$$

It is not hard to find explicit representations of these commutation relations, by observing that (5.1.2) is precisely the definition of $\mathfrak{so}(6)$: For any given representation $\mathcal{H}$ of $\mathfrak{so}(6)$, we can define

$$X^a = r M^{a6}, \qquad a = 1, \dots, 5, \tag{5.1.4}$$

where r is a scale parameter of dimension length. Then the $\mathfrak{so}(6)$ algebra implies (5.1.2), and moreover

$$\Theta^{ab} = r^2 M^{ab}. \tag{5.1.5}$$

However, such a realization will typically not satisfy the radial constraint (5.1.1). Remarkably, there is a series of "minimal irreps" of $\mathfrak{so}(6)$ where this is indeed the case. These are the irreps $\mathcal{H}_\Lambda$ with highest weight $\Lambda = (N, 0, 0)$ or $\Lambda = (0, 0, N)$ in Dynkin label notation,[1] which can be denoted by a Young diagram consisting of one horizontal row with N boxes, e.g. $\mathcal{H}_{(0,0,4)} \cong \;\square\square\square\square$. The crucial property of the $(N, 0, 0)$ and $(0, 0, N)$ irreps is that they remain irreducible if viewed as representations of $\mathfrak{so}(5) \subset \mathfrak{so}(6)$. To see this, we observe that

$$(0, 1)_{\mathfrak{so}(5)} \cong \mathbb{C}^4 \cong (0, 0, 1)_{\mathfrak{so}(6)} \tag{5.1.6}$$

is the spinor representation of either $\mathfrak{so}(5)$ or $\mathfrak{so}(6)$. Then, $\mathcal{H}_\Lambda$ is obtained as the totally symmetric tensor product of these representations,

$$\mathcal{H}_\Lambda = (0, 0, N) \cong (0, 0, 1)^{\otimes_S N} \cong (\mathbb{C}^4)^{\otimes_S N} \cong (0, 1)^{\otimes_S N}_{\mathfrak{so}(5)} \cong (0, N)_{\mathfrak{so}(5)}. \tag{5.1.7}$$

Irreducibility under $\mathfrak{so}(5)$ follows e.g. using Weyl's dimension formula, which states that

$$\dim \mathcal{H}_\Lambda = \frac{1}{6}(N + 1)(N + 2)(N + 3) \tag{5.1.8}$$

---

[1] It is useful to recall here the isomorphism of Lie algebras $\mathfrak{so}(6) \cong \mathfrak{su}(4)$. This implies that the $(0, 0, 1)$ irrep can be viewed either as four-dimensional spinor representation of $\mathfrak{so}(6)$, or as fundamental representation of $\mathfrak{su}(4)$; similarly for the dual irrep $(1, 0, 0)$.

for both $(0, N)_{\mathfrak{so}(5)}$ and $(0, 0, N)_{\mathfrak{so}(6)}$; see Section B.1 for more details. Since $X^a X^b \delta_{ab}$ commutes with $\mathfrak{so}(5)$, (5.1.1) follows by Schur's Lemma. Hence the radial constraint expresses the fact that $\mathcal{H}_\Lambda$ remains irreducible as representation of $\mathfrak{so}(5) \subset \mathfrak{so}(6)$.

The 5 matrices X^a arising from the representation $\mathcal{H}_\Lambda$ (or its dual) constitute the definition of the fuzzy four-sphere, denoted as S_N^4. In particular, we observe that the underlying Hilbert space $\mathcal{H}_\Lambda$ is the same as for fuzzy $\mathbb{C}P_N^3$ (3.3.73), using $\mathfrak{so}(6) \cong \mathfrak{su}(4)$. This means that the space of functions or modes $\mathrm{End}(\mathcal{H}_\Lambda)$ on S_N^4 is the same as for $\mathbb{C}P_N^3$; only the interpretation of the generators is different. This will be understood in more detail later.

The crucial feature of S_N^4 and similar covariant quantum spaces is that the classical isometry group $SO(5)$ is fully realized. This is in marked contrast to the basic quantum spaces such as the Moyal–Weyl quantum plane $\mathbb{R}_\theta^4$, where the Poisson tensor Θ^{ab} breaks the symmetry. The price to pay is that the algebra of "coordinates" X^a does not close, but involves the extra generators Θ^{ab}. Nevertheless, one can define physical theories on such spaces via matrix models, leading to fully covariant higher spin theories with large symmetry groups.

Oscillator construction

Fuzzy S_N^4 can be realized explicitly by an oscillator construction as follows: Consider four bosonic oscillators

$$[Z^\beta, Z_\alpha^\dagger] = \delta_\alpha^\beta, \qquad \alpha, \beta = 1, \dots, 4, \tag{5.1.9}$$

which transform in the spinorial representation of $\mathfrak{so}(5) \subset \mathfrak{so}(6)$. Then define

$$X^c = \frac{r}{2} Z^\dagger \gamma^c Z, \qquad M^{ab} = Z^\dagger \Sigma^{ab} Z \tag{5.1.10}$$

(suppressing spinorial indices), where γ^c are the gamma matrices associated to $SO(5)$ acting on $\mathbb{C}^4$ that satisfy the Clifford algebra

$$\gamma^a \gamma^b + \gamma^b \gamma^a = 2\delta^{ab} \mathbb{1}, \tag{5.1.11}$$

and

$$\Sigma^{ab} = \frac{1}{4i}[\gamma^a, \gamma^b], \qquad a, b = 1, \dots, 5 \tag{5.1.12}$$

are the spinorial $\mathfrak{so}(5)$ generators acting on $\mathbb{C}^4$ which satisfy

$$[\Sigma^{ab}, \gamma^c] = i(\delta^{ac}\gamma^b - \delta^{bc}\gamma^a). \tag{5.1.13}$$

It is then straightforward to see that (5.1.2) holds; for example,

$$\begin{aligned}
[M^{ab}, X^c] &= \frac{r}{2}[Z_\alpha^\dagger (\Sigma^{ab})^\alpha_{\ \beta} Z^\beta, Z_\delta^\dagger (\gamma^c)^\delta_{\ \kappa} Z^\kappa] \\
&= \frac{r}{2} Z_\alpha^\dagger (\Sigma^{ab})^\alpha_{\ \beta} [Z^\beta, Z_\delta^\dagger](\gamma^c)^\delta_{\ \kappa} Z^\kappa + \frac{r}{2} Z_\delta^\dagger (\gamma^c)^\delta_{\ \kappa} [Z_\alpha^\dagger, Z^\kappa](\Sigma^{ab})^\alpha_{\ \beta} Z^\beta \\
&= \frac{r}{2} Z_\alpha^\dagger [\Sigma^{ab}, \gamma^c]^\alpha_{\ \kappa} Z^\kappa \\
&= i(\delta^{ac} X^b - \delta^{bc} X^a)
\end{aligned} \tag{5.1.14}$$

and similarly

$$[X^a, X^b] = \frac{r^2}{4} Z_\alpha^\dagger [\gamma^a, \gamma^b]^\alpha_\kappa Z^\kappa = ir^2 M^{ab} \tag{5.1.15}$$

as it must be. In particular, $r^{-1}X^a \equiv M^{a6}$ together with M^{ab} define an irreducible representation of $\mathfrak{so}(6) \cong \mathfrak{su}(4)$, which acts on the N-particle Fock space

$$\mathcal{H}_N := \{ \underbrace{Z^+ \ldots Z^+}_{N} |0\rangle \} \cong (0, 0, N)_{\mathfrak{so}(6)} \tag{5.1.16}$$

identifying Z_α^+ as $(0, 0, 1) \cong \mathbb{C}^4$. This construction is equivalent to the oscillator construction of $\mathbb{C}P_N^3$ in Section 3.3.2, and the only difference between $\mathbb{C}P_N^3$ and S_N^4 is the choice of a specific set of generators that define the corresponding matrix configuration. We can now compute the radius as

$$\begin{aligned}
X^a X_a &= \frac{r^2}{4} Z_\alpha^\dagger (\gamma^b)^\alpha_\beta Z^\beta Z_\delta^\dagger (\gamma_b)^\delta_\kappa Z^\kappa \\
&= \frac{r^2}{4} Z_\alpha^\dagger (\gamma^b)^\alpha_\beta ([Z^\beta, Z_\delta^\dagger] + Z_\delta^\dagger Z^\beta)(\gamma_b)^\delta_\kappa Z^\kappa \\
&= \frac{r^2}{4} Z_\alpha^\dagger (\gamma^b)^\alpha_\beta (\gamma_b)^\beta_\kappa Z^\kappa + \frac{r^2}{4} Z_\alpha^\dagger Z_\delta^\dagger (\gamma^b)^\alpha_\beta (\gamma_b)^\delta_\kappa Z^\beta Z^\kappa \\
&= \frac{r^2}{4} Z_\alpha^\dagger (\gamma^b)^\alpha_\beta (\gamma_b)^\beta_\kappa Z^\kappa + \frac{r^2}{8} Z_\alpha^\dagger Z_\delta^\dagger (Z^\alpha Z^\delta + Z^\delta Z^\alpha) \\
&= \frac{r^2}{4} \hat{N}(\hat{N} + 4) \tag{5.1.17}
\end{aligned}$$

using the identity (B.2.8) for gamma matrices. Here $\hat{N} = Z_\alpha^\dagger Z^\alpha$ is the number operator, which satisfies

$$[\hat{N}, Z_\alpha] = -Z_\alpha. \tag{5.1.18}$$

We conclude that the radius of S_N^4 is given by

$$R_N^2 = X^a X_a = \frac{r^2}{4} \hat{N}(\hat{N} + 4). \tag{5.1.19}$$

We will see in the following that the oscillator construction (5.1.10) reduces in the semi-classical limit to the well-known Hopf map $S^7 \to S^4$ (5.1.41).

Constraints and $\mathfrak{so}(6)$ structure

The algebraic properties of S_N^4 become more transparent by observing its hidden $\mathfrak{so}(6)$ structure. Consider $\mathbb{C}^4$ as spinorial representation of $\mathfrak{so}(6)$, which is given in terms of the following generators Σ^{ab}, $a, b = 1, \ldots, 6$

$$\Sigma^{ab} = \frac{1}{4i} [\gamma^a, \gamma^b], \qquad \Sigma^{a6} = -\frac{1}{2} \gamma^a \tag{5.1.20}$$

for $a, b = 1, \ldots, 5$. Here γ^a are the $\mathfrak{so}(5)$ Gamma matrices (B.2.4), which are hermitian. Then (5.1.10) can be subsumed as oscillator realization of $\mathfrak{so}(6)$ acting on $\mathcal{H}_N$:

$$M^{ab} = Z^\dagger \Sigma^{ab} Z, \qquad a, b = 1, .., 6. \tag{5.1.21}$$

Recalling $\mathfrak{so}(6) \cong \mathfrak{su}(4)$ and comparing with the construction of fuzzy $\mathbb{C}P_N^3$ in Section 3.3.2, we conclude that S_N^4 is nothing but fuzzy $\mathbb{C}P_N^3$ with a specific choice of generators, which define the matrix configuration X^a under consideration.

Using this insight, we can derive further constraints satisfied by the generators. Consider the following symmetric $\mathfrak{so}(6)$ tensor operator:

$$\mathcal{T}^{bc} := \frac{1}{2} \sum_{a,a'=1}^{6} (M^{ab} M^{a'c} + M^{a'c} M^{ab}) \delta_{aa'} = \mathcal{T}^{cb}, \qquad \in \mathrm{End}(\mathcal{H}_N) \tag{5.1.22}$$

which transforms in the symmetric part of

$$(6) \otimes (6) = (0, 1, 0) \otimes (0, 1, 0) \cong (0, 2, 0)_S \oplus (1, 0, 1)_{AS} \oplus (0, 0, 0)_S \tag{5.1.23}$$

(cf. Appendix B.1), noting that the indices a, b transform as $(6) = (0, 1, 0)$. Since the decomposition (3.3.77) of $\mathrm{End}(\mathcal{H}_N)$ does not contain any $(0, 2, 0)$ modes, it follows that $\mathcal{T}^{ab} \sim \delta^{ab}$. The normalization is obtained by computing the trace

$$\mathcal{T}^{ab} \delta_{ab} = 2C^2[\mathfrak{so}(6)] = \frac{3}{2} N(N + 4). \tag{5.1.24}$$

Therefore,

$$\mathcal{T}^{ab} = \frac{1}{3} C^2[\mathfrak{so}(6)] \delta^{ab} = R_N^2 \delta^{ab}. \tag{5.1.25}$$

Another way to derive this is to use the characteristic equation (3.3.80) for $\mathbb{C}P_N^3$. Rewriting this relation in terms of $\mathfrak{so}(5)$ operators, we obtain

$$\frac{1}{2} \sum_{a,a'=1}^{5} (M^{ab} M^{a'c} + M^{a'c} M^{ab}) \delta_{aa'} = R_N^2 \delta^{bc} - \frac{1}{2} (X^b X^c + X^c X^b). \tag{5.1.26}$$

There is a number of further identities for S_N^4, which are summarized here for convenience:

$$X^a X_a = R^2 = r^2 R_N^2 \mathbb{1}, \qquad R_N^2 = \frac{1}{4} N(N + 4),$$

$$[X_a, \Theta^{ab}]_+ = 0,$$

$$\frac{1}{2} [\Theta^{ab}, \Theta^{a'c}]_+ \delta_{aa'} = r^2 R^2 \Big(\delta^{bc} - \frac{1}{2R^2} [X^b, X^c]_+ \Big),$$

$$\epsilon_{abcde} \Theta^{ab} X^c = (N + 2) r \Theta_{de}$$

$$\epsilon_{abcde} \Theta^{ab} \Theta^{cd} = 4(N + 2) r^3 X_e, \tag{5.1.27}$$

for indices $a, b, \ldots = 1, \ldots, 5$, where $[\cdot, \cdot]_+$ denotes the anti-commutator. They can be proved using the oscillator construction in terms of suitable identities for the gamma matrices, which will be spelled out explicitly for the analogous case of H_n^4 in the next section. Alternatively, they can be seen using uniqueness of the various tensor operators;

for example, the last relation follows from the fact that X^a is the only vector operator in End($\mathcal{H}$) (which follows from (5.1.30) together with (5.1.32)), and the normalization on the rhs can be obtained via an explicit computation using the oscillator construction. Finally, it is easy to verify

$$\square X^a = 4r^2 X^a, \tag{5.1.28}$$

where

$$\square = [X^a, [X_a, .]] \tag{5.1.29}$$

is the Matrix Laplacian. This will imply that the matrix configuration X^a is a solution of Yang–Mills matrix models (7.0.1) with mass term.

Harmonics and Casimir operator

The space of functions End($\mathcal{H}_N$) on S_N^4 or $\mathbb{C}P^3$ decomposes into $\mathfrak{so}(6) \cong \mathfrak{su}(4)$ harmonics as follows (3.3.77):

$$\text{End}(\mathcal{H}_N) = (0,0,N) \otimes (N,0,0) = \bigoplus_{n=0}^{N}(n,0,n). \tag{5.1.30}$$

This is a truncation of the classical algebra of (polynomial) functions on $\mathbb{C}P^3$,

$$L^2(\mathbb{C}P^3) = \bigoplus_{n=0}^{\infty}(n,0,n). \tag{5.1.31}$$

Each of these $(n,0,n)$ decompose into the $SO(5)$ harmonics as follows:

$$(n,0,n) = \bigoplus_{s=0}^{n}(n-s,2s). \tag{5.1.32}$$

The $(n,0)$ modes correspond to the (totally symmetrized traceless) polynomial functions $P_n(X_i)$ on S^4 of degree n. The $(n-s,2s)$ modes with $s \neq 0$ have a nontrivial dependence along the S^2 fiber, i.e. they transform nontrivially under the local stabilizer group $SU(2)_L \times SU(2)_R$ at fixed $p \in S^4$, hence they correspond to higher spin modes. Thus, all irreps of $SO(5)$ arise precisely once in End($\mathcal{H}_N$). A geometric understanding of this decomposition will be obtained later, in terms of functions on S^4 taking values in harmonics Y_m^s on the internal S^2 fiber.

The decomposition into $SO(5)$ harmonics provides in particular a map that maps any function on $\mathbb{C}P_N^3$ to a spin 0 function on S_N^4, defined by

$$[.]_0: \quad \text{End}(\mathcal{H}_N) \to \text{End}(\mathcal{H}_N) \tag{5.1.33}$$

$$(n,s) \mapsto \begin{cases} (n,0), & s=0 \\ 0 & s \neq 0. \end{cases} \tag{5.1.34}$$

This is an $SO(5)$ intertwiner, which amounts to integrating over the local fiber S^2.

It is instructive to work out the Casimir operators of $\mathfrak{so}(5)$ and $\mathfrak{so}(6)$ for this decomposition:

$$C^2[\mathfrak{so}(6)] = \sum_{a<b\leq 6} M_{ab}M_{ab},$$

$$C^2[\mathfrak{so}(5)] = \sum_{a<b\leq 5} M_{ab}M_{ab}. \qquad (5.1.35)$$

Their eigenvalues are obtained using the standard formula for semi-simple Lie algebras $C^2[\mathfrak{g}] = \langle \Lambda, \Lambda + 2\rho \rangle = n_i \langle \Lambda_i, \Lambda_j \rangle (n_j + 2)_j$, where $\rho = \sum_i \Lambda_i$ is the Weyl vector, and n_i are the Dynkin labels $\Lambda = \sum n_i \Lambda_i$ of the highest weight. In the present case, this gives (cf. Appendix B.1)

$$C^2[\mathfrak{so}(6)](n, 0, n) = 2n(n + 3) \qquad (5.1.36)$$

$$C^2[\mathfrak{so}(5)](n - s, 2s) = n(n + 3) + s(s + 1). \qquad (5.1.37)$$

Using the definition (5.1.4), we can express the matrix model Laplacian as

$$\Box = [X_i, [X_i, .]] = r^2\big(C^2[\mathfrak{so}(6)] - C^2[\mathfrak{so}(5)]\big) \qquad (5.1.38)$$

$$\Box\big|_{(n-s,2s)} = r^2\big(n(n + 3) - s(s + 1)\big), \qquad s \leq n, \qquad (5.1.39)$$

This is manifestly positive except for the trivial mode $(0, 0) \sim \mathbb{1}$. Furthermore, the following "spin Casimir"

$$\mathcal{S}^2 := C^2[\mathfrak{so}(5)] - r^{-2}\Box, \qquad \mathcal{S}^2\big|_{(n-s,2s)} = 2s(s + 1) \qquad (5.1.40)$$

measures the spin s of the internal harmonics.

In Section 5.2, we will develop a more explicit characterization of the space of harmonics on fuzzy H_n^4 in terms of totally symmetric traceless tangential tensor fields, which can be literally adapted to S_N^4. The present approach in terms of polynomial functions only works on S_N^4, since polynomials are integrable only on compact spaces.

5.1.1 Semiclassical limit and bundle structure

Hopf map and bundle structure

A more intuitive understanding of the geometry underlying S_N^4 is obtained in the semi-classical limit, where the algebra $\mathrm{End}(\mathcal{H})$ becomes commutative, and commutators reduce to Poisson brackets. Then the oscillator construction (5.1.10) reduces to the well-known Hopf map

$$x^a : \quad S^7 \to S^4$$

$$z \mapsto \frac{r}{2} z^\dagger \gamma^a z, \qquad (5.1.41)$$

which generalizes (3.3.50). Here $(z^\alpha) \in \mathbb{C}^4$ is assumed to be normalized to one, and therefore describes a point on $S^7 \subset \mathbb{C}^4 \cong \mathbb{R}^8$. Since the overall phase drops out, this can be viewed as a map

$$\mathbb{C}P^3 \cong S^7/U(1) \to S^4 \subset \mathbb{R}^5. \qquad (5.1.42)$$

To understand the fiber of this map, consider a reference point $x^a = (0, 0, 0, 0, R) \in S^4$. Recalling the explicit gamma matrices (B.2.4), this point on S^4 clearly arises from $z^\alpha = (1, 0, 0, 0)$, and also from any other $z^\alpha = (z^1, z^2, 0, 0)$ with $|z^1|^2 + |z^2|^2 = 1$. Upon factoring out the $U(1)$ phase, it follows that the fiber over any point is a two-sphere $S^2 = S^3/U(1)$, recalling the basic Hopf map (3.3.50). Another way to see this is to note that the stabilizer of γ^5 is $SU(2)_L \times SU(2)_R$, where $SU(2)_L$ acting on $z^\alpha = (1, 0, 0, 0)$ sweeps out the S^2 fiber, while $SU(2)_R$ acts trivially on $\mathbb{C}P^3$. Therefore $\mathbb{C}P^3$ is an S^2 bundle over S^4,

$$\mathbb{C}P^3 \cong S^2 \,\tilde\times\, S^4. \tag{5.1.43}$$

Here $\tilde\times$ indicates the local bundle structure, and that the action of $SO(5)$ (or rather of its universal covering group) on $\mathbb{C}P^3$ induces via (5.1.41) an action on the base manifold S^4. In view of definition 2.5, this means that $\mathbb{C}P^3$ is an $SO(5)$ – *equivariant* equivariant bundle[2] over S^4.

The above observation is characteristic for all covariant quantum spaces under consideration. It implies that harmonics on the internal S^2 fiber transform nontrivially under the local stabilizer on the base. Therefore, these harmonics behave like higher-spin modes rather than Kaluza–Klein. This will lead to higher-spin theories arising on covariant quantum spaces within Yang–Mills matrix models.

Coadjoint orbit point of view

To better understand the geometrical meaning of Θ^{ab}, it is useful to view the fuzzy sphere as quantization of the six-dimensional coadjoint orbit $\mathbb{C}P^3$ of $SO(6)$. We recall the general construction from Section 3.3.3: For any given (finite-dimensional) irrep $\mathcal{H}_\Lambda$ of $SO(6)$ with highest weight Λ, the generators $M^{ab} \in \mathrm{End}(\mathcal{H}_\Lambda)$ of its Lie algebra $\mathfrak{so}(6)$ can be viewed as quantized embedding functions

$$M^{ab} \sim m^{ab}: \quad \mathcal{O}_\Lambda \hookrightarrow \mathbb{R}^{15} \cong \mathfrak{so}(6) \tag{5.1.44}$$

of the homogeneous space (i.e. the coadjoint[3] orbit)

$$\mathcal{O}_\Lambda = \{g \cdot \Lambda \cdot g^{-1}; \, g \in SO(6)\} \cong SO(6)/\mathcal{K} \subset \mathbb{R}^{15}, \tag{5.1.45}$$

with $\mathcal{K}$ denoting the stabilizer of Λ in $\mathfrak{so}(6)$. One can identify Λ with a Cartan generator of $\mathfrak{so}(6)$ via

$$\Lambda \in \mathfrak{g}_0^* \leftrightarrow H_\Lambda \in \mathfrak{g}_0, \tag{5.1.46}$$

where $\mathfrak{g}_0$ denotes the Cartan subalgebra of $\mathfrak{so}(6)$. For $\Lambda = (N, 0, 0)$, this gives

$$\mathcal{O}_\Lambda \cong SU(4)/S(U(3) \times U(1)) \cong \mathbb{C}P^3, \tag{5.1.47}$$

which is a S^2-bundle over S^4 via the Hopf map (5.1.42)

$$x^a = r \, m^{a6}: \quad \mathbb{C}P^3 \to S^4 \hookrightarrow \mathbb{R}^5. \tag{5.1.48}$$

---

[2] More precisely, the equivariance group is the universal covering group of $SO(5)$ which acts on $\mathbb{C}^4$.

[3] For simplicity we identify the Lie algebra with its dual.

This in turn is a special case of the Jordan–Schwinger construction

$$m^{ab}: \quad \mathbb{C}P^3 \mapsto \mathbb{R}^{15} \cong \mathfrak{so}(6)$$
$$(z^{\alpha}) \mapsto m^{ab} = z^* \Sigma^{ab} z, \tag{5.1.49}$$

which is the semi-classical version of the oscillator construction (5.1.10). We will denote the resulting $SO(5)$-equivariant bundle with $S^4 \cong \mathbb{C}P^3$. Defining

$$\theta^{ab} = r^2 m^{ab}, \tag{5.1.50}$$

one obtains the following semi-classical analogs of (5.1.27):

$$x^a x_a = R^2,$$
$$x_a \theta^{ab} = 0,$$
$$\theta^{ab} \theta^{a'c} \delta_{aa'} = \frac{L_{NC}^4}{4} P^{bc},$$
$$\epsilon_{abcde} \theta^{ab} x^c = 2R\theta_{de}$$
$$\epsilon_{abcde} \theta^{ab} \theta^{cd} = \frac{2}{R} L_{NC}^4 x_e, \tag{5.1.51}$$

for $a, b = 1, \ldots, 5$. Here

$$L_{NC}^2 = 2rR \tag{5.1.52}$$

is the scale of noncommutativity, and

$$P^{ab} = \delta^{ab} - \frac{1}{R^2} x^a x^b \tag{5.1.53}$$

is the tangential projector to $S^4 \subset \mathbb{R}^5$ which satisfies $P^{bc} x_c = x^b$ and $P^2 = P$. In particular, θ^{ab} is recognized as antisymmetric tangential self-dual tensor field on S^4.

Poisson structure

As any coadjoint orbit, $\mathcal{O}_\Lambda$ carries a (Kirillov–Kostant) Poisson structure,

$$\{\theta^{ab}, \theta^{cd}\} = r^2 \left(\delta^{ac} \theta^{bd} - \delta^{ad} \theta^{bc} - \delta^{bc} \theta^{ad} + \delta^{bd} \theta^{ac} \right),$$
$$\{\theta^{ab}, x^c\} = r^2 \left(\delta^{ac} x^b - \delta^{bc} x^a \right),$$
$$\{x^a, x^c\} = \theta^{ac} \tag{5.1.54}$$

for $a, b, c, d = 1, \ldots, 5$, which is invariant under $SO(5)$, and in fact under $SO(6)$. This can also be obtained from the Jordan–Schwinger construction (5.1.49) with Poisson structure $\{z^\beta, z_\alpha^\dagger\} = -i\delta_\alpha^\beta$, cf. Section 2.1. In particular, we note

$$\{\theta^{ab}, x_b\} = 4r^2 x^a, \tag{5.1.55}$$

which are the equations of motion of the Poisson-sigma model (7.1.7), which is a semi-classical version of a Yang–Mills matrix model. Consequently, S^4 generated by x^a is a solution thereof.

For an arbitrary point $p \in S^4$ such as the "north pole" $p = R(0, 0, 0, 0, 1)$, we can decompose $\mathfrak{so}(5)$ into rotation generators $M^{\mu\nu}$ ($\mu, \nu = 1, \ldots, 4$) and translation generators P^μ, given by

$$P^\mu = \frac{1}{R\theta}\theta^{\mu 5}, \qquad \mu = 1, \ldots, 4. \tag{5.1.56}$$

Although the latter vanish as classical functions due to $P^\mu \propto \{x^\mu, x^a\}x_a = 0$, they do not vanish as generators in the Poisson algebra, and hence cannot be dropped.

Coherent states

The present construction in terms of $\mathcal{O}_\Lambda$ provides a natural definition of coherent states in terms of the $SO(6)$ orbit of the highest weight state $|\Lambda\rangle \in \mathcal{H}_\Lambda$:

$$|x\rangle \equiv |x; \xi\rangle := g_x \cdot |\Lambda\rangle, \qquad g_x \in SO(6). \tag{5.1.57}$$

Here we labeled the points on $\mathcal{O}_\Lambda \cong \mathbb{C}P^3$ by $x \in S^4$ and the fiber coordinate $\xi \in S^2$, where the north pole p corresponds to the highest weight state $|\Lambda\rangle$. It is easy to verify using the oscillator construction (5.1.10) that

$$\langle\Lambda|X^a|\Lambda\rangle = \frac{Nr}{2}\delta^{a,5}, \tag{5.1.58}$$

which implies

$$\langle x|X^a|x\rangle = \frac{N}{\sqrt{N(N+4)}}x^a, \tag{5.1.59}$$

where $|x| = R^2 = r^2 R_N^2$. Hence the expectation values of the coherent states sweep out a sphere whose radius is slightly smaller than the radius of S_N^4. Coherent states are optimally localized, i.e. they minimize the uncertainty in position space. Using the radial constraint, their uncertainty is computed as

$$\Delta^2 := \sum_{a=1}^{5}\langle(X^a - \langle X^a\rangle)^2\rangle = \sum_{a=1}^{5}\langle(X^a)^2\rangle - (\langle X^a\rangle)^2$$
$$= r^2\left(R_N^2 - \frac{N^2}{4}\right) \approx \frac{4}{N}R^2 \approx 2rR =: L_{\text{NC}}^2, \tag{5.1.60}$$

which is the scale of noncommutativity L_{NC} introduced in (5.1.52).

Length scales

There are three distinct length scales which arise on fuzzy S_N^4: A large (IR) scale R is clearly set by the radius of the sphere. A shorter scale is set by

$$L_{\text{NC}} = \frac{2R}{\sqrt{N}} = \sqrt{2rR}, \tag{5.1.61}$$

which is the scale where noncommutative effects become relevant. This is manifest e.g. in terms of a star product, which would have the form

$$f \star g = f \cdot g + O\left((L_{NC} \cdot \partial)^2(f, g)\right). \tag{5.1.62}$$

The third scale is the UV cutoff scale $L_{UV} = \frac{R}{N}$ for fluctuation modes on S_N^4. These three scales arise similarly on fuzzy S^2, or on any other fuzzy space defined in terms of finite-dimensional matrices.

5.2 The fuzzy hyperboloid H_n^4

The fuzzy hyperboloid H_n^4 is a noncompact analog to the fuzzy four-sphere S_N^4. Accordingly, the space of (square-integrable) functions is now infinite-dimensional, and the associated group theory is more subtle. We will therefore develop a more intuitive approach based on tensor fields, which is better adapted to field-theoretical consideration.

We start with some irreducible unitary representation $\mathcal{H}$ of Lie algebra $\mathfrak{so}(4, 2)$:

$$[M^{ab}, M^{cd}] = i(\eta^{ac} M^{bd} - \eta^{ad} M^{bc} - \eta^{bc} M^{ad} + \eta^{bd} M^{ac}). \tag{5.2.1}$$

Indices will be raised and lowered with $\eta_{ab} = \text{diag}(-1, 1, 1, 1, 1, -1)$. Then define the following hermitian generators:

$$X^a := r M^{a5}, \qquad \Theta^{ab} := -i[X^a, X^b] = -r^2 M^{ab}, \qquad a = 0, \dots, 4 \tag{5.2.2}$$

where r is a scale parameter of dimension length. Note the minus in the last relation, which is different from (5.1.5) due to $\eta_{55} = -1$. It follows that

$$\Box X^a = -4r^2 X^a. \tag{5.2.3}$$

where

$$\Box = [X^a, [X_a, .]] = \Box_H \tag{5.2.4}$$

is the Matrix Laplacian. This will imply that the matrix configuration X^a is a solution of Yang–Mills matrix models (7.0.1) with mass term. In order to describe a hyperboloid $H^4 \subset \mathbb{R}^{4,1}$, these generators should satisfy the $SO(4, 1)$ invariant constraint

$$X^a X^b \eta_{ab} = -X_0^2 + \sum_{i=1}^{4} X_i^2 = -R^2 \, \mathbb{1}. \tag{5.2.5}$$

As in the case of S_N^4, this holds only for very special representations of $\mathfrak{so}(4, 2)$ that remain irreducible upon reduction to $\mathfrak{so}(4, 1)$. These representations are known as doubleton representations $\mathcal{H}_n$ labeled by $n \in \mathbb{N}$, and they are most easily constructed using an oscillator construction as described in the next section. Fuzzy H_n^4 is then defined through the algebra of functions $\text{End}(\mathcal{H}_n)$ generated by the corresponding X^a.

Oscillator construction

The doubleton representations $\mathcal{H}_n$ can be obtained from an oscillator construction as follows. Consider bosonic creation and annihilation operators a_i, b_j that satisfy

$$[a_i, a_j^\dagger] = \delta_i^j, \qquad [b_i, b_j^\dagger] = \delta_i^j \qquad \text{for } i, j = 1, 2. \tag{5.2.6}$$

Using the a_i, b_j, we form spinorial operators

$$Z := \begin{pmatrix} a_1^\dagger \\ a_2^\dagger \\ b_1 \\ b_2 \end{pmatrix} = Z^\alpha \tag{5.2.7}$$

with Dirac conjugates

$$\bar{Z} \equiv Z^\dagger \gamma^0 = \left(-a_1, -a_2, b_1^\dagger, b_2^\dagger \right). \tag{5.2.8}$$

Here $\gamma^0 = \begin{pmatrix} -\mathbb{1} & 0 \\ 0 & \mathbb{1} \end{pmatrix}$ is a gamma matrix associated to $SO(4, 1)$ as given in (B.3.7). They satisfy

$$[Z^\alpha, \bar{Z}_\beta] = \delta_\beta^\alpha, \tag{5.2.9}$$

and it is an easy exercise based on the Clifford algebra to see that the

$$M^{ab} := \bar{Z} \Sigma^{ab} Z = \left(-a_1, -a_2, b_1^\dagger, b_2^\dagger \right) \Sigma^{ab} \begin{pmatrix} a_1^\dagger \\ a_2^\dagger \\ b_1 \\ b_2 \end{pmatrix} \tag{5.2.10}$$

satisfy the $\mathfrak{so}(4, 2)$ commutation relations (5.2.1). The conventions of the Σ^{ab} and γ^a matrices are detailed in Appendix B.3. The M^{ab} commute with the $SO(4, 2)$- invariant number operator

$$\hat{N} = \bar{Z}Z = -N_a + N_b - 2, \tag{5.2.11}$$

where $N_a = a_i^\dagger a_i$ and $N_b = b_j^\dagger b_j$. The generators of the $SU(2)_L$ and $SU(2)_R$ subgroups are explicitly

$$L_i^k := a_k^\dagger a_i - \frac{1}{2} \delta_i^k N_a$$

$$R_j^i := b_i^\dagger b_j - \frac{1}{2} \delta_j^i N_b,$$

while the non-compact generators $\mathcal{L}^+$, $\mathcal{L}^-$ are given by linear combinations of quadratic terms of the form $a_i^\dagger b_j^\dagger$ and $a_i b_j$, respectively. Explicitly,

$$X^a = \frac{r}{2} \bar{Z} \gamma^a Z = (X^a)^\dagger,$$

$$M^{ab} = \bar{Z} \Sigma^{ab} Z = (M^{ab})^\dagger \tag{5.2.12}$$

using (B.3.9) and (B.3.8) for $a, b = 0, \ldots, 4$. In particular, the time-like generator X^0 is given by

$$r^{-1} X^0 = M^{05} = \bar{Z} \Sigma^{05} Z = \frac{1}{2} Z^\dagger Z = \frac{1}{2}(N_a + N_b + 2) \equiv E,\qquad(5.2.13)$$

which is clearly positive definite.

Minireps

We now consider the Fock space

$$\mathcal{F} = \oplus a^\dagger \ldots b^\dagger |0\rangle \qquad(5.2.14)$$

spanned by all polynomials in the creation operators $a_i^\dagger, b_j^\dagger$ acting on the Fock vacuum $a_i |0\rangle = 0 = b_i |0\rangle$. This is clearly a unitary representation of $\mathfrak{so}(4,2)$ as explained in the previous section; however, it is reducible. To obtain irreducible representations, we note that this Fock space contains the following special states

$$|\Omega_n\rangle := \left| E, 0, \frac{n}{2} \right\rangle := b_{i_1}^\dagger \ldots b_{i_n}^\dagger |0\rangle, \qquad E = 1 + \frac{n}{2}, j_L = 0, j_R = \frac{n}{2}$$

$$|\Omega_{-n}\rangle := \left| E, \frac{n}{2}, 0 \right\rangle := a_{i_1}^\dagger \ldots a_{i_n}^\dagger |0\rangle, \qquad E = 1 + \frac{n}{2}, j_L = \frac{n}{2}, j_R = 0,$$

which are lowest weight states of $\mathfrak{so}(4,2)$ with the given weight, i.e. they are anihilated by all lowering operators $\mathcal{L}^-$ that have form $a_i b_j$ (B.3.4),

$$\mathcal{L}^- |\Omega_n\rangle = 0 = a_i b_j |\Omega_n\rangle. \qquad(5.2.15)$$

This implies that they generate lowest-weight representations, which are easily seen to be irreducible. Acting with all positive root operators $\mathcal{L}^+$ of the form $a_i^\dagger b_j^\dagger$ on $|\Omega_n\rangle$, one obtains positive energy discrete series unitary irreps of $SU(2,2)$ with lowest weight $\Lambda_n = \left(E, \frac{n}{2}, 0 \right)$ and $\Lambda_{-n} = \left(E, 0, \frac{n}{2} \right)$, which are called *doubleton* or minireps, denoted by $\mathcal{H}_n$. These are known as *minireps* of $\mathfrak{so}(4,2)$, because they are free of multiplicities in weight space.[4] Since the $\mathcal{H}_{-n}$ are related by some Weyl reflection to the $\mathcal{H}_n$, we will restrict ourselves to the $\mathcal{H}_{n>0}$ to be specific. Hence the full Fock space decomposes as

$$\mathcal{F} = \bigoplus_{n \in \mathbb{Z}} \mathcal{H}_n, \qquad \hat{N}|_{\mathcal{H}_n} = (N_b - N_a - 2)\mathbb{1} = (n - 2)\mathbb{1}. \qquad(5.2.16)$$

All this is completely analogous to the case of S_N^4, and more generally to the discussion of $\mathbb{C}P_N^n$ in Section 3.3.2. Moreover, the minireps remain irreducible under the subgroups $SO(4,1)$ as well as $SO(3,2)$ of $SO(4,2)$. This can be seen along the lines of (5.1.19), by verifying that the X^a satisfy the relation of a Euclidean hyperboloid in $\mathbb{R}^{4,1}$

$$\boxed{X_a X^a = -\frac{r^2}{4} \hat{N}(\hat{N} + 4) = -R^2 \, \mathbb{1}} \qquad(5.2.17)$$

---

[4] This can be seen e.g. from the characters given in [98].

(also known as Euclidean Anti–de Sitter space), with quantized radius given by

$$R^2 := \frac{r^2}{4}(n^2 - 4), \qquad n = 0, 1, 2, \ldots. \tag{5.2.18}$$

This justifies the name fuzzy hyperboloid H_n^4. However, there is some extra structure, since the lowest-weight state $\left|E, 0, \frac{n}{2}\right\rangle$ of $\mathcal{H}_n$ generates a $(n+1)$-dimensional irreducible representation of $SU(2)_R$ with fixed (minimal) $X^0 = rE$. As for S_N^4, this is naturally interpreted as fuzzy S_n^2, which will be understood in more detail later. Moreover, the following $SO(4,2)$-invariant constraint holds

$$\sum_{a,b=0,1,2,3,4,5} \eta_{ab} M^{ac} M^{bd} + (c \leftrightarrow d) = \frac{1}{2}(n^2 - 4)\eta^{cd} = 2\frac{R^2}{r^2}\eta^{cd}, \tag{5.2.19}$$

which is the analog of (5.1.25). This implies the first two of the $SO(4,1)$-invariant constraints

$$\eta_{cc'}\Theta^{ac}\Theta^{bc'} + (a \leftrightarrow b) = r^2\left(2R^2\eta^{ab} + \left(X^a X^b + X^b X^a\right)\right) \tag{5.2.20}$$

$$X_b\Theta^{ab} + \Theta^{ab}X_b = 0 \tag{5.2.21}$$

$$\epsilon_{abcde}\Theta^{ab}\Theta^{cd} = 4nr^3 X_e \tag{5.2.22}$$

$$\epsilon_{abcde}\Theta^{ab}X^c = nr\Theta_{de}. \tag{5.2.23}$$

The constraints involving the antisymmetric tensor ϵ_{abcde} can be obtained similarly as for S_N^4. Finally, it is worth mentioning that from a field theory point of view, the minireps $\mathcal{H}_n$ can be interpreted as massless fields on AdS^4, or as conformal fields on $\mathbb{R}^{3,1}$ with helicity $\frac{n}{2}$.

Coherent states

Since $\mathcal{H}_n$ is a lowest weight representation, coherent states are defined naturally in terms of the $SO(4,2)$ or rather $SU(2,2)$ orbit of the lowest weight state $|\Lambda\rangle \in \mathcal{H}_\Lambda$:

$$|x\rangle \equiv |x; \xi\rangle := g_x \cdot |\Lambda\rangle, \qquad g_x \in SU(2,2). \tag{5.2.24}$$

Due to the highest weight property the orbit is 6-dimensional, spanned locally by 4 of the non-compact generators identified with X^a for $a = 1, 2, 3, 4$, together with 2 compact generators. This orbit will be identified in Section 5.2.1 as $\mathcal{O}_\Lambda \cong \mathbb{CP}^{2,1}$ which is a S^2 bundle over H^4, the points on which are labeled by $x \in H^4$ and $\xi \in S^2$. The "south pole" $p = R(1, 0, 0, 0, 0) \in H^4$ now corresponds to the lowest weight state $|\Lambda\rangle$. It is easy to verify using the oscillator construction that

$$\langle\Lambda|X^a|\Lambda\rangle = \frac{nr}{2}\delta^{a,5}, \tag{5.2.25}$$

which implies

$$\langle x|X^a|x\rangle = \frac{n}{\sqrt{n^2 - 4}}x^a, \tag{5.2.26}$$

where $|x| = R^2$. Hence, the expectation values of the coherent states sweep out a hyperboloid whose radius is slightly bigger than the radius of H_n^4.

In particular, we can define the coherent state quantization map (4.4.1) as

$$\mathcal{Q}: \quad \mathcal{C}(\mathbb{C}P^{2,1}) \to \mathrm{End}(\mathcal{H}_n)$$

$$\phi \mapsto \int_{\mathbb{C}P^{2,1}} \phi(x)|x\rangle\langle x|, \tag{5.2.27}$$

where the integral is defined as in (4.2.37). This defines a one-to-one map from $L^2(\mathbb{C}^{2,1})$ to the Hilbert–Schmidt operators $\mathrm{HS}(\mathcal{H}_n) \subset \mathrm{End}(\mathcal{H}_n)$, but it makes sense also for non-integrable functions such as polynomials. Surjectivity to $\mathrm{HS}(\mathcal{H}_n)$ can be shown using the fact that $\mathbb{C}P^{2,1}$ is a Kähler manifold, which implies that the coherent states can be lifted to holomorphic sections as in Theorem 4.3. Moreover, this map allows to identify the fiber over each point in H^4 as fuzzy S^2_{n+1}, which admits only finitely many modes. This follows by noting that the lowest eigenspace of X^0 is given by $\mathbb{C}^{n+1}$ (5.2.13), which leads precisely to the structure of fuzzy S^2_{n+1} acting with $SU(2)_R$ on $|\Lambda\rangle = |\Omega_n\rangle$ (5.2.15). This leads to a truncated tower of higher-spin modes in the decomposition of harmonics on fuzzy H^4_n given in the next paragraph, which will be very important for the gauge theory emerging on such a background in Yang–Mills matrix models.

Harmonics and Casimir operator

As for S^4_N, there is a $SO(4,1)$-invariant Casimir operator for H^4_n which measures the internal spin along the S^2 fiber, given by

$$\mathcal{S}^2 := C^2[\mathfrak{so}(4,1)] + r^{-2}\Box = \sum_{a<b\leq 4} [M^{ab},[M_{ab},\cdot]] + r^{-2}[X_a,[X^a,\cdot]] \tag{5.2.28}$$

cf. (5.1.40). These are compatible by construction,

$$[\mathcal{S}^2,\Box] = 0. \tag{5.2.29}$$

For polynomial functions, it follows from $SO(4,2)$ group theory (in complete analogy to the case of S^4_N) that the polynomial functions in $\mathrm{End}(\mathcal{H}_n)$ decomposes into $SO(4,1)$ modes $\mathcal{C}^s$ labeled by spin s, characterized by the eigenvalue $\mathcal{S}^2 = 2s(s+1)$. However, these are not unitarizable. For unitary modes i.e. square-integrable Hilbert–Schmidt operators, the quantization map $\mathcal{Q}$ (5.2.27) allows to understand the modes in terms of quantized square-integrable functions on $\mathbb{C}P^{2,1}$, which in turn decompose into

$$\boxed{\mathcal{C} := \mathrm{End}(\mathcal{H}_n) = \bigoplus_{s=0}^{n} \mathcal{C}^s, \qquad \mathcal{S}^2|_{\mathcal{C}^s} = 2s(s+1).} \tag{5.2.30}$$

This decomposition[5] is truncated at n, and holds for all modes that can be obtained as image of the coherent state quantization map $\mathcal{Q}$ (see Appendix C.1 in [175] for more details),

---

[5] The $\mathcal{C}^s$ are invariant under symmetrized multiplication with functions on H^4, i.e. $\mathcal{S}^2(\{f,X^a\}_+) = \{\mathcal{S}^2 f, X^a\}_+$ [175].

which includes all Hilbert–Schmidt operators. The intuitive reason is that the structure of the lowest eigenspace of X_0 is that of a fuzzy sphere, based on $\mathbb{C}^{n+1}$.

Minimal H_0^4

It is interesting to consider the minimal fuzzy hyperboloid H_0^4. In that case, the subspace of $\mathcal{H}_0$ with lowest eigenvalue of $r^{-1}X_0 = E = 1$ is one-dimensional, i.e. the internal fuzzy sphere collapses to a point. Then the coherent state quantization map identifies the algebra of operators with functions on H^4,

$$\mathcal{Q}: \quad C(H^4) \to \text{End}(\mathcal{H}_0). \tag{5.2.31}$$

This provides an example of a quantum space that is *not* the quantization of a symplectic space. There is no semi-classical analog, because there exists no $SO(4, 1)$-invariant symplectic structure on H^4.

5.2.1 Semiclassical limit and Poisson structure

Hopf map and bundle structure

Now consider the semi-classical limit of fuzzy H_n^4, which is obtained for large n, indicated by $\sim$. Then $X^a \sim x^a$ and $\Theta^{ab} \sim \theta^{ab}$, and the oscillator construction (5.2.12) reduces to the definition of a non-compact Hopf map

$$x^a: \quad \mathbb{C}P^{2,1} \to H^4 \hookrightarrow \mathbb{R}^{4,1}$$
$$z^\alpha \mapsto \frac{r}{2}\bar{z}\gamma^a z \tag{5.2.32}$$

from $\mathbb{C}P^{2,1}$ to H^4. Here we define

$$\mathbb{C}P^{2,1} := \{z \in \mathbb{C}^4; \ \bar{z}z = 1\}/_{U(1)} \ \cong \ SU(2,2)/_{S(U(2,1)\times U(1))} \tag{5.2.33}$$

to be the $U(1)$ equivalence classes of normalized spinors $z \in \mathbb{C}^4$. This can be viewed as an open sector of $\mathbb{C}P^3 \cong \mathbb{C}^4/_{\mathbb{C}^*}$, defined by $\bar{z}z > 0$.

Comparing the oscillator construction (5.2.10) with the above semi-classical version, it is manifest that for each $\mathcal{H}_n$ with large n, the M_{ab} generators can be interpreted as quantized embedding functions

$$M_{ab} \sim m^{ab}: \quad \mathbb{C}P^{2,1} \mapsto \mathfrak{so}(4,2) \cong \mathbb{R}^{15}$$
$$(z^\alpha) \mapsto m^{ab} = \bar{z}\Sigma^{ab}z. \tag{5.2.34}$$

In particular, $\mathbb{C}P^{2,1}$ is a S^2-bundle over H^4 via the $SO(4, 1)$-equivariant Hopf map

$$x^a = r \, m^{a5}: \quad \mathbb{C}P^{2,1} \to H^4 \hookrightarrow \mathbb{R}^{4,1}. \tag{5.2.35}$$

The constraints implicit in this construction will be elaborated later. The local fiber over each point in H^4 is understood as follows: by homogeneity, we can pick the reference point $p = (R, 0, 0, 0, 0)$ on H^4, and consider the spinor $z = (1, 0, 0, 0)^T$ over p. The base point is clearly invariant under the action of $SU(2)_R$ on the upper two spinor components, which sweeps out a two-sphere S^2 (up to phase). Therefore, $\mathbb{C}P^{2,1}$ is a six-dimensional bundle over H^4

$$\mathbb{C}P^{2,1} \cong S^2 \,\tilde{\times}\, H^4. \tag{5.2.36}$$

Here $\tilde{\times}$ indicates the local bundle structure, and that the isometry group $SO(4,1)$ of the base manifold acts nontrivially on the local fiber, in a way that respects the bundle map. Therefore, $\mathbb{C}P^{2,1}$ is an equivariant bundle over H^4.

Coadjoint orbit point of view

It is again useful to view fuzzy H_n^4 as quantization of the six-dimensional coadjoint orbit $\mathbb{C}P^{2,1}$ of $SO(4,2)$ or $SU(2,2)$, which works as follows. For any given lowest weight[6] irrep $\mathcal{H}_\Lambda$ of $SO(4,2)$ with lowest weight Λ, the generators $M^{ab} \in \text{End}(\mathcal{H}_\Lambda)$ of its Lie algebra $\mathfrak{so}(4,2)$ can be viewed as quantized embedding functions

$$M^{ab} \sim m^{ab}: \quad \mathcal{O}_\Lambda \hookrightarrow \mathbb{R}^{15} \cong \mathfrak{so}(4,2) \tag{5.2.37}$$

of the homogeneous space (coadjoint[7] orbit)

$$\mathcal{O}_\Lambda = \{g \cdot \Lambda \cdot g^{-1}; \; g \in SO(4,2)\} \cong SO(4,2)/\mathcal{K} \subset \mathbb{R}^{15}, \tag{5.2.38}$$

with $\mathcal{K}$ denoting the stabilizer of Λ in $SO(4,2)$. One can identify Λ with a Cartan generator of $\mathfrak{so}(4,2)$ via

$$\Lambda \in \mathfrak{g}_0^* \;\leftrightarrow\; H_\Lambda \in \mathfrak{g}_0, \tag{5.2.39}$$

where $\mathfrak{g}_0$ denotes the Cartan subalgebra of $\mathfrak{so}(4,2) \cong \mathfrak{su}(2,2)$. For $\Lambda = (E,0,0)$ in Dynkin label notation, this gives explicitly

$$\mathcal{O}_\Lambda \cong SU(2,2)/_{S(U(2,1)\times U(1))} \cong \mathbb{C}P^{2,1}, \tag{5.2.40}$$

which is a S^2 bundle over H^4. The classical functions

$$\theta^{ab} = -r^2 m^{ab} \tag{5.2.41}$$

(note the sign difference to (5.1.50)) then satisfy some constraints, which are obtained from those of H_n^4 the previous section in the semi-classical limit:

$$x_a x^a = -R^2, \tag{5.2.42a}$$

$$\theta^{ab} x_b = 0, \tag{5.2.42b}$$

$$\epsilon_{abcde}\theta^{ab}x^c = 2R\theta_{de}, \tag{5.2.42c}$$

$$\epsilon_{abcde}\theta^{ab}\theta^{cd} = 8Rr^2 x_e, \tag{5.2.42d}$$

$$\gamma^{bb'} := \eta_{aa'}\theta^{ab}\theta^{a'b'} = \frac{L_{NC}^4}{4}P^{bb'}. \tag{5.2.42e}$$

---

[6] It is convenient to use lowest weight irreps rather than highest weight irreps, but the two are essentially equivalent. The important point is that $\mathcal{H}_\Lambda$ is spanned by all rising operators acting on $|\Lambda\rangle$.

[7] For simplicity we identify the Lie algebra with its dual.

Here the scale of noncommutativity is

$$L_{NC}^4 := \theta^{ab}\theta_{ab} = 4r^2R^2 \tag{5.2.43}$$

in x^a coordinates, and

$$P^{ab} = \eta^{ab} + \frac{1}{R^2}x^a x^b \quad \text{with} \quad P^{ab}x_b = 0 \quad \text{and} \quad P^{ab}P^{bc} = P^{ac} \tag{5.2.44}$$

is the Euclidean projector on H^4; recall that H^4 is a Euclidean space. This implies that the algebra $\text{End}(\mathcal{H}_n)$ of functions on fuzzy H_n^4 reduces for large n to the algebra $\mathcal{C}$ of functions on $\mathbb{C}P^{2,1}$,

$$\boxed{\text{End}(\mathcal{H}_n) \sim \mathcal{C}(\mathbb{C}P^{2,1}) = \oplus_s \mathcal{C}^s,} \tag{5.2.45}$$

where the higher spin modes $\mathcal{C}^s$ are again characterized by the Casimir $\mathcal{S}^2$ (5.2.28). In physical terms, this is the spin of the local $SO(3)$ acting on the S^2 fiber. In the commutative limit, these $\mathcal{C}^s$ are modules over the algebra $\mathcal{C}^0$ of functions on H^4, which can be represented as

$$\boxed{\mathcal{C} = \bigoplus_{s=0}^{\infty} \mathcal{C}^s \ni \phi_{a_1\ldots a_s;b_1\ldots b_s}^s(x)\,\theta^{a_1b_1}\ldots\theta^{a_sb_s} \equiv \phi_{\underline{\alpha}}^s(x)\,\Xi^{\underline{\alpha}}.} \tag{5.2.46}$$

This means that $\mathcal{C}$ is a bundle over H^4, whose structure is determined by the constraints (5.2.42). In the fuzzy case, the noncommutative space of functions $\mathcal{C}^s$ (5.2.30) is the quantization of such tensor fields $\phi_{\underline{\alpha}}(x)$ taking values in the vector space spanned by $\Xi^{\underline{\alpha}}$,

$$\mathcal{C}^s \ni \mathcal{Q}(\phi_{a_1\ldots a_s;b_1\ldots b_s}(x)\theta^{a_1b_1}\ldots\theta^{a_sb_s}) \equiv \mathcal{Q}(\phi_{\underline{\alpha}}(x)\,\Xi^{\underline{\alpha}}), \quad s \leq n \tag{5.2.47}$$

where $\Xi^{\underline{\alpha}}$ denotes both the polynomials in M^{ab} and m^{ab}. It is worth pointing out that even though only $SO(4,1)$ is manifest here, the quantization map $\mathcal{Q}$ (5.2.27) actually respects $SO(4,2)$, which will be very useful in the following.

Poisson structure

Due to the description as coadjoint orbit, the bundle space $\mathbb{C}P^{2,1}$ is endowed with a $SO(4,2)$-invariant (Kirillov–Kostant) Poisson structure, which is the seed of the quantization map $\mathcal{Q}$. It is given explicitly by

$$\{\theta^{ab},\theta^{cd}\} = -r^2\left(\eta^{ac}\theta^{bd} - \eta^{ad}\theta^{bc} - \eta^{bc}\theta^{ad} + \eta^{bd}\theta^{ac}\right),$$
$$\{\theta^{ab},x^c\} = -r^2\left(\eta^{ac}x^b - \eta^{bc}x^a\right),$$
$$\{x^a,x^c\} = \theta^{ac} \tag{5.2.48}$$

for $a,b,c,d = 0,\ldots,4$, recalling that $\theta^{ab} = -r^2m^{ab}$ (5.2.41); these formulas arise from the Lie algebra $\mathfrak{so}(4,2)$ via (2.2.9). Thus $\mathcal{C}$ acquires the structure of a Poisson algebra.

Integration and measure

As for any quantized coadjoint orbit, the trace on $\mathrm{End}(\mathcal{H})$ corresponds to the integral over the underlying symplectic space:

$$\mathrm{Tr}\mathcal{Q}(\phi) = \int_{\mathbb{C}P^{2,1}} \frac{\Omega}{(2\pi)^3} \phi = \int_{H^4} \Omega_H[\phi]_0. \tag{5.2.49}$$

Here Ω is the $SO(4,2)$-invariant volume form on $\mathbb{C}P^{2,1}$ arising from the (Kirillov–Kostant) symplectic form, and $[.]_0$ denotes the average over the internal S^2 fiber. The $SO(4,1)$-invariant volume form Ω_H on H^4 can be written explicitly as

$$\Omega_H = \frac{1}{x_4} dx_0 \dots dx_3 \tag{5.2.50}$$

using the Cartesian x_μ, $\mu = 0, \dots, 3$ as local coordinates.

Derivatives

It is useful to define the following derivations on the Poisson algebra $\mathcal{C}$

$$\eth^a \phi := -\frac{1}{r^2 R^2} \theta^{ab}\{x_b, \phi\} = \frac{1}{r^2 R^2} x_b \{\theta^{ab}, \phi\}, \qquad \phi \in \mathcal{C}, \tag{5.2.51}$$

which are tangential due to $x^a \eth_a = 0$, satisfy the Leibniz rule, and are $SO(4,1)$-covariant. These tangential derivatives satisfy

$$\eth^a x^c = P^{ac} = \eta^{ab} + \frac{1}{R^2} x^a x^b$$

$$\eth^a \theta^{cd} = \frac{1}{R^2}(-\theta^{ac} x^d + \theta^{ad} x^c)$$

$$\{x^a, \cdot\} = \theta^{ab} \eth_b. \tag{5.2.52}$$

due to (5.2.42). The first identity means that they act as expected on functions in x. Furthermore, the following identities hold

$$\eth^a(\{x_a, \phi\}) = 0 \tag{5.2.53}$$

$$\{x_a, \eth^a \phi\} = 0 \tag{5.2.54}$$

$$\Box \phi = -r^2 R^2 \eth^a \eth_a \phi \tag{5.2.55}$$

$$[\eth_a, \eth_b]\phi = -\frac{1}{r^2 R^2}\{\theta_{ab}, \phi\} \tag{5.2.56}$$

for any $\phi \in \mathcal{C}$. The first identity (5.2.53) is seen as follows:

$$\eth^a(\{x_a, \phi\}) = \frac{1}{r^2 R^2}\theta_{ad}\{x^a, \{x^d, \phi\}\} = \frac{1}{2r^2 R^2}\theta_{ad}\{\theta^{ad}, \phi\} = 0 \tag{5.2.57}$$

using (5.2.43). Similarly, (5.2.54) is seen as follows:

$$\{x_a, \eth^a \phi\} = \frac{1}{r^2 R^2} \{x^a, \theta_{ad}\{x^d, \phi\}\}$$
$$= \{x^a, \theta_{ad}\}\{x^d, \phi\} + \theta_{ad}\{x^a, \{x^d, \phi\}\} = 0 \tag{5.2.58}$$

using $\{x^a, \theta_{ad}\} = 4r^2 x_d$ and $x_a\{x^a, .\} = 0$. The identity (5.2.55) is proved as follows:

$$\Box \phi = -\{x_a, \{x^a, \phi\}\} = -\{x_a, \theta^{ab} \eth_b \phi\}$$
$$= -\{x_a, \theta^{ab}\} \eth_b \phi - \theta^{ab} \{x_a, \eth_b \phi\}$$
$$= r^2 \{M^{ab}, x_a\} \eth_b \phi - \theta^{ab} \theta^{ac} \eth_b \partial_c \phi$$
$$= -r^2 R^2 \eth^a \eth_a \phi \tag{5.2.59}$$

for any $\phi \in \mathcal{C}$. The last identity (5.2.56) is left as an exercise, and can be found in [175]. Finally for scalar functions $f \in \mathcal{C}^0$, we have

$$\{M^{ab}, f\} = -(x^a \eth_b - x^b \eth_a) f$$
$$\{Y, f\} = \{Y, x^a\} \eth_a f \tag{5.2.60}$$

for any $Y \in \mathcal{C}$. The first relation is easily verified for the generators x^c,

$$\{M^{ab}, x^c\} = -(x^a \eth_b - x^b \eth_a) x^c = -(x^a P^{bc} - x^b P^{ac}) = -(x^a \eta^{bc} - x^b \eta^{ac}) \tag{5.2.61}$$

since both sides are derivations. The second relation follows similarly.

Gradation and Poisson brackets

The gradation $\mathcal{C} = \oplus \mathcal{C}^s$ into spin s sectors is not respected by the multiplication in $\mathcal{C}$. Using the standard addition law for angular momenta, it follows that the gradation into $\mathcal{C}^s$ is respected by the algebra in the following weaker sense

$$\mathcal{C}^s \cdot \mathcal{C}^t \subset \mathcal{C}^{|s-t|} \oplus \mathcal{C}^{|s-t|+2} \oplus \ldots \oplus \mathcal{C}^{s+t}. \tag{5.2.62}$$

We will denote the projection on $\mathcal{C}^s$ as follows:

$$[\phi]_s \in \mathcal{C}^s. \tag{5.2.63}$$

Explicitly, (5.2.42) implies the following projection

$$\left[\theta^{ab} \theta^{cd} \right]_0 = \frac{r^2 R^2}{3} \left(P^{ac} P^{bd} - P^{bc} P^{ad} + \varepsilon^{abcde} \frac{1}{R} x^e \right), \tag{5.2.64}$$

where P^{ab} is the projector on H^4 (5.2.44); more general formulas can be found in [188]. The Poisson bracket respects the gradation as follows:

$$\{\mathcal{C}^0, \mathcal{C}^0\} \subset \mathcal{C}^1$$
$$\{\mathcal{C}^0, \mathcal{C}^1\} \subset \mathcal{C}^0 \oplus \mathcal{C}^2$$
$$\{\mathcal{C}^0, \mathcal{C}^k\} \subset \mathcal{C}^{k-1} \oplus \mathcal{C}^{k+1}$$
$$\{\mathcal{C}^s, \mathcal{C}^t\} \subset \mathcal{C}^{|s-t|-1} \oplus \mathcal{C}^{|s-t|} \oplus \ldots \oplus \mathcal{C}^{s+t+1} \tag{5.2.65}$$

noting that $\{\theta^{\mu\nu}, \theta^{\rho\sigma}\} \in \mathcal{C}^1$. These relations can be understood using the derivation property from the case $\{\mathcal{C}^0, \mathcal{C}^1\}$, which we consider explicitly. For $f(x) \in \mathcal{C}^0$ and $\phi^{(1)} \in \mathcal{C}^1$, this reduces to

$$\{f(x), \phi^{(1)}\} = \eth_a f \{x^a, \phi^{(1)}\}. \tag{5.2.66}$$

Hence we only need to work out

$$\{x^a, \mathcal{C}^1\} \subset \mathcal{C}^2 \oplus \mathcal{C}^0 \tag{5.2.67}$$

explicitly, which will play an important role in the following. This is computed as follows: for

$$\phi^{(1)} = \{x^a, \phi_a\} = \theta^{ab} \eth_b \phi_a = \frac{1}{2} F_{ab} \theta^{ab} \quad \in \mathcal{C}^1$$

$$F_{ab} := \eth_a \phi_b - \eth_b \phi_a \tag{5.2.68}$$

(we will see in Section 5.2.2 that every element of $\mathcal{C}^1$ can be written in this way), we have

$$\begin{aligned}
2\{x_a, \phi^{(1)}\} &= \{x_a, F_{bc}\theta^{bc}\} = F_{bc}\{x_a, \theta^{bc}\} + \{x_a, F_{bc}\}\theta^{bc} \\
&= r^2 F_{bc}(\eta_{ab}x^c - \eta_{ca}x^b) + \{x_a, F_{bc}\}\theta^{bc} \\
&= r^2 (F_{ac}x^c - F_{ba}x^b) + \{x_a, F_{bc}\}\theta^{bc} \\
&= 2r^2 \phi_a + \{x_a, F_{bc}\}\theta^{bc} \\
&= 2r^2 \phi_a + \eth_d F_{bc}\theta^{ad}\theta^{bc},
\end{aligned} \tag{5.2.69}$$

using $\theta^{bc} = -r^2 M^{bc}$. The projection of the second term is evaluated in (5.2.77), and we obtain

$$-\{x_a, \phi^{(1)}\}_0 = \frac{1}{3}(\Box - 2r^2)\phi_a. \tag{5.2.70}$$

This will be generalized in Lemma 5.1.

Square-integrable and admissible modes

In order to have finite kinetic energy and inner products for functions $\phi \in \mathcal{C}$, we need to impose some integrability conditions on ϕ. Consider for example

$$0 \leq \int \{x_a, \phi\}\{x^a, \phi\} = \int \phi \Box \phi \tag{5.2.71}$$

using partial integration. The first integral is positive since $\{x^a, \phi\}$ is tangential to H^4 due to $x_a\{x^a, .\} = 0$, so that

$$\Box_H \equiv \Box > 0 \tag{5.2.72}$$

should be positive definite. It is clear that this does not hold e.g. for polynomial functions; rather, this defines a class of integrable functions. The same argument carries over to $\text{End}(\mathcal{H}_n)$ for Hilbert–Schmidt operators. However, we will need a slightly stronger bound, which can be obtained from group theory. A heuristic argument for a bound on the Casimir of such admissible modes is as follows: consider the $SO(4, 1)$ invariant expression

$$- \int \{M_{ab}, \phi\}\{M^{ab}, \phi\} = \int \phi\{M^{ab}, \{M_{ab}, \phi\}\} = -2 \int \phi \, C^2[\mathfrak{so}(4, 1)]\phi \qquad (5.2.73)$$

for $a, b = 0, \ldots, 4$. At the reference point $p = (R, 0, 0, 0, 0) \in H^4$, the sum on the lhs separates as

$$-\{M_{ab}, \phi\}\{M^{ab}, \phi\} \overset{p}{=} 2 \sum_a \{M^{a0}, \phi\}\{M^{a0}, \phi\} - \sum_{a,b=1}^4 \{M^{ab}, \phi\}\{M^{ab}, \phi\}. \qquad (5.2.74)$$

The first term is manifestly positive, while the second term is negative and involves the local stabilizer $SO(4)$ acting on ϕ. Hence the second term measures the spin, and we expect heuristically that it contributes $-2s(s + 1)$, if we forget about curvature corrections for the moment. This would give the bound $-C^2[\mathfrak{so}(4, 1)] \geq -s(s + 1)$ for admissible modes.

However, it turns out that this is still not quite strong enough. On the slightly different background $\mathcal{M}^{3,1}$ with Minkowski signature discussed in Section 5.4, the requirements to obtain a ghost-free field theory (i.e. with positive kinetic term) are discussed in detail in [188]. Essentially this boils down to a class of modes known in representation theory as *principal series* of unitary representations. They describe the normalizable fluctuation modes in the present context, corresponding to a continuous basis for square-integrable wavefunctions on the hyperboloids, analogous to plane waves in the flat case. The (bosonic) principal series of unitary representations $\Pi_{\nu,s}$ of $SO(4, 1)$ are determined by the spin $s \in \mathbb{N}_0$ and the real ("kinetic") parameter $\nu \in \mathbb{R}$. They can be identified with spin s wavefunctions on H^4. For these representations, the quadratic Casimir satisfies the following bound [60]

$$-C^2[\mathfrak{so}(4, 1)] = 9/4 + \nu^2 - s(s + 1) > 9/4 - s(s + 1) \qquad (5.2.75)$$

assuming $\nu \neq 0$. These will be denoted as **admissible modes**. This is clearly a refined version of the above heuristic argument, and it entails via (5.2.28) the following bound for $\Box = \Box_H$

$$r^{-2}\Box\phi^{(s)} = 2s(s + 1) - C^2[\mathfrak{so}(4, 1)] > s^2 + s + 9/4, \qquad (5.2.76)$$

which is slightly stronger than (5.2.72). These are basically lower bounds on the wavelength of integrable fluctuations with scale set by the curvature of the space.

Technical supplements

Consider the following term which arises in (5.2.69). Let $F_{ab} = \eth_a\phi_b - \eth_b\phi_a$. Then

$$\eth_d F_{bc}[\theta^{ad}\theta^{bc}]_0 = \frac{R^2 r^2}{3} \left(\eth_d F_{bc}(P_{ab}P_{cd} - P_{ac}P_{bd} + \frac{x^e}{R}\varepsilon_{adbce}) \right)$$

$$= \frac{2R^2 r^2}{3} \left(P_{ab}\eth^c F_{bc} + \frac{1}{2R}x^e\varepsilon_{adbce}\eth_d F_{bc} \right)$$

$$
\begin{aligned}
&= \frac{2R^2 r^2}{3} \left(P_{ab}\eth^c(\eth_c\phi_b - \eth_b\phi_c) + \frac{1}{R}x^e \varepsilon_{adbce}\eth_d\eth_c\phi_b \right) \\
&= \frac{2R^2 r^2}{3}\,\eth^c\eth_c\phi_b - \frac{1}{3}\left(2P_{ab}\{\theta_{bc},\phi_c\} - \frac{x^e}{R}\varepsilon_{adcbe}\{\theta_{dc},\phi_b\} \right) \\
&= -\frac{2}{3}\left(\Box\phi_a + r^2\phi_a \right)
\end{aligned}
\tag{5.2.77}
$$

using (5.2.56) and $\eth^c\phi_c = 0$, and the following identity (5.2.78) which is based on self-duality (5.2.42c):

$$
\begin{aligned}
P_{ab}\{\theta_{bc},\phi_c\} - \frac{x^e}{2R}\varepsilon_{adcbe}\{\theta_{dc},\phi_b\} &= \frac{1}{2R}P_{aa'}\varepsilon_{a'dcbe}\theta_{dc}\{x^e,\phi_b\} \\
&= \frac{1}{2R}P_{aa'}\varepsilon_{a'dcbe}\theta_{dc}\theta^{ef}\eth_f\phi_b \\
&= -r^2 P_{aa'}(g^{a'f}x^b - g^{bf}x^{a'})\eth_f\phi_b \\
&= -r^2 P_{af}x^b \eth_f\phi_b = r^2\phi_a
\end{aligned}
\tag{5.2.78}
$$

for any divergence-free, tangential $\phi_b \in \mathcal{C}$. Here we used irreducibility, (5.2.42), and the self-duality relations (5.2.42c) and

$$
\varepsilon_{abcde}\theta^{cd}\theta^{ef} = -2r^2 R(\eta_{af}x^b - \eta_{bf}x^a).
\tag{5.2.79}
$$

The structure of the last equation follows from $SO(4,1)$ invariance (see [175] for direct derivation), and the pre-factor are obtained by contracting with η^{af} which reproduces (5.2.42d).

5.2.2 Irreducible tensor fields on H^4

Having established the basic structures, we turn to a more explicit description of the degrees of freedom arising on H_n^4, namely the semi-classical space of modes $\mathcal{C}^s$ (5.2.46) for $s \in \mathbb{N}$. This is somewhat technical, but will be the basis for the physically more interesting space $\mathcal{M}^{3,1}$.

We claim that $\mathcal{C}^s$ can be described explicitly in terms of totally symmetric, traceless, divergence-free tangential rank s tensor fields $\Gamma^{(s)}H^4 \ni \phi_{a_1\ldots a_s}(x)$ on H^4, via the one-to-one maps

$$
\boxed{
\begin{aligned}
\Gamma^{(s)}H^4 &\to \mathcal{C}^s \\
\phi_{a_1\ldots a_s}(x) &\mapsto \phi^{(s)} = \{x^{a_1},\ldots \{x^{a_s},\phi_{a_1\ldots a_s}\}\ldots\} \\
\{x_{a_1},\ldots,\{x_{a_s},\phi^{(s)}\}\ldots\}_0 &\leftarrow \phi^{(s)}
\end{aligned}
}
\tag{5.2.80}
$$

recalling $\{x^a,\mathcal{C}^k\} \in \mathcal{C}^{k-1}\oplus\mathcal{C}^{k+1}$ (5.2.65). Here the subscript 0 indicates projection to $\mathcal{C}^0$. To see this, it is useful to refine these maps as series of maps

$$
\begin{array}{ccccccc}
\mathcal{C}^0 & \to & \mathcal{C}^1 & \to & \cdots & \to & \mathcal{C}^s \\
\phi_{a_1\ldots a_s} & \mapsto & \{x^{a_s},\phi_{a_1\ldots a_s}\} & \mapsto & \cdots & \mapsto & \{x^{a_1},\ldots\{x^{a_s},\phi_{a_1\ldots a_s}\}\ldots\} =: \phi^{(s)}
\end{array}
\tag{5.2.81}
$$

all of which are $SO(4, 1)$ intertwiners, considering the indices as vectors. Conversely, we have the chain of maps

$$
\begin{array}{ccccccc}
\mathcal{C}^0 & \leftarrow & \cdots & \leftarrow & \mathcal{C}^{s-1} & \leftarrow & \mathcal{C}^s
\end{array}
$$
$$
\tilde{\phi}_{a_1\ldots a_s} := \{x_{a_1}, \ldots, \{x_{a_s}, \phi^{(s)}\}\ldots\}_0 \quad \leftarrow \quad \cdots \quad \leftarrow \quad \{x_{a_s}, \phi^{(s)}\}_- \leftarrow \quad \phi^{(s)}, \qquad (5.2.82)
$$

which are inverse to (5.2.81) up to normalization, as we will see. Here and in the following, the subscripts $\pm$ denote the projection of

$$
\{x_{a_s}, \phi^{(s)}\} \in \mathcal{C}^{s+1} \oplus \mathcal{C}^{s-1} \qquad (5.2.83)
$$

onto $\mathcal{C}^{s\pm 1}$, using (5.2.67). All the tensors in this chain are automatically symmetric, traceless, divergence-free and tangential[8].

As a first check of this correspondence, let us count the number of independent dof (i.e. independent scalar functions) on both sides. Since the spin s functions on the internal S_n^2 span a $2s + 1$-dimensional vector space, $\mathcal{C}^s$ comprises $2s + 1$ dof. This is consistent with the characterization in terms of rank s traceless divergence-free vector fields on H^4, which are well-known to comprise $2s + 1$ dof.

Moreover, we observe that symmetry follows from the projection on $\mathcal{C}^0$, since the antisymmetric part of e.g.

$$
\{x_{d_1}, \{x_{d_2}, \phi^{(2)}\}\} - \{x_{d_2}, \{x_{d_1}, \phi^{(2)}\}\} = \{\theta_{d_1 d_2}, \phi^{(2)}\} \qquad (5.2.84)
$$

is in $\mathcal{C}^2$ rather than $\mathcal{C}^0$. An analogous argument applies for the intermediate steps. Divergence-free and tangential follows from the identities (5.2.53) and $x^a\{x_a, .\} = 0$ for the first index, which by symmetry extends to all indices.

Conceptually, the above isomorphism follows from group theory. Irreps of $SO(4, 1)$ are labeled by two Casimirs, the spin s and the quadratic Casimir $C^2[\mathfrak{so}(4, 1)]$. Hence, the totally symmetric traceless, divergence-free and tangential rank s tensor fields are uniquely classified by the quadratic Casimir. Thus fixing s and C^2, the above chains relate isomorphic irreps, as the starting point and the endpoint are irreducible. In particular, the intermediate maps must also be inverse of each other up to normalization, since irreducible rank k tensors with overall spin s can only live in $\mathcal{C}^{s'}$ for $s' \geq s - k$. It only remains to show that the maps do not vanish, which follows using the following result:

Lemma 5.1 *Given an irreducible tangential tensor* $\phi_{a_1\ldots a_{s-1}a} \in \mathcal{C}^0$ *and defining*

$$
\phi_a^{(s)} := \{x^{a_1}, \ldots, \{x^{a_{s-1}}, \phi_{a_1\ldots a_{s-1}a}\}\ldots\} \in \mathcal{C}^{s-1}, \qquad (5.2.85)
$$

we have

$$
-\{x_a, \phi^{(s)}\}_{s-1} = \alpha_s(\Box - 2r^2)\phi_a^{(s)}, \qquad \phi^{(s)} := \{x^a, \phi_a^{(s)}\}. \qquad (5.2.86)
$$

for

$$
\alpha_s = \frac{s}{2s + 1} \qquad (5.2.87)
$$

---

[8] However, this does not suffice to characterize the intermediate objects. For example, not all divergence-free vector fields are of this form, as we will see.

Proof The structure of this relation follows again from group theory: the lhs is an intertwiner on $\phi_a^{(s)}$, and must therefore be proportional to $\mathbb{1}$ on irreps. Since it is a second-order derivation, it must be a combination of $\mathbb{1}$ and C^2 or the matrix Laplacian $\square$. The coefficients α_s can be determined by a somewhat lengthy computation. The most important case $s = 1$ was obtained in (5.2.70). The case $s = 2$ is derived in [174], and the general case in [188]. □

Using this lemma, invertibility of (5.2.81) and (5.2.82) follow by considering the composition

$$\phi \mapsto \{\phi, x^a\}_{s-1} \mapsto -\alpha_s^{-1}(\square - 2r^2(s+1))^{-1}\{\{\phi, x^a\}_{s-1}, x_a\}_s = \phi \qquad (5.2.88)$$

for $\phi \in C^s$ using Lemma 5.5. Hence the maps are invertible provided $\square - 2r^2(s+1)$ is invertible on C^s. This holds indeed for all admissible unitary irreps of $\mathfrak{so}(4, 1)$ and for all finite-dimensional irreps, cf. (5.2.76). Then surjectivity also follows, since C^s encodes one and only one irreducible spin s field on H^4. Injectivity of (5.2.82) also follows from the following easy result:

Lemma 5.2 *If $\{x^a, \phi\} = 0$ for all $a = 1, .., 5$, then $\phi = const.$*

This follows from the Jacobi identity together with $\{x^a, x^b\} \sim M^{ab}$, and the fact that $\mathbb{C}P^{2,1}$ is symplectic.

Young diagrams and $\mathfrak{hs}$ field strength tensors

An alternative organization of C^s in terms of Young tableaux works as follows. Any $\phi^{(s)} \in C^s$ can be written as a polynomial in θ^{ab},

$$\phi^{(s)} = F_{a_1...a_s;b_1...b_s}(x)\,\theta^{a_1 b_1}\ldots\theta^{a_s b_s} \equiv F_{\underline{\beta}}(x)\,\Xi^{\underline{\beta}}. \qquad (5.2.89)$$

The generators $\Xi^{\underline{\beta}}$ form a basis of irreducible totally symmetric polynomials in θ^{ab}, or of Young tableaux with two rows and s columns:

$$\mathfrak{hs} := \bigoplus_s \Xi^s \cong \bigoplus_s \boxed{\quad}. \qquad (5.2.90)$$

We shall denote this vector space with $\mathfrak{hs}$ henceforth, called "higher spin algebra" in a loose sense. However, the coefficients of such a representation are not unique, due to the constraints satisfied by the θ^{ab}. A unique representation of C^s in terms of the field strength associated to $\phi^{(s)}$ is obtained as follows. For $s = 1$, the space C^1 consists divergence-free vector fields

$$x^a \phi_a = 0 = \eth^a \phi_a. \qquad (5.2.91)$$

The field strength of the vector field ϕ_a is obtained as

$$\phi^{(1)} = \{x^a, \phi_a\} = \theta^{ab}\eth_b\phi_a = \frac{1}{2}F_{ab}(x)\theta^{ab} \quad \in C^1,$$
$$F_{ab} = \eth_b\phi_a - \eth_a\phi_b. \qquad (5.2.92)$$

Conversely, the "potential" ϕ_a is recovered from $\phi^{(1)}$ using Lemma 5.1 as

$$\phi_a = -3(\Box - 2r^2)^{-1}\{x_a, \phi^{(1)}\}_0. \tag{5.2.93}$$

More generally for any totally symmetric, traceless, divergence-free rank s tensor fields $\phi_{a_1...a_s}$, we can write

$$\begin{aligned}
\phi^{(s)} &= \{x^{a_1}, \ldots \{x^{a_s}, \phi_{a_1...a_s}\} \ldots\} \\
&= \theta^{a_1 b_1} \eth_{b_1} \ldots \theta^{a_s b_s} \eth_{b_s} \phi_{a_1...a_s} = \theta^{a_1 b_1} \ldots \theta^{a_s b_s} \eth_{b_1} \ldots \eth_{b_s} \phi_{a_1...a_s} \\
&=: \mathcal{R}_{a_1...a_s; b_1...b_s}(x) \, \theta^{a_1 b_1} \ldots \theta^{a_s b_s} \equiv \mathcal{R}_{\underline{\alpha}}(x) \, \Xi^{\underline{\alpha}} \quad \in \mathcal{C}^s.
\end{aligned} \tag{5.2.94}$$

Here $\mathcal{R}_{\underline{\alpha}}(x)$ can be viewed as generalized field strength of the tensor fields $\phi_{a_1...a_s}$. Note that due to the self-duality constraint (5.2.42c), $\phi^{(1)}$ encodes only the self-dual field strength, and similarly $\phi^{(s)}$ encodes a sort of self-dual or chiral higher spin field strength. This could be used to define some type of chiral higher spin theory. Needless to say that the same construction could be applied to S_N^4.

Exercise 5.2.1 Show that the field strength F_{ab} (5.2.92) associated to $\phi_a = \eth_a \phi$ vanishes:

$$\phi^{(1)} = \{x^a, \eth_a \phi\} = 0. \tag{5.2.95}$$

This expresses the irreducibility of $\phi^{(1)}$, and can be interpreted as gauge invariance.

Exercise 5.2.2 Verify the commutation relations (5.2.56) for the derivations $\eth_a$.

5.2.3 $SO(4, 1)$ intertwiners

This section contains a number of technical results which will be used in various places, and in particular to establish the reconstruction problem in Section 5.2.4. We will discuss a number of maps which are compatible with the $SO(4, 1)$ group action. First, recall that the $\mathfrak{so}(4, 1)$ Lie algebra acts on vector-valued modes $\mathcal{A}^a \in \mathbb{R}^5 \otimes \mathcal{C}$ as

$$i\{M_{ab}, \cdot\} + M_{ab}^{(5)} \tag{5.2.96}$$

where

$$(M_{ab}^{(5)})_d^c = i(\delta_b^c \eta_{ad} - \delta_a^c \eta_{bd}) \tag{5.2.97}$$

is the five-dimensional representation acting on the vector index. As a check, we note that x^a transforms trivially,

$$i\{M_{ab}, x^c\} + M_{ab}^{(5)} x^c = i(\delta_a^c x_b - \delta_b^c x_a) + i(\delta_b^c x_a - \delta_a^c x_b) = 0 \tag{5.2.98}$$

and so does θ^{ab}. In particular, the following $SO(4, 1)$ intertwiners will play an important role:

$$\begin{aligned}
\mathcal{A}^{\pm}: \quad & \mathcal{C} \to \mathbb{R}^4 \otimes \mathcal{C} \\
& \phi^{(s)} \mapsto \mathcal{A}_a^{\pm}[\phi^{(s)}] := \{x_a, \phi^{(s)}\}_{\pm}.
\end{aligned} \tag{5.2.99}$$

$\mathcal{I}$ intertwiner

The quadratic Casimir $C^2 \equiv C^2[\mathfrak{so}(4,1)]$ acts on general vector fields $\mathcal{A}^a$ on H^4 as follows:

$$C^2 \mathcal{A}^a = \frac{1}{2}(i\{M_{cd}, \cdot\} + M_{cd}^{(5)})^2 \mathcal{A}^a = \frac{1}{r^2}(-\Box - 2\mathcal{I} + r^2(\mathcal{S}^2 + 4))\mathcal{A}^a \tag{5.2.100}$$

using (5.2.28). Here $C^2 = 4$ on the vector irrep (5), and $\mathcal{I}$ is an $SO(4,1)$ intertwiner defined as follows:

$$\mathcal{I}(\mathcal{A}^a) := \{\theta^{ab}, \mathcal{A}_b\}. \tag{5.2.101}$$

For divergence-free vector fields $\phi_a \in \mathcal{C}^0$, $\mathcal{I}$ is given explicitly as

$$\mathcal{I}(\phi_a) = r^2 \phi_a. \tag{5.2.102}$$

This can be verified using (5.2.60),

$$\{\theta^{ab}, \phi_b\} = -r^2\{M^{ab}, \phi_b\} = r^2(x^a \bar{\partial}_b - x^b \bar{\partial}_a)\phi_b = r^2 \phi_a. \tag{5.2.103}$$

A useful result to evaluate $\mathcal{I}$ more generally is the following:

Lemma 5.3

$$2\mathcal{I}(\{\phi^{(s)}, x_a\}) = r^2\left(\frac{1}{2}\mathcal{S}^2 - s(s+1) + 4\right)\{\phi^{(s)}, x_a\} \tag{5.2.104}$$

for $\phi^{(s)} \in \mathcal{C}^s$. In particular,

$$\begin{aligned} \mathcal{I}(\mathcal{A}^-[\phi^{(s)}]) &= r^2(2-s)\mathcal{A}^-[\phi^{(s)}] \\ \mathcal{I}(\mathcal{A}^+[\phi^{(s)}]) &= r^2(3+s)\mathcal{A}^+[\phi^{(s)}]. \end{aligned} \tag{5.2.105}$$

Proof   Consider for $\mathcal{A}_a := \{x_a, \phi^{(s)}\}$

$$\begin{aligned} \mathcal{I}(\mathcal{A}_a) &= \{\theta^{ab}, \mathcal{A}_b\} = \{\{x^a, x^b\}, \mathcal{A}_b\} \\ &= -\{\{x^b, \mathcal{A}_b\}, x_a\} - \{\{\mathcal{A}_b, x_a\}, x_b\}. \end{aligned} \tag{5.2.106}$$

The last term can be rewritten as

$$\begin{aligned} -\{\{\mathcal{A}_b, x_a\}, x_b\} &= -\{\{\{x_b, \phi\}, x_a\}, x_b\} \\ &= \{\{\{\phi, x_a\}, x_b\}, x_b\} + \{\{\theta^{ab}, \phi\}, x_b\} \\ &= \Box \mathcal{A}_a - \{\{\phi, x_b\}, \theta^{ab}\} - \{\{x_b, \theta^{ab}\}, \phi\} \\ &= (\Box + 4r^2)\mathcal{A}_a - \mathcal{I}(\mathcal{A}_b). \end{aligned} \tag{5.2.107}$$

Combining these and using $\{x^b, \mathcal{A}_b\} = -\Box \phi^{(s)}$ leads to

$$2\mathcal{I}(\mathcal{A}_a) = -\{x^a, \Box \phi^{(s)}\} + (\Box + 4r^2)\mathcal{A}_a. \tag{5.2.108}$$

On the other hand, relating the quadratic Casimir (5.2.100) with the intertwiner $\{x_a, \cdot\}$ gives

$$\begin{aligned} \{x_a, (\Box - 2r^2 s(s+1))\phi^{(s)}\} = \{x_a, (\Box - r^2 \mathcal{S}^2)\phi^{(s)}\} &= -r^2\{x_a, C^2 \phi^{(s)}\} = -r^2 C^2 \mathcal{A}_a \\ &= (\Box + 2\mathcal{I} - r^2(\mathcal{S}^2 + 4))\mathcal{A}_a. \end{aligned} \tag{5.2.109}$$

Together with (5.2.108), this gives

$$2\mathcal{I}(\mathcal{A}_a) = -\{x_a, \Box\phi^{(s)}\} + (\Box + 4r^2)\mathcal{A}_a = (4r^2 - 2\mathcal{I} + r^2(\mathcal{S}^2 + 4) - 2r^2 s(s+1))\mathcal{A}_a \tag{5.2.110}$$

which is (5.2.104). $\Box$

Intertwining properties of the Laplacian $\Box$

We can establish the following intertwiner relations for the Laplacian on H_n^4 [188]:

Lemma 5.4

$$(\Box + 2r^2 s)\mathcal{A}_a^-[\phi^{(s)}] = \mathcal{A}_a^-[\Box\phi^{(s)}],$$
$$(\Box - 2r^2(s+1))\mathcal{A}_a^+[\phi^{(s)}] = \mathcal{A}_a^+[\Box\phi^{(s)}]. \tag{5.2.111}$$

Proof   For any intertwiner $\mathcal{A}_a[.]$, the quadratic Casimir C^2 acts as

$$(\Box + 2\mathcal{I} - r^2(\mathcal{S}^2 + 4))\mathcal{A}^a[\phi^{(s)}] = -r^2 C^2 \mathcal{A}^a[\phi^{(s)}] = -\mathcal{A}^a[r^2 C^2 \phi^{(s)}]$$
$$= \mathcal{A}^a[(\Box - 2r^2 s(s+1))\phi^{(s)}]. \tag{5.2.112}$$

Evaluating the lhs using (5.2.105) and (5.2.30) we obtain (5.2.111).

As a check, consider $\phi = x^c$. Then

$$(\Box - 2r^2)\{x^c, x_a\}_+ = \{\Box x^c, x_a\}_+$$
$$(\Box - 2r^2)\theta^{ca} = -4r^2\theta^{ca}$$
$$\Box\theta^{ca} = -2r^2\theta^{ca}, \tag{5.2.113}$$

which is easily verified. $\Box$

Finally, we have

Lemma 5.5

$$-\{x^a, \{x_a, \phi^{(s)}\}_-\}_+ = \alpha_s(\Box - 2r^2(s+1))\phi^{(s)}, \tag{5.2.114}$$
$$-\{x^a, \{x_a, \phi^{(s)}\}_+\}_- = ((1 - \alpha_s)\Box + 2r^2\alpha_s(s+1))\phi^{(s)}. \tag{5.2.115}$$

Proof   The first relation follows from Lemma 5.1 using Lemma 5.4. The second relation follows immediately from the first, noting that $\{x^a, \{x_a, \phi^{(s)}\}_-\}_- = 0 = \{x^a, \{x_a, \phi^{(s)}\}_+\}_+$ by irreducibility (or because $\Box$ commutes with $\mathcal{S}^2$ (5.2.29)). $\Box$

$\mathcal{D}^+$ intertwiner

Consider the following intertwiner

$$\mathcal{D}^+: \quad \mathbb{R}^5 \otimes \mathcal{C}^0 \to \mathcal{C}^1$$

$$\mathcal{A}^a \mapsto \mathcal{D}^+(\mathcal{A}) = -\frac{1}{\alpha_1}\{(\Box - 2r^2)^{-1}\mathcal{A}^b, x_b\} \tag{5.2.116}$$

(recall that $\alpha_1 = \frac{1}{3}$). Using (5.2.111) and (5.2.92), this takes the following explicit form

$$\mathcal{D}^+(\mathcal{A}) = -\frac{1}{\alpha_1}(\Box - 4r^2)^{-1}\{\mathcal{A}^b, x_b\} = -\frac{1}{2\alpha_1}(\Box - 4r^2)^{-1}\left(F_{ab}\theta^{ab}\right) \qquad (5.2.117)$$

in terms of the self-dual part of $F_{ab} = \eth_b\mathcal{A}_a - \eth_a\mathcal{A}_b$. Together with (5.2.54), this implies that

$$\mathcal{D}^+\mathcal{A} = 0, \qquad \text{for } \mathcal{A}^a = \eth^a\phi. \qquad (5.2.118)$$

Moreover, Theorem 5.6 states that

$$\{\mathcal{D}^+(\mathcal{A}), \phi\}_0 = \mathcal{A}[\phi] = \mathcal{A}^a\eth_a\phi \qquad (5.2.119)$$

for any divergence-free vector field $\mathcal{A}^a$ and $\phi \in \mathcal{C}^0$. This will be the basis of the reconstruction of frames in Section 5.4.6. We also note the following asymptotic relation

$$\int \mathcal{D}^+(\mathcal{A}^a)\mathcal{D}^+(\mathcal{B}^a) = 9\int\{(\Box - 2r^2)^{-1}\mathcal{A}^a, x_a\}_+\{(\Box - 2r^2)^{-1}\mathcal{B}^a, x_a\}_+$$

$$= -9\int (\Box - 2r^2)^{-1}\mathcal{A}^a\{\{(\Box - 2r^2)^{-1}\mathcal{B}^a, x_a\}_+, x_a\}_-$$

$$= 6\int \mathcal{A}^a\Box(\Box - 2r^2)^{-2}\mathcal{B}^a$$

$$\approx 6\int \mathcal{A}^a\Box^{-1}\mathcal{B}^a \qquad (5.2.120)$$

for $\mathcal{A}_a, \mathcal{B}_b \in \mathcal{C}^0$ provided $r^{-2}\Box \gg 1$, which will be used in the following.

$\mathcal{D}^{++}$ intertwiner

Finally, consider the following intertwiner

$$\mathcal{D}^{++}: \quad \mathbb{R}^5 \otimes \mathcal{C}^0 \to \mathbb{R}^5 \otimes \mathcal{C}^2$$

$$\mathcal{A}^a \mapsto (\mathcal{D}^{++}\mathcal{A})^a = \{\mathcal{D}^+\mathcal{A}, x^a\}_+ \qquad (5.2.121)$$

which will play a role in the reconstruction of vector fields. For $\mathcal{A}_a, \mathcal{B}_b \in \mathcal{C}^0$ in the regime[9] with $r^{-2}\Box \gg 1$, this satisfies

$$\int \mathcal{D}^{++}(\mathcal{A}^a)\mathcal{D}^{++}(\mathcal{B}^a) = \int\{\mathcal{D}^+\mathcal{A}, x^b\}_+\{\mathcal{D}^+\mathcal{B}, x^b\}_+$$

$$= -\int \mathcal{D}^+\mathcal{A}\{\{\mathcal{D}^+\mathcal{B}, x^b\}_+, x^b\}_-$$

$$= \frac{2}{3}\int \mathcal{D}^+\mathcal{A}(\Box + 2r^2)(\mathcal{D}^+\mathcal{B})$$

$$= 4\int \mathcal{A}(\Box + 2r^2)\Box(\Box - 2r^2)^{-2}\mathcal{B}$$

$$\approx 4\int \mathcal{A}^a\mathcal{B}^a \qquad (5.2.122)$$

---

[9] This regime will be identified as effectively four-dimensional regime in Section 5.4.3.

using (5.2.120). Hence the intertwiner $\mathcal{D}^{++}$ is an approximate isometry (up to a factor 4). In this regime, it also satisfies the following approximate compatibility with derivatives and multiplications:

$$\mathcal{D}^{++}(x^a \mathcal{A}) \approx x^a(\mathcal{D}^{++}\mathcal{A}),$$
$$\mathcal{D}^{++}(\eth^a \mathcal{A}) \approx \eth^a(\mathcal{D}^{++}\mathcal{A}). \tag{5.2.123}$$

The first follows from $r^{-2}\square \gg 1$, which together with the intertwining property implies the second. This will be important for the reconstruction of vector fields and frames in the matrix model.

5.2.4 Divergence-free vector fields on H^4 and reconstruction

Consider the following (possibly $\mathfrak{hs}$-valued) vector fields on H^4

$$V := V^a \partial_a, \qquad V^a = \{\phi, x^a\} \tag{5.2.124}$$

generated by some $\phi \in \mathcal{C}$. V is clearly tangential to H^4 and divergence-free due to (5.2.53),

$$\eth_a V^a = 0. \tag{5.2.125}$$

Its components $V^a \in \mathcal{C}$ are in general $\mathfrak{hs}$-valued functions on H^4. V can be viewed as push-forward of the Hamiltonian vector field $\{\phi, .\}$ on $\mathbb{C}P^{2,1}$ to H^4 via the bundle projection[10]. For $\phi = \phi^{(s)} \in \mathcal{C}^s$, it decomposes into different $\mathfrak{hs}$ components as

$$V^a = V_{s-1}^a + V_{s+1}^a \qquad \in \mathcal{C}^{s-1} \oplus \mathcal{C}^{s+1}. \tag{5.2.126}$$

One might hope that all divergence-free $\mathfrak{hs}$-valued vector fields can be written in this form, but this is not possible; there are whole towers of divergence-free $\mathfrak{hs}$ vector fields which are not of this form. The reason is simply that $\mathcal{C}^{s+1}$ encodes fewer degrees of freedom[11] than $3 \times \mathcal{C}^s$ i.e. divergence-free $\mathcal{C}^s$-valued vector fields, except for $s = 0$. In that case, all *ordinary* divergence-free vector fields $V^a \in \mathcal{C}^0$ can indeed be obtained in this way, up to $\mathfrak{hs}$ corrections. This will allow to reconstruct classical geometries through matrix configurations in the context of gravity, based on the following result:

Theorem 5.6 *Given any divergence-free tangential vector field $\eth_a V^a = 0$ on H^4 with $V^a \in \mathcal{C}^0$, there is a unique generator $Y \in \mathcal{C}^1$ such that*

$$V^a = \{Y, x^a\}_0. \tag{5.2.127}$$

Y is given explicitly in terms of the intertwiner $\mathcal{D}^+$ (5.2.116)

$$Y := \mathcal{D}^+(V) = -3(\square - 4r^2)^{-1}\{V^a, x_a\} \qquad \in \mathcal{C}^1. \tag{5.2.128}$$

---

[10] In general, the push-forward of a vector field via a non-injective map is not well defined. However, the push-forward in the present situation makes sense if interpreted as $\mathfrak{hs}$-valued map.

[11] This is closely related to the isomorphism discussed in Section 5.2.2, which holds only for spin s irreps.

Proof   First, note that for divergence-free tangential $V^a \in \mathcal{C}^0$, the expression $\{V_a, x^a\}$ vanishes only if $(\Box - 2r^2)V^a = 0$ due to Lemma 5.1, which has no solution in $\mathcal{C}^0$. Indeed, it follows from $\mathfrak{so}(4,1)$ representation theory that $\Box$ is bounded from below by $\frac{9}{4}$ on H^4 for unitary irreps (cf. (5.2.76)), while $\Box$ is negative for finite irreps. Similarly, $(\Box - 4r^2)$ is positive-definite on unitary modes in $\mathcal{C}^1$.

We have seen that the vector field on the rhs is always divergence-free using (5.2.53),

$$\eth_a\{Y, x^a\} \equiv 0. \tag{5.2.129}$$

The intertwiner result (5.2.111) implies

$$(\Box - 2r^2)\{\phi^{(1)}, x_a\}_- = \{(\Box - 4r^2)\phi^{(1)}, x_a\}_- \tag{5.2.130}$$

for any $\phi^{(1)} \in \mathcal{C}^1$. Therefore,

$$\{Y, x^a\}_0 = -3\{(\Box - 4r^2)^{-1}\{V^b, x_b\}, x^a\}_0 = -3(\Box - 2r^2)^{-1}\{\{V^b, x_b\}, x^a\}_0 = V^a \tag{5.2.131}$$

using Lemma 5.1 in the last step.

To establish uniqueness, assume that $Y \in \mathcal{C}^1$ satisfies $\{Y, x^a\}_0 = 0$. Then Lemma 5.5 implies that $(\Box_H - 4r^2)Y = 0$, which implies as above that $Y = 0$ via representation theory. □

Hence, given V^a, we define

$$Y^{(1)} := \mathcal{D}^+(V), \qquad \in \mathcal{C}^1 \tag{5.2.132}$$

which encodes (the self-dual component of) the field strength of the vector field. By construction, this satisfies

$$V^a = \{Y^{(1)}, x^a\}_0 = -\mathcal{A}^a_-[Y^{(1)}]. \tag{5.2.133}$$

However, $\{Y^{(1)}, x^a\}$ contains in general also a spin two component

$$V^{(2)a} := \{Y^{(1)}, x^a\}_+ = -\mathcal{A}^a_+[Y^{(1)}] = \mathcal{D}^{++}(V^a) \qquad \in \mathcal{C}^2 \tag{5.2.134}$$

which satisfies

$$\eth_a V^{(2)a} = 0,$$
$$\{V^{(2)a}, x_a\}_+ = \{\{Y^{(1)}, x^a\}_+, x_a\}_+ = 0 \tag{5.2.135}$$

noting that $\eth$ respects $\mathcal{C}^n$, and that there is no spin 1 mode in $\mathcal{C}^3$. The second relation means that we cannot just repeat the above procedure to cancel this component. We can in fact identify via (5.2.134) the generators $Y \in \mathcal{C}^1$, which do *not* induce any components in $\mathcal{C}^2$:

Lemma 5.7   *Assume that $Y \in \mathcal{C}^1$ generates a Hamiltonian vector field $V^a = \{Y, x^a\} \in \mathcal{C}^0$ on H^4 whose $\mathfrak{hs}$ components vanish. Then $Y \in \mathfrak{so}(4,1)$, i.e. Y is a linear combination of θ^{ab}.*

We shall denote these vector fields V^a as *pure* vector fields.

Proof   Since every $Y \in \mathcal{C}^1$ can be written as $Y = \{x^b, y_b\}$ for some $y_b \in \mathcal{C}^0$, Lemma 5.1 gives

$$\{Y, x^a\}_0 = \frac{1}{3}(\Box - 2r^2)y_a. \tag{5.2.136}$$

Then we have

$$\begin{aligned}
\Box Y &= -\{\{Y, x^a\}, x_a\} = -\{\{Y, x^a\}_0, x_a\}_1 + \{\{Y, x^a\}_2, x_a\}_1 \\
&= -\{\{Y, x^a\}_0, x_a\}_1 \\
&= -\frac{1}{3}\{(\Box - 2r^2)y_a, x_a\}_1 = -\frac{1}{3}(\Box - 4r^2)\{y_a, x_a\}_1 \\
&= \frac{1}{3}(\Box - 4r^2)Y \tag{5.2.137}
\end{aligned}$$

using Lemma 5.4. Therefore,

$$\Box Y = -2r^2 Y. \tag{5.2.138}$$

This holds indeed for $Y = \theta^{ab}$ (5.2.113), and these are the only elements in $\mathcal{C}^1$ with that property. The vector fields corresponding to the $\mathfrak{so}(4, 1)$ generators θ^{ab} are given explicitly by

$$V_{(ab)}^c = \{\theta^{ab}, x^c\} = -r^2(\eta^{ac}x^b - \eta^{ab}x^c) \quad \in \mathcal{C}^0. \tag{5.2.139}$$

$\square$

We conclude that the reconstruction of vector fields V^a on H^4 via Theorem 5.6 leads generically to a spin 2 component given by

$$V^{(2)a} = \{Y, x^a\}_2 = \mathcal{D}^{++}(V^a) \quad \in \mathcal{C}^2. \tag{5.2.140}$$

Since a $\mathcal{C}^0$-valued divergence-free vector field V^a is fully characterized by the three degrees of freedom of $Y \in \mathcal{C}^1$, its spin two sibling $V^{(2)a} \in \mathcal{C}^2$ encodes the same information (i.e. $SO(4, 1)$ representation) as V^a. We also recall that the intertwiner $\mathcal{D}^{++}$ is an approximate isometry (5.2.122) and compatible with derivatives for short wavelengths.

Exercise 5.2.3   Verify (5.2.59) for $\phi = \theta^{ab}$, and obtain $\Box \theta^{ab} = -2r^2 \theta^{ab}$ (5.2.113).

Volume-preserving diffeomorphisms

Let $\phi \in \mathrm{End}(\mathcal{H}_n)$ be some field on fuzzy H_n^4, which can be interpreted as quantized function on $\mathbb{C}P^{2,1}$. Consider a transformation of ϕ given by the conjugation with some unitary operator $U \in \mathrm{End}(\mathcal{H}_n)$:

$$\phi \to U\phi U^{-1}. \tag{5.2.141}$$

The infinitesimal transformation for $U = e^{-i\varepsilon\Lambda}$ reduces in the semi-classical limit to

$$\delta_\Lambda \phi = \{\Lambda, \phi\} = \mathcal{L}_\xi \phi, \qquad \Lambda = \Lambda^* \in \mathcal{C}^\infty(\mathbb{C}P^{2,1}), \tag{5.2.142}$$

which is the Lie derivative of $\phi \in C^\infty(\mathbb{C}P^{2,1})$ along the Hamiltonian vector field $\xi = \{\Lambda, .\}$. From the base space point of view i.e. for functions $\phi \in C^0$ on H^4, this reduces to

$$\delta_\Lambda \phi = \mathcal{L}_\xi \phi = \xi^\mu \partial_\mu \phi. \tag{5.2.143}$$

However, ξ generally takes value in $\mathfrak{hs}$. Accordingly, the transformation should be interpreted as Lie derivative along the $\mathfrak{hs}$-valued vector field

$$\xi = \xi^\mu \partial_\mu, \qquad \xi^\mu = \{\Lambda, x^\mu\}, \tag{5.2.144}$$

in local coordinates x^μ on H^4. E.g. for $\Lambda \in C^1$, its components are

$$\xi^\mu = \xi^\mu_{(0)} + \xi^\mu_{(2)} \quad \in C^0 \oplus C^2. \tag{5.2.145}$$

As shown in the previous section, any divergence-free vector field ξ on H^4 can be represented in the form $\xi^\mu = \{\Lambda, x^\mu\}_0$. Then

$$[\delta_\Lambda \phi]_0 = \mathcal{L}_\xi \phi \tag{5.2.146}$$

is precisely the Lie derivative of $\phi \in C^0$ along ξ. In this sense, the gauge transformations arising from fuzzy H^4_n comprise all volume-preserving diffeos on H^4. Generalized to more generic covariant quantum spaces, this will play an important role to understand gravity.

5.3 *Relation to twistor space

There is an interesting relation of the covariant quantum spaces S^4_N and H^4_n to twistor space, which was introduced in the 60s by Roger Penrose; for a nice review on twistor theory see e.g. [1, 126]. Twistor space can be defined to be (an open set of) the projective space $\mathbb{C}P^3 \cong \mathbb{C}^4/\mathbb{C}^*$. It is thus also closely related to the non-compact projective space $\mathbb{C}P^{2,1}$ (5.2.33), which can be identified with an open sector of $\mathbb{C}P^3$ given by $\bar{z}z > 0$. The covariant quantum spaces S^4_N and H^4_n can therefore be viewed in terms of quantized twistor space.

To exhibit the relation between the present point of view and the twistor space formalism, we recall the spinorial construction of $\mathbb{C}P^3$ (5.1.10)

$$z^{\mathcal{A}} = \begin{pmatrix} z^1 \\ z^2 \\ z^3 \\ z^4 \end{pmatrix} = \begin{pmatrix} \lambda^\alpha \\ \mu^{\alpha'} \end{pmatrix} \in \mathbb{C}^4, \ \mathcal{A} = 1, 2, 3, 4 \tag{5.3.1}$$

in terms of a pair of 2-component spinors $\lambda^\alpha, \mu^{\alpha'}$ of opposite chirality, which transform under $\mathfrak{su}(2)_L \times \mathfrak{su}(2)_R \cong \mathfrak{so}(4) \subset \mathfrak{so}(5) \subset \mathfrak{so}(6) \cong \mathfrak{su}(4)$. Twistor space $\mathbb{PT}$ is defined as an open subset of $\mathbb{C}P^3$

$$\mathbb{PT} := \{z^{\mathcal{A}} \in \mathbb{C}P^3 \mid \lambda^\alpha \neq 0\}, \tag{5.3.2}$$

with the projective line $\{\lambda^\alpha = 0\}$ removed. We will see that from the point of view of the Hopf map (5.1.41), this amounts to removing the south pole of S^4, which is mapped to infinity by the stereographic projection as explained in the following. Note that the twistor z^A transforms in the fundamental representation of $\mathfrak{su}(4)$.

Now consider the complex conjugation of the twistor z^A, denoted as

$$z^*_A = (\lambda^*{}_\alpha, \mu^*{}_{\alpha'}), \tag{5.3.3}$$

which transforms in the anti-fundamental of $\mathfrak{su}(4)$. This representation is self-dual if restricted to $\mathfrak{so}(5) \subset \mathfrak{su}(4)$, so that spinorial indices can be raised and lowered with the $\mathfrak{so}(5)$-invariant antisymmetric matrix (B.2.12)

$$z_A \kappa^{BA} = z^B, \quad z^A \kappa_{AB} = z_B, \quad \kappa_{AB} = \begin{pmatrix} \epsilon_{\alpha\beta} & 0 \\ 0 & \epsilon_{\alpha'\beta'} \end{pmatrix} \tag{5.3.4}$$

using the conventions $\epsilon^{\alpha\beta} = \epsilon_{\alpha\beta}$ with $\epsilon^{01} = -\epsilon^{10} = 1$. In particular, we can then define the dual twistor $\hat{z}^A$ of z^A as

$$\hat{z}^A = z^*{}_B \kappa^{AB} = (\hat{\lambda}^\alpha, \hat{\mu}^{\alpha'}), \tag{5.3.5}$$

where

$$\hat{\lambda}^\alpha = \bar{\lambda}_\beta \epsilon^{\alpha\beta}, \qquad \hat{\mu}^{\alpha'} = \bar{\mu}_{\beta'} \epsilon^{\alpha'\beta'}. \tag{5.3.6}$$

One defines the angle and square brackets as

$$\langle uv \rangle = u^\alpha v_\alpha, \qquad [uv] = u^{\alpha'} v_{\alpha'}. \tag{5.3.7}$$

The $SU(4)$-invariant inner product between the twistor z and its complex conjugate or dual twistor can then be written as follows:

$$N := z^*_A z^A = \lambda^*_\alpha \lambda^\alpha + \mu^*_{\alpha'} \mu^{\alpha'} = -\langle \hat{\lambda}\lambda \rangle - [\hat{\mu}\mu] = -z^{*A} z_A, \tag{5.3.8}$$

which defines the seven-sphere $S^7 \subset \mathbb{C}^4$; note that

$$\langle \lambda\hat{\lambda} \rangle = -\langle \hat{\lambda}\lambda \rangle = \lambda^*_\alpha \lambda^\alpha \geq 0 \tag{5.3.9}$$

is positive. The Hopf map (5.1.41)

$$\mathbb{CP}^3 \simeq S^7/_{U(1)} \to S^4,$$
$$z^A \mapsto y^a := \frac{r}{2} z^*{}_A (\gamma^a)^A{}_B z^B = -\frac{r}{2} \hat{z}^A (\gamma^a)_{AB} z^B \tag{5.3.10}$$

(cf. (5.3.8) for the last sign) can be made more explicit in the chiral basis of the γ matrices in (B.2.4), which are manifestly antisymmetric if the first index is lowered with the $\mathfrak{so}(5)$-invariant matrix κ_{AB}. Explicitly,

$$(\gamma_m)_{AB} = \begin{pmatrix} 0 & (\hat{\sigma}_m)_{\alpha\beta'} \\ -(\hat{\sigma}_m)_{\beta'\alpha} & 0 \end{pmatrix}, \qquad m = 1, 2, 3$$

$$(\gamma_4)_{AB} = \begin{pmatrix} 0 & -\epsilon_{\alpha\alpha'} \\ \epsilon_{\alpha\alpha'} & 0 \end{pmatrix}, \quad (\gamma_5)_{AB} = \begin{pmatrix} -\epsilon_{\alpha\beta} & 0 \\ 0 & \epsilon_{\alpha'\beta'} \end{pmatrix} \tag{5.3.11}$$

where $\hat{\sigma}^m_{\alpha\alpha'} = i\epsilon_{\alpha\bullet} (\sigma_m)^\bullet{}_{\alpha'}$. In particular, for $a = 5$ we obtain

$$y^5 = -\frac{r}{2} \hat{z}^A (\gamma^5)_{AB} z^B = \frac{r}{2}([\hat{\mu}\mu] - \langle \hat{\lambda}\lambda \rangle) = -(R + r\langle \hat{\lambda}\lambda \rangle) \geq -R \tag{5.3.12}$$

using (5.3.8), where the radius of S^4 is obtained using the Fierz identity as (5.1.19)

$$y_a y^a = R^2 = \left(\frac{rN}{2}\right)^2. \tag{5.3.13}$$

Similarly, for $\mu = 1, \ldots, 4$ the above Hopf map yields

$$y^\mu = -\frac{r}{2}\hat{Z}^A (\gamma^\mu)_{AB} Z^B = \hat{\sigma}^\mu_{\alpha\alpha'} y^{\alpha\alpha'}, \tag{5.3.14}$$

where

$$\hat{\sigma}^\mu_{\alpha\alpha'} = (\hat{\sigma}^m_{\alpha\alpha'}, \epsilon_{\alpha\alpha'}) \tag{5.3.15}$$

and

$$y^{\alpha\alpha'} = -\hat{\lambda}^\alpha \mu^{\alpha'} + \lambda^\alpha \hat{\mu}^{\alpha'} \tag{5.3.16}$$

using (5.3.12). The relation (5.3.12) implies

$$\langle \lambda \hat{\lambda} \rangle = \frac{N}{2}\left(1 + \frac{y_5}{R}\right). \tag{5.3.17}$$

Now we define the rescaled coordinates x^μ on $\mathbb{R}^4$ as

$$x^\mu = \frac{R}{r\langle \lambda\hat{\lambda} \rangle} y^\mu, \qquad y^\mu = \left(1 + \frac{y_5}{R}\right) x^\mu \tag{5.3.18}$$

which is recognized as a stereographic projection from the south pole corresponding to $\lambda = 0$. Then

$$1 + \frac{y^5}{R} = \frac{2R^2}{x^2 + R^2} \tag{5.3.19}$$

and we obtain

$$y^\mu = \frac{2R^2 x^\mu}{R^2 + x^2}, \qquad y^5 = R\frac{R^2 - x^2}{R^2 + x^2}. \tag{5.3.20}$$

The round metric on S^4 then takes the form

$$ds^2 = \left(\frac{\partial y^a}{\partial x^\mu}\frac{\partial y^b}{\partial x^\nu}\delta_{ab}\right) dx^\mu dx^\nu = \frac{4R^4 dx_\mu dx^\mu}{(R^2 + x^2)^2} := g_{\mu\nu} dx^\mu dx^\nu, \tag{5.3.21}$$

which means that the stereographic projection is conformal. From (5.3.8), we can also deduce

$$[\mu\hat{\mu}] = \frac{N}{2}\left(1 - \frac{y^5}{R}\right) = \frac{Nx^2}{R^2 + x^2}. \tag{5.3.22}$$

The south pole $\lambda = 0$ is mapped to infinity in this projection.

The point is that we can now parametrize twistor space $\mathbb{PT}$ or $\mathbb{C}P^3$ in terms of coordinates x^μ on $\mathbb{R}^4$, and a normalized Weyl spinor λ (modulo phase) which describes the internal S^2 fiber. The $SO(5)$ isometry group on S^4 is recognized as part of the conformal group $SO(5,1)$ on $\mathbb{R}^4$. Moreover, the fuzzy spaces S^4_N and H^4_n are now recognized as quantizations of twistor space, and the Weyl spinor λ then realizes the oscillator or Jordan–Schwinger construction of the internal fuzzy sphere S^2_N.

To make the relation with standard twistor space constructions more explicit, we recast the spinorial coordinates (5.3.16) in terms of the dimensionless re-scaled generators

$$x^{\alpha\alpha'} := \frac{y^{\alpha\alpha'}}{\langle\lambda\hat\lambda\rangle} = \frac{\hat\lambda^\alpha \mu^{\alpha'} - \lambda^\alpha \hat\mu^{\alpha'}}{\langle\hat\lambda\lambda\rangle}. \tag{5.3.23}$$

Multiplying with λ_α, this gives

$$x^{\alpha\alpha'}\lambda_\alpha = \frac{\hat\lambda^\alpha \lambda_\alpha \mu^{\alpha'} - \lambda^\alpha \lambda_\alpha \hat\mu^{\alpha'}}{\langle\hat\lambda\lambda\rangle} = \mu^{\alpha'}, \tag{5.3.24}$$

which is known as *incidence relation* in twistor theory. Therefore, the incidence relation of twistor space is understood in terms of the Hopf map $\mathbb{C}P^3 \to S^4$ followed by the stereographic projection $S^4 \to \mathbb{R}^4$. For some field-theoretic explorations of these structures in the present framework see e.g. [182, 191].

5.4 Cosmological $k = -1$ quantum spacetime and higher spin

Now we make a big step toward real physics, and discuss a solution or background of the model (4.0.2) that describes a quantized cosmological FLRW spacetime $\mathcal{M}^{3,1}$. The fluctuation modes on this background will lead to a consistent higher-spin gauge theory, with very interesting and near-realistic physical properties. We will discuss the basic properties of this solution following [174] and elaborate some aspects of the resulting gauge theory, including the reconstruction of generic divergence-free vector fields, which will be essential to describe generic geometries in the emergent gravity sector.

Before giving the explicit mathematical realization, let us briefly discuss the idea. As illustrated in the previous examples, noncommutative or quantum geometries are described by two structures: one is a (noncommutative) *algebra*, interpreted as quantized algebra of functions $\mathcal{C}^\infty(\mathcal{M})$ of a classical manifold. This algebra encodes the abstract manifold, and it is always $\mathrm{End}(\mathcal{H})$ in the present framework, for some separable Hilbert space $\mathcal{H}$. In addition we need to define a *metric* structure, corresponding to a Riemannian or Lorentzian manifold. This is defined here through a matrix Laplacian or d'Alembertian $\Box$ that acts on $\mathrm{End}(\mathcal{H})$, such as (5.2.4). Thus, the same algebra $\mathrm{End}(\mathcal{H})$ can describe very different geometries even with different signatures, as the same abstract manifold can have different metrics.

In this spirit, the cosmological spacetime $\mathcal{M}_n^{3,1}$ under consideration coincides with the fuzzy hyperboloid H_n^4 as a (quantized) manifold, but it inherits a Lorentzian effective metric through a different matrix d'Alembertian $\Box$ (5.4.3), which governs the fluctuations around the background in the matrix model. An intuitive picture of $\mathcal{M}_n^{3,1}$ is obtained in terms of quantized embedding functions from the manifold into target space, as in the case of fuzzy S_N^2 and fuzzy H_n^4. More specifically, we consider *four* generators X^μ of H_n^4 as quantized embedding functions into target space

$$X^\mu \sim x^\mu: \quad \mathcal{M}^{3,1} \hookrightarrow \mathbb{R}^{3,1}, \qquad \mu = 0,\dots,3, \tag{5.4.1}$$

Figure 5.1   Projection Π from H^4 to $\mathcal{M}^{3,1}$ with Minkowski signature.

where Greek indices μ, ν will run from 0 to 3 from now on. This can be interpreted as a brane $\mathcal{M}^{3,1}$ embedded in $\mathbb{R}^{3,1}$. Dropping X^4 means that the same abstract manifold is now embedded in $\mathbb{R}^{3,1}$ (rather than $\mathbb{R}^{4,1}$ as for H_n^4), so that $\mathcal{M}_n^{3,1}$ can be interpreted as squashed hyperboloid projected to $\mathbb{R}^{3,1}$, as sketched in Figure 5.1. Accordingly, (5.2.17) is rewritten as

$$\eta_{\mu\nu}X^\mu X^\nu = -R^2\mathbb{1} - X_4^2 \le -R^2\mathbb{1}. \tag{5.4.2}$$

It is easy to see that the X^μ alone generate the full algebra $\mathrm{End}(\mathcal{H}_n)$ of (quantized) functions on $\mathbb{C}P^{2,1}$, which is now viewed as quantized algebra of functions on an S^2-bundle over $\mathcal{M}^{3,1}$. The $X^\mu \sim x^\mu$ define an embedding of this bundle in $\mathbb{R}^{3,1}$ which is degenerate along the fibers. Although we will largely focus on the $3 + 1$-dimensional base manifold, $\mathcal{M}^{3,1}$ is understood to carry this bundle structure.

This picture strongly suggests that $\mathcal{M}^{3,1}$ will carry an effective metric with Lorentzian signature. That metric is encoded in the following matrix d'Alembertian

$$\Box\phi \equiv \Box_T\phi = [T_\mu, [T^\mu, \phi]] = (C^2[\mathfrak{so}(4,1)] - C^2[\mathfrak{so}(3,1)])\phi, \tag{5.4.3}$$

where

$$T^\mu = \frac{1}{R}M^{\mu 4}. \tag{5.4.4}$$

This defines an $SO(3,1)$-invariant d'Alembertian for $\mathcal{M}^{3,1}$ with Lorentzian structure. The underlying metric will be discussed in Section (5.4.2). We will also see that (5.4.4) is indeed a solution of the matrix model, which we simply accept for now and discuss some of its properties.

Note that we now have two vector operators X^μ and T^μ, both of which satisfy the $SO(3,1)$ -covariant commutation relations

$$[X^\mu, X^\nu] =: i\Theta^{\mu\nu} = -ir^2 M^{\mu\nu},$$

$$[T^\mu, T^\nu] = -\frac{i}{r^2 R^2}\Theta^{\mu\nu},$$

$$[T^\mu, X^\nu] = \frac{i}{R}\eta^{\mu\nu}X_4. \tag{5.4.5}$$

This looks like a noncommutative version of phase space, but there is an important difference: the special representations $\mathcal{H}_n$ lead to constraints (5.4.7), which are crucial for the consistency of the resulting gauge theory. Moreover, X^μ and T^μ provide solutions of the following matrix equations of motion with different signs of the inhomogeneous or mass term,

$$\Box_X X^\mu = -3r^2 X^\mu, \qquad \Box_T T^\mu = 3R^{-2} T^\mu. \tag{5.4.6}$$

To simplify the analysis of the constraints, we will focus on the semi-classical (Poisson) limit $n \to \infty$ from now on and work with commutative functions of $x^\mu \sim X^\mu$ and $t^\mu \sim T^\mu$, keeping the Poisson structure $[.,.] \sim i\{.,.\}$ encoded in $\theta^{\mu\nu}$.

5.4.1 Semiclassical structure

Since the full matrix algebra $\text{End}(\mathcal{H}_n)$ coincides with the one for fuzzy H_n^4, its semi-classical (or Poisson) limit $\mathcal{C}$ is again the algebra of functions on the bundle space $\mathbb{C}P^{2,1}$. The subalgebra $\mathcal{C}^0 \subset \mathcal{C}$ of functions on the base space $\mathcal{M}^{3,1}$ is now generated by the x^μ for $\mu = 0, \ldots, 3$. The generators x^μ and t^μ satisfy the following constraints, which arise from (5.2.19):

$$x_\mu x^\mu = -R^2 - x_4^2 = -R^2 \cosh^2 \eta, \qquad R \sim \frac{r}{2} n \tag{5.4.7a}$$

$$t_\mu t^\mu = r^{-2} \cosh^2 \eta, \tag{5.4.7b}$$

$$t_\mu x^\mu = 0. \tag{5.4.7c}$$

Here $x^\mu : \mathcal{M}^{3,1} \hookrightarrow \mathbb{R}^{3,1}$ is interpreted as Cartesian coordinate functions, and η plays the role of a time parameter, defined via

$$x^4 = R \sinh \eta. \tag{5.4.8}$$

Hence $\eta = const$ defines a foliation of $\mathcal{M}^{3,1}$ into spacelike surfaces H^3, which will be recognized as spacelike slice of a FLRW cosmological spacetime with $k = -1$. Note that the sign of η distinguishes the two degenerate (upper and lower) sheets of $\mathcal{M}^{3,1}$, which result from the projection along x^4 indicated in Figure 5.1. The t^μ generators clearly describe the S^2 fiber over $\mathcal{M}^{3,1}$, which is spacelike due to (5.4.7c). These generators satisfy the Poisson brackets

$$\{x^\mu, x^\nu\} = \theta^{\mu\nu} = -r^2 R^2 \{t^\mu, t^\nu\},$$

$$\{t^\mu, x^\nu\} = \frac{x^4}{R} \eta^{\mu\nu},$$

$$\{\theta^{\mu\nu}, x^\gamma\} = -r^2 (\eta^{\mu\gamma} x^\nu - \eta^{\nu\gamma} x^\mu),$$

$$\{\theta^{\mu\nu}, t^\gamma\} = -r^2 (\eta^{\mu\gamma} t^\nu - \eta^{\nu\gamma} t^\mu) \tag{5.4.9}$$

as well as the following useful relations

$$\{x^4, x^\mu\} = r^2 R t^\mu,$$

$$\{x^4, t^\mu\} = \frac{1}{R} x^\mu. \tag{5.4.10}$$

The Poisson tensor $\theta^{\mu\nu}$ satisfies the following constraints which arise from H_n^4

$$t_\mu \theta^{\mu\alpha} = -\sinh \eta \, x^\alpha, \tag{5.4.11a}$$

$$x_\mu \theta^{\mu\alpha} = -r^2 R^2 \sinh \eta \, t^\alpha, \tag{5.4.11b}$$

$$\eta_{\mu\nu} \theta^{\mu\alpha} \theta^{\nu\beta} = R^2 r^2 \eta^{\alpha\beta} - R^2 r^4 t^\alpha t^\beta + r^2 x^\alpha x^\beta. \tag{5.4.11c}$$

As a check for the last equation, consider the contraction with $x_\alpha x_\beta$:

$$\eta_{\mu\nu}\theta^{\mu\alpha}x_\alpha\theta^{\nu\beta}x_\beta = -R^4 r^2 \cosh^2\eta + r^2 R^4 \cosh^4\eta, \tag{5.4.12}$$

which is easily verified using the above relations. Furthermore due to the self-duality relations of θ^{ab} on H^4, $\theta^{\mu\nu}$ can be expressed in terms of t^μ as [175]

$$\theta^{\mu\nu} = \frac{r^2}{\cosh^2\eta}\left(\sinh\eta(x^\mu t^\nu - x^\nu t^\mu) + \epsilon^{\mu\nu\alpha\beta}x_\alpha t_\beta\right). \tag{5.4.13}$$

The commutation relations imply in particular

$$\{t_\mu, \phi\} = \sinh\eta\,\partial_\mu\phi \tag{5.4.14}$$

for $\phi = \phi(x)$, which suggest that $T^\mu \sim t^\mu$ can be viewed as momentum generators on $\mathcal{M}^{3,1}$.

Furthermore, it is important to note that there is a global time-like vector field on $\mathcal{M}^{3,1}$ which is compatible with the spacelike $SO(3,1)$ isometry group. This is given explicitly by

$$\tau = x^\mu\partial_\mu \tag{5.4.15}$$

in the Cartesian coordinates. The vector field τ describes the time evolution on the FLRW cosmological background, and makes perfect physical sense. However, it is of course not invariant under local Lorentz transformations, and as such its presence may lead to some (typically mild) violation of local Lorentz invariance. It will be important in the following to check that this does not have any significant impact on the local physics.

For convenience, we also note the following useful relations

$$\beta^{-1}\partial_\mu\beta = \frac{\beta^2}{R^2}x_\mu, \tag{5.4.16}$$

$$\beta^{-1}\tau\beta = -(\beta^2 + 1), \qquad \tau x_4 = -x_4 + x_4\tau \tag{5.4.17}$$

$$\Box\beta\phi = \beta\Box\phi - \frac{2}{R^2}(\tau + 2)\beta\phi. \tag{5.4.18}$$

Here $\phi \in C^0$ is a scalar function, and

$$\beta = \sinh(\eta)^{-1} = \frac{R}{x_4} \tag{5.4.19}$$

decays with the cosmic time evolution and is proportional to the symplectic density ρ_M (5.4.26).

Fiber projection

As for H_n^4, the projection $[.]_0$ along the S^2 fiber can be evaluated explicitly as follows:

$$[t^\mu t^\nu]_0 =: \frac{1}{3r^2}\kappa^{\mu\nu}. \tag{5.4.20}$$

Here

$$\kappa^{\mu\nu} = \cosh^2\eta\,\eta^{\mu\nu} + \frac{x^\mu x^\nu}{R^2}, \qquad \kappa_{\mu\nu}x^\nu = 0 \tag{5.4.21}$$

is the unique $SO(3, 1)$-invariant positive semi-definite spacelike metric which vanishes along the time-like direction τ. Similarly, one finds

$$[t^\alpha \theta^{\mu\nu}]_0 = \frac{1}{3}\left(\sinh\eta(\eta^{\alpha\nu}x^\mu - \eta^{\alpha\mu}x^\nu) + x_\beta \varepsilon^{\beta 4\alpha\mu\nu}\right),$$

$$[t^{\mu_1}\dots t^{\mu_4}]_0 = \frac{3}{5}\left([t^{\mu_1}t^{\mu_2}][t^{\mu_3}t^{\mu_4}]_0 + [t^{\mu_1}t^{\mu_3}][t^{\mu_2}t^{\mu_4}]_0 + [t^{\mu_1}t^{\mu_4}][t^{\mu_2}t^{\mu_3}]_0\right). \tag{5.4.22}$$

A more general Wick-type formula for the projection was given in [188].

Integration and measure

Since the symplectic space underlying $\mathcal{M}^{3,1}$ is precisely the same as for H_n^4, the invariant measure for the integral corresponding to the trace is the same as for the hyperboloid (5.2.49), given in terms of the $SO(4, 2)$-invariant volume form $d\Omega$ on $\mathbb{C}P^{2,1}$ arising from the (Kirillov–Kostant) symplectic form. We would like to find its explicit form on $\mathcal{M}^{3,1}$ in terms of the cosmological time η, which determines the $SO(3, 1)$-invariant spacelike slice $H^3 \subset H^4 \subset \mathbb{R}^{4,1}$ via $x_\mu x^\mu = -R^2 \cosh^2\eta$. The $SO(4, 1)$-invariant volume form (5.2.50) on H^4 can be written as

$$\Omega_H = \frac{1}{\sinh\eta}dx_0\dots dx_3 = \cosh^3\eta d\eta \sinh^2\chi d\chi \sin\theta d\theta d\varphi \tag{5.4.23}$$

where the second form holds in hyperbolic coordinates

$$\begin{pmatrix} x^0 \\ x^1 \\ x^2 \\ x^3 \end{pmatrix} = R\cosh\eta \begin{pmatrix} \cosh\chi \\ \sinh\chi\sin\theta\cos\varphi \\ \sinh\chi\sin\theta\sin\varphi \\ \sinh\chi\cos\theta \end{pmatrix}, \tag{5.4.24}$$

which are adapted to the projection along $x^4 = R\sinh\eta$. This volume form should also be used on $\mathcal{M}^{3,1}$. Thus,

$$\mathrm{Tr}\mathcal{Q}(\phi) = \int_{\mathbb{C}P^{2,1}} \frac{\Omega}{(2\pi)^3}\phi = \int_{H^4}\Omega_H[\phi]_0 = \int_{\mathcal{M}} d^4x\,\rho_M[\phi]_0 \tag{5.4.25}$$

where $d\Omega$ is the $SO(4, 2)$-invariant volume form on $\mathbb{C}P^{2,1}$, and

$$\Omega_H = \Omega_M = \rho_M(x)d^4x, \qquad \rho_M(x) = \frac{1}{R^4\sinh\eta}. \tag{5.4.26}$$

Since the symplectic volume form is invariant under symplectomorphisms, it satisfies

$$\int \Omega f\{g, h\} = -\int \Omega\{g, f\}h. \tag{5.4.27}$$

As a check, for $g = t^\mu$ this reduces using (5.4.9) to

$$\int_{\mathcal{M}} d^4x\,\rho_M fx_4(\partial_\mu g) = -\int_{\mathcal{M}} d^4\rho_M x_4(\partial_\mu f)\,g, \tag{5.4.28}$$

as it must.

Derivatives

For the subalgebra $\mathcal{C}^0 \subset \mathcal{C}$ which corresponds to functions on the classical base manifold $\mathcal{M}^{3,1}$, the partial derivative operators ∂_μ act on the Cartesian coordinate functions x^μ as $\partial_\mu x^\nu = \delta_\mu^\nu$. A priori, it is not evident how to extend these derivations to general $\mathfrak{h}\mathfrak{s}$-valued functions $\mathcal{C}$. It turns out that this can be achieved using the tangential derivative operators $\eth$ on H^4 as follows. Using (5.2.51), we define the derivations

$$\partial_\mu := \eth_\mu - x_\mu \frac{1}{x_4}\eth_4 \qquad \text{on} \quad \mathcal{C}. \tag{5.4.29}$$

It is easy to check from (5.2.52) that they satisfy

$$\partial_\mu x^\nu = \delta_\mu^\nu$$
$$\partial_\mu(\rho_M \theta^{\mu\nu}) = 0. \tag{5.4.30}$$

We will understand the latter on more general grounds in Section 5.4.5 in the context of conserved vector fields, since $V^a = \{x^\nu, x^a\}$ is conserved on H^4. This identity can also be seen as a manifestation of the Jacobi identity restricted to $\mathcal{C}^0$. It is also interesting to observe

$$\partial^\mu t^\nu = \frac{x_4}{r^2 R^3}(\theta^{\mu\nu} - \frac{r^2}{\cosh\eta}(x^\mu t^\nu - t^\mu x^\nu))$$
$$\overset{\eta\to\infty}{\sim} \frac{1}{R^2 \cosh\eta}\varepsilon^{\mu\nu\alpha\beta}x_\alpha t_\beta \tag{5.4.31}$$

using (5.4.13) in the last step. At the reference point $\xi = (x^0, 0, 0, 0)$ on $\mathcal{M}$, the last relation becomes

$$\partial^\mu t^\nu \overset{\xi}{\sim} \frac{1}{R}\varepsilon^{0\mu\nu k}t_k \tag{5.4.32}$$

which means that ∂^i generates local rotations of t^j, while $\partial^0 t^j \sim 0$.

Late-time regime and noncommutativity scale

Consider the regime of large η, so that $\sinh\eta \gg 1$. Then the Poisson tensor $\theta^{\mu\nu}$ (5.4.13) reduces to

$$\theta^{\mu\nu} \sim \frac{r^2}{\cosh\eta}(x^\mu t^\nu - x^\nu t^\mu), \qquad \eta \to \infty. \tag{5.4.33}$$

More specifically, consider some given reference point $\xi = (x^0, 0, 0, 0)$ on $\mathcal{M}$. Then this reduces to

$$\theta^{0i} \overset{\xi}{=} \frac{r^2}{\cosh^2\eta}\sinh\eta\, x^0 t^i \ \sim\ r^2 R t^i \quad = O(L_{\text{NC}}^2)$$
$$\theta^{ij} \overset{\xi}{=} \frac{r^2}{\cosh^2\eta}x^0 \epsilon^{0ijk} t_k \ \sim\ \frac{1}{\sinh\eta}r^2 R\epsilon^{ijk} t_k \quad = O(rR) \tag{5.4.34}$$

where

$$L_{\text{NC}}^2 := Rr \cosh \eta \qquad (5.4.35)$$

is the effective scale of noncommutativity on $\mathcal{M}^{3,1}$, in x^μ coordinates. Even though this grows with η, it is much shorter than the curvature scale of the spacelike H^3

$$L_{\text{NC}} \sim R\sqrt{\frac{1}{n} \cosh \eta} \ll R \cosh \eta =: L_{\text{H}} \qquad (5.4.36)$$

which in turn is much shorter than the cosmic curvature scale (5.4.49). Hence, there is plenty of room for interesting physics in between. In particular, $\theta^{0i} \sim r^2 R t^i \gg \theta^{ij}$ at late times $\eta \gg 1$. We furthermore note that the time-like generator $t^0 \overset{\xi}{=} 0$ vanishes as function, while the spacelike generators t^i describe the internal fuzzy sphere S_n^2, and satisfy

$$\{t^i, t^j\} \overset{\xi}{=} -\frac{1}{r^2 R^2}\theta^{ij} = -\frac{1}{R \sinh \eta}\epsilon^{ijk} t^k. \qquad (5.4.37)$$

However, t^0 is nontrivial as a generator even at ξ, and induces local time translations via $\{t^0, .\}$.

Exercise 5.4.1   Verify the relation (5.4.18)
Exercise 5.4.2   Verify the characteristic equation (9.1.65)

$$0 = J^4 - \frac{1}{2}(tr J^2)J^2 - \theta^4 \mathbb{1} \qquad (5.4.38)$$

for covariant $\mathcal{M}^{3,1}$, using the identity (5.4.11c).

5.4.2 Effective FLRW metric on $\mathcal{M}^{3,1}$

In the matrix model framework, the effective metric on the background $T^\alpha \sim t^\alpha$ is obtained by rewriting the kinetic term acting on scalar fields $\phi \in \mathcal{C}^0$ in covariant form in terms of the frame

$$E^\alpha = \{t^\alpha, .\}, \qquad E^{\alpha\nu} = \{t^\alpha, x^\nu\} = \frac{x^4}{R}\eta^{\alpha\nu} \qquad (5.4.39)$$

defined by the background. This will be discussed in detail in Sections 9.1.1 and 10.1. Equivalently, the metric can be obtained from the matrix d'Alembertian, which can be written as

$$\Box := [T^\alpha, [T_\alpha, .]] \sim -\{t^\alpha, \{t_\alpha, .\}\} = \sinh^3 \eta \Box_G \qquad (5.4.40)$$

where $\Box_G = -\frac{1}{\sqrt{|G|}}\partial_\mu(\sqrt{|G|}\, G^{\mu\nu}\partial_\nu)$. This effective metric turns out to be (10.1.7)

$$G_{\mu\nu} = \sinh \eta\, \eta_{\mu\nu} \qquad (5.4.41)$$

in Cartesian coordinates. This will be recognized as Friedmann–Lemaître–Robertson–Walker (FLRW) metric, where $\eta = const$ determines $SO(3,1)$-invariant spacelike hyperboloids H_η^3.

FLRW form of the metric

In terms of the hyperbolic coordinates (5.4.24), the $SO(1,3)$-invariant flat metric $\eta_{\mu\nu}$ restricted to the spacelike hyperboloid H^3_η has the form

$$ds^2_\eta|_{H^3} = R^2 \cosh^2\eta \, d\Sigma^2, \tag{5.4.42}$$

where $d\Sigma^2 = d\chi^2 + \sinh^2\chi \, d\Omega^2$ is the metric on the unit hyperboloid H^3. Therefore

$$
\begin{aligned}
ds^2_G = G_{\mu\nu}dx^\mu dx^\nu &= -R^2 \sinh^3\eta \, d\eta^2 + R^2 \sinh\eta \cosh^2\eta \, d\Sigma^2 \\
&= -dt^2 + a^2(t)d\Sigma^2
\end{aligned} \tag{5.4.43}
$$

has the form of a $k = -1$ cosmological FLRW metric, with scale parameter $a(t)$ determined by the following equations:

$$
\begin{aligned}
a(t)^2 &= R^2 \sinh\eta \cosh^2\eta, \\
dt &= R(\sinh\eta)^{\frac{3}{2}}d\eta.
\end{aligned} \tag{5.4.44}
$$

The Einstein tensor of this metric is found to be

$$\mathcal{G}_{\mu\nu} = \mathcal{R}_{\mu\nu} - \frac{1}{2}G_{\mu\nu}\mathcal{R} \sim \frac{5}{2}\frac{1}{a(t)^2}\left(\frac{\tau_\mu\tau_\nu}{a(t)^2} - \frac{1}{2}G_{\mu\nu}\right) \tag{5.4.45}$$

for large η, where

$$\tau = x^\mu\partial_\mu = a(t)\frac{\partial}{\partial t} = \coth\eta\frac{\partial}{\partial\eta} \tag{5.4.46}$$

(here x^μ are Cartesian coordinates) is the $SO(3,1)$-invariant cosmic time-like vector field which measures the cosmic scale,

$$G_{\rho\rho'}\tau^\rho\tau^{\rho'} = -R^2 \cosh^2\eta \sinh\eta = -a(t)^2. \tag{5.4.47}$$

At the reference point $\xi = R\cosh\eta\,(1,0,0,0)$ in Cartesian coordinates x^μ, this has the form

$$\mathcal{G}_{\mu\nu} \sim \frac{1}{R^2\cosh^2\eta}\mathrm{diag}(3,-1,-1,-1) = \mathrm{O}\left(\frac{1}{a(t)^2}G_{\mu\nu}\right). \tag{5.4.48}$$

From a GR point of view, this would be interpreted in terms of pressure $p < 0$ and energy density $\rho > 0$ satisfying $p \sim -\frac{1}{3}\rho$. In particular, the cosmic curvature scale set by the FLRW metric G is

$$L_{\mathrm{cosm}} = a(t) \sim R\cosh^{3/2}\eta. \tag{5.4.49}$$

We note that the scales L_{cosm} and L_{NC} (measured by G) are widely separated at late times,

$$\frac{L^2_{\mathrm{cosm}}}{L^2_{\mathrm{NC}}} \sim \frac{R}{r}\cosh\eta \sim n\cosh\eta, \tag{5.4.50}$$

which leaves plenty of space for interesting physics in between for large η.

To gain some insight into the cosmological evolution, it is interesting to determine $a(t)$ explicitly. The first equation of (5.4.44) gives

$$2a\,da = R^2 \cosh\eta\left(1 + 3\sinh^2\eta\right)d\eta \tag{5.4.51}$$

and combined with the second we obtain

$$2\frac{da}{dt} = 3 + \frac{1}{\sinh^2\eta}. \tag{5.4.52}$$

Hence, at late times, the evolution is that of a "coasting" universe

$$a(t) \approx \frac{3}{2}t, \qquad t \to \infty \tag{5.4.53}$$

as considered e.g. in [123, 140]. This correctly reproduces the accepted age of the universe upon inserting the present Hubble rate, which is quite remarkable since no specific matter content was assumed. For early times, we can approximate

$$R\eta^{\frac{3}{2}}\, d\eta = dt, \qquad \eta \propto t^{\frac{2}{5}} \tag{5.4.54}$$

such that

$$a(t) \propto \eta^{\frac{1}{2}} \propto t^{\frac{1}{5}}. \tag{5.4.55}$$

Hence, we obtain a FLRW cosmology that is asymptotically coasting at late times, with a singularity at $t = 0$. This singularity describes naturally a Big Bounce (BB) rather than a Big Bang, since the propagation of perturbations can be followed through the BB from one sheet of $\mathcal{M}^{3,1}$ to the other [24, 115]. At late times $\eta \to \infty$, the corresponding matrix d'Alembertian takes the following simple form [186]

$$\Box\phi = -\sinh^2\eta\left(\eta^{\alpha\beta}\partial_\alpha\partial_\beta - \frac{1}{R^2\sinh^2\eta}\tau\right)\phi \approx -\sinh^2\eta\,\eta^{\alpha\beta}\partial_\alpha\partial_\beta\phi \tag{5.4.56}$$

in Cartesian coordinates for $\phi = \phi(x)$, as the contribution from $\tau \approx x^0\partial_0$ is negligible.

5.4.3 Algebra of functions and higher spin

To understand the physics arising on such a background, we need to study the algebra of functions $\mathrm{End}(\mathcal{H}_n)$ on $\mathcal{M}^{3,1}$. As pointed out above, $\mathrm{End}(\mathcal{H}_n)$ can be viewed as algebra of functions on H_n^4, which decomposes into various $SO(4,1)$ modes. Now recall the spin Casimir $\mathcal{S}^2$ (5.2.28), which can be written as

$$\mathcal{S}^2 = 2C^2[\mathfrak{so}(4,1)] - C^2[\mathfrak{so}(4,2)]. \tag{5.4.57}$$

This Casimir also commutes with the matrix d'Alembertian $\Box$ on $\mathcal{M}^{3,1}$:

$$[\mathcal{S}^2, \Box] = 0 \tag{5.4.58}$$

because the latter can be written as

$$\Box = \Box_T \equiv [T_\mu, [T^\mu, \cdot]] = R^{-2}\left(C^2[\mathfrak{so}(4,1)] - C^2[\mathfrak{so}(3,1)]\right), \tag{5.4.59}$$

and all these Casimirs arising from subgroups commute. As a consequence, the algebra of functions

$$\mathrm{End}(\mathcal{H}_n) = \mathcal{C}^0 \oplus \mathcal{C}^1 \oplus \ldots \oplus \mathcal{C}^n \tag{5.4.60}$$

decomposes into the same spin s sectors C^s (5.2.30) as on H_n^4, corresponding to spin s harmonics on the S^2 fiber. This decomposition is compatible with the kinematics defined through $\square$.

In the semi-classical limit, the C^s were identified with traceless divergence-free vector fields on H_n^4, which comprise $2s + 1$ dof. Because the isometry group on $\mathcal{M}^{3,1}$ is given by the spacelike $SO(3, 1)$, it is more natural to organize the $\phi \in C^s$ as functions $\phi = \phi(x, t)$ on $\mathbb{C}P^{2,1}$, which can be identified with totally symmetric traceless spacelike[12] (!) rank s tensor fields on $\mathcal{M}^{3,1}$ via

$$\phi^{(s)} = \phi_{\mu_1 \ldots \mu_s}(x) t^{\mu_1} \ldots t^{\mu_s}, \qquad \phi_{\mu_1 \ldots \mu_s} x^{\mu_i} = 0 \qquad (5.4.61)$$

which are not necessarily divergence free. This captures the $2s + 1$ d.o.f. of a spin s field; however, local Lorentz invariance is not manifest. The gauge invariance of the underlying matrix model is expected to effectively restore local Lorentz invariance in the resulting higher-spin gauge theory. Indeed all physical tensor fields – such as the metric $G_{\mu\nu}$ – will arise from $\phi^{(s)}$ via brackets using the map (5.2.80), i.e.

$$\tilde{\phi}_{\mu_1 \ldots \mu_s}(x) = [\{x_{\mu_1}, \ldots \{x_{\mu_s}, \phi^{(s)}\} \ldots\}]_0, \qquad (5.4.62)$$

where $[.]_0$ denotes the projection to C^0. These are totally symmetric tensor fields which are no longer spacelike, but satisfy divergence constraints as discussed in Section 5.4.5.

The constraints (5.4.61) are closely related to the issue of ghosts. In a consistent theory, it is essential that the time-like components of the tensor fields do not lead to negative-norm states, i.e. ghosts. The standard way to define such a consistent Lorentz-invariant quantum theory is as a gauge theory, where the unphysical negative norm modes can be removed while effectively preserving Lorentz invariance. The prime examples are Yang–Mills gauge theory for spin 1 and diffeomorphism-invariant general relativity for spin 2. For higher spin this is achieved by higher spin theory [59, 74, 208], which however becomes very problematic at the level of interactions.

The present framework provides a slightly different resolution of this problem. Now the spacelike constraint (5.4.61) is built into the above organization of modes[13], thus removing explicitly the dangerous time-like components. However, Lorentz invariance is no longer manifest, more precisely only the spacelike isometries $SO(3, 1)$ are manifest, while boosts are not. Again, gauge invariance comes to the rescue: The matrix model enjoys a large (higher-spin type) gauge invariance, which includes volume-preserving diffeomorphisms. Together with the spacelike symmetries of the model, this protects the theory from significant Lorentz violations. In particular, we will see that all modes propagate in the standard relativistic way. These topics will be discussed in more detail in the following.

Higher-spin d'Alembertian and internal harmonics

Consider $\phi \in C^s$, written as (5.4.61)

$$\phi^{(s)} = \phi_{\mu_1 \ldots \mu_s}(x) t^{\mu_1} \ldots t^{\mu_s} \qquad (5.4.63)$$

---

[12] This reflects the constraint (5.4.7c), evaluated at the reference point $x^\mu = (x^0, 0, 0, 0)$.

[13] This is not the case for the tangential fluctuation modes $\mathcal{A}^\mu$, which do have time-like components a priori. These are then removed as in Yang–Mills theory, as explained in Section 7.5.

denoted as "spacelike gauge." We would like to find an explicit formula for the matrix d'Alembertian $\Box\phi^{(s)}$ in terms of the scalar d'Alembertian $\Box$ acting on the $\phi_{\mu_1...\mu_s}(x) \in \mathcal{C}(\mathcal{M}^{3,1})$, taking into account the contributions from the internal modes on S^2, similar as in Kaluza–Klein compactification. To this end, we introduce the normalized spin s harmonics as follows:

$$\hat{Y}_m^s := \beta^s \underbrace{(t \ldots t)}_{s}, \qquad \beta = \frac{1}{\sinh \eta} \tag{5.4.64}$$

where $(t \ldots t)$ stands for the appropriate irreducible rank s polynomial in t^μ on the internal fuzzy sphere, and β^s is the appropriate normalization factor. We claim that

$$\Box\hat{Y}_m^s = \left(\frac{-s(s-4)}{R^2} + \beta^{-s}\Box\beta^s\right)\hat{Y}_m^s = \frac{1}{R^2}\left(-s(s-4) - 4s + s(s+1)\right)\hat{Y}_m^s = \frac{s}{R^2}\hat{Y}_m^s \tag{5.4.65}$$

up to sub-leading terms for large η. This can be verified using (5.4.37), noting that the mixed terms $\{t_\mu, \beta\}\{t^\mu, t^i\} = O(\beta)$ is suppressed, and using (5.4.18) and (5.4.17) which imply

$$\beta^{-s}\Box\beta^s = \beta^{-s}\Box\beta\beta^{s-1} = \beta^{-s+1}\Box\beta^{s-1} - \frac{4}{R^2} - \frac{2}{R^2}\beta^{-s}\tau\beta^s$$

$$= -\frac{4s}{R^2} + \frac{s(s+1)}{R^2}(1 + \beta^2) \tag{5.4.66}$$

after a little algebra. Now we can expand $\phi^{(s)}$ into these harmonics

$$\phi^{(s)} = \sum_{m=-s}^{s} \phi_{sm}(x)\hat{Y}_m^s. \tag{5.4.67}$$

Then

$$\Box\phi^{(s)} = \Box\phi_{lm}\hat{Y}_m^s + \phi_{lm}\Box\hat{Y}_m^s - 2\{t_\nu, \phi_{lm}\}\{t^\nu, \hat{Y}_m^s\}. \tag{5.4.68}$$

Now consider the mixed term

$$\{t_\nu, \phi_{\mu_1...\mu_s}(x)\}\{t^\nu, t^{\mu_1} \ldots t^{\mu_s}\} = -\frac{s \sinh \eta}{r^2 R^2}\partial_\nu\phi_{\mu_1...\mu_s}(x)\theta^{\nu\mu_1}t^{\mu_2} \ldots t^{\mu_s}$$

$$\sim -\frac{s \sinh \eta}{R^2 \cosh \eta}(x^\nu t^{\mu_1} - x^{\mu_1}t^\nu)\partial_\nu\phi_{\mu_1...\mu_s}(x)t^{\mu_2} \ldots t^{\mu_s}$$

$$\sim -\frac{s \sinh \eta}{R^2 \cosh \eta}(x^\nu\partial_\nu + 1)\phi_{\mu_1...\mu_s}(x)t^{\mu_1} \ldots t^{\mu_s} \tag{5.4.69}$$

for large η, using the late-time relation (5.4.33) and the spacelike gauge in the last step. Similarly,

$$\{t_\mu, \phi_{lm}\}\{t^\mu, \beta^s\} = \frac{s}{R^2}\beta^s x^\mu\partial_\mu\phi_{lm} \tag{5.4.70}$$

using

$$\partial_\mu\beta^s = s\beta^{s+2}\frac{1}{R^2}x_\mu, \tag{5.4.71}$$

which follows from (5.4.17). Combining these terms using the late-time approximation $\eta \to \infty$, the mixing term simplifies as

$$\{t_\nu, \phi_{sm}(x)\}\{t^\mu, \hat{Y}_m^s\} \sim -\frac{s}{R^2}\phi^s, \tag{5.4.72}$$

which is the reason for choosing the normalization in (5.4.64). Therefore we obtain

$$\boxed{\Box\phi^{(s)} \sim \hat{Y}_m^s\left(\Box + \frac{3s}{R^2}\right)\phi_{lm}(x)} \tag{5.4.73}$$

at late times $\eta \to \infty$. This has a similar structure as in Kaluza–Klein compactification, with mass scale set by $\frac{1}{R^2}$. Note that these masses are essentially zero at late times, taking into account the conformal factor $\rho^2 = \sinh^3 \eta$ in the matrix Laplacian (5.4.40).

Poisson brackets and effectively four-dimensional regime

Now consider the following Poisson bracket

$$\{x^\nu, \phi^{(s)}\} = \{x^\nu, \phi_{\mu_1...\mu_s}(x)\}t^{\mu_1}\ldots t^{\mu_s} - s\sinh\eta\, \eta^{\nu\mu_1}\phi_{\mu_1...\mu_s}(x)t^{\mu_2}\ldots t^{\mu_s}. \tag{5.4.74}$$

We would like to understand which term dominates under which conditions. It turns out that the second term dominates only in the extreme (cosmological) IR regime, while the first term dominates for shorter wavelengths relevant to the local physics. To determine this scale, it suffices to consider the case of $s = 1$ for $\phi^{(1)} = \phi_\alpha(x)t^\alpha \in \mathcal{C}^1$. Then

$$\begin{aligned}
\{x^\mu, \phi^{(1)}\} &= \{x^\mu, \phi_\alpha\}t^\alpha + \phi_\alpha\{x^\mu, t^\alpha\} = \theta^{\mu\nu}t^\alpha\partial_\nu\phi_\alpha - \sinh\eta\,\phi_\mu \\
&\sim \cosh\eta\big(\cosh\eta\, R\partial\phi - \phi\big) \\
&\sim \{x^\mu, \phi(x)_\alpha\}t^\alpha\left(1 + O\big(\frac{1}{R\cosh\eta\,\partial}\big)\right)
\end{aligned} \tag{5.4.75}$$

using (5.4.33) in the late-time regime and recalling $t \sim r^{-1}\cosh\eta$. Hence, the first (derivative) term dominates for all wavelengths λ shorter than the curvature scale (5.4.36) of H^4 and longer than the scale of noncommutativity:

$$\boxed{L_{NC} \ll \lambda \ll L_H = R\cosh\eta.} \tag{5.4.76}$$

This will be denoted as the **effectively 4D regime**. The upper bound for λ can alternatively be characterized by

$$r^{-2}\Box_H \gg 1, \tag{5.4.77}$$

where $\Box_H$ is the Laplacian (5.2.4) on H_n^4. In this regime, the following approximations hold

$$\begin{aligned}
\{x^\nu, \phi^{(s)}\} &\approx \{x^\nu, \phi_{\mu_1...\mu_s}(x)\}t^{\mu_1}\ldots t^{\mu_s} \\
\{t^\nu, \phi^{(s)}\} &\approx \{t^\nu, \phi_{\mu_1...\mu_s}(x)\}t^{\mu_1}\ldots t^{\mu_s}.
\end{aligned} \tag{5.4.78}$$

The second relation can be seen similarly as above by noting that

$$\{t^\mu, \phi^{(1)}\} \sim \{t^\mu, \phi(x)_\alpha\} t^\alpha \left(1 + O(\frac{1}{R \cosh \eta \partial})\right). \tag{5.4.79}$$

This means that $\phi^{(s)} = \phi_{\mu_1...\mu_s}(x) t^{\mu_1} ... t^{\mu_s} \in \mathcal{C}^s$ can be viewed effectively as four-dimensional $\mathfrak{hs}$-valued function on $\mathcal{M}^{3,1}$, and the Poisson brackets are oblivious to the $\mathfrak{hs}$ generators. We can then use the four-dimensional approximations

$$\{g, f\} \approx \{g, x^\mu\}(\partial_\mu f)$$
$$\{f, g\} \approx \theta^{\mu\nu} \partial_\mu g \partial_\nu f =: \{f, g\}_\mathcal{M}, \tag{5.4.80}$$

which in turn suggests (5.4.30)

$$\partial_\mu(\rho_M \theta^{\mu\nu}) \sim 0, \qquad \rho_M = \sqrt{|\theta^{\mu\nu}|}^{-1} \tag{5.4.81}$$

where $\mu, \nu = 0, .., 3$. This is in fact an exact relation on $\mathcal{M}^{3,1}$ provided the derivatives (5.4.29) are used.

5.4.4 Spacelike $SO(3, 1)$ substructure and associated identities

Consider again the spacelike tensor fields (5.4.61). They decompose further into divergence-free and pure divergence modes, compatible with the spacelike isometry $SO(3, 1)$. This decomposition can be realized algebraically using the underlying $\mathfrak{so}(4, 2)$ structure. Consider the $SO(3, 1)$ invariant derivation

$$D\phi := \{x^4, \phi\} = r^2 R^2 \frac{1}{x^4} t^\mu \{t_\mu, \phi\} = -\frac{1}{x^4} x_\mu \{x^\mu, \phi\}$$
$$= r^2 R \, t^{\mu_1} ... t^{\mu_s} t^\mu \nabla^{(3)}_\mu \phi_{\mu_1...\mu_s}(x), \tag{5.4.82}$$

where $\nabla^{(3)}$ is the covariant derivative along the spacelike $H^3 \subset \mathcal{M}^{3,1}$. Hence D relates the different spin sectors in (5.4.60):

$$D = D^- + D^+ : \mathcal{C}^s \to \mathcal{C}^{s-1} \oplus \mathcal{C}^{s+1}, \qquad D^\pm \phi^{(s)} = [D\phi^{(s)}]_{s\pm1} \tag{5.4.83}$$

where $[.]_s$ denotes the projection to $\mathcal{C}^s$. It is easy to see that $(D^+)^\dagger = -D^-$ w.r.t. the canonical invariant inner product. Explicitly,

$$Dx^\mu = r^2 R \, t^\mu, \qquad Dt^\mu = R^{-1} x^\mu. \tag{5.4.84}$$

In particular, $\mathcal{C}^{(s,0)} \subset \mathcal{C}^s$ is the space of divergence-free traceless spacelike rank s tensor fields on $\mathcal{M}^{3,1}$, which encode 2 dof as in ordinary massless gauge theories. We can then organize the $\mathcal{C}^s$ modes into primals[14] and descendants

$$\mathcal{C}^{(s,0)} = \{\phi \in \mathcal{C}^s; \ D^- \phi = 0\} \qquad \qquad ...\text{primal fields}$$
$$\mathcal{C}^{(s+k,k)} = (D^+)^k \mathcal{C}^{(s,0)} \qquad \qquad ...\text{descendants.} \tag{5.4.85}$$

---

[14] These names and the structures are similar but distinct from the more familiar case of CFT modes, which are organized into primaries and descendants. The algebraic structure arises in both cases from $SO(4, 2)$, albeit in a different way.

This means that the $2s + 1$ dof in $\mathcal{C}^s$ decompose further into those of massless gauge fields with spin $1 + \ldots + s$, all of which have the standard 2 dof, plus a scalar mode. Then

$$D^{\pm}: \quad \mathcal{C}^{(s,k)} \to \mathcal{C}^{(s\pm1,k\pm1)}, \tag{5.4.86}$$

and one can show using the $\mathfrak{so}(4, 2)$ structure and the special properties of the minireps $\mathcal{H}_n$ that the $D^{\pm}$ satisfy ladder-type commutation relations

$$D^+D^-\phi^{(s,k)} = \left(a_{s,k}D^-D^+ + b_{s,k}\right)\phi^{(s,k)}, \qquad \phi^{(s,k)} = (D^+)^k\phi^{(s-k,0)} \tag{5.4.87}$$

with constants $a_{s,k}, b_{s,k}$ given in [188]. This sub-structure encodes two different concepts of spin on the FRW background, which arise from the spacelike foliation: $\mathcal{S}^2 = 2s(s + 1)$ measures the four-dimensional spin on H^4, while $s - k$ measures the three-dimensional spin of $\mathcal{C}^{(s,k)}$ on H^3. The maps $D^{\pm}$ arise also in the following identities for $\phi^{(s)} \in \mathcal{C}^s$

$$\{t^{\mu}, \mathcal{A}_{\mu}^{(+)}[\phi^{(s)}]\} = \frac{s+3}{R}D^+\phi^{(s)},$$

$$\{t^{\mu}, \mathcal{A}_{\mu}^{(-)}[\phi^{(s)}]\} = \frac{-s+2}{R}D^-\phi^{(s)},$$

$$x^{\mu}\{x_{\mu}, \phi^{(s)}\}_{\pm} = -x_4 D^{\pm}\phi^{(s)}, \tag{5.4.88}$$

which follow from the $a = 4$ component in 5.2.104 and (5.4.82), respectively. Using these relations, one can verify the following intertwining properties of $\Box$

$$\Box D^+\phi^{(s)} = D^+\left(\Box + \frac{2s+2}{R^2}\right)\phi^{(s)}, \qquad \Box D^-\phi^{(s)} = D^-\left(\Box - \frac{2s}{R^2}\right)\phi^{(s)}. \tag{5.4.89}$$

Now consider the following $SO(3, 1)$ intertwiner

$$\tilde{\mathcal{I}}^{(4)}\mathcal{A}^{\mu} = \{\theta^{\mu\nu}, \mathcal{A}_{\nu}\} \tag{5.4.90}$$

(cf. (5.2.101)), which arises in the vector d'Alembertian on $\mathcal{M}^{3,1}$ and satisfies

$$C^2[\mathfrak{so}(3, 1)]^{(4)\otimes(\mathrm{ad})} = C^2[\mathfrak{so}(3, 1)]^{(4)} + C^2[\mathfrak{so}(3, 1)]^{(\mathrm{ad})} - \frac{2}{r^2}\tilde{\mathcal{I}}^{(4)}, \tag{5.4.91}$$

where (ad) indicates the adjoint action on $\mathcal{C}$. Using (5.2.105) we obtain

$$\tilde{\mathcal{I}}^{(4)}\mathcal{A}_{\mu}^{(+)}[\phi^{(s)}] = r^2(s + 3)\mathcal{A}_{\mu}^{(+)}[\phi^{(s)}] + r^2 R\{t_{\mu}, D^+\phi^{(s)}\} \tag{5.4.92}$$

$$\tilde{\mathcal{I}}^{(4)}\mathcal{A}_{\mu}^{(-)}[\phi^{(s)}] = r^2(-s + 2)\mathcal{A}_{\mu}^{(-)}[\phi^{(s)}] + r^2 R\{t_{\mu}, D^-\phi^{(s)}\}. \tag{5.4.93}$$

Finally, one can establish an explicit relation between the matrix Laplacians on H_n^4 and $\mathcal{M}^{3,1}$ [188]

$$\Box_H\phi^{(s)} = \left(-R^2 r^2\Box + (2s - 1)D^+D^- - (2s + 3)D^-D^+\right)\phi^{(s)}. \tag{5.4.94}$$

To appreciate the significance of this relation, we note that

$$-D^-D^+\phi^{(0)} = \frac{r^2 R^2}{3}\cosh^2\eta\,\Delta^{(3)}\phi^{(0)}, \tag{5.4.95}$$

where $\Delta^{(3)} = -\nabla^{(3)\alpha}\nabla_{\alpha}^{(3)}$ is the spacelike Laplacian on H^3 w.r.t. the metric $\eta^{\mu\nu}$. Together with (5.4.56), we obtain

$$\Box_H \phi^{(0)} = \left(-R^2 r^2 \Box + r^2 R^2 \cosh^2\eta \Delta^{(3)} \right) \phi^{(0)}$$

$$= R^2 r^2 \left(\sinh^2\eta \, \eta^{\alpha\beta} \partial_\alpha \partial_\beta + \cosh^2\eta \Delta^{(3)} - \frac{1}{R^2 \sinh^2\eta} \tau \right) \phi^{(0)}$$

$$\approx R^2 r^2 \left(-\frac{1}{R^2} \tau^2 + \Delta^{(3)} \right) \phi^{(0)}. \tag{5.4.96}$$

at late times $\eta \to \infty$. This should be compared with the matrix d'Alembertian, which can be written as (cf. (5.4.56), [186])

$$\Box \phi^{(0)} = \sinh^2\eta \left(\Delta^{(3)} + \frac{1}{R^2 \sinh^2\eta} \tau + \frac{1}{R^2 \cosh^2\eta} (2+\tau)\tau \right) \phi^{(0)}. \tag{5.4.97}$$

We observe not only the expected change of signature, but also the dominance of the derivatives τ^2 in the Euclidean operator $\Box_H \approx -r^2 \tau^2$ for on-shell modes at late times.

Example: Spin one sector

Using the above structure, the map (5.4.62) can be refined and computed explicitly. We illustrate this for the spin 1 case. For $\phi^{(1)} = \phi_\alpha t^\alpha$, we obtain using (5.4.22)

$$\tilde{\phi}_\mu[\phi^{(1)}] := \{x^\mu, \phi_\alpha t^\alpha\}_0 = \partial_\nu \phi_\alpha [\theta^{\mu\nu} t^\alpha]_0 + \phi_\alpha \{x^\mu, t^\alpha\}$$

$$= \frac{1}{3} \sinh\eta (x^\mu \partial^\alpha \phi_\alpha - (\tau+3)\phi_\mu) + \frac{1}{3} x_\beta \varepsilon^{\beta 4\alpha\mu\nu} \partial_\nu \phi_\alpha. \tag{5.4.98}$$

This vector field separates as

$$\tilde{\phi}_\mu[\phi^{(1)}] = \tilde{\phi}_\mu[\phi^{(1,0)}] + \tilde{\phi}_\mu[\phi^{(1,1)}] \tag{5.4.99}$$

into a spacelike divergence-free field arising from $\phi^{(1,0)}$

$$\tilde{\phi}_\mu[\phi^{(1,0)}] = -\frac{1}{3} \sinh\eta (\tau+3)\phi_\mu + \frac{1}{3} x_\beta \varepsilon^{\beta 4\alpha\mu\nu} \partial_\nu \phi_\alpha,$$

$$\partial^\mu \tilde{\phi}_\mu = 0 = x^\mu \tilde{\phi}_\mu \tag{5.4.100}$$

using (5.4.82) and (5.4.88), and a scalar mode arising from $\phi^{(1,1)} = D\phi$ for $\phi \in \mathcal{C}^0$

$$\tilde{\phi}_\mu[D\phi] = \{x^\mu, D\phi\}_0 = \frac{r^2 R}{3} \sinh\eta (x_\mu \partial^\alpha \partial_\alpha \phi - (\tau+3)\partial_\mu \phi), \tag{5.4.101}$$

which is neither spacelike nor divergence-free. This illustrates the role of the above substructure for tensor fields.

5.4.5 Divergence-free vector fields on $\mathcal{M}^{3,1}$

Divergence-free vector fields will play an important role in the following. Clearly any vector field V^a on H^4 can be mapped to a vector field V^μ on $\mathcal{M}^{3,1}$, by simply dropping the V^4 component (in Cartesian coordinates). This is nothing but the push-forward via the projection in Figure 5.1. For example, a Hamiltonian vector field $V^a = \{\phi, x^a\}$ is mapped to $V^\mu = \{\phi, x^\mu\}$ in Cartesian coordinates. Conversely, any vector field V^μ on $\mathcal{M}^{3,1}$ can be lifted to H^4 by defining

$$V^4 := -\frac{1}{x_4} x_\mu V^\mu, \qquad (5.4.102)$$

which defines a tangential vector field $V^a x_a = 0$ on H^4. We claim that this correspondence maps divergence-free vector fields $\eth_a V^a = 0$ on H^4 to divergence-free vector fields on $\mathcal{M}^{3,1}$, in the sense that

$$\partial_\mu(\rho_M V^\mu) = 0. \qquad (5.4.103)$$

Here ∂_μ is the derivation on $\mathcal{C}$ defined in (5.4.29), and ρ_M is the symplectic density (5.4.26) on $\mathcal{M}^{3,1}$, which in Cartesian coordinates is given by $\rho_M = (\sinh \eta)^{-1}$. In fact, the following more general result holds:

Lemma 5.8 *Let V^a be a (tangential) vector field on H^4. Then its reduction (or push-forward) V^μ to $\mathcal{M}^{3,1}$ satisfies*

$$\eth_a V^a = \sinh \eta \, \partial_\mu(\rho_M V^\mu). \qquad (5.4.104)$$

Conversely, the lift of V^μ to H^4 defined by (5.4.102) satisfies (5.4.104). In particular if V^a is divergence-free on H^4, then its reduction to $\mathcal{M}^{3,1}$ satisfies

$$\partial_\mu(\rho_M V^\mu) = 0. \qquad (5.4.105)$$

Proof   Using the definition of ∂_μ, we compute

$$\begin{aligned}
\eth_a V^a &= \eth_\mu V^\mu + \eth_4 V^4 \\
&= \left(\partial_\mu + \frac{1}{x_4} x_\mu \eth_4\right) V^\mu + \eth_4 V^4 \\
&= \partial_\mu V^\mu + \frac{1}{x_4} \eth_4(x_\mu V^\mu) - \frac{1}{x_4} V^\mu \eth_4 x_\mu + \eth_4 V^4 \\
&= \partial_\mu V^\mu - \frac{1}{x_4} \eth_4(x_4 V^4) - \frac{1}{R^2} x_\mu V^\mu + \eth_4 V^4 \\
&= \partial_\mu V^\mu - \frac{1}{x_4} V^4 \eth_4 x_4 + \frac{1}{R^2} x_4 V^4 \\
&= \partial_\mu V^\mu - \frac{1}{x_4} V^4 \\
&= \sinh \eta \, \partial_\mu\left(\frac{1}{\sinh \eta} V^\mu\right) \qquad (5.4.106)
\end{aligned}$$

using (5.4.17).   □

As a special case, we recover the identity (5.4.30) by noting that $V^a = \{x^b, x^a\}$ is conserved on H^4. Finally, we note that the divergence constraint (5.4.105) for vector fields on $\mathcal{M}^{3,1}$ can be written in covariant form in terms of the effective metric $G_{\mu\nu}$ on $\mathcal{M}^{3,1}$:

$$\partial_\mu(\rho_M V^\mu) = 0 = \nabla_\mu^{(G)}(\rho^{-2} V^\mu). \qquad (5.4.107)$$

This will be discussed in Section 9.2.1.

This correspondence generalizes immediately to higher-rank tensors on H^4 and $\mathcal{M}^{3,1}$. In particular, two-forms $F_{\mu\nu}$ on $\mathcal{M}^{3,1}$ correspond to (tangential) two-forms on H^4.

5.4.6 Reconstruction of divergence-free vector fields

As for H^4, we can now solve the following "reconstruction" problem on $\mathcal{M}^{3,1}$: Given any $\mathcal{C}^0$-valued divergence-free vector field V^μ,

$$\partial_\mu(\rho_M V^\mu) = 0 \tag{5.4.108}$$

there exists a unique generator $Y \in \mathcal{C}^1$ such that

$$V^\mu = \{Y, x^\mu\}_0. \tag{5.4.109}$$

This can be obtained by lifting V^μ to a divergence-free vector field V^a on H^4 as in Lemma 5.8. Then the result (5.2.127) on H^4 states that $V^a = \{Y, x^a\}_0$ for some $Y \in \mathcal{C}^1$, which implies $V^\mu = \{Y, x^\mu\}_0$. Explicitly, the generator Y is given by

$$Y = \mathcal{D}^+(V) = -3(\Box_H - 4r^2)^{-1}\big(\{V^\mu, x_\mu\} + \{V^4, x_4\}\big)$$

$$= -3(\Box_H - 4r^2)^{-1}\big(\{V^\mu, x_\mu\} + \frac{1}{x_4}\{x_4, x_\mu V^\mu\}\big). \tag{5.4.110}$$

Moreover, Y is uniquely determined by (5.4.109). Indeed, assume that $Y \in \mathcal{C}^1$ satisfies $\{Y, x^\mu\}_0 = 0$. Since x^4 is a function of x^μ, it follows that $\{Y, x^a\}_0 = 0$ on H^4, which implies $Y = 0$ according to Lemma 5.2. This means that the corresponding spin 2 part $\{Y, x^\mu\}_+ \in \mathcal{C}^2$ is uniquely determined by the vector field. In view of Lemma (5.7), this spin 2 component vanishes only for $Y \in \mathfrak{so}(4, 1)$. This will be used in Section 10.2.4 to reconstruct divergence-free frames from suitable deformed backgrounds in the matrix model.

As in the case of fuzzy H_n^4, the realization of a divergence-free vector field V^μ as a Hamiltonian vector field entails the presence of a spin two sibling

$$V^{(2)\mu} = \{Y, x^\mu\}_2 = \mathcal{D}^{++}(V^\mu), \quad \in \mathcal{C}^2 \tag{5.4.111}$$

where $\mathcal{D}^{++}$ is understood in terms of the corresponding vector field V^a on H^4. In other words, the Hamiltonian vector field generated by $Y \in \mathcal{C}^1$ acts on a function $\phi \in \mathcal{C}^0$ as

$$\{Y, \phi(x)\} = \{Y, x^\mu\}\partial_\mu\phi = (V^\mu + V^{(2)\mu})\partial_\mu\phi, \tag{5.4.112}$$

through the sum of the vector fields $V^\mu + V^{(2)\mu} \in \mathcal{C}^0 \oplus \mathcal{C}^2$. Both components are essentially isomorphic and contain the same information, being related by the intertwiner $\mathcal{D}^{++}$. This will apply in particular to the frame in the effective field theory on $\mathcal{M}^{3,1}$ arising from matrix models.

5.4.7 *Local bundle geometry

In Section 5.4.1, the six-dimensional bundle $\mathbb{C}P^{2,1}$ over $\mathcal{M}^{3,1}$ was described in terms of Cartesian x^μ and t_μ generators, subject to some constraints (5.4.7). This description treats these two sets of generators in an asymmetric way, using four x coordinates to describe spacetime and the two independent t coordinates to describe the S^2 fiber. In this section, we will discuss a different and completely symmetric description of the bundle, which helps to understand the structure of the geometry.

We start with the description of the bundle space $\mathbb{C}P^{2,1}$ in terms of the spacelike coordinates x^i and p_j, while x^0 and p_0 are determined by the constraints. This is most transparent in terms of the re-scaled generators

$$p_\mu = rRt_\mu. \tag{5.4.113}$$

Then the constraints (5.4.7) take the form

$$p_0 x^0 = p_k x^k \,,$$
$$-(p_0)^2 + p_i p_i = (x^0)^2 - \vec{x}^2, \tag{5.4.114}$$

where the sum over i is understood. The constraints can be written as

$$(p_0 + x^0)^2 = |\vec{x} + \vec{p}|^2$$
$$(p_0 - x^0)^2 = |\vec{x} - \vec{p}|^2, \tag{5.4.115}$$

which is solved by

$$x^0 = \frac{1}{2}(|\vec{x} + \vec{p}| + |\vec{x} - \vec{p}|) \approx |\vec{p}|$$
$$p_0 = \frac{1}{2}(|\vec{x} + \vec{p}| - |\vec{x} - \vec{p}|) \approx \frac{\vec{p}\vec{x}}{|\vec{p}|} \tag{5.4.116}$$

up to corrections suppressed by $\frac{|\vec{x}|}{|\vec{p}|}$. In these variables, the Poisson brackets take the form

$$\{p_i, x^j\} = rx^4 \delta_i^j, \qquad x^4 = \sqrt{x_\mu x^\mu + R^2}$$
$$\{x^i, x^j\} = \theta^{ij} = -\{p_i, p_j\}. \tag{5.4.117}$$

Thus at late times $\eta \to \infty$, the Poisson structure *almost* reduces to the canonical structure on $\mathbb{R}^3_x \times \mathbb{R}^3_p$, modified by quantized fluxes on the 2-spheres S^2_n.

5.4.8 General $k = -1$ cosmologies

The covariant quantum spacetime $\mathcal{M}^{3,1}$ can be naturally generalized to the following (semi-classical) background

$$T^\mu = \alpha(x^4)t^\mu \tag{5.4.118}$$

for some generic function $\alpha(x^4)$. This is still covariant under $SO(3, 1)$, and hence describes a generic homogeneous and isotropic FLRW-like cosmology with $k = -1$. Such modifications are expected due to quantum corrections in the effective action, but this is also required to obtain a solution of the (semi-) classical matrix model *without mass term*. Indeed, one can find backgrounds with the above structure which satisfy

$$\Box_T T^\mu = 0 \tag{5.4.119}$$

for suitable $\alpha(x^4) \approx x^4$ at late times. This leads to a modified effective geometry, which will not be elaborated here since the salient features of $\mathcal{M}^{3,1}$ are preserved. The appropriate choice of spacetime coordinates $\tilde{x}^\mu$ is discussed in Section 10.2.5.

5.5 Cosmological $k = 0$ quantum spacetime

Using a slight variation of the above ansatz, we obtain a covariant quantum space which describes a flat $k = 0$ FLRW spacetime. Consider the following generators

$$X^a = M^{ac}\alpha_c, \qquad a = 0, \dots, 4$$
$$T^\mu = M^{\mu d}\beta_d, \qquad \mu = 0, \dots, 3 \tag{5.5.1}$$

where M^{ab} are $\mathfrak{so}(4,2)$ generators acting on the minirep $\mathcal{H}_n$, and α, β are given by

$$\alpha_a = r(0,0,0,0,0,1), \qquad \beta_a = \frac{1}{R}(1,0,0,0,1,0) \tag{5.5.2}$$

which satisfy

$$\alpha \cdot \beta = 0 = \beta \cdot \beta, \qquad \alpha \cdot \alpha = -r^2. \tag{5.5.3}$$

X^a is the same as for fuzzy H_n^4 and $\mathcal{M}_n^{3,1}$, but T^μ is different. It is easy to see that these generators satisfy

$$[X^\mu, X^\nu] = -ir^2 M^{\mu\nu}$$
$$[T^\mu, T^\nu] = i\left(-T^\mu\beta^\nu + \beta^\mu T^\nu\right)$$
$$[X^\mu, T^\nu] = i\left(\beta^\mu X^\nu - \eta^{\mu\nu} X^d \beta_d\right)$$
$$= i\left(\beta^\mu X^\nu - \frac{1}{R}\eta^{\mu\nu}(X^0 + X^4)\right) \tag{5.5.4}$$

Explicitly, the T^μ, $\mu = 0, \dots, 3$ generate the relations[15]

$$\boxed{[T^i, T^j] = 0, \qquad [T^0, T^i] = -\frac{i}{R}T^i} \tag{5.5.5}$$

and

$$[T^i, X^j] = -\frac{i}{R}\delta^{ij}(X^0 + X^4)$$
$$[T^i, X^0] = \frac{i}{R}X^i$$
$$[T^0, X^j] = 0$$
$$[T^0, X^0] = -\frac{i}{R}X^4. \tag{5.5.6}$$

---

[15] The commutation relations of the T^μ are sometimes denoted as κ Minkowski space.

Here X^4 is a function of X^μ due to the constraint (5.2.5). These relations simplify for

$$Y^i = X^i, \qquad Y^0 = X^a \beta_a R = X^0 + X^4, \tag{5.5.7}$$

and we obtain

$$\boxed{[T^\mu, Y^\nu] = i\frac{Y^0}{R}\eta^{\mu\nu}.} \tag{5.5.8}$$

Since X^4 is a function of X^μ, the Y^μ generate the same algebra of functions $\mathcal{C}^0$ on quantum spacetime $\mathcal{M}_n^{3,1}$ as the X^μ. The matrix background T^μ is covariant under the Euclidean group $E(3)$ of rotations and translations in three dimensions, generated by $[M^{ij}, .]$ and $[T^i, .]$. In particular, the matrix d'Alembertian

$$\Box\phi = [T_\mu, [T^\mu, \phi]] \tag{5.5.9}$$

is invariant under $E(3)$. This suggests that this space describes a $k = 0$ FLRW cosmology. This can be seen explicitly in the semi-classical limit, where y^μ become spacetime coordinates, and $y^0 \in \mathbb{R}$ is a time-like coordinate. Then t^μ defines again an S^2 bundle over spacetime with nondegenerate Poisson structure, and the matrix background $T^\alpha \sim t^\alpha$ defines a frame

$$E^\alpha = \{t^\alpha, .\}, \qquad E^{\alpha\mu} = \{t^\alpha, y^\mu\} = \frac{y^0}{R}\eta^{\alpha\mu} \tag{5.5.10}$$

(cf. (5.4.39)) which is diagonal due to (5.5.8). Observing that $y^0 = R(\cosh\eta + \sinh\eta) \in \mathbb{R}_{>0}$ in hyperbolic coordinates, we note that the symplectic density $\rho_M = \sinh^{-1}\eta = \rho_M(y^0)$ is $E(3)$ invariant, and takes the asymptotic form

$$\rho_M \sim y_0^{-1} \tag{5.5.11}$$

at late times $y_0 \to \infty$. Then the general results in Section 9.1.1 lead to the $E(3)$–invariant effective metric

$$G_{\mu\nu} = \rho_M y_0^2 \eta_{\mu\nu} \tag{5.5.12}$$

analogous to (5.4.41). This is easily recognized as a cosmological $k = 0$ FLRW metric, featuring a Big Bang at $y^0 = 0$ rather than a Big Bounce. In the late-time regime, the metric becomes

$$G_{\mu\nu} \sim y_0 \eta_{\mu\nu}, \tag{5.5.13}$$

which can be written in FLRW form as

$$ds_G^2 \sim y_0 dy_0^2 + y_0 d\Sigma^2 = dt^2 + a(t)^2 d\Sigma^2, \tag{5.5.14}$$

where $d\Sigma^2 = dr^2 + r^2 d\Omega^2$ is the Euclidean flat metric on $\mathbb{R}_y^3$. This has the form of a $k = 0$ cosmological FLRW metric, with scale parameter $a(t)$ determined by the following equations:

$$a(t)^2 = y_0, \qquad dt = y_0^{1/2} dy_0. \tag{5.5.15}$$

The solution is

$$t = \frac{2}{3}y_0^{3/2} \tag{5.5.16}$$

and therefore

$$a(t) \propto t^{1/3}. \tag{5.5.17}$$

Therefore, the late-time expansion of the universe is quite different from the $k = -1$ case, which fits better to the observational data.

Further comments and reading

The first examples of covariant quantum spaces (in a wider sense) are due to Snyder [172], based on the Poincare algebra, and Yangs variant thereof [214] based on $\mathfrak{so}(4, 1)$. There is considerable literature on these and similar spaces, often using an ad-hoc reduction to the base space and dropping the extra-dimensional modes. Covariant quantum spaces with compact internal spaces and a truncation of the internal modes were first considered in [83]. They were applied in various contexts including string theory and higher-dimensional quantum Hall effect; for some early references see [42, 100, 111, 161, 217], and e.g. [94, 96, 97, 206] for related work.

In view of the covariant FLRW quantum spacetimes with $k = 1$ and $k = 0$, one may wonder about the case $k = +1$. One can indeed find suitable matrix configurations based on either S_N^4 or H_n^4 [185], which lead to an $SO(4)$–invariant matrix d'Alembertian. This requires a matrix background with five rather than four matrices. Then there seems to be no algebra of functions $\mathcal{C}^0$ on spacetime which is respected by the frame, and the general discussion in Chapter 9 no longer applies. This is the reason why the case $k = +1$ is missing in the above discussion. It is not clear if this gap is significant, or simply reflects a lack of inventiveness.

Part III

Noncommutative field theory and matrix models

6 Noncommutative field theory

Given the correspondence between matrix configurations and quantized symplectic spaces, it is natural to generalize field theory within this framework. This leads to noncommutative field theory (NCFT), formulated in terms of matrix models. Note that "noncommutative" should be distinguished from "quantum" here: a NCFT is a deformation or analog of a classical field theory on a quantized space(time). Such a NCFT can then be quantized in a second step, leading to a noncommutative quantum field theory (NCQFT). This second step is naturally defined in terms of a matrix integral, which has a clear-cut definition.

Due to the intrinsic quantization of area and volume, one might expect that the characteristic UV divergences in QFT should not arise in (quantum) field theory on noncommutative spaces. However, such an expectation is too naive in general. Within perturbation theory, it turns out that even though some (suitably generalized) Feynman diagrams become UV finite, a new type of IR divergence arises, which is linked to UV modes running in the loops. This phenomenon is known as UV/IR mixing [141]; it violates the Wilsonian picture of QFT, and typically destroys renormalizability of NCQFT in four dimensions. However, there is an important exception: the maximally supersymmetric noncommutative $\mathcal{N} = 4$ super-Yang–Mills theory is spared from that problem, because its commutative analog is UV finite. This model is nothing but the IKKT matrix model in disguise.

The original computations leading to the discovery of UV/IR mixing were carried out in a traditional momentum-space formulation on the Moyal–Weyl quantum plane. On noncommutative spaces with sufficient symmetry such as quantized coadjoint orbits, analogous computations can be performed using a harmonic mode expansion, cf. [51] for fuzzy S_N^2. Some basic examples of such loop computations will be discussed. However, this approach is cumbersome, and it completely obscures the origin and significance of UV/IR mixing. We will then discuss a different approach based on coherent states and string modes, which makes the underlying mechanism much more transparent, and exhibits the stringy nature of noncommutative field theory. This approach is computationally very simple and powerful, albeit at the expense of rigor and precision: it amounts to a particular approximation, which is very plausible but not yet fully under control. Nevertheless, meaningful approximations are very valuable in physics. In particular, this tool will allow to derive the induced gravity action of the IKKT model in Section 12.

6.1 Noncommutative scalar field theory

As emphasized in the introductory chapters, the algebra of functions on some given quantum space is given by the operator algebra $\mathrm{End}(\mathcal{H})$. To make the presentation more transparent, we shall use the notation

$$\mathcal{A} := \mathrm{End}(\mathcal{H}) \cong \mathcal{H} \otimes \mathcal{H}^* \qquad (6.1.1)$$

in this section. This is viewed as quantization of the algebra of functions $L^2(\mathcal{M})$ via some quantization map $\mathcal{Q}$, which could be the Weyl quantization map (3.1.30) or the coherent state quantization map (3.1.76). As emphasized in Section 4.4, this correspondence is only appropriate for modes in the semi-classical regime, with wavelengths longer than L_{NC}. In that regime, NC field theory behaves very similarly as field theory on a continuous space. However, the vast majority of modes in $\mathcal{A} = \mathrm{End}(\mathcal{H})$ is in the deep quantum regime spanned by the string modes $|x\rangle\langle y|$ (6.2.1), with completely nonlocal behavior unlike anything in ordinary field theory.

This observation already explains the major difference between classical and NC field theory: Classical field theory contains only local fields, which are functions on the underlying manifold. In contrast, most of the functions or eigenmodes in NC field theory are completely nonlocal, and can be viewed as bi-local objects. Hence, an almost-local behavior of NCFT can only be expected in the IR limit. Due to the UV divergences in QFT, these nonlocal modes running in the loops lead to significant nonlocal behavior in NCQFT even in the IR. This is responsible for UV/IR mixing, which is suppressed only in very special, (maximally) supersymmetric models.

The definition of toy models of NCFT is simple. Given some matrix configuration $X^a \in \mathcal{A}$ corresponding to a quantized symplectic space $(\mathcal{M}, \omega)$, we can define a noncommutative analog of scalar field theory on that space through an action of the form

$$S = \mathrm{Tr}\left(-\frac{1}{2}[X^a, \phi][X_a, \phi] + \frac{1}{2}m^2\phi^2 + \frac{\lambda}{4!}\phi^4 \right), \qquad \phi \in \mathcal{A}. \qquad (6.1.2)$$

To be specific, we consider a ϕ^4 theory. The indices in the first term are contracted with the Euclidean metric δ_{ab}, so that the action is positive definite. A prototype for such a model for the fuzzy two-sphere S_N^2 is obtained for X^a, $a = 1, 2, 3$ as in (3.3.13), where S^2 should be viewed as a submanifold of $\mathbb{R}^3$. The generalization to higher-dimensional coadjoint orbits $\mathcal{O} \subset \mathfrak{g} \cong \mathbb{R}^D$ is straightforward, and the formulation applies equally well to non-compact spaces such as the Moyal–Weyl quantum plane $\mathbb{R}^{2n}_\theta$, where X^a, $a = 1, \ldots, 2n$ are given by (3.1.12).

The commutator term in (6.1.2) should be viewed as kinetic part of the action. In the semi-classical regime, it reduces to the standard form of a kinetic term in Euclidean field theory. This is obvious in the Moyal–Weyl case, where

$$-[X^a, \phi][X_a, \phi] = \gamma^{ab}\partial_a\phi\partial_b\phi \qquad (6.1.3)$$

using (3.1.64) and (3.1.68), and γ^{ab} is recognized as effective metric. The effective metric can be extracted similarly for more general quantum spaces, which for symmetric spaces such as S_N^2 is invariant under the symmetry group.

Given such an action, noncommutative analogs of n-point functions are defined via

$$\langle \phi_{\Lambda_1} \ldots \phi_{\Lambda_n} \rangle := \frac{1}{Z} \int D\phi \, \phi_{\Lambda_1} \ldots \phi_{\Lambda_n} e^{-S[\phi]}, \tag{6.1.4}$$

where Λ_i label an ON basis of modes in $\mathcal{A}$, such that

$$\phi = \sum_{\Lambda} \phi_{\Lambda} \mathbf{T}_{\Lambda}, \qquad \mathbf{T}_{\Lambda} \in \mathcal{A}. \tag{6.1.5}$$

Typically the $\mathbf{T}_{\Lambda}$ will be chosen as eigenmodes of the matrix Laplacian $\Box = \delta_{ab}[X^a, [X^b, .]]$ (3.1.67), and on spaces with sufficient symmetries they can be chosen as ONB of irreducible modes which respect these symmetries, such as spherical harmonics on S_N^2, or plane waves on $\mathbb{R}_\theta^{2n}$. The integral $\int D\phi$ is defined to be the integral over the Euclidean vector space $\mathcal{A} = \mathrm{End}(\mathcal{H})$, which is invariant under shifts $\phi \to \phi + \psi$ in field space. This is the natural analog of the Feynman path or functional integral, which is completely well-defined and finite for compact quantum spaces with finite-dimensional $\mathcal{H}$, so that no divergences can arise. On noncompact spaces such as the Moyal–Weyl quantum plane $\mathbb{R}_\theta^{2n}$, UV-divergences typically do arise, though their meaning is different from standard QFT as discussed later.

On the other hand, it is often more useful to expand ϕ in terms of string modes, as discussed in Section 6.2.

Given a quantization map $\mathcal{Q} : \mathcal{C}(\mathcal{M}) \to \mathcal{A}$, we can map the modes $\mathbf{T}_{\Lambda}$ to classical functions $T_{\Lambda}(x)$ via $\mathbf{T}_{\Lambda} = \mathcal{Q}(T_{\Lambda}(x))$, which span the classical space of functions via $\phi(x) = \sum_{\Lambda} \phi_{\Lambda} T_{\Lambda}(x) \in \mathcal{C}(\mathcal{M})$. This allows to define classical n-point functions

$$\langle \phi(x_1) \ldots \phi(x_n) \rangle := \sum T_{\Lambda_1}(x_1) \ldots T_{\Lambda_1}(x_n) \langle \phi_{\Lambda_1} \ldots \phi_{\Lambda_n} \rangle, \tag{6.1.6}$$

which are totally symmetric functions in the classical coordinates x_i. They are a convenient and intuitive way to encode the correlation functions (6.1.4). One might hope that in some suitable limit where quantum space becomes commutative, these n-point functions should reduce to the corresponding n-point functions of the corresponding QFT on commutative $\mathcal{M}$. We will see that this is true for free theories, but typically not for interacting theories, where some nonlocal effects of noncommutative field theory persist even in the commutative limit. This will be referred to as "noncommutative anomaly": quantization and (noncommutative) deformation of field theory do not commute [51], and NCQFT is distinct from ordinary QFT.

It is straightforward to extend the standard formalism of QFT in terms of sources, generating functions and effective action into the noncommutative setting. The correlators or n-point functions are obtained as usual from a generating function

$$Z[J] = \int D\phi \, e^{-S[\phi] + \mathrm{Tr}\phi J} =: e^{-W[J]}. \tag{6.1.7}$$

Then the correlators can be obtained as

$$\langle \phi_{\Lambda_1} \ldots \phi_{\Lambda_n} \rangle = \frac{1}{Z[0]} \frac{\partial^n}{\partial J_{\Lambda_1} \ldots \partial J_{\Lambda_n}} Z[J] \Big|_{J=0}, \tag{6.1.8}$$

while $W[J]$ generates the connected correlators

$$\langle \phi_{\Lambda_1} \ldots \phi_{\Lambda_n} \rangle_c := -\frac{\partial^n}{\partial J_{\Lambda_1} \ldots \partial J_{\Lambda_n}} W[J] \Big|_{J=0}. \tag{6.1.9}$$

For example,

$$\langle \phi_{\Lambda_1} \phi_{\Lambda_2} \rangle_c = \langle \phi_{\Lambda_1} \phi_{\Lambda_2} \rangle - \langle \phi_{\Lambda_1} \rangle \langle \phi_{\Lambda_2} \rangle \tag{6.1.10}$$

and so on. Defining the *classical matrix* in the presence of a source J via

$$\phi_{\mathrm{cl}} := \langle \phi \rangle_J = -\frac{\partial}{\partial J} W[J] \qquad \in \mathrm{End}(\mathcal{H}) \tag{6.1.11}$$

or more explicitly

$$(\phi_\Lambda)_{\mathrm{cl}} = -\frac{\partial}{\partial J_\Lambda} W[J] = \frac{1}{Z[J]} \int D\phi \, \phi_\Lambda e^{-S[\phi] + \mathrm{Tr}(\phi J)} \tag{6.1.12}$$

for any matrix basis $\mathbf{T}_\Lambda$, the *effective action* is defined as Legendre transform of $W[J]$

$$\Gamma[\phi_{\mathrm{cl}}] := W[J] + \mathrm{Tr}(\phi_{\mathrm{cl}} J) \Big|_{J=J(\phi_{\mathrm{cl}})} \tag{6.1.13}$$

expressing J in terms of ϕ_{cl} through (6.1.11). Repeating the standard arguments in field theory or thermodynamics, we conclude that the vacuum expectation values $\phi_{\mathrm{cl}} = \langle \phi \rangle_{J=0}$ are critical points of the effective action,

$$\frac{\partial \Gamma[\phi_{\mathrm{cl}}]}{\partial \phi_{\mathrm{cl}}} = J = 0. \tag{6.1.14}$$

Analogous formulas hold in Minkowski signature, with the appropriate $i\varepsilon$ prescription discussed in Section (7.2). The shift invariance of the integral allows to derive analogs of the Schwinger–Dyson equation in close analogy to the commutative case, as well as Ward identities in the presence of symmetries. For compact quantum spaces, all these manipulations are well-defined and essentially rigorous.

The perturbative analysis of such a NCQFT then proceeds by expanding the integral over ϕ perturbatively using Gaussian integration, based on the standard formulas collected in the appendix. It is straightforward to derive Feynman rules for a perturbative expansion of these correlation functions. Consider first a free theory, defined by an action of the form

$$S_0[\phi] = \frac{1}{2} \mathrm{Tr} \phi \mathcal{K} \phi, \tag{6.1.15}$$

where $\mathcal{K}$ is a positive definite hermitian operator, which typically has the form

$$\mathcal{K} \phi = (\Box + m^2)\phi, \qquad \Box = [X_a, [X^a, .]]. \tag{6.1.16}$$

Here $\phi = \phi^\dagger \in \mathcal{A}$ should be viewed as a real scalar field on the noncommutative space under consideration. We can then compute the free generating function $Z_0[J]$ as usual via a Gaussian integral,

$$Z_0[J] = \int d\phi \, e^{-\text{Tr}(\frac{1}{2}\phi\mathcal{K}\phi - \phi J)} = e^{\frac{1}{2}\text{Tr}J\mathcal{K}^{-1}J} \tag{6.1.17}$$

dropping an irrelevant normalization constant, so that

$$W_0(J) = -\frac{1}{2}\text{Tr}J\mathcal{K}^{-1}J. \tag{6.1.18}$$

In particular, the two-point function is

$$\langle \phi_{\Lambda_1}\phi_{\Lambda_2}\rangle_0 = -\frac{\partial^2}{\partial J_1 \partial J_2} W_0[J]\Big|_{J=0} = \mathcal{K}^{-1}_{\Lambda_1\Lambda_2}. \tag{6.1.19}$$

For complex scalar fields, the action takes the form

$$S_0[\phi] = \text{Tr}\phi^\dagger \mathcal{K}\phi \tag{6.1.20}$$

which leads to

$$Z_0[J] = \int d\phi \, e^{-\text{Tr}(\phi^\dagger \mathcal{K}\phi - \phi^\dagger J - J^\dagger \phi)} = \frac{1}{\mathcal{N}} e^{\text{Tr}J^\dagger \mathcal{K}^{-1}J}. \tag{6.1.21}$$

Then

$$\langle \phi_{\Lambda_1}\phi^*_{\Lambda_2}\rangle_0 = \langle \mathbf{T}_{\Lambda_1}|\mathcal{K}^{-1}|\mathbf{T}_{\Lambda_2}\rangle \tag{6.1.22}$$

can be viewed as matrix elements of the propagator

$$\mathcal{K}^{-1} = (\Box + m^2)^{-1} = \sum \mathbf{T}_{\Lambda_1}\langle \phi_{\Lambda_1}\phi^*_{\Lambda_2}\rangle_0 \mathbf{T}^\dagger_{\Lambda_2} \quad \in \mathcal{A} \otimes \mathcal{A}^*. \tag{6.1.23}$$

Even for real (i.e. hermitian) scalar fields ϕ, it is often useful to consider a complex (i.e. non-hermitian) ON basis $\mathbf{T}_\Lambda$ of matrices such as a plane waves e^{ikX}, spherical harmonics $\hat{Y}^l_m$, or the over-complete basis of string modes $\left|\begin{smallmatrix}x\\y\end{smallmatrix}\right\rangle$ discussed later. Then the expansion coefficients ϕ_Λ satisfy some reality constraint, which will be understood from now on, and (6.1.22) holds also for real fields.

By construction, the free propagator reduces to the classical one in the semi-classical or commutative limit. Let us illustrate this in the example of the two-dimensional fuzzy sphere S^2_N. Then the free action in the basis of spherical harmonics is

$$S_0[\phi] = \frac{1}{2}\sum \phi_{l,-n}(l(l+1) + m^2)\phi_{l,n} \tag{6.1.24}$$

and the two-point function is given by

$$\langle \phi_{l_1 m_1}\phi_{l_2 m_2}\rangle = \frac{\delta_{l_1 l_2}\delta_{m_1,-m_2}}{l(l+1) + m^2}. \tag{6.1.25}$$

This coincides with the commutative result up to the cutoff in l. In this sense, free NCQFT typically reduces to the commutative theory.

We can now write the interacting generating function $Z[J]$ in terms of the free one as follows

$$Z[J] = \int D\phi\, e^{-S_0[\phi]-S_{\text{int}}[\phi]+\text{Tr}\phi J}$$

$$= e^{-S_{\text{int}}[\partial_J]} \int D\phi\, e^{-S_0[\phi]+\text{Tr}\phi J} = e^{-S_{\text{int}}[\partial_J]} Z_0[J]$$

$$= e^{-S_{\text{int}}[\partial_J]} e^{-W_0[J]} = e^{-W[J]}. \tag{6.1.26}$$

Then the n-point functions are obtained as

$$\langle \phi_{\Lambda_1} \cdots \phi_{\Lambda_{2n}} \rangle = \frac{1}{Z[0]} \frac{\partial^n}{\partial J_1 \ldots \partial J_{2n}} e^{-S_{\text{int}}[\partial_J]} Z_0[J]|_{J=0}$$

$$= \sum_{\text{contractions}} \langle \phi\phi \rangle \ldots \langle \phi\phi \rangle. \tag{6.1.27}$$

This is nothing but Wicks theorem, and the factor $Z[0]$ cancels as usual in the connected n-point functions. Expanding $e^{-S_{\text{int}}}$ into a power series e.g. for

$$S_{\text{int}} = \frac{\lambda}{4!} \text{Tr}\phi^4 \tag{6.1.28}$$

yields immediately the Feynman rules for NCQFT in terms of Wick contractions of the vertices with the propagator:

Feynman diagrams and ribbon graphs

As in ordinary QFT, the perturbative expansion can be represented in terms of suitably adapted Feynman diagrams. Following the standard arguments, $W[J]$ generates the connected diagrams, while $\Gamma[\phi_c]$ generates the one-particle irreducible diagrams. However, there is an important difference to ordinary QFT, which is best illustrated in the ϕ^4 model. Since the fields $\phi \in \mathcal{A} \cong \text{End}(\mathcal{H})$ are matrices and the vertices are defined by a matrix product, the ordering of the lines within a vertex does matter, apart from cyclic re-ordering which is trivial due to the trace property. This leads to a distinction between planar and nonplanar diagrams, which is best incorporated by making the index structure of matrices explicit, and representing the propagator by a double line rather than a single line, as familiar from the t'Hooft large N expansion of nonabelian gauge theory [196].

To make this explicit, consider some basis $(\mathbf{T}_\Lambda)^i_j$ of the space of matrices $\mathcal{A}$ (6.1.5), such as the fuzzy spherical harmonics on S^2_N (3.3.21). Since the vertices are defined through the matrix product, we should keep track of the matrix indices i, j and contract them appropriate in the vertices. This can be achieved using the t'Hooft double line representation. For example, a quartic vertex is represented in that basis as

$$= \tfrac{\lambda}{4!} \text{Tr}(\mathbf{T}_{\Lambda_1}\mathbf{T}_{\Lambda_2}\mathbf{T}_{\Lambda_3}\mathbf{T}_{\Lambda_4}) = \tfrac{\lambda}{4!}(\mathbf{T}_{\Lambda_1})^i_j(\mathbf{T}_{\Lambda_2})^j_k(\mathbf{T}_{\Lambda_3})^k_l(\mathbf{T}_{\Lambda_4})^l_i,$$

$$\tag{6.1.29}$$

where upper matrix indices are represented by lines with out-going arrows. These are contracted with the propagator

$$= \langle \mathbf{T}_{\Lambda_1} | (\Box + m^2)^{-1} | \mathbf{T}_{\Lambda_2} \rangle = \langle \phi_{\Lambda_1} \phi^*_{\Lambda_2} \rangle \qquad (6.1.30)$$

using complex modes, summing over the labels Λ_i which are suppressed. Alternatively, one can view the propagator as element in $\mathcal{A} \otimes \mathcal{A}^*$ (6.1.23) represented by a double-line, and then the vertices would only contain the factor $\frac{\lambda}{4!}\delta$, where δ implements the appropriate contraction of indices. More abstractly, the vertex can be viewed as element in $\mathcal{A}^{\otimes 4}$, which is contracted with the propagator $\mathcal{K}^{-1} \in \mathcal{A} \otimes \mathcal{A}^*$ using the canonical identification $\mathcal{A} \cong \mathcal{A}^*$ via the trace. Either way, the matrix indices i, j are often suppressed, and the Wick expansion (6.1.27) is represented graphically through **ribbon graphs**, where the sum over modes is understood. In particular, the ordering of the legs in the vertices must be respected.

As familiar from the t'Hooft double line representation of nonabelian gauge theory, diagrams can now be organized in terms of the minimal genus of a Riemann surface on which the diagrams can be drawn without crossing. This arises by defining "faces" in a diagram through closed single lines; then the Euler character of a graph can be defined as $\chi = V - E + F$, with V, E, and F the numbers of vertices, edges and faces respectively. Planar diagrams are those which can be drawn in a plane without crossing lines. A diagram with $\chi = 2 - 2g$ for $g \geq 1$ is nonplanar of genus g and can be drawn on a surface of genus g without crossings. If the propagators were $\propto \delta^{i_1 i_2} \delta_{j_1 j_2}$ as in the single-matrix models discussed in Section 7.2.1, the diagrams would scale like $(\dim(\mathcal{H}))^\chi$ (upon appropriate rescaling of the action), which means that planar diagrams dominate. Even though the propagators are more complicated in NCFT, the distinction between planar and nonplanar diagrams remains highly relevant here, in contrast to ordinary scalar field theory.

Let us illustrate this by the one-loop corrections to the propagator, which is given by the following two different types of diagrams

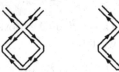

The planar diagram is given by

$$= \frac{\lambda}{4!} \operatorname{Tr}(\mathbf{T}_{\Lambda_1} \mathbf{T}_{\Lambda_3} \mathbf{T}^\dagger_{\Lambda_4} \mathbf{T}_{\Lambda_2}) \langle \phi_{\Lambda_3} \phi^*_{\Lambda_4} \rangle,$$

where only adjacent legs of the vertex are contracted. The nonplanar diagram is given by

$$= \frac{\lambda}{4!} \operatorname{Tr}(\mathbf{T}_{\Lambda_1} \mathbf{T}_{\Lambda_3} \mathbf{T}_{\Lambda_2} \mathbf{T}^\dagger_{\Lambda_4}) \langle \phi_{\Lambda_3} \phi^*_{\Lambda_4} \rangle,$$

where nonadjacent legs are contracted.

To evaluate such diagrams, we must choose some basis $\mathbf{T}_\Lambda$ of $\mathcal{A}$, and compute the vertices by evaluating their matrix product. If the quantum space admits a large symmetry group such as $\mathbb{R}^{2n}_\theta$ or S^2_N, it seems natural to choose a basis consisting of irreducible representations of that group, i.e. plane waves e^{ikX} or $\hat{Y}^l_m$, respectively. Then the propagator manifestly respects that symmetry, so that the label Λ of the irrep is preserved along the (double) lines. To evaluate the vertices, we need to know the multiplication law

$$\mathbf{T}_\Lambda \mathbf{T}_{\Lambda'} = \sum_K C^K_{\Lambda \Lambda'} \mathbf{T}_K \tag{6.1.31}$$

in $\mathcal{A}$. This is easy for $\mathbb{R}^{2n}_\theta$ but nontrivial for curved spaces such as S^2_N, where the $C^K_{\Lambda\Lambda'}$ boil down to $6J$ symbols. In the semi-classical or IR regime, these group-theoretical structure constants $C^K_{\Lambda\Lambda'}$ approximately coincide with the classical ones, according to the correspondence principle (3.0.7). In the deep quantum regime they are typically highly oscillatory, and the evaluation of the corresponding sums or integrals in the diagrams becomes very problematic. We shall illustrate such computations for $\mathbb{R}^{2n}_\theta$ and S^2_N, and discuss a better approach in Section 6.2.

$\mathbb{R}^{2n}_\theta$ and plane waves

These diagrams can be evaluated straightforwardly on $\mathbb{R}^{2n}_\theta$ using a basis of plane waves, expanding the (real) field as

$$\phi = \int \frac{d^{2n}k}{(2\pi)^{2n}} \phi_k \mathbf{T}_k, \qquad \mathbf{T}_k = e^{ikX} \tag{6.1.32}$$

with

$$\phi^*_k = \phi_{-k}, \qquad \mathbf{T}^\dagger_k = \mathbf{T}_{-k}. \tag{6.1.33}$$

There is no UV cutoff in $k \in \mathbb{R}^{2n}$ on the Moyal–Weyl quantum plane. Then the action for scalar ϕ^4 theory can be written as[1]

$$S = \int_{\mathbb{R}^{2n}_\theta} \frac{1}{2} \delta^{ab} \partial_a \phi \partial_b \phi + \frac{1}{2} m^2 \phi^2 + \frac{\lambda}{4!} \phi^4$$

$$= \int \frac{d^{2n}k}{(2\pi)^{2n}} \frac{1}{2}(k^2 + m^2)\phi_k \phi^*_k + \int \frac{d^{2n}k_1}{(2\pi)^{2n}} \cdots \frac{d^{2n}k_4}{(2\pi)^{2n}} \frac{\lambda}{4!} \Big(\int_{\mathbb{R}^{2n}_\theta} \mathbf{T}_{k_1} \dots \mathbf{T}_{k_4} \Big) \phi_{k_1} \dots \phi_{k_4}. \tag{6.1.34}$$

Here the integral over $\mathbb{R}^{2n}_\theta$ is defined by $\int_{\mathbb{R}^{2n}_\theta} \equiv (2\pi\theta)^n \mathrm{Tr}_{\mathcal{H}}$, which leads to the standard formula $\int e^{ikX} = (2\pi)^{2n} \delta^{2n}(k)$ (3.1.9). The propagator in the momentum basis then coincides exactly with the usual one, given by

$$\langle \phi_k \phi^*_{k'} \rangle = \quad\xrightarrow{\hspace{2cm}}\quad = \frac{\delta(k-k')}{k^2+m^2} \tag{6.1.35}$$

---

[1] we assume that the effective metric (6.1.3) and the scaling constants on $\mathbb{R}^{2n}_\theta$ are normalized accordingly.

for external momenta k and k' in Euclidean signature. The vertices can be evaluated using the Weyl algebra (3.1.20)

$$\mathbf{T}_k \mathbf{T}_{k'} = e^{-ik_a\theta^{ab}k'_b} \mathbf{T}_{k'} \mathbf{T}_k = e^{-\frac{i}{2}k_a\theta^{ab}k'_b} \mathbf{T}_{k+k'}. \tag{6.1.36}$$

In the planar diagram, the contraction of the adjacent vertices leads to $\mathbf{T}_{k_3}\mathbf{T}_{k_4}^\dagger = e^{ik_3 X}e^{-ik_4 X} = \mathbb{1}$ due to the momentum conservation $k_3 = k_4$ in the propagator. This means that the vertex $\int \mathbf{T}_{k_1}\mathbf{T}_{k_2}\mathbf{T}_{k_3}\mathbf{T}_{k_4}^\dagger = (2\pi)^{2n}\delta(k_1 + k_2)$ does not contribute any phase factor. Hence the planar diagram is given by

$$\vcenter{\hbox{}} = \delta(k-k')\tfrac{\lambda}{3} \int \frac{d^{2n}p}{(2\pi)^{2n}} \frac{1}{p^2+m^2} =: \delta(k-k')\Gamma_{\mathrm{P}} \tag{6.1.37}$$

taking into account the multiplicity factor 8, which is the number of different contractions leading to the same planar diagram. This amounts to a divergent mass renormalization which is the same as in ordinary QFT, except for the combinatorial factor. The fact that it is UV-divergent seems strange at first sight, since $\mathbb{R}_\theta^{2n}$ is quantized; this will be understood in Section 6.2 in terms of a general UV/IR relation on quantum spaces. That divergence disappears on compact quantum spaces, as illustrated in the following.

Now consider the nonplanar diagram. Here the contraction of the nonadjacent vertices leads upon reordering $\mathbf{T}_{k_2}\mathbf{T}_{k_4}^\dagger = e^{ik_2\theta k_4}\mathbf{T}_{k_4}^\dagger\mathbf{T}_{k_2}$ to a phase factor using (6.1.36). Then

$$\vcenter{\hbox{}} = \delta(k-k')\tfrac{\lambda}{6} \int \frac{d^{2n}p}{(2\pi)^{2n}} \frac{e^{ik_a\theta^{ab}p_b}}{p^2+m^2} =: \delta(k-k')\Gamma_{\mathrm{NP}}(k) \tag{6.1.38}$$

taking into account the multiplicity factor 4. This result is very strange: for non-vanishing external momentum $k \neq 0$ the integral is oscillatory and finite, but it becomes divergent as $k \to 0$. Once again this is a manifestation of a general UV/IR relation on quantum spaces. The result (6.1.38) becomes more transparent in position space, where it amounts to the following nonlocal contribution to the one-loop effective action

$$\begin{aligned}
\Gamma_{\mathrm{NP}} &= \frac{\lambda}{12(2\pi)^{2n}} \int \frac{d^{2n}k}{(2\pi)^{2n}} \phi_{-k}\phi_k \int \frac{d^{2n}p}{(2\pi)^{2n}} \frac{e^{ik_a\theta^{ab}p_b}}{p^2+m^2} \\
&= \frac{\lambda}{12(2\pi)^{2n}L_{\mathrm{NC}}^{4n}} \int \frac{d^{2n}k}{(2\pi)^{2n}} \phi_{-k}\phi_k \int \frac{d^{2n}\tilde{p}}{(2\pi)^{2n}} \frac{e^{ik_a\tilde{p}^a}}{\tilde{p}^a\tilde{p}^b\gamma_{ab}+m^2} \\
&= \frac{\lambda}{12} \frac{1}{(2\pi L_{\mathrm{NC}})^{4n}} \int d^{2n}x\, d^{2n}y \frac{\phi(x)\phi(y)}{|x-y|_\gamma^2 + m^2},
\end{aligned} \tag{6.1.39}$$

where γ denotes the metric (3.1.68). The universal nature and origin of this nonlocal contribution will become clear in terms of string modes in Section 6.2.

Even though these computations are fairly simple, the existence of UV divergences leaves some lingering doubt on the consistency of the model. These concerns can be resolved by carrying out an analogous computation for compact quantum spaces, starting with the example of S_N^2.

S_N^2 and spherical harmonics

Consider the analogous one-loop contributions on the fuzzy sphere S_N^2. Then a natural basis of $\mathcal{A}$ is given by the spherical harmonics,

$$\phi = \sum \phi_\Lambda \mathbf{T}_\Lambda, \qquad \mathbf{T}_\Lambda = \hat{Y}_n^l. \tag{6.1.40}$$

(3.3.21), (3.3.25). Here $\Lambda = (l, n)$ labels the $\mathfrak{su}(2)$ irreps which comprise $l = 0, 1, \ldots,$ $N - 1$ and $-l \leq n \leq l$, with reality constraints

$$\phi_{l,n}^* = (-1)^n \phi_{l,-n}, \qquad \hat{Y}_n^{l\dagger} = (-1)^n \hat{Y}_{-n}^l. \tag{6.1.41}$$

Then the action for scalar ϕ^4 theory can be written as

$$S = \int_{S_N^2} \frac{1}{2}\phi \Box \phi + \frac{1}{2}m^2\phi^2 + \frac{\lambda}{4!}\phi^4$$

$$= \sum_{l,n} \frac{1}{2}\big(l(l+1) + m^2\big)\phi_{l,n}^* \phi_{l,n} + \sum \frac{\lambda}{4!}\Big(\int_{S_N^2} \mathbf{T}_{\Lambda_1} \ldots \mathbf{T}_{\Lambda_4}\Big) \phi_{\Lambda_1} \ldots \phi_{\Lambda_4} \tag{6.1.42}$$

Here the integral over S_N^2 is defined by $\int_{S_N^2} \equiv \frac{4\pi}{N}\mathrm{Tr}_{\mathcal{H}}$, and $\Box = \sum_a J_a^2$ is the $SU(2)$ Casimir. Then the propagator takes the form

$$\langle \phi_\Lambda \phi_{\Lambda'}^* \rangle = \quad\underrightarrow{\rule{3cm}{0pt}}\quad = \frac{\delta_{l,l'}\delta_{n,n'}}{l(l+1)+m^2}, \tag{6.1.43}$$

which coincides with the commutative theory up to the UV cutoff in l. The product of the $\mathbf{T}_\Lambda$ is more complicated, given in terms of $6J$ symbols (3.3.26). In the planar diagram, the contraction of adjacent vertices simplifies again using

$$\sum_{n=-l}^{l} \hat{Y}_n^l \hat{Y}_n^{l\dagger} = \frac{2l+1}{4\pi}\mathbb{1} \tag{6.1.44}$$

as operators acting on $\mathcal{H}$, due to $SU(2)$ invariance. Thus the vertex does not contribute any phase factor, and the planar diagram reduces to

$$\includegraphics = \delta^{ll'}\delta_{nn'} \frac{\lambda}{3}\frac{1}{4\pi}\sum_{j=0}^{N-1} \frac{2j+1}{j(j+1)+m^2} =: \delta^{ll'}\delta_{nn'}\Gamma_{\mathrm{P}}. \tag{6.1.45}$$

This is independent of the external momentum l, and amounts to a finite mass renormalization which scales like $\ln(N)$ with the cutoff, as expected in a two-dimensional QFT. This is clearly a regular analog of (6.1.37), which is recovered in the flat limit $S_N^2 \to \mathbb{R}_\theta^2$ as will be shown in the following.

The evaluation of the nonplanar diagram is more complicated, and requires to use the $6J$ symbols in the product formula (3.3.26). After unwinding these ingredients, the nonplanar diagram can be written as [51, 205]

$$
= \delta^{ll'} \delta_{nn'} \frac{\lambda}{6} \frac{1}{4\pi} \sum_{j=0}^{N-1} (-1)^{l+j+2\alpha} \frac{(2j+1)(2\alpha+1)}{j(j+1)+m^2} \left\{ \begin{array}{ccc} \alpha & \alpha & l \\ \alpha & \alpha & j \end{array} \right\}
$$

$$
=: \delta^{ll'} \delta_{nn'} \Gamma_{\mathrm{NP}}(l), \tag{6.1.46}
$$

where $\alpha = (N-1)/2$. This is a nontrivial and rather obscure function of the external momentum l, which becomes more transparent using the asymptotic formula

$$
(-1)^{l+j+2\alpha} \left\{ \begin{array}{ccc} \alpha & \alpha & l \\ \alpha & \alpha & j \end{array} \right\} \approx \frac{1}{2\alpha} P_l \left(1 - \frac{j^2}{2\alpha^2} \right) \tag{6.1.47}
$$

due to Racah, where P_l are the Legendre polynomials. The approximation is very good for all $0 \leq j \leq 2\alpha$, provided α is large and $l \ll \alpha$. Then

$$
\Gamma_{\mathrm{NP}}(l) \approx \frac{\lambda}{6} \frac{1}{4\pi} \sum_{j=0}^{N-1} \frac{(2j+1)(2\alpha+1)}{j(j+1)+m^2} \frac{1}{2\alpha} P_l \left(1 - \frac{j^2}{2\alpha^2} \right)
$$

$$
\approx \frac{\lambda}{6} \frac{1}{4\pi} \int_0^2 du \, \frac{2u + \frac{1}{\alpha}}{u^2 + \frac{u}{\alpha} + \frac{m^2}{\alpha^2}} P_l \left(1 - \frac{u^2}{2} \right)
$$

$$
= \frac{\lambda}{6} \frac{1}{4\pi} \int_{-1}^1 dt \, \frac{P_l(t)}{1 - t + \frac{2m^2}{N^2}} + O(1/N), \tag{6.1.48}
$$

where $u = j/N$ and $t = 1 - \frac{u^2}{2}$. The Legendre polynomials $P_l(t)$ are highly oscillatory, which obscures the physical significance of these nonplanar graphs. As for the Moyal–Weyl quantum plane, this contribution to the effective action becomes much more transparent in position space:

Commutative limit $S_N^2 \to S^2$ and noncommutative anomaly

The nonplanar contributions of the one-loop effective action for scalar field theory deserves further discussion. We claim that it can be written in position space for large $N \to \infty$ as follows

$$
\Gamma_{\mathrm{NP}} = \frac{\lambda}{6} \frac{1}{4\pi^2} \int_{S^2 \times S^2} dx dy \frac{\phi(x)\phi(y)}{|x-y|^2 + \frac{4m^2}{N^2}} \tag{6.1.49}
$$

for real fields $\phi = \phi^*$, in complete analogy to (6.1.39). To see this, it suffices to evaluate this for $\phi = Y_n^l$ and sum over n, which gives

$$
(2l+1)\Gamma_{\mathrm{NP}}(l) = \frac{\lambda}{6}\frac{1}{4\pi^2}\int\limits_{S^2\times S^2} dx dy\, \frac{\sum_n Y_n^l(x)Y_n^{l*}(y)}{|x-y|^2 + \frac{4m^2}{N^2}}
$$

$$
= \frac{\lambda}{6}\frac{2l+1}{4\pi}\frac{1}{4\pi^2}\int\limits_{S^2\times S^2} dx dy\, \frac{P_l(\cos\vartheta)}{|x-y|^2 + \frac{4m^2}{N^2}}
$$

$$
= \frac{\lambda}{6}\frac{2(2l+1)}{4\pi}\int_0^\pi d\vartheta\,\sin\vartheta\,\frac{P_l(\cos\vartheta)}{(1-\cos\vartheta)^2 + \sin\vartheta^2 + \frac{4m^2}{N^2}}
$$

$$
= \frac{\lambda}{6}\frac{2l+1}{4\pi}\int\limits_{-1}^{1} dt\, \frac{P_l(t)}{1-t+\frac{2m^2}{N^2}} = (2l+1)\Gamma_{\mathrm{NP}}(l), \tag{6.1.50}
$$

where $x\cdot y = \cos\vartheta$, using the spherical harmonics addition theorem

$$
P_l(\cos\vartheta) = \frac{4\pi}{2l+1}\sum_n Y_n^l(x)Y_m^{l*}(y). \tag{6.1.51}
$$

This reproduces the result (6.1.48), which is recognized as a nonlocal contribution to the effective action. The universal origin of this contribution will become clear later.

The most striking fact is that the nonlocal term (6.1.49) has no analog in the 1-loop effective action of QFT on the classical 2-sphere or on $\mathbb{R}^2$. It is hence a genuine new effect of noncommutativity. This may not seem too disturbing at first sight but it should be, since this term survives in the commutative limit $N \to \infty$ on S_N^2 for fixed radius. In other words: quantizing the scalar field theory on S_N^2 and then taking the commutative limit $N \to \infty$ gives a different effective action $\Gamma[S^2]$ than first taking the commutative limit $S_N^2 \to S^2$ and then quantizing the theory. Hence the diagram

$$
\begin{array}{ccc}
S_N^2 & \longrightarrow & S^2 \\
\downarrow & & \downarrow \\
\Gamma[S_N^2] & \longrightarrow & \Gamma[S^2]
\end{array}
\tag{6.1.52}
$$

does not commute, and quantum field theory feels the "shadow" of noncommutativity even in the commutative limit. This is somewhat reminiscent of anomalies in QFT, and may hence be denoted as **noncommutative anomaly** [51]. It is a manifestation of the more general UV/IR mixing due to nonlocal modes running in the loops, which will be understood better in the following sections.

This observation is rather disturbing, since scalar ϕ^4 field theory on classical S^2 is a renormalizable quantum field theory. Conventional wisdom then suggests that the resulting quantum field theory should be independent of the specific UV regularization of the theory. The fuzzy sphere S_N^2 seems like a perfectly fine UV regularization, yet we do not recover the standard QFT. How is this possible?

The origin of the anomaly is the fact that the space of functions on quantum spaces is vastly bigger than the space of commutative functions in the semi-classical regime, and comprises a huge pool of highly nonlocal modes in the deep quantum regime. Since the loops are dominated by the UV contributions, these UV modes have a significant impact even in the commutative limit. Hence, the UV regularization provided by compact quantum spaces is more intricate than a naive UV cutoff. NCQFT is clearly well-defined on compact quantum spaces, but it differs significantly from ordinary QFT, at least for generic, nonsupersymmetric models.

These observations may be interpreted in different ways. One may simply conclude that QFT on noncommutative spaces is not related to ordinary QFT. A more cautious view is to take the noncommutative anomaly as a warning, and as an indication that a clean construction of a theory comprising quantized fields may lead to surprising and significant physical implications. This is relevant since no fully satisfactory UV regularization and construction of ordinary QFT is available in more than two dimensions, notably in the context of gauge theories. We will see that gauge theories do fit very well into the noncommutative framework of matrix models.

Quantum plane limit $S_N^2 \to \mathbb{R}_\theta^2$

Recall from Section 3.3.1 that the two-dimensional Moyal–Weyl quantum plane $\mathbb{R}_\theta^2$ can be approximated by taking the scaling limit of the fuzzy sphere with radius

$$R = \sqrt{\frac{N}{2}\theta} \tag{6.1.53}$$

keeping θ constant as $N \to \infty$. To describe a scalar field theory with fixed mass $\tilde{m}$, we should also scale

$$m^2 = \tilde{m}^2 R^2 \sim N \tag{6.1.54}$$

such that $\tilde{m}$ is fixed. Matching the Laplacian on the plane with that on the sphere $j(j+1)/R^2 = p^2$ in the large radius limit, it follows that

$$p = j/R. \tag{6.1.55}$$

Then the planar diagram (6.1.45) on S_N^2 becomes

$$\Gamma_P = 2 \int_0^\Lambda dp \frac{p}{p^2 + \tilde{m}^2} \tag{6.1.56}$$

with UV cutoff $\Lambda = \sqrt{2N\theta^{-1}}$, which recovers the planar contribution (6.1.37) to the two point function on $\mathbb{R}_\theta^2$. It is not hard to see that the nonplanar contribution (6.1.38) to the two point function on $\mathbb{R}_\theta^2$ is also recovered from S_N^2 in this way, consistent with the universal form in position space (6.1.39), (6.1.49). Hence the rigorous one-loop calculation on S_N^2 provides a justification for the simpler but formal computations on $\mathbb{R}_\theta^2$.

Even though the bases of $\mathcal{A}$ discussed so far are suggested by the underlying symmetry, other choices are often more useful. One choice is a basis of rank one elementary matrices given by $(E_a^b)^i{}_j = \delta_a^i \delta_j^b$. Then the vertices are simple, but the propagator is typically complicated and does not respect the matrix indices ij; nevertheless this is useful in some models, such as the Grosse–Wulkenhaar model discussed in Section 6.4. However, the most useful but over-complete basis is given by the string modes discussed in the next section, for which both the vertices and the propagator are quite simple. Then the evaluation of such diagrams simplifies greatly, even on generic quantum spaces without any symmetries. The oscillatory behavior of nonplanar diagrams is avoided altogether, and the nonlocal action(s) will be obtained directly in position space.

One-loop effective action

As in standard QFT, the perturbative expansion of n-point functions and of the effective action can be organized in terms of a loop expansion. This amounts to an expansion in the Planck constant $\hbar$, replacing $S \to \frac{1}{\hbar}S$. There is a particularly simple and useful formula for the 1-loop contribution to the effective action

$$\Gamma_{\text{eff}}[\phi_{\text{cl}}] = \frac{1}{\hbar}S[\phi_{\text{cl}}] + \Gamma_{1\,\text{loop}}[\phi_{\text{cl}}] + \sum_{n>0} O(\hbar^n) \tag{6.1.57}$$

which is obtained by approximating the integral (6.1.12) over the fluctuations $\phi = \phi_{\text{cl}} + \delta\phi$ around some classical background by a Gaussian integral. This leads to

$$\Gamma_{1\,\text{loop}}[\phi_{\text{cl}}] = \frac{1}{2}\text{Tr}_{\text{End}(\mathcal{H})} \log\left(S''[\phi]\right)$$
$$(\delta\phi, S''[\phi_{\text{cl}}]\delta\phi) = \frac{1}{N}\text{tr}\left(\delta\phi(\Box + m^2)\delta\phi + \frac{\lambda}{3}\phi_{\text{cl}}^2\delta\phi^2 + \frac{\lambda}{6}\delta\phi\phi_{\text{cl}}\delta\phi\phi_{\text{cl}}\right), \tag{6.1.58}$$

where $S''[\phi_{\text{cl}}]$ is the quadratic form for fluctuations $\delta\phi$ around the background ϕ_{cl}. This will allow to reproduce the present Feynman graph computations in a more transparent way, and it will be applied to Yang–Mills matrix models in Section 7.6.

6.2 String modes and trace formulas

We will now use coherent states to define string modes, which provide crucial new insights into the UV regime of noncommutative field theory. UV/IR mixing will be understood as an essential feature of NCQFT, which captures the nonlocal nature of quantum spaces and the breakdown of the Wilsonian paradigm of QFT.

The central role in the quantization of a symplectic space $(\mathcal{M}, \omega)$ is played by the quantization map $\mathcal{Q}\colon L^2(\mathcal{M}) \to \text{End}(\mathcal{H})$, which allows to interpret observables in $\text{End}(\mathcal{H})$ as quantized functions on $\mathcal{M}$. This is most transparent in the example of the fuzzy 2-sphere S_N^2, where every $\text{End}(\mathcal{H})$ is spanned by the fuzzy spherical harmonics $\hat{Y}_m^l$.

However, it is actually misleading to think of modes with $l > l_{\text{NC}} = \sqrt{N}$ as quantized functions, or more generally for modes with energy above the semi-classical regime L_{NC}^{-1} on any quantized space. Instead, it is more adequate to describe this regime in terms of nonlocal **string modes**. String modes are nonclassical bi-local modes in the algebra of "functions" $\text{End}(\mathcal{H})$ on some quantum space $\mathcal{M}$, defined as[2]

---

[2] A quantum mechanical analog of string modes would be density matrices built from coherent states.

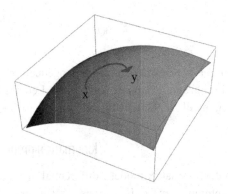

Figure 6.1

Visualization of a string mode $\left| {x \atop y} \right)$ on a quantum space $\mathcal{M}$.

$$\left| {x \atop y} \right) := |x\rangle\langle y| \qquad \in \mathrm{End}(\mathcal{H}), \tag{6.2.1}$$

where $|x\rangle, |y\rangle$ are (quasi-) coherent states on $\mathcal{M}$. We use the quantum mechanical "ket" notation to emphasize that this is a normalized vector in the Hilbert space $\mathrm{End}(\mathcal{H})$, equipped with the (Hilbert–Schmidt) inner product (3.0.17). Accordingly, the dual or adjoint string "bra" is denoted with $\left({x \atop y} \right|$. These modes have several remarkable properties. One obvious property is bi-locality in x and y, which is manifest in the inner product as

$$\left({x \atop y} \Big| {x' \atop y'} \right) = \mathrm{Tr}\big(|y\rangle\langle x|x'\rangle\langle y'|\big) = \langle x|x'\rangle\langle y'|y\rangle \approx e^{-\frac{1}{4}|x-x'|_g^2 - \frac{1}{4}|y-y'|_g^2}\, e^{i\varphi} \tag{6.2.2}$$

using the localization properties of (quasi-) coherent states (4.2.24). Here g is the quantum metric and φ is a gauge-dependent phase, which can locally be chosen as on the Moyal–Weyl quantum plane (3.1.113)

$$\varphi = \varphi(x,x';y,y') = -\frac{1}{2}(x^\mu \theta_{\mu\nu}^{-1} x'^\nu - y^\mu \theta_{\mu\nu}^{-1} y'^\nu). \tag{6.2.3}$$

Note that

$$\delta_{\mathrm{NC}}(x,y) = |\langle x|y\rangle|^2 \sim e^{-\frac{1}{2}|x-x'|_g^2} \tag{6.2.4}$$

defines an effective delta function on $\mathcal{M}$ for the semi-classical regime.

It will be useful to take a more abstract point of view, identifying the space of string modes with

$$\left\{ \left| {x \atop y} \right), \ x \in \mathcal{M}, y \in \mathcal{M} \right\} \cong \mathcal{M}_x \times \mathcal{M}_y. \tag{6.2.5}$$

This space inherits the symplectic (product) measure from the symplectic measure Ω on $\mathcal{M}$. Most importantly, the string modes provide an over-complete basis of $\mathrm{End}(\mathcal{H})$. We assume that the (quasi-) coherent states on $\mathcal{M}$ satisfy the completeness relation (4.2.36) with (4.2.37). Noting that $\mathrm{End}(\mathcal{H}) \cong \mathcal{H} \otimes \mathcal{H}^*$, we obtain the following (approximate or exact) completeness relation for string modes in $\mathrm{End}(\mathcal{H})$:

$$\mathbb{1}_{\mathrm{End}(\mathcal{H})} \approx \int_{\mathcal{M}_x \times \mathcal{M}_y} \frac{\Omega_x \Omega_y}{(2\pi)^m} \left| \begin{smallmatrix} x \\ y \end{smallmatrix} \right) \left(\begin{smallmatrix} x \\ y \end{smallmatrix} \right|, \tag{6.2.6}$$

where $\dim \mathcal{M} = m$. For quantized homogeneous spaces the relation is exact, and follows directly from group invariance and Schur's Lemma. This relation is crucial to understand the UV-regime of NC field theory and the quantization of matrix models.

Kinematical properties

In the context of matrix models, differential operators on quantum spaces are realized in terms of commutators i.e. derivations. Any given matrix configuration X^a naturally defines the following derivative or "matrix momentum" operators acting on $\mathrm{End}(\mathcal{H})$:

$$\mathcal{P}^a \phi := [X^a, \phi], \qquad \phi \in \mathrm{End}(\mathcal{H})$$
$$\Box \phi := \mathcal{P}^a \mathcal{P}_a \phi. \tag{6.2.7}$$

These can be viewed as quantized differential operators on $\mathcal{M}$, which will play an important role in the following. Acting on (quantized) function on $\mathcal{M}$ in the semi-classical regime i.e. with wavelength $> L_{\mathrm{NC}}$, the $\mathcal{P}^a \sim i\{x^a, .\}$ are adequately interpreted as (Hamiltonian) vector fields on $\mathcal{M}$. For example if $\phi = e^{i\tilde{p}X}$ is a plane wave on $\mathbb{R}_\theta^{2n}$ where $\tilde{p}_a = p^b \theta_{ba}^{-1}$, then

$$\mathcal{P}^a e^{i\tilde{p}X} = [X^a, e^{i\tilde{p}X}] = -\theta^{ab} \tilde{p}_b e^{i\tilde{p}X} = p^a e^{i\tilde{p}X}, \tag{6.2.8}$$

cf. (3.1.23). Outside of the semi-classical regime, this interpretation is inappropriate, and operators such as $\mathcal{P}^a \in \mathrm{End}(\mathrm{End}(\mathcal{H}))$ are in general *nonlocal*. Although this nonlocality is only significant for functions in the UV regime, it plays a crucial role in quantum field theory on quantum spaces, due to virtual modes propagating in the loops. This is manifest in terms of the string representation, i.e. for matrix elements w.r.t. string modes. For example, the matrix elements

$$\left(\begin{smallmatrix} x \\ y \end{smallmatrix} \right| \mathcal{P}^a \left| \begin{smallmatrix} x' \\ y' \end{smallmatrix} \right) \approx (\mathbf{x}^{\mathbf{a}}(x) - \mathbf{x}^{\mathbf{a}}(y)) \langle x|x' \rangle \langle y'|y \rangle \tag{6.2.9}$$

$$\left(\begin{smallmatrix} x \\ y \end{smallmatrix} \right| \mathcal{P}^a \left| \begin{smallmatrix} x \\ y \end{smallmatrix} \right) = \mathbf{x}^{\mathbf{a}}(x) - \mathbf{x}^{\mathbf{a}}(y) \approx x^a - y^a \tag{6.2.10}$$

of $\mathcal{P}^a$ are approximately diagonal, using $\langle x|X^a|x' \rangle \approx \mathbf{x}^{\mathbf{a}}(x) \langle x|x' \rangle$ in terms of the embedding map (4.2.7) $\mathbf{x}^{\mathbf{a}} \colon \mathcal{M} \to \mathbb{R}^D$. Similarly the Laplacian is also approximately diagonal, with

$$\left(\begin{smallmatrix} x \\ y \end{smallmatrix} \right| \Box \left| \begin{smallmatrix} x' \\ y' \end{smallmatrix} \right) = \left(\begin{smallmatrix} x \\ y \end{smallmatrix} \right| \mathcal{P}^a \mathcal{P}_a \left| \begin{smallmatrix} x' \\ y' \end{smallmatrix} \right) \approx E_{xy} \langle x|x' \rangle \langle y'|y \rangle, \tag{6.2.11}$$

$$\left(\begin{smallmatrix} x \\ y \end{smallmatrix} \right| \mathcal{P}^a \mathcal{P}_a \left| \begin{smallmatrix} x \\ y \end{smallmatrix} \right) = E_{xy}. \tag{6.2.12}$$

Here

$$E_{xy} = (\vec{\mathbf{x}}(x) - \vec{\mathbf{x}}(y))^2 + \Delta_x^2 + \Delta_y^2 \tag{6.2.13}$$

can be interpreted as energy of a string mode, which is given by its length square plus the intrinsic uncertainty Δ_x^2 (4.1.6) of the coherent states. These results can be summarized as follows:

$$\mathcal{P}^a \big|_y^x \rangle \approx (x-y)^a \big|_y^x \rangle, \qquad \Box \big|_y^x \rangle \approx E_{xy} \big|_y^x \rangle. \tag{6.2.14}$$

Hence the string mode $\big|_x^{x+y}\rangle$ is an approximate eigenstate of $\mathcal{P} \approx y$ localized near x. This suggests that it should be interpreted as plane wave packet. To make this precise, we should distinguish between a semi-classical and a "deep quantum regime" for string modes:

Short string modes as localized wave packets

We will denote string modes $\big|_y^x\rangle$ with $|x-y| \le L_{NC}$ as **short string modes**. They provide the NC analog of optimally localized wave packets around $x \approx y$, with characteristic size L_{NC} and linear momentum determined by $x-y$. They also mark the boundary between the classical IR and stringy UV regime in $\mathrm{End}(\mathcal{H})$. To see this, consider a four-dimensional quantum space $\mathcal{M}$, which we assume to satisfy the almost-Kähler condition (4.2.28). We can then identify it locally with a quantum plane $\theta^{\mu\nu} \approx const$,

$$\mathcal{M} \cong \mathbb{R}_\theta^4, \tag{6.2.15}$$

such that their quantum metrics g coincide. Using the Wigner map (3.1.118), we obtain the following isometric identification of short string modes on $\mathcal{M} \approx \mathbb{R}_\theta^4$ with Gaussian wave-packets on $\mathbb{R}^4$:

$$\boxed{e^{\frac{i}{2}k^a \theta_{ab}^{-1} y^b} \big|_{y-\frac{k}{2}}^{y+\frac{k}{2}}\rangle =: \Psi_{\tilde{k};y} \cong \psi_{\tilde{k};y}(x) = 4\, e^{i\tilde{k}x} e^{-|x-y|_g^2}} \tag{6.2.16}$$

recalling $\tilde{k}_a = k^b \theta_{ba}^{-1}$. To verify this, we compute the inner product for the Gaussian modes as

$$\langle \psi_{\tilde{k};x}, \psi_{\tilde{l};y}\rangle = \int \frac{d^4z}{(2\pi L_{NC}^2)^2}\, \psi_{\tilde{k};x}^*(z)\psi_{\tilde{l};y}(z) = e^{\frac{i}{2}(\tilde{l}-\tilde{k})(x+y)} e^{-\frac{1}{2}|x-y|_g^2 - \frac{1}{8}|k-l|_g^2} \tag{6.2.17}$$

using the almost-Kähler relation $|\tilde{k}|_g^2 = |k|_g^2$ (3.1.109). This agrees with the inner product (6.2.2) of the short string modes

$$\langle \Psi_{\tilde{k};x}, \Psi_{\tilde{l};y}\rangle = e^{-\frac{1}{4}|x+\frac{1}{2}k-y-\frac{1}{2}l|_g^2 - \frac{1}{4}|x-\frac{1}{2}k-y+\frac{1}{2}l|_g^2}\, e^{\frac{i}{2}l^a \theta_{ab}^{-1} y^b}\, e^{-\frac{i}{2}k^a \theta_{ab}^{-1} x^b}$$

$$\cdot\, e^{-\frac{i}{2}\left((x+\frac{k}{2})^a(y+\frac{l}{2})^b + (y-\frac{l}{2})^a(x-\frac{k}{2})^b\right)\theta_{ab}^{-1}}$$

$$= e^{\frac{i}{2}(\tilde{l}-\tilde{k})(x+y)}\, e^{-\frac{1}{2}|x-y|_g^2 - \frac{1}{8}|k-l|_g^2}, \tag{6.2.18}$$

confirming the identification (6.2.16) which is equivalent to Weyl quantization.[3] It may be useful to make the link with Quantum Mechanics, where coherent states can be identified with Gaussian wave-functions in position space. In contrast, the string modes $\big|_y^x\rangle \in \mathrm{End}(\mathcal{H})$ are operators, which can be interpreted as Gaussian functions on phase space. However, the symbol of the string modes

$$\langle x|\Psi_{\tilde{k};y}|x\rangle = e^{\frac{i}{2}k^a \theta_{ab}^{-1} y^b}\langle x|y+k/2\rangle\, \langle y-k/2|x\rangle = e^{\frac{i}{2}\tilde{k}_a(x^a+y^a)}\, e^{-\frac{1}{2}|x-y|_g^2}\, e^{-\frac{1}{8}|k|_g^2} \tag{6.2.19}$$

---

[3] Recall that Weyl quantization is an isometry from $L^2(\mathbb{R}^{2n})$ with measure $\frac{\Omega}{(2\pi)^n}$ to Hilbert–Schmidt operators.

does not quite agree with the Gaussian functions $\psi_{\tilde{k};y}$. The reason is that the short string modes are not in the space of semi-classical or almost-local functions $\mathrm{Loc}(\mathcal{H})$ (4.4.6), since the size of the coherent states is set by the scale L_{NC}. This is reflected in the peculiar algebraic properties of the string modes, which satisfy relations such as

$$|x\rangle\langle y||y\rangle\langle z| = |x\rangle\langle z|, \tag{6.2.20}$$

which are not reproduced by their symbol.[4] This issue will be remedied shortly in terms of semi-classical modes. In particular, the short string modes with $k = 0$ should be identified with fuzzy delta functions:

$$\hat{\delta}_y := \left|{}^y_y\right) \cong 4\,e^{-|y-x|^2_g} \cong \delta_y(x), \tag{6.2.21}$$

which is justified for test-functions which are approximately constant on the length scale L_{NC}:

$$\delta_y[f] = \int \frac{d^4x}{(2\pi L_{\mathrm{NC}}^2)^2} e^{-|y-x|^2_g} f(x) \approx f(y). \tag{6.2.22}$$

Refining the identification (6.2.5), we can now identify the space of short string modes with the cotangent bundle

$$(\mathcal{M} \times \mathcal{M})_{\mathrm{short}} \cong T^*\mathcal{M}$$
$$(x,y) \mapsto (x,\tilde{k}), \qquad \tilde{k}_\nu = (y-x)^\mu \theta_{\mu\nu}^{-1}, \tag{6.2.23}$$

assuming that $\mathcal{M}$ is approximately flat on scales $|y-x| < L_{\mathrm{NC}}$. The symplectic (product) measure on $\mathcal{M} \times \mathcal{M}$ then reduces to the canonical measure on $T^*\mathcal{M}$

$$\Omega_x\Omega_y = \sqrt{\theta_{\mu\nu}^{-1}}\,d^m x\sqrt{\theta_{\mu\nu}^{-1}}\,d^m y = d^m x\,d^m\,\tilde{k}. \tag{6.2.24}$$

Note that the quantum metric is an extra structure on $\mathcal{M}$ and $T_x\mathcal{M}$, which is encoded in the inner product (6.2.2) of the short string modes.

Semi-classical Gaussian modes

We have seen that short string modes can be viewed as localized Gaussian wave packets, which are just outside of the semi-classical regime. This suggests that semi-classical wavefunctions can be obtained as Gaussian averages of short string modes over larger scales $L \gg L_{\mathrm{NC}}$:

$$\Psi^{(L)}_{k;y} := \int \frac{d^4z}{(2\pi L_{\mathrm{NC}}^2)^2}\, e^{-|y-z|^2/L^2}\,\Psi_{k;z} \tag{6.2.25}$$

(dropping the tilde on k), where

$$|y-z|^2 = |y-z|^2_g\, L_{\mathrm{NC}}^2 = (y-z)^a(y-z)^b\delta_{ab}, \tag{6.2.26}$$

---

[4] Of course, this could be reconciled by using the star product (3.2.1) on the classical space of functions. The point here is to show that they are not in the semi-classical regime.

using coordinates adapted to the quantum metric. They provide a useful (over-complete) basis of semi-classical modes, assuming that $\mathcal{M} \sim \mathbb{R}^4_\theta$ is valid over the scale L. According to (6.2.16), $\Psi^{(L)}_{k;y}$ is identified isometrically with the wavefunction

$$\psi^{(L)}_{k;y}(x) = \int \frac{d^4z}{(2\pi L^2_{NC})^2} e^{-|y-z|^2/L^2} \psi_{k;z}(x) \overset{L \gg L_{NC}}{\approx} e^{ikx} e^{-|x-y|^2/L^2}. \tag{6.2.27}$$

For $L \gg L_{NC}$, this agrees indeed with the symbol of these semi-classical modes

$$\langle x|\Psi^{(L)}_{k;y}|x\rangle = \int \frac{d^4z}{(2\pi L^2_{NC})^2} e^{-|y-z|^2/L^2} \langle x|\Psi_{k;y}|x\rangle \overset{L \gg L_{NC}}{\approx} e^{ikx} e^{-|x-y|^2/L^2} \tag{6.2.28}$$

using (6.2.19), assuming that $\mathcal{M}$ is sufficiently flat near y and $k^2 \ll \frac{1}{L^2_{NC}}$. In particular, plane waves on $\mathbb{R}^4_\theta$ are recovered for $L = \infty$ as

$$e^{ik_aX^a} = \int \frac{d^4z}{(2\pi L^2_{NC})^2} \Psi_{k;z}. \tag{6.2.29}$$

As a check, we recover the symbol of the plane wave

$$\langle x|e^{ik_aX^a}|x\rangle = \int \frac{d^4y}{(2\pi L^2_{NC})^2} e^{\frac{i}{2}k_a(x^a+y^a)} e^{-\frac{1}{2}|x-y|^2_g} e^{-\frac{1}{8}|k|^2_g} = e^{ik_ax^a} e^{-\frac{1}{4}|k|^2_g} \tag{6.2.30}$$

consistent with (4.6.34). We conclude that the $\Psi^{(L)}_{k;y}$ can be interpreted for $k^2 \ll \frac{1}{L^2_{NC}}$ as semi-classical Gaussian modes with size $L \gg L_{NC}$ on a generic quantum space which approximate plane waves $e^{ikX} \sim e^{ikx}$. Their momentum or wavenumber is measured by

$$\mathcal{P}^a\Psi^{(L)}_{k;y} \approx k^a\Psi^{(L)}_{k;y}, \qquad i\{x^a, \Psi^{(L)}_{k;y}\} \approx k^a\Psi^{(L)}_{k;y}. \tag{6.2.31}$$

Since (6.2.16) is an isometry, the inner products of $\Psi^{(L)}_{k;y}$ and $\psi^{(L)}_{k;y}$ coincide, and are given for large L by

$$\langle \Psi^{(L)}_{k;x}, \Psi^{(L)}_{l;y}\rangle = \int \frac{d^4z}{(2\pi L^2_{NC})^2} \psi^{(L)}_{k;x}(z)\psi^{(L)}_{l;y}(z)$$

$$\overset{L \gg L_{NC}}{\approx} \frac{L^4}{16 L^4_{NC}} e^{\frac{i}{2}(l-k)_a(x+y)^a} e^{-\frac{1}{2}|x-y|^2/L^2 - \frac{1}{8}|k-l|^2L^2}. \tag{6.2.32}$$

For large L, this approaches the $\delta(k - l)$ orthogonality relation for plane waves, and it implies the completeness relation

$$\mathbb{1} = \int \frac{d^4y}{(2\pi L^2_{NC})^2} d^4k \, |\Psi^{(L)}_{k;y}\rangle\langle\Psi^{(L)}_{k;y}| \tag{6.2.33}$$

This confirms the identification

$$T_x\mathcal{M} \sim (\mathbb{R}^4, \theta^{\mu\nu}) \tag{6.2.34}$$

in the semi-classical regime, where $\theta^{\mu\nu}$ is the (local) Poisson tensor. Since these definitions in terms of short string modes make sense for any almost-local quantum space and generalize straightforwardly to any even dimensions, we have obtained an explicit description of the space $\mathrm{Loc}(\mathcal{H}) \subset \mathrm{End}(\mathcal{H})$ (4.4.6) of almost-local operators: This space is spanned by

$$\mathrm{Loc}(\mathcal{H}) = \langle \Psi^{(L)}_{k;y}| \quad y \in \mathcal{M}, \ L^{-1} \ll k \ll L^{-1}_{\mathrm{NC}} \rangle, \tag{6.2.35}$$

which is isometrically mapped to $\mathcal{C}_{IR}(\mathcal{M})$:

$$\boxed{\begin{aligned} \mathrm{End}(\mathcal{H}) \supset \mathrm{Loc}(\mathcal{H}) &\cong \mathcal{C}_{\mathrm{IR}}(\mathcal{M}) \subset L^2(\mathcal{M}, \theta^{\mu\nu}) \\ \Psi^{(L)}_{k;y} &\leftrightarrow \psi^{(L)}_{k;y}. \end{aligned}} \tag{6.2.36}$$

This description will be very useful to compute traces of differential operators for the one-loop effective action in Sections 12 and 6.2.3.

Laplacian and effective metric

Generalizing the momentum identification (6.2.31) to the matrix d'Alembertian $\Box = \eta_{ab}\mathcal{P}^a\mathcal{P}^b$, we obtain

$$\Box \Psi^{(L)}_{\tilde{k};y} \approx |k|^2_{\eta} \Psi^{(L)}_{\tilde{k};y} \approx \mathcal{Q}(|\tilde{k}|^2_{\gamma} \psi^{(L)}_{\tilde{k};y}) \approx \mathcal{Q}(-\gamma^{ab}\partial_a\partial_b \psi^{(L)}_{\tilde{k};y}) \approx \mathcal{Q}(\Box_{\gamma} \psi^{(L)}_{\tilde{k};y}) \tag{6.2.37}$$

provided $\frac{1}{L_{\mathrm{NC}}} \gg k \gg 1/L$. Here we recall (3.1.68)

$$\begin{aligned} \gamma^{ab} &:= \theta^{aa'}\theta^{bb'}\eta_{a'b'}, \\ |k|^2_{\eta} = \eta_{ab}k^a k^b &= \gamma^{ab}\tilde{k}_a\tilde{k}_b = |\tilde{k}|^2_{\gamma}, \end{aligned} \tag{6.2.38}$$

assuming that $\partial\theta^{ab}$ is negligible. Hence in that regime, γ^{ab} is the effective metric underlying the Matrix d'Alembertian $\Box$. This will be generalized and refined in Section 9.1), where η_{ab} is replaced by the pullback of the metric on target space $(\mathbb{R}^D, \eta)$ to the brane $\mathcal{M}$, and the effective metric G^{ab} is recognized as a conformal rescaling of γ^{ab}. In string theory, this is known as open string metric on a brane with B field, which governs the gauge theory on the brane.

Long string modes

Finally, the **long string modes** $|^x_y)$ for $|x - y| > L_{\mathrm{NC}}$ are completely nonlocal objects on $\mathcal{M}$, which have no analog on classical spaces. They are naturally viewed as dipoles or strings linking x with y, which are bi-local in position space due to (6.2.2), and have good localization properties in momentum space as well. This makes them novel and very interesting objects from a QFT point of view. They provide the appropriate description of the UV or deep quantum regime of NC field theory, and are responsible for UV/IR mixing. The long string modes comprise the vast majority of $\mathrm{End}(\mathcal{H})$: e.g. on the fuzzy sphere S^2_N, the dividing line between the semi-classical regime and the deep quantum regime is given by the angular momentum $l \sim \sqrt{N}$, which is far below the UV cutoff at $l_{UV} = N - 1$. The short string modes define the boundary between these regimes.

6.2.1 The off-diagonal string symbol and regularity

We have seen that for $\Phi \in \text{End}(\mathcal{H})$ which is not in the semi-classical regime, the diagonal "string" representation in (4.5.17) involves a highly oscillatory function and is therefore rather useless. In the UV regime, it is better to use a non-diagonal string representation of Φ. This is a priori no longer unique, but we can turn this into a virtue and find a representation that is not oscillatory and well-behaved:

Let $\Phi \in \text{End}(\mathcal{H})$ be normalized as

$$\|\Phi\|^2 = \text{Tr}(\Phi^\dagger \Phi) = 1. \tag{6.2.39}$$

Using the completeness relation we can write[5]

$$\Phi = \int\limits_{\mathcal{M} \times \mathcal{M}} |x\rangle \phi(x,y) \langle y|, \tag{6.2.40}$$

where $\phi(x,y)$ is the **off-diagonal string symbol**

$$\phi(x,y) := \langle x|\Phi|y\rangle. \tag{6.2.41}$$

As opposed to the diagonal kernel in (4.5.17), we claim that $\phi(x,y)$ is always in the IR regime, i.e. its derivative is bounded by the scale of noncommutativity

$$|\nabla \phi(x,y)| \le \frac{1}{L_{\text{NC}}}. \tag{6.2.42}$$

This follows from the definition of the quantum metric (4.2.19) and the scale of noncommutativity (4.2.26), (4.2.40), which imply

$$\|\partial_\mu|y\rangle\|^2 = (\partial_\mu \langle y|)(\partial_\mu|y\rangle) = \frac{1}{2}g_{\mu\mu} \le \frac{1}{2L_{\text{NC}}^2} \tag{6.2.43}$$

(in a gauge where $A|_y$ vanishes) where $\|.\|$ denotes the norm in $\mathcal{H}$. Therefore, the Cauchy–Schwarz inequality implies

$$\left|\frac{\partial}{\partial y^\mu}\phi(x,y)\right| = |\langle x|\Phi\partial_\mu|y\rangle| \le \|\Phi^\dagger|x\rangle\| \|\partial_\mu|y\rangle\| \le \frac{1}{\sqrt{2}\,L_{\text{NC}}} \tag{6.2.44}$$

since

$$\|\Phi^\dagger|x\rangle\|^2 = \langle x|\Phi\Phi^\dagger|x\rangle \le \|\Phi\|^2 = 1 \tag{6.2.45}$$

using the normalization condition (6.2.39), where $\|.\|$ denotes the Hilbert–Schmidt norm on $\text{End}(\mathcal{H})$. This bound is essentially saturated by $\Phi = |y\rangle\langle y|$, as it is easy to verify

$$\left|\frac{\partial}{\partial y^\mu}\langle z|y\rangle\right| \lesssim \frac{1}{L_{\text{NC}}}. \tag{6.2.46}$$

The reason for this mild behavior of the string symbol is of course the fact that coherent states are spread over an area L_{NC}^2, and average out any finer oscillations. This may seem inconsistent with the fact that the diagonal symbol faithfully captures the full UV structure

---

[5] This can be viewed as coherent state version of Weyl quantization and Wigner's inverse map.

which includes much shorter wavelengths, as short as $L_{UV} = \frac{R}{N} = \frac{1}{\sqrt{N}} L_{NC}$ in the example of fuzzy S_N^2. This puzzle is resolved noting that the amplitude of such UV modes is strongly suppressed, due to the factor $e^{-\frac{1}{4}|k|_g^2}$ in the string symbol (6.2.19). In this sense, the extreme UV wavelengths are indeed smoothed out on quantum spaces, but the price to pay is long-range nonlocality mediated by the string modes. The off-diagonal string symbol (6.2.40) allows to capture the full range of modes on quantum spaces in terms of semi-classical *bi-local* functions.

6.2.2 The string representation of the propagator

A particularly interesting application of the string modes is a novel representation of the propagator, which is only possible on noncommutative spaces, and exhibits the difference between commutative and noncommutative field theory in a most striking way.

Given these string modes, we can consider the following object

$$K(xy;x'y') := \left(\begin{smallmatrix}x\\y\end{smallmatrix}\middle|(\Box+m^2)^{-1}\middle|\begin{smallmatrix}x'\\y'\end{smallmatrix}\right) \qquad (6.2.47)$$

interpreted as string two-point function. It provides the kernel of the following exact string representation of the propagator

$$(\Box+m^2)^{-1} = \int \frac{\Omega_x\Omega_y\Omega_{x'}\Omega_{y'}}{(2\pi)^{2m}} \left|\begin{smallmatrix}x\\y\end{smallmatrix}\right) K(xy;x'y') \left(\begin{smallmatrix}x'\\y'\end{smallmatrix}\right| \qquad (6.2.48)$$

where $\dim \mathcal{M} = m$. Using similar estimates as in the last section, it follows that the kernel $K(xy;x'y')$ is well-behaved without significant oscillatory behavior. We claim that it takes the following approximate form:

$$K(xy;x'y') \approx \begin{cases} G_\Lambda(x,x') & x \approx y \neq x' \approx y' \\ \tilde{G}(x,y) & x \approx x' \neq y \approx y' \\ 0 & \text{otherwise.} \end{cases} \qquad (6.2.49)$$

Here

$$\tilde{G}(x,y) := \left(\begin{smallmatrix}x\\y\end{smallmatrix}\middle|(\Box+m^2)^{-1}\middle|\begin{smallmatrix}x\\y\end{smallmatrix}\right) \approx \frac{1}{|x-y|^2+m^2+2\Delta^2}, \qquad |x-y| \geq L_{NC} \quad (6.2.50)$$

governs the stringy or UV regime, where

$$|x-y|^2 = |x-y|_\eta^2 := (x^a-y^a)(x^b-y^b)\eta_{ab} \qquad (6.2.51)$$

is the distance in target space $\mathbb{R}^D$. On the other hand, the semi-classical regime is governed by

$$G_\Lambda(x,y) := \left(\begin{smallmatrix}x\\x\end{smallmatrix}\middle|(\Box+m^2)^{-1}\middle|\begin{smallmatrix}y\\y\end{smallmatrix}\right) \approx G(x,y), \qquad |x-y| \geq L_{NC}$$

$$(\Box+m^2)G_\Lambda(x,y) = e^{-\frac{1}{2}|x-y|_g} \qquad (6.2.52)$$

which is a regularization of the commutative Greens function $G(x,y)$ for the massive Laplacian $(\Box_\gamma + m^2)$ corresponding to the effective metric $\gamma^{\mu\nu}$ (6.2.38), with UV cutoff $\Lambda = \Lambda_{NC}$ and a nonstandard normalization.

To justify these claims, consider first the UV regime of long-string modes with $x \approx x' \neq y \approx y'$, and $|x - y| \gg L_{NC}$. This regime has no commutative analog. Then $\Box$ is approximately diagonal with eigenvalues given by (6.2.13), which directly yields (6.2.50).

Now consider the semi-classical regime with $x \approx y \neq x' \approx y'$. We can then approximate $\mathcal{M}$ by a local quantum plane[6] $\mathbb{R}^{2n}_\theta$ near $x \approx y$. Then (4.3.24) gives

$$
\begin{aligned}
-\gamma^{\mu\nu} \partial_\mu \partial_\nu \left| {}^x_x \right) &= -\gamma^{\mu\nu} \partial_\mu \partial_\nu (|x\rangle\langle x|) \\
&\approx \gamma^{\mu\nu} \theta^{-1}_{\mu\alpha} \theta^{-1}_{\nu\beta} [X^\alpha, [X^\beta, |x\rangle\langle x|]] \\
&= \eta_{\alpha\beta} [X^\alpha, [X^\beta, |x\rangle\langle x|]] = \Box(|x\rangle\langle x|)
\end{aligned}
\tag{6.2.53}
$$

(cf. (6.2.37)), so that

$$
\begin{aligned}
(-\gamma^{\mu\nu} \partial^x_\mu \partial^x_\nu + m^2) \left({}^y_x \middle| (\Box + m^2)^{-1} \middle| {}^x_x \right) &\approx \left({}^y_x \middle| (\Box + m^2)^{-1} (\Box + m^2) \middle| {}^x_x \right) = \left({}^y_x \middle| {}^x_x \right) \\
&= \delta_{NC}(x, y) = e^{-\frac{1}{2}|x-y|^2_g}
\end{aligned}
\tag{6.2.54}
$$

recalling the effective delta function (6.2.4). Therefore,

$$
G_\Lambda(x, y) = \int d^{2n}z\, G(x, z) e^{-\frac{1}{2}|z-y|_g} \overset{y \to x}{\to} \int d^{2n}z\, G(x, z) e^{-\frac{1}{2}|z-x|_g}
\tag{6.2.55}
$$

reduces for $|x - y| \geq L_{NC}$ to the classical Greens function

$$
G(x, y) = \int \frac{d^{2n}k}{(2\pi)^{2n}} \frac{1}{k \cdot k + m^2} e^{ik(x-y)} \overset{m=0}{\sim} \frac{1}{|x - y|^{2n-2}_\gamma},
\tag{6.2.56}
$$

in terms of the effective metric $\gamma^{\mu\nu}$ which is encoded in the dotted product

$$
k \cdot k = \gamma^{\mu\nu} k_\mu k_\nu.
\tag{6.2.57}
$$

The coincidence limit of the semi-classical two-point function remains regular, given by

$$
\begin{aligned}
G_\Lambda(x, x) &= \int d^{2n}z \int \frac{d^{2n}k}{(2\pi)^{2n}} \frac{1}{k \cdot k + m^2} e^{ik(x-z)} e^{-\frac{1}{2}|x-z|^2_g} \\
&= (2\pi L^2_{NC})^n \int \frac{d^{2n}k}{(2\pi)^{2n}} \frac{1}{|k|^2_g L^2_{NC} + m^2} e^{-\frac{1}{2}|k|^2_g} \\
&= L^{2n-2}_{NC} \frac{1}{2^{n-1}(n-1)!} \int_0^\infty dk \frac{k^{2n-1}}{k^2 L^2_{NC} + m^2/L^2_{NC}} e^{-\frac{1}{2}k^2 L^2_{NC}} \\
&= \frac{1}{2(n-1)} \frac{1}{L^2_{NC}} \left(1 + O(m^2/L^2_{NC}) \right)
\end{aligned}
\tag{6.2.58}
$$

using the area of the $2n - 1$-dimensional unit sphere[7]

$$
A(S^{2n-1}) = \frac{(2\pi)^n}{2^{n-1}(n-1)!}.
\tag{6.2.59}
$$

---

[6] This should be done in a local frame where $\partial \gamma^{\mu\nu} = 0$, or otherwise the effective metric $G^{\mu\nu}$ should be used.

[7] in the Lorentzian case, an appropriate contour integration is understood.

We assume that $m^2/L_{NC}^2 \ll 1$ for physical mass m, which has dimension length here. This agrees indeed approximately with the short string limit of the stringy two-point function

$$G_\Lambda(x,x) \approx \frac{1}{2\Delta^2} \approx \tilde{G}(x,x) \tag{6.2.60}$$

using (3.1.111). The standard normalization of the Greens function is recovered by replacing

$$\delta_{NC}(x,y) = e^{-\frac{1}{2}|x-y|_g} \;\to\; \frac{1}{(2\pi L_{NC}^2)^n} e^{-\frac{1}{2}|x-y|_g} \approx \delta(x,y) \tag{6.2.61}$$

in (6.2.55), and taking the factor Λ_{NC}^4 out of $\Box \sim \gamma^{\mu\nu}\partial_\mu\partial_\nu \approx \Lambda_{NC}^4 \eta^{\mu\nu}\partial_\mu\partial_\nu$. Then

$$\frac{L_{NC}^4}{(2\pi L_{NC}^2)^n} G_\Lambda(x,y) \sim \int \frac{d^{2n}k}{(2\pi)^{2n}} \frac{e^{ik(x-z)}}{k^2 + \frac{m^2}{L_{NC}^4}} \sim \begin{cases} \frac{1}{2(n-1)} \frac{1}{L_{NC}^{2n-2}}, & x \to y \\ \frac{1}{|x-y|^{2n-2}}, & |x-y| < \frac{L_{NC}^4}{m^2} \end{cases} \tag{6.2.62}$$

which reduces to the ordinary Greens function in $\mathbb{R}^{2n}$, with UV cutoff L_{NC}.

To summarize, the string two-point function (6.2.49) is a regular function of 4 variables. It is dominated by two regimes with simple behavior, one interpreted as semi-classical IR channel with standard decay behavior $\frac{1}{|x-y|_\gamma^{2n-2}}$ in terms of the effective metric for sufficiently small m, and the other as UV stringy (or crossed) channel which scales like $\frac{1}{|x-y|^2+2\Delta^2}$ in the target space metric. Both channels merge at the noncommutativity scale L_{NC}. This representation of the propagator is very useful to evaluate loop integrals in NCFT, which can thus be approximated by simple geometrical non-oscillatory integrals, generalizing the discussion in Section 6.1. Feynman diagrams become ribbon graphs represented via t'Hooft's double line notation, where the lines correspond to the positions x,y of the strings, which are easily evaluated in position space. More details and some explicit loop computations can be found in [190]. This provides a very effective approach to compute loops on quantum spaces, leading directly to local contributions similar as on classical spaces with UV cutoff L_{NC}, as well as novel nonlocal contributions characteristic for quantum spaces reflecting UV/IR mixing. However, we will restrict ourselves to one-loop computations in this book, which can be conveniently organized in terms of the trace formulas discussed in the next section.

6.2.3 Trace formulas on classical and quantum spaces

To understand better the relation between the commutative and noncommutative setting, we consider (pseudo-)differential operators, and develop some tools to compute their trace in the classical and noncommutative case. This will be essential for the one-loop computations in Chapter 12.

Trace formulas for classical spaces

In classical field theory, differential operators on configuration space can be viewed as functions on phase space. This suggests that their traces can be computed in terms of an integral on phase space. Choosing $\mathbb{R}^{2n}$ as configuration space, such a trace formula can be obtained using the following Gaussian wave packets

$$\psi_{k;y}^{(L)}(x) = e^{ikx}e^{-(x-y)^2/L^2} \tag{6.2.63}$$

(cf. (6.2.27)), which are centered at y with size L and characteristic wave vector k. In this section, $\mathbb{R}^{2n}$ is equipped with the standard Euclidean metric, and inner product of functions is defined as

$$\langle f, g \rangle = \int \frac{d^{2n}x}{(2\pi)^n} f^*(x)g(x). \tag{6.2.64}$$

Then their inner products are obtained as

$$\langle \psi_{k;x}^{(L)}, \psi_{l;y}^{(L)} \rangle = \frac{L^{2n}}{2^{2n}} e^{-\frac{i}{2}(x+y)(k-l)} e^{-\frac{1}{2}(x-y)^2/L^2 - \frac{1}{8}L^2(k-l)^2} \tag{6.2.65}$$

cf. (6.2.17). These $\psi_{k;y}^{(L)}$ form an over-complete set of functions in $\mathcal{H} \cong L^2(\mathbb{R}^{2n})$ for any fixed L, and their quantum counterparts $\Psi_{k;y}^{(L)}$ (6.2.25) will be used in the quantum case.

We would like to compute the trace $\mathrm{Tr}_{\mathcal{H}}(\mathcal{O})$ for (pseudo-)differential operators $\mathcal{O} = \mathcal{O}(x, \partial)$ acting on $\mathcal{H} \cong L^2(\mathbb{R}^{2n})$. In principle, one could use the plane wave basis e^{ikx} of $\mathcal{H}$ to compute the trace; however, that would be hard to evaluate for differential operators with non-constant coefficients. A more useful approach is to use the over-complete basis $\psi_{k;y}^{(L)}$, choosing L such that the coefficients of $\mathcal{O}$ are approximately constant on the length scale L. We claim that the following general trace formula holds:

$$\boxed{\mathrm{Tr}_{\mathcal{H}}(\mathcal{O}) = \frac{1}{(2\pi)^m} \int d^m y \int d^m k \, \frac{\langle \psi_{k;y}^{(L)}, \mathcal{O}\psi_{k;y}^{(L)} \rangle}{\langle \psi_{k;y}^{(L)}, \psi_{k;y}^{(L)} \rangle}} \tag{6.2.66}$$

for any L, where $m = \dim \mathcal{M} = 2n$. There are various ways to show this formula; one possibility is to view the $\psi_{k;y}^{(L)}$ as position space wavefunctions of coherent states on the doubled Moyal–Weyl space $\mathbb{R}_\theta^{4n}$, for $\theta = \frac{1}{4}L^2$. Then (6.2.66) is nothing but (3.1.73). This is basically a consequence of the invariance of (6.2.66) under the translations in position and momentum space $\mathbb{R}_y^{2n} \times \mathbb{R}_k^{2n}$: Since $\mathcal{H}$ is irreducible under this group, Schur's Lemma implies that there is only one invariant functional on $\mathrm{End}(\mathcal{H})$ up to normalization, which is thus proportional to the above trace formula. The normalization is obtained e.g. for $\mathcal{O}_0 = \frac{|\psi_{0;0}^{(L)}\rangle\langle\psi_{0;0}^{(L)}|}{\langle\psi_{0;0}^{(L)}, \psi_{0;0}^{(L)}\rangle}$ using

$$\mathrm{Tr}_{\mathcal{H}}(\mathcal{O}_0) = \frac{1}{(2\pi)^m} \int d^m y \int d^m k \, \frac{|\langle\psi_{k;y}^{(L)}, \psi_{0;0}^{(L)}\rangle|^2}{\langle\psi_{k;y}^{(L)}, \psi_{k;y}^{(L)}\rangle\langle\psi_{0;0}^{(L)}, \psi_{0;0}^{(L)}\rangle} = 1. \tag{6.2.67}$$

The most transparent proof is via the Wigner transform of the quantum trace formula (6.2.77).

Note that the scale L of the enveloping function in (6.2.66) is arbitrary. If L is sufficiently large, the $\psi^{(L)}$ are approximately plane waves, which will be useful to compute traces of differential operators with non-constant coefficients. On the other hand, we can choose L to be very short, to the point where the $\psi^{(L)}$ are approximately delta functions. Then the trace of integral operators $\phi(x) \to \int dy G(x,y)\phi(y)$ reduce to the integral $\int dx G(x,x)$ for the integral kernel. Therefore, the trace formula (6.2.66) allows to interpolate between the position and momentum point of view. The trace diverges for the propagator $(\Box + m^2)G(x,y) = \delta(x,y)$, but it converges for the UV-regulated propagator $G_\Lambda(x,y)$ (6.2.52).

The heat kernel

As an application of this trace formula, we compute the leading contribution to the heat kernel

$$\text{Tr} e^{-\alpha\Delta} \tag{6.2.68}$$

which will be motivated in Section 12.1. Here Δ is the Laplacian on a Riemannian manifold $(\mathcal{M}, g)$, which is assumed to be four-dimensional to be specific. In order to use (6.2.66), we consider $\mathcal{M}$ as a homogeneous manifold $\mathbb{R}^4$, equipped with a non-trivial metric $g_{\mu\nu}$; accordingly, the homogeneous measure and inner product on $\mathbb{R}^4$ will be used in this formula. To evaluate the inner product in (6.2.66), we choose the length scale

$$L \gg \frac{1}{k} \tag{6.2.69}$$

such that the relevant $\psi^{(L)}_{k;y}(x)$ are approximate local eigenfunctions of Δ_g, i.e.

$$\Delta_g \psi^{(L)}_{k;y} \approx g^{\mu\nu} k_\mu k_\nu \psi^{(L)}_{k;y} \tag{6.2.70}$$

up to corrections suppressed by a factor $O(\frac{1}{kL})$ from the derivatives of the enveloping function in (6.2.63). We also assume that the metric $g_{\mu\nu}$ is approximately constant over the length scale L; this is easily satisfied in the context of gravity. Then we have more generally

$$f(\Delta_g)\psi^{(L)}_{k;y} \approx f(k \cdot k)\psi^{(L)}_{k;y}, \tag{6.2.71}$$

where

$$k \cdot k = k_\mu k_\nu g^{\mu\nu}(x) \tag{6.2.72}$$

may depend on x, and a simple Gaussian integration using $\langle \psi^{(L)}_{k;x}, \psi^{(L)}_{k;x} \rangle = \frac{L^4}{16}$ (6.2.65) gives

$$\text{Tr} e^{-\alpha\Delta} = \frac{1}{(2\pi)^4} \int d^4x \int d^4k\, e^{-\alpha k \cdot k} = \frac{1}{\alpha^2} \frac{1}{(4\pi)^2} \int d^4x \sqrt{g}. \tag{6.2.73}$$

This will be recognized as leading term $n = 0$ in the heat kernel expansion (12.1.1), or as leading Seeley–de Witt coefficient (12.1.3). Using a Laplace transformation in α, this allows to obtain traces over other functions of Δ, including UV-regulated traces as required in the context of QFT. For example, the leading term of the one-loop effective action induced by integrating out a scalar field coupled to a background metric g is obtained as

$$
\mathrm{Tr}_\Lambda \ln(\Delta) := -\frac{1}{2}\mathrm{Tr}\int_0^\infty \frac{d\alpha}{\alpha} e^{-\alpha\Delta - \frac{1}{\alpha\Lambda^2}}
$$

$$
= -\frac{1}{2}\frac{1}{(4\pi)^2}\int d^4x \sqrt{g}\int_0^\infty \frac{d\alpha}{\alpha^3} e^{-\frac{1}{\alpha\Lambda^2}}
$$

$$
= -\frac{\Lambda^4}{2(4\pi)^2}\int d^4x \sqrt{g} \tag{6.2.74}
$$

up to sub-leading corrections[8] in the UV cutoff Λ. That result is familiar from field theory, where it can be interpreted as induced cosmological constant. This computation can be interpreted in terms of the volume of phase space truncated at Λ,

$$
\mathrm{Vol}_\Lambda T^*\mathcal{M} \sim \int_{|p|<\Lambda} \frac{d^4k}{(2\pi)^2}\int \frac{d^4x}{(2\pi)^2}\sqrt{g} \sim \Lambda^4 \int \frac{d^4x}{(2\pi)^2}\sqrt{g}. \tag{6.2.75}
$$

However, the dominant contribution arises from the UV region near $|k| = \Lambda$, since the integral over k is UV divergent. That issue will be avoided later by SUSY.

It will be useful to observe that the $\psi_{k;y}^{(L)}$ are also approximate local eigenfunctions of generic vector fields $v[.]$, in the sense

$$
v[\psi_{k;y}^{(L)}] = v^\mu(x)\partial_\mu \psi_{k;y}^{(L)} \approx ik_\mu v^\mu \psi_{k;y}^{(L)}
$$

$$
v[v[\psi_{k;y}^{(L)}]] \approx -(k_\mu k_\nu v^\mu v^\nu)\psi_{k;y}^{(L)} \tag{6.2.76}
$$

etc., noting that $\partial_x \psi_{k;y}^{(L)} \sim ik\psi_{k;y}^{(L)}$ for $|k| \geq \frac{1}{L}$.

Trace formulas on quantum spaces and UV/IR mixing

Now consider quantum spaces $\mathcal{M}$ arising as quantized symplectic spaces. The analogs of differential operators on such quantum space are defined in terms of commutators $-i[X^a, .]$ with the matrices of the given matrix configuration underlying $\mathcal{M}$. These are derivations of the algebra $\mathrm{End}(\mathcal{H})$, which in the semi-classical regime reduce to vector fields $\{x^a, .\}$ which are derivations of $\mathcal{C}(\mathcal{M})$. Nevertheless, there is a crucial difference. Even though the semi-classical sector is closely related to the commutative case via (6.2.36), the space of quantum functions $\mathrm{End}(\mathcal{H})$ contains nonlocal objects such as the long string modes, which have no classical counterpart. This means that the trace over $\mathrm{End}(\mathcal{H})$ is in general very different from the classical trace over some space of functions.

---

[8] The specific form of the smooth UV-cutoff chosen here is convenient but not crucial.

The trace over $\text{End}(\mathcal{H})$ can be computed using the completeness relation (6.2.6) for the string modes in $\text{End}(\mathcal{H})$, which implies the following geometric trace formula for $\mathcal{O} \in \text{End}(\text{End}(\mathcal{H}))$:

$$\text{Tr}_{\text{End}(\mathcal{H})}\mathcal{O} = \int_{\mathcal{M} \times \mathcal{M}} \frac{\Omega_x \Omega_y}{(2\pi)^m} \left(\begin{smallmatrix} x \\ y \end{smallmatrix} \middle| \mathcal{O} \middle| \begin{smallmatrix} x \\ y \end{smallmatrix}\right), \tag{6.2.77}$$

where $\dim \mathcal{M} = m$. In view of (6.2.16), this can be viewed as quantum counterpart of the classical trace relation (6.2.66). For quantized coadjoint orbits $\mathcal{M} \cong G/H$ such as $\mathbb{C}P^n$ and for $\mathbb{R}_\theta^{2n}$, this trace formula is exact and follows from invariance under the the following action of $G \times G$ on $\text{End}(\mathcal{H})$:

$$(G \times G) \times \text{End}(\mathcal{H}) \to \text{End}(\mathcal{H})$$
$$((g_1, g_2), \Phi) \to U_{g_1} \Phi U_{g_2}^{-1} \tag{6.2.78}$$

which acts on the string modes as

$$\left| \begin{smallmatrix} x \\ y \end{smallmatrix} \right) \to \left| \begin{smallmatrix} g_1 x \\ g_2^{-1} y \end{smallmatrix} \right). \tag{6.2.79}$$

In particular, this applies to H_n^4 and $\mathcal{M}^{3,1}$. For more generic quantum spaces, its validity is tied to the completeness relation (6.2.6), which is expected to hold quite generally to a good approximation.

For example, consider the propagator associated to the Matrix Laplacian $\Box$. Its trace can be computed as

$$\text{Tr}(\Box + m^2)^{-1} = \int_{\mathcal{M} \times \mathcal{M}} \frac{\Omega_x \Omega_y}{(2\pi)^m} \left(\begin{smallmatrix} y \\ x \end{smallmatrix} \middle| (\Box + m^2)^{-1} \middle| \begin{smallmatrix} y \\ x \end{smallmatrix}\right) = \int_{\mathcal{M} \times \mathcal{M}} \frac{\Omega_x \Omega_y}{(2\pi)^m} \tilde{G}(x,y), \tag{6.2.80}$$

where $\tilde{G}(x,y)$ is the diagonal string propagator (6.2.50). As discussed in Section 6.2.2, there are two regimes which should be distinguished. For $|x - y| > L_{\text{NC}}$, we can approximate this by $\tilde{G}(x,y) \approx \frac{1}{|x-y|^2 + 2\Delta^2 + m^2}$ (6.2.50), leading to a highly non-local integral in position space. For $|x - y| < L_{\text{NC}}$, the propagator approximates the classical regulated propagator (6.2.60)

$$\tilde{G}(x,x) \approx G_\Lambda(x,x), \tag{6.2.81}$$

which is governed by the effective metric $\gamma^{\mu\nu}$ encoded in $\Box$.

Therefore, in general, there are two scenarios which should be distinguished. In the first, UV-divergent scenario, the integral over $y - x$ is divergent or *not* localized to a small region near the origin. This is the case for the propagator, where

$$\text{Tr}(\Box + m^2)^{-1} \sim \int \frac{\Omega_x \Omega_y}{(2\pi)^m} \frac{1}{|x - y|^2 + 2\Delta^2 + m^2} \tag{6.2.82}$$

is divergent in dimensions ≥ 2. Nevertheless, similar integrals can make sense as a nonlocal interaction. This nonlocal behavior is a hallmark for UV/IR mixing: a UV-divergent momentum integral is understood as an IR-divergent integral in position space.

On the other hand, if for some operator $\mathcal{O}$ the integral over $y - x$ is convergent and localized to a small region near the origin, we can relate its trace to a semi-classical localized trace as follows:

UV-convergent traces on quantum spaces

Now assume that the trace is UV-finite, which is the case e.g. for $\mathcal{O} = (\Box + m^2)^{-n}$ for sufficiently large n, and assume that $\mathcal{M}$ is almost-Kähler. Using the correspondence (6.2.16), this implies that the dominant part of the y integral (6.2.77) originates from some ball $B_L(x)$ with radius L around x, which we assume to be sufficiently small so that the identification (6.2.24) and (6.2.23) with $T^*\mathcal{M}$ can be used. Then

$$
\mathrm{Tr}\mathcal{O} \approx \frac{1}{(2\pi)^m} \int_{\mathcal{M}} \Omega_x \int_{B_L(x)} \Omega_y \left({}_x^y \middle| \mathcal{O} \middle| {}_x^y \right)
$$

$$
\approx \frac{1}{(2\pi)^m} \int_{T^*\mathcal{M}} d^m x\, d^m k \langle \Psi_{k,x}, \mathcal{O}\Psi_{k,x}\rangle = \frac{1}{(2\pi)^m} \int_{\mathcal{M}} d^m x \sqrt{G} \int \frac{d^m k}{\sqrt{G}} \langle \Psi_{k,x}, \mathcal{O}\Psi_{k,x}\rangle
$$

$$
(6.2.83)
$$

(dropping the tilde on k), which is made more transparent in terms of an arbitrary metric $G_{\mu\nu}$ on $\mathcal{M}$. No cutoff in k is needed due to the assumed UV finiteness. We can then effectively work on the local $\mathcal{M} \approx \mathbb{R}^{2n}_\theta$, and replace the short string modes $\Psi_{k,x}$ with the semi-classical modes $\Psi^{(L)}_{k;y}$ (6.2.27) with characteristic size $L \gg 1/k \gg L_{\mathrm{NC}}$. Then the $G \times G'$ action (6.2.79) reduces locally to translations $\mathbb{R}^{2n}_y \times \mathbb{R}^{2n}_k$ in position and momentum space, which acts on the short string modes $\Psi_{k,x}$ as follows:

$$
\Psi_{k,y} \sim \left| {}^{y+k/2}_{y-k/2} \right) \rightarrow \left| {}^{y+a+(k+l)/2}_{y+a-(k+l)/2} \right) = \Psi_{k+l,y+a}
\tag{6.2.84}
$$

and identically on the semi-classical string modes $\Psi^{(L)}_{k,y}$. Since the trace is the unique invariant functional under this group action, the geometric trace formula generalizes immediately to the semi-classical string modes

$$
\mathrm{Tr}\mathcal{O} \approx \frac{1}{(2\pi)^m} \int_{\mathcal{M}} d^m x \int d^m k \langle \Psi_{k,x}, \mathcal{O}\Psi_{k,x}\rangle
$$

$$
= \frac{1}{(2\pi)^m} \int_{\mathcal{M}} d^m x \sqrt{G} \int \frac{d^m k}{\sqrt{G}} \frac{\langle \Psi^{(L)}_{k,x}, \mathcal{O}\Psi^{(L)}_{k,x}\rangle}{\langle \Psi^{(L)}_{k,x}, \Psi^{(L)}_{k,x}\rangle}
\tag{6.2.85}
$$

for any L. This corresponds precisely to the classical trace formula (6.2.66), but arises within the present quantum framework. All these formulas can be understood as a consequence of the isometry arising from the equivalent Gaussian inner product structures (6.2.18) and (6.2.32). They are equivalent to a completeness relation, which for homogeneous spaces[9] follows directly from translation invariance as in (6.2.66). The normalization can be obtained for $\mathcal{O}_0 = |\Psi_{0,0}\rangle\langle\Psi_{0,0}|$,

$$1 = \frac{1}{(2\pi)^m} \int_{\mathcal{M}} d^m x \int d^m k \, |\langle\Psi_{k,x}, |\Psi_{0,0}\rangle|^2 \qquad (6.2.86)$$

using the inner products (6.2.17) and a Gaussian integration, and similarly for $\Psi_{k,x}^{(L)}$.

This semi-classical version of the trace formula is useful, because the $\Psi_{k;y}^{(L)}$ are approximate local eigenmodes of $\Box$ for sufficiently large L, with

$$\Box\Psi_{k;y}^{(L)} \sim \gamma^{\mu\nu} k_\mu k_\nu \Psi_{k;y}^{(L)}, \qquad (6.2.87)$$

cf. (6.2.37). This will allow to evaluate the relevant traces in the one-loop effective action explicitly, and in particular to obtain the induced Einstein–Hilbert action on covariant quantum spaces in Section 12.4.1. Furthermore, IIB supergravity in $\mathbb{R}^{9,1}$ will be understood in this way in Section 11.1.7 as 1-loop induced (inter)action.

6.2.4 String modes on fuzzy S_N^2

To illustrate the power of the exact trace formula (6.2.77), we use it to compute the trace of the Laplacian on the fuzzy sphere S_N^2. Using (6.2.12) for the normalization $X^a X_a = \mathbb{1}$, we obtain

$$\begin{aligned}
\mathrm{Tr}_{\mathrm{End}(\mathcal{H})}[X^a, [X_a, .]] &= \int_{S^2 \times S^2} \frac{\Omega_x \Omega_{x'}}{(2\pi)^2} \mathrm{tr}(|x\rangle\langle x'|)\big(|\mathbf{x}'^a - \mathbf{x}^a|^2 + 2\Delta^2\big)(|x'\rangle\langle x|) \\
&= \int_{S^2 \times S^2} \frac{\Omega_x \Omega_{x'}}{(2\pi)^2}\big(|\mathbf{x}'^a - \mathbf{x}^a|^2 + 2\Delta^2\big) \\
&= N \int_{S^2} \frac{\Omega_x}{2\pi}\big(|\mathbf{n}^a - \mathbf{x}^a|^2 + 2\Delta^2\big), \qquad (6.2.88)
\end{aligned}$$

where $\int \frac{\Omega}{2\pi} = N$. Here $|.|$ is the Euclidean distance in target space $\mathbb{R}^3$, and $\mathbf{n}$ denotes the north pole of the sphere given by the expectation values $\mathbf{x}^a = \langle x|X^a|x\rangle$ with radius (4.6.7)

$$r_N^2 = \frac{N-1}{N+1}. \qquad (6.2.89)$$

Recalling $\Delta^2 = 1 - r_N^2 = \frac{2}{N+1}$ (4.6.14), we obtain

$$\begin{aligned}
\mathrm{Tr}_{\mathrm{End}(\mathcal{H})}[X^a, [X_a, .]] &= N \int_{S^2} \frac{\Omega_x}{2\pi}\Big(|\mathbf{n}^a - \mathbf{x}^a|^2 + \frac{4}{N+1}\Big) \\
&= \frac{N(N-1)}{N+1} \int_{S^2} \frac{\Omega_x}{2\pi}\Big(|e_3{}^a - x^a|^2 + \frac{4}{N-1}\Big), \qquad (6.2.90)
\end{aligned}$$

where x is normalized to 1 in the last line. Evaluating the integral over the unit sphere

$$\int_{S^2} \frac{\Omega_x}{2\pi}|e_3 - x|^2 = \frac{N}{2}\int_0^\pi d\theta \sin\theta((1 - \cos\theta)^2 + \sin^2\theta) = 2N \qquad (6.2.91)$$

---

[9] If $\mathcal{M}$ is not homogeneous, the argument generalizes using a partition of unity argument, for suitable scales.

results in

$$\mathrm{Tr}_{\mathrm{End}(\mathcal{H})}[X^a, [X_a, \cdot]] = 2N^2. \tag{6.2.92}$$

By restoring the normalization $J^a = C_N X^a$ (4.6.1), we recover the exact result

$$\mathrm{Tr}_{\mathrm{End}(\mathcal{H})}[J^a, [J_a, \cdot]] = \sum_{j=0}^{N-1} j(j+1)(2j+1) = \frac{1}{2}N^2(N^2 - 1). \tag{6.2.93}$$

It is remarkable that this result from representation theory is recovered exactly by the new geometric trace formula. In fact, it is not hard to show the following exact "string representation" for the Matrix Laplacian on S_N^2:

$$[J^a, [J_a, \cdot]] = \frac{(N+1)^2}{4} \int \frac{\Omega_x \Omega_{x'}}{(2\pi)^2} \left| \begin{matrix} x \\ y \end{matrix} \right) \left(\begin{matrix} x \\ y \end{matrix} \right| \left(|x - y|^2 - \frac{4}{N+1} \right), \tag{6.2.94}$$

which has no classical analog; for a systematic discussion see [190]. Similar formulas can be derived for any quantized coadjoint orbit. The main merit of such formulas is that the leading properties of operators and traces can be understood in a very transparent way. It is also possible to derive a similar formula for the propagator, but in that case the off-diagonal string representation (6.2.48) turns out to be better behaved.

Exercise 6.2.1   Show (6.2.94) on S_N^2 using the coherent state representation (4.6.26) for X^a.

6.2.5 String modes on covariant quantum spaces

Now consider the covariant quantum spaces discussed in Section 5. All these spaces – including the spacetime $\mathcal{M}_n^{3,1}$ with Minkowski signature – are based on certain quantized coadjoint orbits, specifically $\mathbb{C}P^3$ and $\mathbb{C}P^{1,2}$, defined in terms of some highest (or lowest) weight representation $\mathcal{H}_\Lambda$ of some Lie group G. These are naturally quantum Kähler manifolds, equipped with coherent states $|x\rangle$ for $x \in \mathcal{M} = \mathbb{C}P^3$ or $\mathcal{M} = \mathbb{C}P^{1,2}$ defined as G orbit of the highest (or lowest) weight state in $\mathcal{H}_\Lambda$. We can therefore use these to define the string modes

$$\left| \begin{matrix} x \\ y \end{matrix} \right) := |x\rangle\langle y| \qquad \in \mathrm{End}(\mathcal{H}). \tag{6.2.95}$$

Then the geometric trace formulas apply, keeping in mind that the underlying symplectic manifold to be integrated over is the six-dimensional bundle space. This is useful e.g. to compute UV-divergent traces. However, the definition of the semi-classical modes $\Psi_{k;y}^{(L)}$ (6.2.25) is not adapted to the local structure as a S^2 bundle over spacetime. Hence, to compute UV-convergent local traces, it may be more appropriate to use a local factorization of the semi-classical modes into S^2 harmonics and functions on spacetime. This will be used in Section 12.3 to compute the one-loop effective action on covariant quantum spaces.

6.2.6 Factorized string modes for product spaces

Consider a quantum space which has a product structure $\mathcal{M} \times \mathcal{K}$; this will arise in the context of fuzzy extra dimensions. Assuming that $|x\rangle$ and $|\xi\rangle$ are coherent states on $\mathcal{M}$ and $\mathcal{K}$, respectively, it is natural to define coherent states on the product space as as $|x\rangle \otimes |\xi\rangle \in \mathcal{H}_{\mathcal{M}} \otimes \mathcal{H}_{\mathcal{K}} = \mathcal{H}$. Then the string modes (6.2.1) on this background are naturally defined in a factorized form

$$\left| {x \atop y} \right) \otimes \left| {\xi \atop \zeta} \right) \quad \in \ \mathrm{End}(\mathcal{H}) \cong \mathrm{End}(\mathcal{H}_{\mathcal{M}}) \otimes \mathrm{End}(\mathcal{H}_{\mathcal{K}}), \tag{6.2.96}$$

where ξ, ζ are points on the internal space $\mathcal{K}$. With this definition, the trace formula (6.2.77) generalizes as follows:

$$\mathrm{Tr}_{End(\mathcal{H})} \mathcal{O} = \int\limits_{\mathcal{M} \times \mathcal{M}} \Omega_x \Omega_y \int\limits_{\mathcal{K} \times \mathcal{K}} \Omega_\xi \Omega_\zeta \, \left({x \atop y} \right| \left(\left. {\xi \atop \zeta} \right| \mathcal{O} \left| {\xi \atop \zeta} \right) \right) \left| {x \atop y} \right) \tag{6.2.97}$$

in self-explanatory notation. This is again useful to compute the one-loop effective action, although it may sometimes be more appropriate to use a hybrid form of the trace formula, e.g. summing over some finite basis for $\mathcal{K}$ and integrating over string modes on $\mathcal{M}$.

6.3 String modes at one loop in NC field theory

In this section, we indicate how the one-loop effective action in noncommutative field theory can be computed directly in position space, using the geometric trace formulas for string modes. This leads directly to the local form (6.1.49) of the nonplanar contribution, avoiding the use of oscillatory integrals in momentum space. We shall illustrate this in the example of the scalar ϕ^4 theory on fuzzy S^2.

One-loop effective action and nonlocality on S_N^2

As an application of this formalism, we want to compute the one-loop effective action for scalar ϕ^4 theory on S_N^2, with hermitian scalar field $\phi^\dagger = \phi$ and action

$$S[\phi] = \frac{1}{N} \mathrm{tr}\left(\frac{1}{2} \phi(\square + m^2)\phi + \frac{\lambda}{4!} \phi^4 \right). \tag{6.3.1}$$

The result will agree with the (more complicated and less transparent) original computation in [51]. We use the following normalization

$$X^a = J_{(N)}^a, \qquad X^a X_a = \frac{1}{4}(N^2 - 1) = R_N^2$$

$$\square \phi = [X^a, [X_a, \phi]] \tag{6.3.2}$$

so that the spectrum of $\square$ is $l(l+1)$, and $\mathcal{M}$ is a sphere with radius R_N. Then the effective action including one-loop quantum corrections can be written as (6.1.58)

$$\Gamma_{\text{eff}}[\phi] = S[\phi] + \frac{1}{2}\text{Tr}_{\text{End}(\mathcal{H})} \log\left(S''[\phi]\right)$$

$$(\delta\phi, S''[\phi]\delta\phi) = \frac{1}{N}\text{tr}\left(\delta\phi(\Box + m^2)\delta\phi + \frac{\lambda}{3}\phi^2\delta\phi^2 + \frac{\lambda}{6}\delta\phi\phi\delta\phi\phi\right), \qquad (6.3.3)$$

where $S''[\phi]$ is the quadratic form for fluctuations $\delta\phi$ around the background ϕ. The one-loop contribution can be expanded follows

$$\Gamma_{1-loop}[\phi] = \text{Tr}\log(.(\Box + m^2). + \frac{\lambda}{3}.\phi^2. + \frac{\lambda}{6}.\phi.\phi)$$

$$= \text{Tr}\log(\Box + m^2) + \text{Tr}\left(.\frac{1}{\Box + m^2}(\frac{\lambda}{3}\phi^2. + \frac{\lambda}{6}\phi.\phi)\right) + O(\phi^4), \qquad (6.3.4)$$

where the dots indicate the position of the $\delta\phi$ to be traced over. We assume that the background field

$$\phi = \int_{\mathcal{M}} \phi(y)|y\rangle\langle y| \qquad (6.3.5)$$

is slowly varying on the scale of noncommutativity. Then ϕ acts almost-diagonally on the string basis $\psi_{yx} = \left|{x \atop y}\right) = |y\rangle\langle x|$, and we can replace

$$\phi\psi_{yx} \approx \phi(y)\psi_{yx} \qquad (6.3.6)$$

and similarly for ϕ^2. As a warm-up, consider

$$\text{Tr}(.\phi^2.) = \int_{\mathcal{M}\times\mathcal{M}} \frac{\Omega_x\Omega_y}{(2\pi)^2}\text{tr}(\psi_{y,x}\phi^2\psi_{x,y})$$

$$= \dim(\mathcal{H})\int_{\mathcal{M}} \frac{\Omega_x}{2\pi} \langle x|\phi^2|x\rangle \approx N\int_{\mathcal{M}} \frac{\Omega_x}{2\pi} \phi(x)^2 \qquad (6.3.7)$$

using (6.2.77). Similarly, using the property (6.2.12) or (6.2.14) of the propagator we find

$$\text{Tr}(.\Box^{-1}\phi^2.) = \int_{\mathcal{M}\times\mathcal{M}} \frac{\Omega_x\Omega_y}{(2\pi)^2} \text{tr}(\psi_{y,x}(\Box + m^2)^{-1}(\phi^2\psi_{x,y}))$$

$$\approx \int_{\mathcal{M}\times\mathcal{M}} \frac{\Omega_x\Omega_y}{(2\pi)^2} \frac{1}{|x - y|^2 + 2\Delta^2 + m^2}\text{tr}(\psi_{y,x}\phi^2\psi_{x,y})$$

$$= \int_{\mathcal{M}\times\mathcal{M}} \frac{\Omega_x\Omega_y}{(2\pi)^2} \frac{1}{|x - y|^2 + \tilde{m}^2}\langle x|\phi^2|x\rangle$$

$$\approx \frac{m_N^2}{N}\int_{\mathcal{M}} \frac{\Omega_x}{2\pi} \phi^2(x), \qquad (6.3.8)$$

where $|x - y|^2$ is the distance of $x, y \in \mathbb{R}^3$ on the sphere with radius R_N in target space,

$$\tilde{m}^2 = m^2 + 2\Delta^2 \qquad (6.3.9)$$

and m_N^2 is the one-loop planar mass renormalization

$$m_N^2 = N \int_{\mathcal{M}} \frac{\Omega_y}{2\pi} \frac{1}{|x_0 - y|^2 + \tilde{m}^2}$$

$$= \frac{N^2}{2R_N^2} \int_0^\pi d\vartheta \sin\vartheta \frac{1}{(1 - \cos\vartheta)^2 + \sin\vartheta^2 + \frac{\tilde{m}^2}{R_N^2}}$$

$$= 2 \int_{-1}^1 du \frac{1}{2 - 2u + \frac{\tilde{m}^2}{R_N^2}}$$

$$=: I^P \approx \sum_{j=0}^{N-1} \frac{2j+1}{j(j+1) + \tilde{m}^2}. \tag{6.3.10}$$

Here x_0 is taken to be the north pole on S^2, and we recall $\Omega = \frac{N}{2} \sin\vartheta d\vartheta d\varphi$. The approximation in (6.3.8) amounts to replacing $(\Box + m^2)^{-1}$ by its diagonal matrix elements, as discussed in Section 6.2.2. Note also that $\Box^{-1}$ has bounded matrix elements in the string basis, so that the computation makes sense even for $m = 0$. This "planar" contribution is schematically depicted in Figure 6.2. It can be interpreted in terms of an open string stretching from the field at x to $y \in \mathcal{M}$, integrated over y. Now consider the "nonplanar" contribution

$$\text{Tr}(.(\Box + m^2)^{-1}\phi.\phi) = \int_{\mathcal{M}\times\mathcal{M}} \frac{\Omega_x\Omega_y}{(2\pi)^2} \text{tr}(\psi_{y,x}(\Box + m^2)^{-1}(\phi\psi_{x,y}\phi))$$

$$= \int_{\mathcal{M}\times\mathcal{M}} \frac{\Omega_x\Omega_y}{(2\pi)^2} \langle x|(\Box + m^2)^{-1}\phi|x\rangle \langle y|\phi|y\rangle$$

$$\approx \int_{\mathcal{M}\times\mathcal{M}} \frac{\Omega_x\Omega_y}{(2\pi)^2} \frac{1}{|x - y|^2 + \tilde{m}^2}\phi(x)\phi(y) \tag{6.3.11}$$

depicted in Figure 6.3. In contrast to the planar contribution, this results in a nonlocal term. The diagram can be interpreted in terms of an open string[10] connecting the fields at x and y. Note that the loop integral has a direct physical meaning here as integral over position space $\mathcal{M}$. As discussed before, this is justified as long as ϕ^2 is in the IR regime, i.e. it varies only slowly at the NC scale.

Finally, we compute the "vacuum energy" contribution

$$\text{Tr}\log(\Box + m^2) = \int_{\mathcal{M}} \frac{\Omega_x}{2\pi} \int_{\mathcal{M}} \frac{\Omega_y}{2\pi} \log\left(|x - y|^2 + \tilde{m}^2\right)$$

$$= N \int_{\mathcal{M}} \frac{\Omega_y}{2\pi} \log\left(|x_0 - y|^2 + \tilde{m}^2\right)$$

$$= \frac{N^2}{2} \int_0^2 du \log\left(2u + \frac{\tilde{m}^2}{R_N^2}\right) + N^2 \log(R_N^2)$$

---

[10] In string theory, this diagram could alternatively be interpreted in terms of a closed string propagating from x to y.

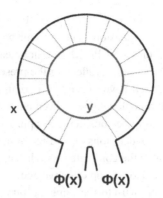

Figure 6.2

Planar one-loop contribution. The string modes $\left|\begin{smallmatrix}x\\y\end{smallmatrix}\right)$ are indicated as dashed lines.

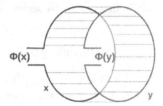

Figure 6.3

Nonplanar 1-loop contribution. The string modes $\left|\begin{smallmatrix}x\\y\end{smallmatrix}\right)$ are indicated as dashed lines.

$$= N^2\left(2\ln N - 1 + O\left(\frac{\tilde{m}^2}{R_N^2}\right)\right)$$

$$=: \Gamma_{vac}, \tag{6.3.12}$$

where x_0 is the north pole. Collecting these results, the one-loop contribution to the effective action up to quadratic order in ϕ is obtained as

$$\boxed{\Gamma_{1-loop} = \Gamma_{vac} + \frac{\lambda}{3}\frac{1}{N}\int_{\mathcal{M}}\frac{\Omega_x}{2\pi}\,m_N^2\phi^2(x) + \frac{\lambda}{6}\int_{\mathcal{M}\times\mathcal{M}}\frac{\Omega_x\Omega_y}{(2\pi)^2}\frac{\phi(x)\phi(y)}{|x-y|^2+\tilde{m}^2} + O(\phi^4),}$$

$$\tag{6.3.13}$$

where $|x-y|^2 = (x-y)^\mu(x-y)^\nu\eta_{\mu\nu}$ is the distance in target space. This should be added to the bare action (6.3.1), which can be written in the semi-classical regime as

$$S[\phi] = \frac{1}{N}\int_{\mathcal{M}}\frac{\Omega_x}{2\pi}\left(\frac{1}{2}\phi(\Box+m^2)\phi + \frac{\lambda}{4!}\phi^4\right). \tag{6.3.14}$$

The planar contribution is local and leads to a standard mass renormalization, which can easily be obtained using a more traditional mode expansion [51]. The non-planar loop contribution coincides with our previous result (6.1.49) (noting that the present $\mathcal{M}$ is a sphere with radius R_N), which was obtained with considerable effort using a group-theoretical mode expansion. The present approach is clearly much simpler and more transparent, and provides directly the nonlocal action in position space. This effect is of distinctly stringy nature,[11] reflecting the presence of virtual long strings described by the string modes, which have no counterpart in standard quantum field theory. Hence, the model describes a nonlocal theory even on scales much longer than the noncommutativity scale, and should *not* be considered as approximation or deformation of some local QFT.

The present considerations generalize immediately to any higher-dimensional quantum space $\mathcal{M}$, provided the completeness relation (4.2.36) holds, where $|x - y|$ is the distance in target space. The one-loop effective action has always the *same universal form* (6.3.13). This would be very hard to see using a group-theoretical mode expansion.

In the case of Minkowski signature, the spectrum of on-shell modes will include long and therefore nonlocal string modes, which are extreme UV modes. Their contribution through loop effects as above clearly leads to a significant clash with causality and locality. This is avoided only in the maximally supersymmetric model as discussed in Chapter 11.

UV/IR mixing on the Moyal–Weyl plane

Although the nonlocal term (6.3.13) is perfectly well-defined on compact fuzzy spaces for finite N, it corresponds in the flat limit to an IR-divergent oscillatory term in momentum space, as discussed in Section 6.1. It is worth revisiting this issue from the present point of view. According to the general result, the nonplanar contribution to the one-loop effective action of the ϕ^4 theory on $\mathbb{R}^{2n}_\theta$ is given by

$$\Gamma_{NP} = \frac{\lambda}{6} \int\limits_{\mathbb{R}^{2n} \times \mathbb{R}^{2n}} \frac{\Omega_x \Omega_y}{(2\pi)^{2n}} \frac{\phi(x)\phi(y)}{|x - y|^2 + m^2}. \tag{6.3.15}$$

Even though it is nonlocal, this term is invariant under translations, and we can compute it in a plane wave basis $\phi(x) = \int \frac{d^{2n}k}{(2\pi)^{2n}} \phi_k \, e^{ikx}$. For real $\phi = \phi^*$, this leads to

$$\begin{aligned}
\Gamma_{NP} &= \frac{\lambda}{6} \int \frac{d^{2n}k}{(2\pi)^{2n}} \phi(-k)\phi(k) \int \frac{d^{2n}z}{(2\pi L_{NC}^2)^{2n}} \frac{1}{|z|^2 + m^2} e^{ik_a z^a} \\
&= \frac{\lambda}{6} \int \frac{d^{2n}k}{(2\pi)^{2n}} \phi(-k)\phi(k) \int \frac{d^{2n}p}{(2\pi)^{2n}} \frac{1}{p_i p_j \gamma^{ij} + m^2} e^{ik_a \theta^{ab} p_b}.
\end{aligned} \tag{6.3.16}$$

(cf. (6.1.39)), replacing $z^a = \theta^{ab} p_b$ in the last step. Here $|.|$ is the target space metric, and

$$\gamma^{ab} = \theta^{aa'} \theta^{bb'} \delta_{a'b'} \tag{6.3.17}$$

is the effective or "open string" metric (3.1.68) which governs noncommutative field theory. This is the familiar form[12] for the nonplanar contribution to the propagator on

---

[11] The present low-dimensional model can be viewed as noncritical string theory. The connection to critical (super)string theory will be discussed in Chapter 11.

[12] In many papers on NC field theory [63, 194], the kinetic term is defined as $\partial^a \phi \partial_a \phi$ rather than $[X^a, \phi][X_a, \phi]$. Then the closed string metric rather than the open string one appears in the last line of (6.3.16), as in (6.1.39). The present form of the action arises naturally for transversal modes in the matrix model approach.

$\mathbb{R}_\theta^{2n}$ in the momentum basis. In this form, the nonlocality leads to an IR divergence as $k \to 0$, which entails a failure[13] of the standard renormalization procedure with local counterterms. Note that the interpretation of the loop variables is transmuted from position to momentum space.

In particular, the IR divergence in (6.3.16) suggests that the standard translation-invariant vacuum may be inappropriate here. Indeed, the nonlocal form of the equation of motion resulting from (6.3.13) suggests the presence of a novel phase featuring a nonuniform or "striped" vacuum structure in position space, as verified in numerical simulations [33, 87, 152]. This novel phase structure can also be seen in terms of the eigenvalue distribution of the matrix ϕ, which will be discussed briefly in Section 7.2.1.

The important message is that quantum effects in NC field theory generically lead to long-range nonlocality. This means that these models should not be considered as deformations or regularization of standard QFT, but they may be useful e.g. as effective description of physics in strong magnetic fields, and possibly in other contexts. However, there is an important exception to this conclusion, given by the maximally supersymmetric IKKT matrix model. Nontrivial backgrounds such as fuzzy $\mathcal{M}^{3,1}$ arise in this model as branes in target space, leading to a much milder form of nonlocality at one loop, understood in terms of 10-dimensional IIB supergravity. This leads to a noncommutative gauge theory on quantum spacetime with mild nonlocal effects at the quantum level, which will be computed in Chapter 11 using the present method.

6.4 Higher loops and pathological UV/IR mixing

At higher loops, it was observed by Minwalla, van Raamsdonk and Seiberg [141] that certain types of nonplanar contributions lead to ever stronger IR divergences on $\mathbb{R}_\theta^{2n}$. This can again be understood in terms of even stronger nonlocality in position space, and computed conveniently using the string representation of the propagator discussed in Section 6.2.2. Here we will only explain some qualitative features, and refer to the literature for more details, notably [190] on the string mode formalism and [49, 63, 72, 194] for the momentum space approach on $\mathbb{R}_\theta^{2n}$.

Consider the nonplanar two-loop contribution to the propagator in the present scalar model corresponding to the diagram in Figure 6.4. This leads to a contribution to the effective action of the form

$$\int_{\mathcal{M}\times\mathcal{M}} \Omega_x \Omega_y \Omega_z \frac{\phi(x)\phi(y)}{(|x-z|^2 + m^2)(|y-z|^2 + m^2)} \tag{6.4.1}$$

and hence to a non-local kinetic term of the form

$$K(x,y) \sim \int \Omega_z \frac{1}{|x-z|^2 + m^2} \frac{1}{|y-z|^2 + m^2}. \tag{6.4.2}$$

---

[13] This may be circumvented by adding additional terms to the action which strongly modify the noncommutative geometry, as discussed in Section 6.4. However, then translational invariance is lost.

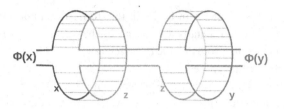

Nonplanar two-loop contribution. The string modes are indicated as dashed lines.

The internal integral over z is divergent on $\mathbb{R}_\theta^d$ in $d \geq 4$ dimensions, so that the action and the propagator do not make sense. Such terms cannot be absorbed by local counterterms, so that the model is not renormalizable. Note that the divergence arises from long high-energy strings for large separations of $|z - x|$ and $|z - y|$. The problem persists and gets amplified in higher loops. This means that it is not possible to separate the UV and IR, and the Wilsonian paradigm of QFT breaks down.

For compact noncommutative spaces (or in two dimensions), the integrals are actually finite, but lead to a stronger nonlocality of the effective action at higher loops. Hence, the models do make sense at least for finite-dimensional matrices, but they lead to physics which is very different from standard local field theory. This problem is avoided in the maximally supersymmetric IKKT matrix model, which will be the main focus in the following.

The Grosse–Wulkenhaar model

It is interesting to briefly discuss a somewhat different model for scalar field theory on $\mathbb{R}_\theta^{2n}$, where the UV/IR mixing is suppressed by adding a confining harmonic potential $V(X) \sim X^2$. In matrix model language, the model takes the form

$$S = \mathrm{Tr}\left(- [X^a, \phi][X_a, \phi] + \Omega [X^a, \phi]_+ [X_a, \phi]_+ + \frac{\lambda}{4!}\phi^4 \right), \qquad (6.4.3)$$

where $[.,.]_+$ denotes the anti-commutator. In terms of $\mathcal{P}^a = [X^a, .]$ (6.2.7), this can be cast in the form

$$S = \mathrm{Tr}\left(- (1 - \Omega)\mathcal{P}^a \phi \mathcal{P}_a \phi + 4\Omega X^a X_a \phi^2 + \frac{\lambda}{4!}\phi^4 \right). \qquad (6.4.4)$$

This model turns out to be renormalizable in four dimensions [85], and exhibits the full complexity of a renormalizable QFT in four dimensions. This is quite plausible after having understood UV/IR mixing as a nonlocal effect of long strings running in the loops, since the long string modes are suppressed the X^2 potential. In the proof of renormalizability it is convenient to use a certain matrix basis that is somewhat reminiscent of our basis of string modes, but is local only near the origin.

Moreover, this model admits a remarkable "Langmann-Szabo" duality between momentum space and position space [129], suggested by the form of the kinetic term in

(6.4.3). The invariance of the interaction term under this duality follows from the intrinsic nature of the operator product in End($\mathcal{H}$), which is oblivious to any specific (position or momentum) representation. For the special "self-dual" case $\Omega = 1$, the model takes a particularly simple form $\mathrm{Tr}(2X^a X_a \phi^2 + \frac{\lambda}{4!}\phi^4)$. This is quite close to the simple matrix models discussed in Section 7.2.1, and the model can indeed be solved exactly at the quantum level [86], applying matrix model techniques in the spirit of a field-theoretic model. However, translational invariance is lost, and the model should not be considered as a deformation of the commutative ϕ^4 QFT. Nevertheless, it demonstrates how the framework of matrix models allows to introduce new techniques into this generalized realm of field theoretic models.

6.5 Diffeomorphisms and symplectomorphisms on quantum spaces

Consider a scalar field $\phi \in \mathrm{End}(\mathcal{H})$ on some quantum space $\mathcal{M}$, and transformations of ϕ of the form

$$\phi \to U^{-1}\phi U, \tag{6.5.1}$$

where $U \in U(\mathcal{H})$ is an unitary operator. This is indeed a symmetry in the context of Yang–Mills matrix models, which will play the role of gauge transformation. An infinitesimal transformation for $U = e^{i\Lambda}$ then has the form

$$\delta_\Lambda \phi = -i[\Lambda, \phi], \tag{6.5.2}$$

where $\Lambda = \Lambda^\dagger$ is hermitian. In the semi-classical limit, such an infinitesimal gauge transformation has the form

$$\delta_\Lambda \phi = \{\Lambda, \phi\} = \mathcal{L}_\xi \phi = \xi^\mu \partial_\mu \phi, \qquad \Lambda = \Lambda^* \in \mathcal{C}^\infty(\mathcal{M}). \tag{6.5.3}$$

This is nothing but the Lie derivative of $\phi \in \mathcal{C}^\infty(\mathcal{M})$ along the Hamiltonian vector field

$$\xi = \{\Lambda, .\} = \xi^\mu \partial_\mu, \qquad \xi^\mu = \{\Lambda, x^\mu\} \tag{6.5.4}$$

in local coordinates x^μ. The extension to tensor fields will be discussed in Section 9.1.3.

However, not every infinitesimal diffeomorphism on $\mathcal{M}$ is of this form. The diffeomorphisms corresponding to Hamiltonian vector fields are symplectomorphisms or symplectic diffeos, because they preserve the symplectic form, as discussed in Section 1.1. In two dimensions, a symplectic form can be viewed as a volume form, so that symplectomorphisms are essentially volume-preserving diffeos. In dimensions larger than 2, they provide only a specific class of volume-preserving diffeos. We will thus view the transformations (6.5.2) as infinitesimal symplectomorphisms on quantum spaces, or as quantized symplectomorphisms.

6.6 Noncommutative Yang–Mills theory I

Having discussed scalar field theory, the next step is to define a Yang–Mills gauge theory on quantum spaces. We first consider the case of the Moyal–Weyl quantum plane $\mathbb{R}_\theta^{2n}$, and show that Yang–Mills gauge theories are naturally formulated as certain types of multi-matrix models. This is the starting point for the more general discussion in Section 7.

Inspired by ordinary Yang–Mills gauge theory, it is natural to write down the following action on $\mathbb{R}_\theta^{2n}$,

$$S_{\mathrm{YM}} = -\frac{1}{g^2}\mathrm{Tr}\big(F_{ab}F_{a'b'}\delta^{aa'}\delta^{bb'}\big), \tag{6.6.1}$$

where

$$F_{ab} = \partial_a A_b - \partial_b A_a - i[A_a, A_b] \tag{6.6.2}$$

is the field strength of the real $\mathfrak{u}(k)$-valued gauge field

$$A_a = A_{a,\alpha}\lambda^\alpha. \tag{6.6.3}$$

Here λ_α is a basis of the Lie algebra $\mathfrak{u}(k)$ and $A_{a,\alpha} = A_{a,\alpha}^\dagger \in \mathrm{End}(\mathcal{H})$ are functions on $\mathbb{R}_\theta^{2n}$. It is easy to check that this action is invariant under the following gauge transformations

$$A_b \to U^{-1}A_b U + U^{-1}i\partial_b U \tag{6.6.4}$$

for any unitary $U \in \mathrm{End}(\mathcal{H}) \otimes \mathfrak{u}(k)$, noting that the field strength transforms as

$$F_{ab} \to U^{-1}F_{ab}U. \tag{6.6.5}$$

Even though this action looks simple enough, it becomes even simpler in terms of

$$\mathcal{A}^a = \theta^{ac}A_c, \tag{6.6.6}$$

which has dimension length. Recalling the crucial relation $[X^a, .] = i\theta^{ab}\partial_b$ (3.1.64), the gauge transformations take the form

$$\mathcal{A}^a \to U^{-1}\mathcal{A}^a U + i\theta^{ab}U^{-1}\partial_b U = U^{-1}(X^a + \mathcal{A}^a)U - X^a \tag{6.6.7}$$

or equivalently

$$X^a + \mathcal{A}^a \to U^{-1}(X^a + \mathcal{A}^a)U. \tag{6.6.8}$$

Infinitesimal gauge transformation can accordingly be written as

$$\delta_\Lambda \mathcal{A}^a = i[X^a + \mathcal{A}^a, \Lambda], \qquad U = e^{i\Lambda}. \tag{6.6.9}$$

This suggests to define the following "covariant coordinates" [133] or dynamical matrix variables

$$T^a := X^a + \mathcal{A}^a \to U^{-1}T^a U, \tag{6.6.10}$$

where $X^a = X^a \mathbb{1}_k$ is understood. Then we observe

$$-i[T^a, T^b] = \theta^{ab} + \theta^{aa'}\theta^{bb'}([\partial_{a'}A_{b'} - \partial_{b'}A_{a'} - i[A_{a'}, A_{b'}])$$
$$= \theta^{ab} + \theta^{aa'}\theta^{bb'}F_{a'b'}$$
$$= \theta^{aa'}\theta^{bb'}(-\theta^{-1}_{a'b'} + F_{a'b'}), \tag{6.6.11}$$

so that the Yang–Mills action (6.6.1) assumes the form of a **Yang–Mills matrix model**

$$S_{\text{YM}} = \frac{1}{g^2}\text{Tr}\big([T^a, T^b][T^{a'}, T^{b'}]\gamma_{aa'}\gamma_{bb'} + c\mathbb{1}\big), \qquad c = \theta^{ab}\theta^{a'b'}\gamma_{aa'}\gamma_{bb'} \tag{6.6.12}$$

after dropping a "surface" term $\text{Tr}[.,.] = 0$. Here

$$\gamma^{ab} = \theta^{aa'}\theta^{aa'}\delta_{a'b'} \tag{6.6.13}$$

is the effective metric (3.1.68). This action is trivially invariant under the gauge transformations

$$T^a \to U^{-1}T^a U. \tag{6.6.14}$$

Since the Lie algebra $\mathfrak{u}(k)_{\mathbb{C}} \cong \text{End}(\mathbb{C}^k)$ is nothing but the space of complex $k \times k$ matrices, the T^a and the gauge generator Λ take values in

$$T^a \in \text{End}(\mathcal{H}) \otimes \mathfrak{u}(k)_{\mathbb{C}} \cong \text{End}(\mathcal{H} \otimes \mathbb{C}^k). \tag{6.6.15}$$

Dropping the irrelevant constant c, we can thus define $U(k)$ Yang–Mills theory on $\mathbb{R}^{2n}_\theta$ through the matrix model[14]

$$S_{\text{YM}} = \frac{1}{g^2}\text{Tr}[T^a, T^b][T^{a'}, T^{b'}]\delta_{aa'}\delta_{bb'} = -\frac{1}{g^2}\int \frac{\Omega}{(2\pi)^n}\,\text{tr}(F_{ab}F_{a'b'})\gamma^{aa'}\gamma^{bb'} \tag{6.6.16}$$

recalling that $\text{Tr} = \int \frac{\Omega}{(2\pi)^n}$ (3.1.39). It is straightforward to add fermions to the present NC gauge theories, provided they transform either in the adjoint or in the (anti-) fundamental representation.[15] For example, the covariant derivative of a scalar field $\phi \to U^{-1}\phi U$ transforming in the adjoint of the gauge group is given by the commutator

$$[X^a, \phi] = i\theta^{ab}D_b\phi, \qquad D_b = \partial_b - i[A_b, \phi] \tag{6.6.17}$$

which obviously transforms as $D_b\phi \to U^{-1}D_b\phi U$. This automatic for supersymmetric gauge theories, some of which will be discussed in more detail in Section 11.1.2.

Hence, after the dust has settled, Yang–Mills gauge theory turns out to be much simpler to formulate on noncommutative spaces than in the commutative case, and the concepts of fiber bundles and connections emerge automatically from the formulation in terms of matrix models. The crucial ingredient is the fact that derivations on quantum spaces are generated by commutators, as discussed in Section 3.1.4. However, the formulation of Yang–Mills theory as a matrix model leads to a puzzle, since the action is invariant under the scaling transformation

---

[14] The metric γ^{ab} can be absorbed by a suitable re-definition of T^a.

[15] It is notoriously difficult to define noncommutative gauge theory for fields in different representations and for gauge groups different from $U(n)$. However, the more general effective theories required in physics can be obtained via spontaneous symmetry breaking, e.g. on intersecting branes.

$$T \mapsto \alpha T,$$
$$g^2 \mapsto \alpha^4 g^2 \tag{6.6.18}$$

for $\alpha \in \mathbb{R}$. This transformation entails a change of the noncommutativity scale $L_{\mathrm{NC}} \to \alpha L_{\mathrm{NC}}$. Since T^a has dimension length due to (6.6.10), it follows that g^2 has dimension length^4. Hence, g^2 cannot be the physical Yang–Mills coupling constant. To understand this, it suffices to observe that the physical scale L_{NC} arises only after choosing some background, which could be any $\mathbb{R}^{2n}_\theta$ with *any* noncommutativity tensor θ^{ab}, since all these are solutions of the same model. To identify the physical Yang–Mills coupling constant, we rewrite the action taking into account the contribution from the symplectic volume form $\Omega = L_{\mathrm{NC}}^{-2n} d^{2n}x$ (3.1.18) on the given background:

$$S_{\mathrm{YM}} = -\frac{1}{(2\pi)^n g^2 L_{\mathrm{NC}}^{2n-8}} \int d^{2n}x \, \mathrm{tr}(F_{ab}F_{a'b'}) \tilde{\gamma}^{aa'} \tilde{\gamma}^{bb'}, \tag{6.6.19}$$

where

$$\tilde{\gamma}^{ab} := L_{\mathrm{NC}}^{-4} \gamma^{ab} \tag{6.6.20}$$

is dimensionless with $\det \tilde{\gamma} = 1$. We can now read off the Yang–Mills coupling constant:

$$\boxed{g_{\mathrm{YM}}^2 = g^2 L_{\mathrm{NC}}^{2n-8}.} \tag{6.6.21}$$

This is dimensionless in $2n = 4$ dimensions as it should be. Hence, the physical coupling depends not only on g but also on the scale of noncommutativity L_{NC} which is set by the background X^a. In particular, it depends on the background (rather than just on g) whether the resulting gauge theory is weakly or strongly coupled. On the covariant quantum space $\mathcal{M}^{3,1}$, this relation will be somewhat modified, but the conclusion is the same.

A related observation is in order, which is based on the mathematical fact that all infinite-dimensional separable Hilbert spaces are equivalent. Since the Hilbert space $\mathcal{H}$ underlying Moyal–Weyl quantum space is such an infinite-dimensional separable Hilbert space, we can conclude that

$$\mathcal{H} \otimes \mathbb{C}^k \cong \mathcal{H}. \tag{6.6.22}$$

This means that all $U(k)$-valued Yang–Mills theories on $\mathbb{R}^{2n}_\theta$ are equivalent for all k, and the rank k makes sense only[16] for perturbative fluctuations around a given background $X^a \mathbb{1}_k$ (6.6.10). On such a background, the action reduces to the standard $U(k)$ Yang–Mills action on $\mathbb{R}^{2n}$ in the commutative limit $\theta \to 0$, or more properly in the regime where all wavelengths are much larger than L_{NC}. This equivalence underscores the fundamental difference between noncommutative and commutative gauge theories. It is obvious from the matrix model point of view, since we simply expand the same model around different "backgrounds" X^a or $X^a \otimes \mathbb{1}_k$, which lead to different perturbative physics for the fluctuations. However, the explicit relation between the backgrounds

$$X^a \otimes \mathbb{1}_k = X^a \otimes \mathbb{1}_l + \mathcal{A}^a \tag{6.6.23}$$

---

[16] This can be stated in terms of a more abstract mathematical notion called Morita equivalence. We will refrain from such abstraction.

involves "large" gauge fields $\mathcal{A}^a$ which have no commutative analog. Similar features also arise in finite-dimensional matrix models of noncommutative gauge theory, such as the Alekseev–Recknagel–Schomerus model (7.3.2) for the fuzzy sphere. These topics – and related aspects such as noncommutative instantons – will be discussed more generally in Section (7.4.5) in terms of brane solutions of the underlying matrix model.

While these observations are very satisfactory and natural, the present construction of NC gauge theory only works for $U(k)$ gauge groups. Any attempts to define noncommutative gauge theories for more general gauge groups run into serious obstacles, since constraints are typically not compatible with gauge transformations.[17] Fortunately, the standard model of elementary particles requires only (products of) unitary gauge groups, which could be realized in the matrix model framework via spontaneous symmetry breaking on suitable (brane-type) backgrounds. This will be briefly discussed in Section 8.

$U(1)$ sector, gauge transformations, and geometry

There is one small but crucial detail which will play an important role later: the $U(1)$ sector does not decouple, even though all fields in the matrix model action are in the adjoint. In the commutative case, $U(1)$ components of fields in the adjoint always decouple from the $SU(k)$ components. This is not the case in the NC setting, since commutators reduce to nontrivial Poisson brackets. This leads to a universal coupling of the $U(1)$ fields to the $SU(k)$ fields, which is precisely the reason why one cannot define a pure $SU(k)$ NC gauge theory. Since the $U(1)$ component of the background $X^a \otimes \mathbb{1}_k$ describes the geometry of the (stack of) branes, that coupling is nothing but gravity. Indeed the separation into background plus fluctuation in (6.6.10) is completely artificial for the $U(1)$ component, and adding a $U(1)$ gauge field $\mathcal{A}^a$ is tantamount to changing the (noncommutative) geometry of the background. Therefore, noncommutative $U(1)$ gauge theory – and more generally the $U(1)$ sector of NC $U(k)$ gauge theory [178] – should be understood as a gauge theory of geometry. That observation is the starting point for emergent gravity, which has been discussed by various authors including [162, 181, 215, 216]. It will be developed systematically in Chapter 9, where the general relation between background and geometry will be derived.

The special role of the $U(1)$ sector is also reflected in the gauge transformations (6.6.4) of the $U(k)$-valued gauge fields, which take the infinitesimal form

$$A_b \rightarrow -\partial_b \Lambda + i[A_b, \Lambda]. \tag{6.6.24}$$

The last term vanishes on classical spaces, but is essential to maintain gauge invariance on a noncommutative space, even if $\Lambda = \Lambda_0(x)\mathbb{1}_k$ contains no $\mathfrak{su}(k)$ components. The reason is that NC $U(1)$ gauge transformations correspond to diffeomorphisms generated by Hamiltonian vector fields $\{\Lambda, .\}$, which are *nonlocal* transformations. In particular, the abelian field strength $F_{\mu\nu}$ transforms non-trivially under such gauge transformations. This

---

[17] The approaches in the literature toward more general gauge groups typically rely on formal star product constructions, see e.g. [110]. However, their quantization is very problematic.

can be handled by absorbing any $U(1)$ contributions in the symplectic form $\theta_{\mu\nu}^{-1} = \bar{\theta}_{\mu\nu}^{-1} + F_{\mu\nu}$, which is then invariant under symplectic diffeos. This point of view will be very useful in the context of more general geometries.

From now on we will simply start from the matrix model formulation, which is much simpler and more natural for the noncommutative setting. This formulation provides immediately a definition of Yang–Mills gauge theory on any quantum space which can arise as a background (or solution) of such matrix models. The appropriate structures including derivatives, connections, metric and calculus then emerge automatically on the given background.

6.7 *Relation to string theory

Even though the approach of this book is not built on string theory, it is appropriate to discuss how noncommutative field theory and gauge theory arise in that context. We shall assume some familiarity with string theory in this section, which is not essential in the later sections.

The basic observation is that open strings can end on D-branes, which are submanifolds embedded in target space. In the presence of a Neveu–Schwarz B-field in target space, the worldsheet action of the string takes the form

$$S_\Sigma = \frac{1}{4\pi\alpha'} \int_\Sigma \left(g_{ij}\partial_a x^i \partial^a x^j - 2\pi i\alpha' B_{ij}\epsilon^{ab}\partial_a x^i \partial_b x^j \right). \tag{6.7.1}$$

Here $\alpha' = \ell_s^2$ is the Regge slope, Σ is the string worldsheet, and x^i is the embedding function of the strings into target space with "closed string" metric g_{ij}. The term involving the B-field is a total derivative, which for open strings can be written as an integral over the boundary of the string worldsheet

$$S_{\partial\Sigma} = -\frac{i}{2} \oint_{\partial\Sigma} B_{ij} x^i \partial_t x^j. \tag{6.7.2}$$

In a suitable decoupling limit of the parameters, the worldsheet theory is topological. All that remains are the boundary degrees of freedom of the open strings, which are governed by the action (6.7.2). This one-dimensional action coincides with the action of charged particles in a strong magnetic field, as discussed in Section 4.8. Taking into account the results of that section, this means that the endpoints of the open strings live effectively on a D-brane with intrinsic coordinate noncommutativity[18] $[x^i, x^j] = i\theta^{ij}$ with $\theta^{ij} = \left(\frac{1}{B}\right)^{ij}$, hence they are governed by a noncommutative field theory on $\mathbb{R}^{2n}_\theta$. Moreover, the Σ-model for the string world sheet is coupled to $U(1)$ gauge field degrees of freedom which live

[18] A more elaborate version of this argument can be given in the context of deformation quantization [43], following results of Kontsevich. A similar argument shows that the D-branes in the Wess–Zumino–Witten model are given by fuzzy spheres governed by the ARS matrix model (7.3.2).

on the worldsheet boundary i.e. on the D-brane. The open strings accordingly induce a noncommutative Yang–Mills gauge theory on the D-brane, with the same structure as discussed in this chapter. That noncommutative gauge theory – defined by the open strings – is governed by a metric denoted as *open string metric*, which corresponds to the effective metric (7.4.24), (9.1.9) identified in the present framework of matrix models. The remaining massive closed string modes decouple in a suitable limit, so that the string theory reduces the non-commutative gauge theory described by the endpoints of the open strings. These observations were made precise in a seminal work by Seiberg and Witten [170], following earlier observations [50, 168].

Further discussion and reading

There is a large body of literature on field theory on noncommutative spaces, which is impossible to properly discuss here, so that only a few representative topics and citations will be selected.

Early ideas about a possible quantum structure of spacetime in physics and quantum field theory were discussed by pioneers including Schrödinger, Heisenberg, Oppenheimer and Snyder, leading to a first specific proposal in [172]. These ideas were overshadowed by the success of the renormalization programme in QFT, but revived following the development of noncommutative geometry [52]. After the first proposals to formulate NC field theories on various noncommutative spaces [62, 72, 81, 84], the subject became highly active following the paper of Seiberg and Witten [170], exhibiting the noncommutative gauge theory that arises on D-branes in a *B*-field background, due to open strings. The developments during this period are covered in the excellent reviews [63, 194]. Similar relations were observed on D-brane solutions in Yang–Mills matrix models [3, 153, 154]. However, the discovery of UV/IR mixing in [141] showed that the quantization of generic NC field theories is highly problematic, and the Wilsonian view of QFT does not apply. This basically doomed the early hopes that generic models of NCFT might replace or supersede ordinary QFT as a description of particle physics at all scales. This problem also arises in noncommutative gauge theories [138], with the notable exception of maximally supersymmetric or possibly other UV-finite models [107].

Nevertheless, novel approaches led to new insights to NC QFT. For example, the phase structure has been related to the eigenvalue distribution [85, 125, 159, 177, 201]. The entanglement entropy in NC field theory was studied e.g. in [114, 142, 151]. The relevance of string-like modes in NC field theory was realized already in [34, 105], and provides a natural understanding of UV/IR mixing in NCFT [184, 190]. In the framework of matrix models and NCFT, their use to evaluate loop integrals was pointed out in [105] and elaborated further in [184], providing a simple derivation of IIB supergravity amplitudes and propagators on curved NC geometries. Noncommutative Chern–Simons gauge theory was applied in the context of the quantum Hall effect [112, 113, 158, 193]. Finally, the observations of features in NC gauge theory reminiscent of gravity [162, 178, 215] and the geometric interpretation of NC gauge theory formulated through matrix models [3, 11, 176] lead to a better understanding of the dynamical geometry on D-branes in matrix models [181]. Together with the UV finiteness of maximally supersymmetric gauge theories, this suggests to focus on the IKKT model, in order to resolve the pathological UV/IR mixing and to make the link with gravity.

7 Yang–Mills matrix models and quantum spaces

A comprehensive theory including gravity should generalize the description of NC Yang–Mills gauge theory in the previous chapter to more generic quantum spaces. It may be tempting to follow the standard construction in terms of gauge fields, gauge transformations and field strength, generalizing the approach of Section 6.6. However, defining a consistent differential calculus on more general quantum spaces is problematic, and following old ideas in a new world is not always the best strategy.

There is a much simpler and more profound way to formulate noncommutative gauge theories, through certain matrix models denoted as Yang–Mills matrix models. These models are best viewed as models for dynamical quantum spaces defined through matrix configurations, which is precisely what gravity should accomplish. At the same time, they also define Yang–Mills-type gauge theory for the fluctuations around such quantum spaces. This construction is simpler and more compelling than the classical construction of Yang–Mills theories, and provides a novel way to understand gauge theory.

Yang–Mills matrix models are defined through an action of the form

$$S[T^a] = \frac{1}{g^2} \text{Tr}\big([T^a, T^b][T^{a'}, T^{b'}]\eta_{aa'}\eta_{aa'} - 2m^2 T^a T^b \eta_{ab}\big). \qquad (7.0.1)$$

Here $T^a \in \text{End}(\mathcal{H})$, $a = 0, \ldots, D-1$ are hermitian matrices acting on a (separable) Hilbert space $\mathcal{H}$, and g^2 is a coupling constant that will often be absorbed in the matrices via $T^a \to g^{-1/2}T^a$, which are then dimensionless. The indices are contracted with the $SO(D-1, 1)$ invariant target space metric $\eta_{ab} = \text{diag}(-1, 1, \ldots, 1)$ to specific, while the Euclidean model is defined by replacing $\eta_{ab} \to \delta_{ab}$; both cases will be understood henceforth, raising and lowering Latin indices $a, b \ldots$ with either η_{ab} or δ_{ab}. This action has the same form as (6.6.12), except that we include a quadratic mass term with a scale parameter m^2.

A comment on the choice of sign is in order. We will see that in the Minkowski case, the action has the appropriate structure $S = \int \mathcal{L}$ of a field-theoretic action with $\mathcal{L} = T - V$, with kinetic term T and potential V. To maintain a uniform notation in both signatures, we adopt the convention that the action is negative definite in Euclidean signature in this chapter,

$$S_E < 0. \qquad (7.0.2)$$

It is straightforward to include fermions in the adjoint[1] to these matrix models as follows:

$$S[T, \Psi] = \frac{1}{g^2} \text{Tr}\big([T^a, T^b][T_a, T_b] - 2m^2 T_a T^a + \overline{\Psi}\Gamma_a[T^a, \Psi]\big), \qquad (7.0.3)$$

where Ψ are spinors of $SO(D)$ or $SO(D-1,1)$ whose entries are (Grassmann-valued) matrices. The configuration space of Yang–Mills matrix models is thus given by the space of such matrices T^a and possibly Ψ.

As the name suggests, Yang–Mills matrix models are closely related to Yang–Mills gauge theory. In fact, these models can be obtained by dimensionally reducing Yang–Mills gauge theory on flat space $\mathbb{R}^d$ with $U(N)$ gauge group to a point, i.e. dropping any x-dependence of the fields. To see this, consider the standard YM action

$$S = -\frac{1}{g^2} \int d^D x \, F^{\mu\nu} F_{\mu\nu}, \qquad F_{\mu\nu} = i[\partial_\mu - iA_\mu, \partial_\nu - iA_\nu], \qquad (7.0.4)$$

where $A_\mu = (A_\mu(x))^i_j$ is a $\mathfrak{u}(N)$ valued gauge field, i.e. a hermitian matrix valued gauge field. Assuming that $(A_\mu(x))^i_j \to (A_\mu)^i_j$ are constant, this reduces to the first term in (7.0.1). In the same way, the fermionic term in (7.0.3) arises by dimensional reduction from Yang–Mills gauge field theories with fermions in the adjoint. The mass term does not arise in this way, and has no commutative analog. This term may be useful to stabilize some classical backgrounds, and as an infrared (IR) regularization for the quantization.

We will consider these matrix models as fundamental starting point, rather than as derived or effective models. This is appropriate, since it is hard to conceive anything simpler than a matrix model. They will be recognized as models for dynamical quantum geometry, whose quantization turns out to be well-behaved for a unique such model, namely the maximally supersymmetric IKKT model. We are hence led to a uniquely selected candidate for a quantum theory of gravity.

Symmetries

The essential property of matrix models is the invariance under the transformations

$$T^a \to U T^a U^{-1}, \qquad \Psi \to U \Psi U^{-1} \qquad (7.0.5)$$

for any unitary matrices $U \in U(\mathcal{H})$. They will be treated as gauge transformations, which means that the points of such an orbit are identified. The models are also invariant under the "global" symmetry

$$T^a \to \Lambda^a_{\ b} T^b \qquad (7.0.6)$$

for $\Lambda \in SO(D)$ or $\Lambda \in SO(D-1,1)$. To emphasize this transformation, we will sometimes put a dot over such indices $\dot{a}$ etc. In the case $m^2 = 0$, the model also enjoys the shift invariance

$$T^a \to T^a + c^a \mathbb{1}, \qquad (7.0.7)$$

---

[1] It is also possible to include also (anti-)fundamental fermions, but these do not acquire the appropriate kinetic term on quantum spaces.

which amounts to a non-compact flat direction of the model. The matrix component along $\mathbb{1}$ will be denoted as trace-$U(1)$ sector. That flat direction is lifted in the presence of a mass term with $m^2 \neq 0$. It can also be removed by requiring that the matrices are traceless.

In addition to these symmetries, the model is also invariant under the following scaling relation:

$$
\begin{aligned}
T &\mapsto \alpha T, \\
m^2 &\mapsto \alpha^2 m^2 \\
g^2 &\mapsto \alpha^4 g^2.
\end{aligned}
\tag{7.0.8}
$$

This means that the coupling constant is redundant, and we will sometimes drop g. An meaningful coupling constant for fluctuations can only be identified once a specific background is chosen.

Supersymmetry

For an appropriate choice of fermions Ψ and $m^2 = 0$, Yang–Mills matrix models also enjoy supersymmetry (SUSY), relating the bosonic and fermionic matrices. This is obtained rather trivially via dimensional reduction from the standard $\mathcal{N} = 1$ supersymmetric Yang–Mills theories on $\mathbb{R}^D$, which exist for $D = 2, 4, 6$ and $D = 10$. The most interesting case of $D = 10$ (or rather $D = 9 + 1$) leads to the IKKT or IIB matrix model, which will be discussed in detail in Chapter 11. Here we merely indicate the basic structure of the SUSY transformations, which take the form

$$
\begin{aligned}
\delta \Psi &= i[T^a, T^b] \Sigma_{ab}^{(\psi)} \epsilon \\
\delta T^a &= i \bar{\epsilon} \Gamma^a \Psi,
\end{aligned}
\tag{7.0.9}
$$

where the SUSY generator ϵ is a (Grassmann–) matrix-valued spinor and $\Sigma_{ab}^{(\psi)} = \frac{i}{4}[\Gamma_a, \Gamma_b]$. This matrix supersymmetry gives rise to spacetime supersymmetry on the Moyal–Weyl background $\mathbb{R}_\theta^{2n}$, which reduces to ordinary super-Yang–Mills theory on $\mathbb{R}^{2n}$ for $\theta \to 0$. In particular, the square of the SUSY transformation on $\mathbb{R}_\theta^{2n}$ generates translations, which is nothing but the shift symmetry (7.0.7) in the matrix model. The SUSY generators ϵ must be proportional to the unit matrix $\mathbb{1}_N$, as the shift symmetry only works along a unit matrix.

Equations of motion and matrix Laplacian

Given an action, we can consider solutions of the matrix models defined as critical points $\delta S = 0$ of the action. For the bosonic model, this leads to the equations of motion

$$
[T^b, [T_b, T_a]] + m^2 T_a = 0, \qquad \text{or} \qquad (\Box + m^2) T_a = 0
\tag{7.0.10}
$$

where the *matrix Laplacian* is defined as

$$
\Box := [T_a, [T^a, .]]: \quad \mathrm{End}(\mathcal{H}) \to \mathrm{End}(\mathcal{H}).
\tag{7.0.11}
$$

This matrix Laplacian governs the kinetic term and hence the fluctuations in Yang–Mills matrix models. In the presence of fermions, the equations of motion are extended as follows:

$$(\Box + m^2)T_a = -\frac{1}{4}[\overline{\Psi}, \Gamma_a \Psi]$$

$$\displaystyle{\not{D}}\Psi = 0, \qquad \not{D} = \Gamma_a[X^a, .], \tag{7.0.12}$$

where the commutator in the first line pertains only to the matrix part of the spinor, and $\not{D}$ is the matrix Dirac operator for adjoint fermions.

The presence of a mass term $m^2 \neq 0$ deserves some discussion. One motivation for adding a mass term to the model is the fact that purely bosonic Yang–Mills matrix models with finite-dimensional matrices and $m^2 = 0$ have only trivial i.e. commutative $[T^a, T^b] = 0$ classical solutions, even for Lorentzian signature (cf. Appendix A in [185]). The parameter m^2 also provides a scale for the theory, and without mass term there would be no scale in the model. Adding a mass term to the model can also be seen as a way to regularize possible IR divergences.

Moreover, a nonvanishing mass term breaks the supersymmetry (7.0.9) of the models. From the point of view of the emergent gauge theories, such a mass term is usually considered as "soft" in the QFT sense, i.e. it does not significantly affect the UV behavior of the model. Nevertheless, it would suffice to spoil the UV finiteness of loops in the IKKT model. A mass term might also capture some quantum effects on some given background. Indeed a nontrivial background typically defines a scale, which may lead effectively to a mass term. These issues will be discussed in more depth in Section 7.3.

7.1 Almost-commutative matrix configurations and branes

A priori, the model knows nothing about geometry, except for the target space metric δ_{ab} or η_{ab} that enters the action. Clearly it does not make much sense to attach geometric significance to generic matrix configurations $\{T^a\}$, and most of the configuration space is basically "white noise." However, since the Yang–Mills action is defined as square of commutators, it prefers matrix configurations that are almost-commutative. This is clear in Euclidean signature, where the action is defined in terms of δ_{ab}. Then the dominant contributions are matrices that are almost-commutative, which means that the T^a can "almost" be simultaneously diagonalized. In the case Minkowski signature with target space metric η_{ab}, that argument seems to have a loop-hole, since there can be configurations whose spacelike commutators $[T^i, T^j]$ are large but canceled by equally large mixed contributions $[T^0, T^i]$. However, the same conclusion is reached using the conserved matrix energy momentum tensor T^{ab} discussed in Section 7.4.2. Then these configurations would have high energy, as measured by

$$E = T^{00} = [T^0, T^i][T^0, T_i] + \frac{1}{2}[T^i, T^j][T_i, T_j] \tag{7.1.1}$$

for $i, j = 1, \ldots, D - 1$. The dominant configurations of the model should be those with lowest energy, which are almost-commutative.

Now we recall from Chapter 4 that almost-commutative matrix configuration can be interpreted as a quantization of some underlying symplectic space $\mathcal{M}$. This is achieved in terms of quasi-coherent states $|t\rangle \in \mathcal{H}$, which are approximate common eigenstates of the T^a localized at some point in target space

$$t^a = \langle t | T^a | t \rangle \quad \in \mathbb{R}^D. \tag{7.1.2}$$

These states sweep out some variety $\mathcal{M}$, which allows to associate classical functions to the matrices, and in particular an embedding

$$T^a \sim t^a : \quad \mathcal{M} \hookrightarrow \mathbb{R}^D. \tag{7.1.3}$$

Then $\mathcal{M}$ – or its image under this map – is interpreted geometrically as quantized "brane" in target space $\mathbb{R}^D$. Moreover, the following semi-classical identification was established in (4.4.8)

$$\mathrm{End}(\mathcal{H}) \supset \mathrm{Loc}(\mathcal{H}) \;\overset{\cong}{\longleftrightarrow}\; \mathcal{C}_{\mathrm{IR}}(\mathcal{M}) \subset \mathcal{C}(\mathcal{M}). \tag{7.1.4}$$

We will mostly restrict ourselves to this semi-classical regime, and simply write $\mathrm{End}(\mathcal{H})$ and $\mathcal{C}(\mathcal{M})$; the restriction to almost-local operators and IR functions is understood. In this way, a fuzzy notion of geometry can be extracted from almost-commuting matrix configurations. In the irreducible case, the T^a will generate the full matrix algebra

$$\mathrm{End}(\mathcal{H}) \sim \mathcal{C}(\mathcal{M}), \tag{7.1.5}$$

which is interpreted as algebra of function on $\mathcal{M}$ in the spirit of noncommutative geometry [53]. This brane may be a sharply defined submanifold, or it may be fuzzy in all directions. It may also carry internal extra dimensions that are not resolved by the T^a. In any case, one can view the commutator

$$\Theta^{ab} = -i[T^a, T^b] \sim \{t^a, t^b\} = \theta^{ab} \tag{7.1.6}$$

as an antisymmetric tensor field on $\mathcal{M}$. More generally, $[.,.] \sim i\{.,.\}$ defines an antisymmetric bracket on the space of "fuzzy functions" $\mathrm{End}(\mathcal{H}) \sim \mathcal{C}(\mathcal{M})$, which is a derivation and satisfies the Jacobi identity. In other words, $\mathcal{M}$ can be viewed as a Poisson manifold in the IR regime, where the functions are approximately commuting. Moreover, the matrix Laplacian or d'Alembertian (7.0.11) encodes an effective metric in the spirit of spectral geometry. This geometry will be understood more directly using the embedding (7.1.3), leading to a frame underlying this metric.

In the following, we will develop a suitable framework to describe the effective geometry of $\mathcal{M}$ and its dynamics in the semi-classical regime, in terms of familiar objects such as frames and a metric. One can then view the matrix model as a sort of (higher-dimensional) Poisson-sigma model for functions $t^a : \mathcal{M} \to \mathbb{R}^D$ on a symplectic space or brane $\mathcal{M}$ with target space $\mathbb{R}^D$:

$$S \sim \frac{1}{g^2} \int_{\mathcal{M}} -\{t^a, t^b\}\{t_a, t_b\} - 2m^2 t^a t_a. \tag{7.1.7}$$

Since $\mathcal{M}$ may have any even dimension, these models are highly nontrivial and contain rich physics. However, this is justified only in the IR regime and does not capture the full matrix model, whose saddle points correspond to different symplectic spaces. Note that all the IR physics lives on the brane $\mathcal{M}$ here, while $\mathbb{R}^D$ is an auxiliary space; this is in contrast to string theory, where target space assumes the central role.

To obtain real physics, however, we face a problem: There is no known physical quantity in spacetime which could be interpreted as a Poisson tensor, i.e. an antisymmetric rank 2 tensor field. If such an object exists in nature, it would break Lorentz invariance, which holds to an excellent approximation in nature. This conundrum is resolved on certain **covariant quantum spaces** $\mathcal{M}$, which are interpreted as higher-dimensional bundles over spacetime with a particular structure. That part of the story will be developed in Chapter 10, leading to higher-spin gauge theory and gravity.

The stability of such backgrounds is a nontrivial question that can only be answered in the full quantum theory, where such backgrounds are supposed to arise as nontrivial vacua of the matrix model, through spontaneous symmetry breaking. This will be addressed in the next section.

Of course, there are also commutative matrix configurations $[T^a, T^b] = 0$, which are clearly solutions for $m = 0$. They can be simultaneously diagonalized, and their eigenvalues define some lattice in target space, which one might want to interpret as discretized space(time). However, this does not work and these configuration are negligible, for reasons explained in Section 7.3.1.

We conclude that the dominant configurations in the matrix model are given by noncommutative but almost-commutative matrices, which should be interpreted in terms of an underlying symplectic space.

7.2 Quantization: Matrix integral and $i\varepsilon$ regularization

Perhaps the most important feature of matrix models is that they provide a natural notion of quantization, defined by integrating over the space of all matrices. This is the matrix model analog of the Feynman path integral. For finite-dimensional hermitian matrices, the integral is defined by the obvious translational invariant measure on a vector space, which is invariant under the gauge transformations (7.0.5). In the Euclidean case, the matrix integral[2]

$$Z = \int dT \; e^{S_E[T]} \tag{7.2.1}$$

is well-defined for traceless T^a [15, 127]. In the case of Minkowski signature, the analogous integral

$$Z = \int dT \; e^{iS[T]} \tag{7.2.2}$$

---

[2] Recall that our conventions are such that the Euclidean action is negative definite (7.0.2).

is oscillating and not well-defined a priori. It can be regularized by adding a suitable imaginary mass term

$$S_\varepsilon[T] = \frac{1}{g^2} \mathrm{Tr}\Big([T^a, T^b][T_a, T_b] - 2m^2 T^a T_a + i\varepsilon \sum_a T_a^2 \Big) \qquad (7.2.3)$$

with $\varepsilon \to 0$, where $a, b = 0, \ldots, D-1$. Then the integral

$$Z_\varepsilon = \int dT e^{iS_\varepsilon[T]} \qquad (7.2.4)$$

is absolutely convergent and gauge invariant for $\varepsilon \in (0, \frac{\pi}{2})$, at least for finite-dimensional matrices[3]. It turns out that this imposes at the same time Feynman's $i\varepsilon$ prescription for the propagator, so that (7.2.4) provides a physically sensible and nonperturbative definition for the quantized model. We can then define correlators as

$$\langle \mathcal{O}(T) \rangle := \frac{1}{Z_\varepsilon} \int dT \mathcal{O}(T) \, e^{iS_\varepsilon[T]}, \qquad (7.2.5)$$

where $\mathcal{O}(T)$ is some matrix observable, i.e. some function of the matrices T^a. As in quantum field theory, these correlators encode all physical information of the model. Of course, Planck's constant $\hbar$ enters in the path integral (7.2.4) via $\int dT e^{\frac{i}{\hbar} S}$, but is usually dropped using natural units. Recall that $\hbar$ should not be confused with the scale of noncommutativity of $\mathcal{M}$.

Another possible way to make sense of the oscillatory integral is to perform a generalized Wick rotation $T^0 \to e^{is} T^0$, $T^i \to e^{-it} T^i$. For a suitable combination of s and t, this even allows to relate the Minkowski model to the Euclidean one [145]. However, this is not very useful for studying the physics arising from fluctuations on some background, since it does not map backgrounds (i.e. spacetime branes) which are saddle-points of the Minkowski action to Euclidean ones. The regularization in (7.2.3) is more physical, since it reduces on a given background to the usual $i\varepsilon$ procedure familiar from QFT.

Given the definition of quantization in terms of a matrix integral, many of the standard steps in quantum field theory can be straightforwardly adapted. For example, the Schwinger–Dyson equation is obtained by noting that the integral measure is invariant under arbitrary translations $T^a \to T^a + \delta T^a$ in matrix space. Using this change of variables in the definition of Z and noting that

$$\delta S = -\frac{4}{g^2} \mathrm{Tr} \delta T^a \big([T^b, [T_b, T_a]] + m^2 T_a \big), \qquad (7.2.6)$$

we obtain the simplest form of the **Schwinger–Dyson equation**

$$\langle [T^b, [T_b, T_a]] + m^2 T_a \rangle = 0. \qquad (7.2.7)$$

This can easily be generalized in the presence of matrix correlators and fermions.

---

[3] This also works in the presence of fermions, since the determinant det $\not{D}$ (11.1.9) or the Pfaffian (11.1.14) do not spoil the convergence.

7.2.1 *Single-matrix models

It is appropriate to recall here some aspects of random matrix theory, defined through the single-matrix model

$$Z = \int dT e^{-\frac{N}{g^2} \text{Tr} V(T)} \tag{7.2.8}$$

that is invariant under the symmetry

$$T \to U^{-1} T U, \tag{7.2.9}$$

and its invariant correlators. Here T is a hermitian $N \times N$ matrix, and $V(T)$ is a function called potential which should be bounded from below, such as[4] $V(T) = \frac{1}{2} T^2 + \frac{1}{4} T^4$. Such models were introduced long ago by Wigner as a statistical description of the Hamiltonians governing the energy levels of complicated nuclei, and subsequently found numerous other applications including 2D quantum gravity. These models have many remarkable and analytically accessible features, which lead to important insights. In particular, it was realized that the typical spacing of energy levels in these models display a certain universal behavior, which reproduces the correlations of energy levels in chaotic quantum systems. These results and insights are often subsumed by the term "random matrix theory," and reviewed e.g. in [58, 88, 139].

From a perturbative point of view, the expansion around the Gaussian term leads to ribbon-type Feynman graphs as explained in Section 6.1. Using the basis of elementary matrices, the propagator is diagonal and given by $\frac{g^2}{N} \delta^i_j \delta^k_l$. Let E be the number of edges i.e. propagators, V the number of vertices, and F the number of faces in such a graph. Each vertex contributes a factor of N, each propagator contributes a factor of N^{-1}, and each face contributes a factor of N due to the associated index summation. Taking into account also the (t'Hooft) coupling g^2, each vacuum diagram contributes an overall factor

$$g^{E-V} N^{V+F-E} = g^{L-1} N^\chi \tag{7.2.10}$$

in terms of their Euler character $\chi = V + F - E = 2(1 - \text{genus})$, starting with planar graphs. Here $L = E - V + 1$ is the number of independent loops.

On the other hand, the integral $\int dX$ over the space of hermitian matrices is well-defined and under nonperturbative control, which serves as an analytically accessible prototype of a path integral. This is possible due to the symmetry (7.2.9) of the model, which is respected by the matrix integral. Writing the matrices in diagonalized form

$$T = U^{-1} D U, \qquad D = \text{diag}(\lambda_1, \dots, \lambda_N) \tag{7.2.11}$$

for some $U \in SU(N)$, the path integral measure decomposes into the Haar measure for $SU(N)$ and a suitable measure for the eigenvalues, which is easily seen to be

$$dT = dU \prod_i d\lambda_i \Delta^2(\lambda), \qquad \Delta^2(\lambda) = \prod_{i<j} (\lambda_i - \lambda_j)^2. \tag{7.2.12}$$

---

[4] The coupling constant for the quartic term is absorbed in T and the t'Hooft coupling g^2.

Here $\Delta^2(\lambda_i)$ is recognized as the Vandermonde determinant

$$\Delta^2(\lambda) = \det(\lambda_i^{j-1}) = \exp\left(2\sum_{i<j}\ln|\lambda_i - \lambda_j|\right). \tag{7.2.13}$$

Therefore, the matrix model reduces to an interacting model for the eigenvalues λ_i, with a repulsive potential $-\ln|\lambda_i - \lambda_j|$ between the eigenvalues. The ordered eigenvalues $\lambda_1 \leq \lambda_2 < \cdots < \lambda_N$ can be viewed as a distribution of eigenvalues with eigenvalue density $\rho(\lambda) = \frac{1}{N}\sum_i \delta(\lambda - \lambda_i)$. The repulsive interaction due to the Vandermonde determinant suppresses eigenvalue distributions with coinciding $\lambda_i = \lambda_j$ and favors uniform distributions, so that $\rho(\lambda)$ becomes smooth in the large N limit. This model for the eigenvalues can then be analyzed with various methods, such as the saddle point approximation which nicely captures the large N or planar limit, or more sophisticated methods such as orthogonal polynomials or loop equations which allow to capture also the nonplanar sector.

It is illuminating to interpret the change of variables (7.2.12) in terms of a coadjoint orbit of $SU(N)$ through $\mathrm{diag}(\lambda_1, \ldots, \lambda_N)$: the space of hermitian matrices decomposes into leaves

$$\mathcal{O}[\lambda] = \{U^{-1}\mathrm{diag}(\lambda_1, \ldots, \lambda_N)U, \quad U \in SU(N)\}. \tag{7.2.14}$$

These are nothing but the symplectic leaves for the Poisson structure (2.2.9) on $\mathfrak{su}(n)$ as discussed in Section 2.2, and the volume of $\mathcal{O}[\lambda]$ is proportional to $\Delta^2(\lambda)$. Therefore the orbits with coinciding eigenvalues have vanishing volume, and are insignificant in the matrix model. This can be interpreted as an effective repulsion of the eigenvalues λ_i, which therefore tend to assume a specific smooth distribution. In the context of Yang–Mills matrix models, analogous considerations will imply that commutative matrix configurations $[T^a, T^b] = 0$ lead to a reduction of the dimension of their $U(N)$ orbit. Therefore, commutative matrix configurations will be suppressed upon quantization, and the path integral will be dominated by generic, noncommutative matrix configurations whose action is small.

The shape of the eigenvalue density $\rho(\lambda)$ in single-matrix models depends of course on the potential $V(T)$. In the simplest case of a quadratic potential $V(T) = \frac{1}{2}T^2$, the eigenvalue density turns out to be a semicircle,

$$\rho(\lambda) = \frac{2}{\pi}\sqrt{1 - \lambda^2} \tag{7.2.15}$$

and $\rho(\lambda) = 0$ for $|\lambda| > 1$ (after suitable rescaling). This is known as Wigner's semicircle law. Many properties – such as the level spacing between adjacent eigenvalues – turn out to be universal, i.e. essentially independent of the choice of the potential $V(T)$. For more complicated potentials, the support of $\rho(\lambda)$ may split up into disconnected pieces, which marks a phase transition of the matrix model. Zooming into some critical point via a suitable double-scaling limit, the divergent perturbative expansion leads to large Feynman diagrams, with significant contributions of any genus. Hence, the Feynman graphs describe discretized two-dimensional surfaces of any genus, which leads to a description of two-dimensional quantum gravity [58].

However, we should not get carried away too far with the example of single-matrix models. The mechanism of 2D gravity arising from large Feynman diagrams in the "deep quantum" regime of the single-matrix models is completely different from the case of multi-matrix models under consideration here, where the quantum space $\mathcal{M}$ arises as a nontrivial but essentially classical vacuum condensate. Nevertheless, the main message is that matrix integrals can serve as a quantum mechanical path integral, which is well-defined and under control. Similarly, the matrix integral (7.2.4) in the Yang–Mills matrix models is well-defined as long as N is finite, even though it is much harder to evaluate. Of course, this does not suffice, and most models will turn out to be too nonlocal. This should be expected since four-dimensional gravity is not renormalizable, and we will see that an almost-local four-dimensional quantum theory can emerge only from one specific supersymmetric such model, known as IKKT model.

Eigenvalue distribution for noncommutative scalar field theory

The idea of an eigenvalue distribution or density can also be applied in the context of noncommutative scalar field theory defined through matrix models such as (6.1.2). Even though they are not invariant under $SU(N)$, there is a unique eigenvalue distribution associated to any hermitian matrix, which provides a novel observable for scalar fields in NCFT. For free scalar NCFT, the eigenvalue density $\rho(\lambda)$ turns out to be given again by the Wigner semicircle law (7.2.15). This can be seen by computing the expectation values [143, 177]

$$\frac{\langle \mathrm{Tr}(\Phi^{2n}) \rangle}{\langle \mathrm{Tr}(\Phi^2) \rangle^n} \sim N_{\mathrm{planar}}(2n), \tag{7.2.16}$$

up to nonplanar corrections that are suppressed. Here N_{planar} is the number of planar diagrams [40], just like in the $SU(N)$-invariant matrix models. This implies that the eigenvalue distribution is given by a Wigner semicircle. The eigenvalue distribution is particularly useful to study the phase structure and phase transitions in scalar NCFT [159, 201], and it appears that the novel nonuniform or striped phase indicated in Section 6.3 corresponds to an eigenvalue density $\rho(\lambda)$ with disconnected support.

7.3 Nontrivial vacua and emergent quantum spaces

We recall the familiar concept of spontaneous symmetry breaking (SSB) in QFT and statistical mechanics: In a nontrivial vacuum or phase, some scalar fields acquire a nontrivial VEV $\langle \phi^a \rangle \neq 0$, and fluctuations around that background give rise to massless fields or modes known as Goldstone bosons. Moreover if the ϕ^a transform nontrivially under some gauge group, that gauge symmetry is spontaneously broken and realized nonlinearly on the fluctuations. The mechanism of emergent space or spacetime in matrix models is completely analogous: the nontrivial vacuum defines the quantum space(time)[5]

---

[5] In the following, "spacetime" will be used irrespective of signature.

or brane $\mathcal{M}$, and the massless modes turn out to be gauge fields on $\mathcal{M}$ which transform nonlinearly under gauge transformations.

The basic assumption or hypothesis is that for sufficiently large N, there are nontrivial matrix configurations of the model that are stable or at least sufficiently metastable at the quantum level. We shall indicate this by writing

$$\langle T^a \rangle = \overline{T}^a,$$
(7.3.1)

where $\overline{T}^a \in \text{End}(\mathcal{H})$ is some nontrivial matrix configuration; the bar will often be omitted. Here $\langle . \rangle$ denotes the quantum expectation value (7.2.5) or its Euclidean analog. Such a matrix configuration will be considered as "vacuum" or **background**, which can be interpreted as spacetime in suitable cases. Hence, spacetime is not given a priori, but it **emerges** dynamically from the matrix model; it can be viewed as a condensate in matrix model, much like in condensed matter physics. Such vacua are given by critical points of the quantum effective action (7.6.6) of the matrix model. If the quantum effects are negligible, then the Schwinger–Dyson equation $\langle [T^b, [T_b, T_a]] + m^2 T_a \rangle = 0$ (7.2.7) essentially reduces to the eom of the classical matrix model (7.0.10), which determines $\overline{T}^a$. We shall denote such a configuration as classical solution, even though it typically describes a quantum spacetime. In general, quantum effects arising in the matrix integral (7.2.5) will modify such a background, but $\overline{T}^a$ may still be perturbatively close to some classical solution. We will accordingly focus on the **weak coupling regime** of the gauge theory that governs the fluctuations around the background $\overline{T}^a$, assuming that such a regime exists and is physically interesting. This is the basic idea of the present approach.

On the other hand, in the deep quantum regime of the matrix model, many different unstable matrix configurations may contribute. This regime is hard to grasp analytically except for very special models, but it is accessible to numerical simulations, since the matrix models are well-defined.

This mechanism can be illustrated and justified in a simple toy-model with Euclidean signature. Consider the Alekseev–Recknagel–Schomerus (ARS) model [3]

$$S_{\text{ARS}} = \frac{1}{g^2} \text{Tr} \left([T^a, T^b][T_a, T_b] + \frac{4}{3} i \alpha \varepsilon_{abc} T^a [T^b, T^c] - 2m^2 T^a T_a \right)$$
(7.3.2)

involving three matrices T^a, $a = 1, 2, 3$; supersymmetric extensions of this model are also available [30, 106]. The equations of motion

$$\Box T^a = i \alpha \varepsilon^{abc} [T_b, T_c] - m^2 T^a$$
(7.3.3)

clearly admit fuzzy spheres with suitable radius r as solution. Moreover, any block-matrix configuration consisting of several fuzzy spheres $S_{n_i}^2$ with $\sum n_i = N$ is a solution.

Given these backgrounds, we should ask about stability under small perturbations. The stabilization of a fuzzy sphere in the ARS model is rather obvious for a choice of parameters where the action takes the form

$$S_{\text{ARS}} = -\frac{1}{g^2} \text{Tr} \left(\left(T^a - \frac{i}{r} \varepsilon^{abc} T_b T_c \right) \left(T_a - \frac{i}{r} \varepsilon_{ade} T^d T^e \right) \right).$$
(7.3.4)

Then $-S$ is positive definite[6], and the global minimum $S = 0$ is assumed for backgrounds $T^a = \oplus r J^a_{(n)}$, corresponding to a fuzzy sphere with radius r or direct sums thereof. More generally, the cubic term is essential for the classical stability of the fuzzy spheres. Fuzzy spheres are classical solutions even for $\alpha = 0$ and $m^2 < 0$, but then some unstable modes typically appear.

At the quantum level, stability of these backgrounds is less obvious. In numerical simulations and consistent with one-loop computations, three phases have been identified in these and similar models, depending on the choice of parameters [17, 149]:

- a disordered phase without any nontrivial matrix background
- an ordered phase which describes a single maximal fuzzy sphere $T^a = r J^a_{(N)}$ in accord with the hypothesis (7.3.1)
- a third phase which appears to describe a strongly interacting or "striped" phase on a fuzzy sphere, triggered by UV/IR mixing of the transversal fluctuations as in scalar NCFT.

We will mainly be interested in the weakly interacting ordered phase, and analogs thereof in other matrix models, notably the IKKT model.

In the absence of a cubic term and in higher dimensions, the existence of nontrivial vacua (7.3.1) is not obvious. If the model is not supersymmetric, one-loop effects typically lead to an attractive interaction mediated by string modes, which induce strongly nonlocal effects and tend to invalidate the classical considerations. As discussed in Section 7.6, this can be interpreted in terms of UV/IR mixing in the corresponding noncommutative gauge theory. Numerical simulations [18, 19] also seem to indicate that higher-dimensional fuzzy spaces are typically only metastable, although they might become long-lived at large N.

In the case of the maximally supersymmetric IKKT model, the situation is much more interesting and promising. The long-range interactions mediated by string modes then largely cancel, leading to a residual short-range interaction that decays like r^{-8}, and can be interpreted in terms of IIB supergravity in target space. Then the hypothesis (7.3.1) is much more plausible. In fact, the model has many "valleys" or almost-flat directions given by commutative or almost-commutative matrix configurations, which can be interpreted as branes in string theory language. For infinite-dimensional matrices, some backgrounds are protected from quantum corrections by supersymmetry and therefore stable, notably the flat Moyal–Weyl quantum planes $\mathbb{R}^{3,1}_\theta$ embedded in target space via

$$T^a = \begin{cases} X^a, & a = 0, \ldots, 3, \\ 0, & a = 4, \ldots, 9. \end{cases} \qquad (7.3.5)$$

On such a background, the model reduces to noncommutative $\mathcal{N} = 4$ SYM, as discussed in Section 11.1.2.

More generally, this matrix model can be considered at weak coupling[7] on some given background where it is perturbatively under control. It then remains to elaborate the

---

[6] Recall that our conventions are such that the action is negative definite in the Euclidean case.

[7] Note that the coupling constant g can be formally absorbed by a rescaling of the matrices. However, this does not invalidate the argument, because such a change of g entails a change of the scale L_{NC} of the resulting noncommutative gauge theory, which is clearly not scale invariant.

physics of the fluctuations on the background. In particular, it is then manifest that all fluctuations on the background are *propagating on the brane*, and do not escape into the bulk of target space. This perturbative approach is no longer justified in the strong coupling regime, where a holographic point of view in terms of target space geometry may be more appropriate. We will restrict ourselves in the following to the weak coupling regime around a quantum space or brane embedded in target space.

7.3.1 Commutative backgrounds

Now consider commutative matrix configurations $[T^a, T^b] = 0$, which are clearly solutions for $m = 0$. These can be simultaneously diagonalized, and their eigenvalues define some lattice in target space, which one might want to interpret as discretized space(time). However, this does not work for several reasons:

- Commutative configurations are suppressed for entropic reasons: their $U(\mathcal{H})$ gauge orbit $U^{-1}T^a U$ is a manifold whose dimension is smaller than the gauge orbit of noncommutative matrix configurations. The reason is that the stabilizer group of a commutative background is at least $U(1)^N$ assuming $\mathcal{H} = \mathbb{C}^N$, while the stabilizer of a noncommutative background is generically $U(1)$. Since different points of any gauge orbits are to be identified, the commutative matrix configurations have measure zero under the matrix integral, and are negligible.
- The momentum operators $\mathcal{P}^a = [T^a, .]$ (6.2.7) acting on $\mathrm{End}(\mathcal{H})$ have a large space of zero modes, given by any functions $f(T^a)$ of the background. Therefore, these modes acquire no kinetic energy and will be degenerate, and the matrix model will not define a meaningful theory for the fluctuations. Therefore, such a background does not lead to a local field theory on the lattice, but to some degenerate field theory on the underlying phase space, or a pathological higher spin theory.

The second issue is particularly transparent in the discrete case, where we can assume that

$$T^a = \mathrm{diag}(\lambda_i^a) \qquad (7.3.6)$$

with $\lambda_i \in \mathbb{R}^D$, which form some lattice $L \subset \mathbb{R}^D$. This can be viewed as a collection of point branes, and one might expect naively that the matrix model leads to some lattice field theory. However, this does not work, not only because of the degeneracy discussed above, but also because $\mathrm{End}(\mathcal{H})$ is much bigger than the space of functions on the lattice. $\mathrm{End}(\mathcal{H})$ is better described in terms of (quantized) functions on the Cartesian product $L \times L$ of the lattice, spanned by the string-like modes $|i\rangle \langle j|$.

Similarly, consider the simplest one-dimensional case of a continuous commutative background given by the matrix configuration

$$T^a = \begin{cases} X & a = 1 \\ 0 & a \neq 1 \end{cases} . \qquad (7.3.7)$$

Here $X\phi(s) = s\phi(s)$ and $P\phi(s) = i\frac{d}{ds}\phi(s)$ are position and momentum operators acting on $\phi(s) \in \mathcal{H} \cong L^2(\mathbb{R})$. Again, any $f(X)$ is a zero mode of the momentum operator $\mathcal{P} = [X, .]$

and the matrix Laplacian $\Box = [X, [X, .]]$. The point is that $\text{End}(\mathcal{H})$ describes (quantized) functions on two-dimensional phase space, rather than functions on $\mathbb{R}$, and the momentum operators act only in one direction. Similarly, D commuting operators define a Hilbert space $\mathcal{H}$ whose operator algebra $\text{End}(\mathcal{H})$ describes quantized functions on phase space $\mathbb{R}_\theta^{2D}$, rather than position space $\mathbb{R}^D$.

We conclude that commutative matrix configurations have measure zero and are pathological as backgrounds. We will therefore not consider them any further.

7.4 Yang–Mills gauge theory on quantum spaces from SSB

Now we consider some nontrivial vacuum given by the matrix configuration $\bar{T}^a$, which can be interpreted in terms of an underlying quantum space $\mathcal{M} \hookrightarrow \mathbb{R}^D$. Stability requires that the spectrum of fluctuations around the background $\bar{T}^a$ does not contain any unstable modes. We should therefore study the fluctuations around such a background.

7.4.1 Fluctuations and gauge fields

Consider the most general perturbations $\mathcal{A}^a$ of such a background

$$T^a = \bar{T}^a + \mathcal{A}^a, \tag{7.4.1}$$

which describes the physics arising from the matrix model on the background. We shall drop the bar if no confusion can arise. The gauge transformations $T^a \to U^{-1}T^a U$ of the matrix model clearly do not preserve a nontrivial background $\bar{T}^a$, which means that the gauge symmetry is *spontaneously broken*. Such a situation is familiar from quantum field theory, where it describes nontrivial vacua of the theory. The important point is that the underlying symmetry still holds, but takes on a different guise: for the perturbation $\mathcal{A}^a$ of the background, the gauge symmetry acts in the inhomogeneous form

$$\mathcal{A}^a \to U^{-1}\mathcal{A}^a U + U^{-1}[\bar{T}^a, U]. \tag{7.4.2}$$

For infinitesimal gauge transformations $U = e^{i\Lambda}$, this takes the form

$$\delta_\Lambda \mathcal{A} = i[\bar{T}^a + \mathcal{A}^a, \Lambda] \sim \{\Lambda, \bar{t}^a\} + \{\Lambda, \mathcal{A}^a\}. \tag{7.4.3}$$

This is familiar from gauge theory on the Moyal–Weyl quantum plane (6.6.9), and we will see that in general, (the nonabelian components of) tangential fluctuations $\mathcal{A}^a$ play the role of gauge fields in a Yang–Mills type gauge theory on the generic quantum space $\mathcal{M}$. Hence, quantum spacetime arises from spontaneous symmetry breaking in matrix model, and gauge fields arise as Goldstone bosons, which are governed by a noncommutative gauge theory. It is then natural to define a matrix field strength as follows:

$$\mathcal{F}^{ab} := i[T^a, T^b] = i[\bar{T}^a + \mathcal{A}^a, \bar{T}^b + \mathcal{A}^b], \tag{7.4.4}$$

which transforms homogeneously under gauge transformations

$$\mathcal{F}^{ab} \to U^{-1}\mathcal{F}^{ab}U. \tag{7.4.5}$$

Then for $m^2 = 0$, the matrix model action takes the familiar form

$$S = -\frac{1}{g^2}\text{Tr}\mathcal{F}^{ab}\mathcal{F}_{ab}. \tag{7.4.6}$$

On a Moyal–Weyl background $T^a = X^a$, the matrix field strength can be written as

$$\mathcal{F}^{ab} = -\theta^{ab} - \theta^{aa'}\theta^{bb'}F_{a'b'}, \tag{7.4.7}$$

where $F_{ab} = \partial_a A_b - \partial_b A_a - i[A_a, A_b]$ is the familiar field strength tensor, and $A_a = \theta^{-1}_{ab}A^b$ transforms as

$$A_a \to U^{-1}A_a U + U^{-1}i\partial_a U. \tag{7.4.8}$$

Then the action is nothing but the noncommutative gauge theory as discussed in Section 6.6, which will be elaborated in the nonabelian case in some detail. Notice that there is no need to introduce any ad-hoc mathematical structures by hand.

On more generic backgrounds, the separation between a $U(1)$ field strength tensor and the background is more subtle, because the background field strength $\mathcal{F}^{ab} = i[T^a, T^b]$ transforms nontrivially under gauge transformations. The relation between $\mathcal{F}^{ab}$ and ordinary $U(1)$ gauge fields A_μ is the subject of the Seiberg–Witten map, which will be briefly discussed in Section 9.1.4. However, the trace-$U(1)$ sector of the matrix model is more properly understood in terms of deformations of the underlying symplectic structure, and should be viewed as a gravity-like theory of (symplectic) geometry rather than a Maxwell-like theory. This geometric point of view will be discussed in detail in Chapter 9. In contrast, nonabelian $SU(n)$ fluctuations indeed correspond to gauge fields of a nonabelian Yang–Mills theory.

7.4.2 *The matrix energy–momentum tensor

A fundamental property of Yang–Mills matrix models

$$S = \text{Tr}\mathcal{L}, \qquad \mathcal{L} = [T^a, T^b][T_a, T_b] - 2m^2 T^a T_a \tag{7.4.9}$$

is the invariance under gauge transformations

$$\delta_\Lambda T^a = i[T^a, \Lambda]. \tag{7.4.10}$$

As usual, this should imply a conservation law, which turns out to be a matrix version of the conservation of the energy–momentum tensor. To derive it, we observe that the matrix Lagrangian $\mathcal{L}$ transforms as

$$\begin{aligned}
\delta_\Lambda \mathcal{L} = i[\mathcal{L}, \Lambda] &= 2[\delta T^a, T^b][T_a, T_b] + 2[T^a, T^b][\delta T_a, T_b] - 2m^2(\delta T^a T_a + T^a \delta T_a) \\
&= 2[(\delta T_a[T^a, T^b] + [T^a, T^b]\delta T_a), T_b] \\
&\quad - 2\delta T_a([[T^a, T^b], T_b] + m^2 T^a) - 2([[T^a, T^b], T_b] + m^2 T^a)\delta T_a.
\end{aligned} \tag{7.4.11}$$

The last line vanishes using the equations of motion (7.0.10), which gives the on-shell relation

$$0 = [\mathcal{L}, \Lambda] - 2[[T^a, \Lambda][T_a, T_b] + [T_a, T_b][T^a, \Lambda], T^b]. \tag{7.4.12}$$

For $\Lambda = T^c$, we obtain

$$0 = [\mathcal{L}, T^c] - 2[[T^a, T^c][T_a, T_b] + [T_a, T_b][T^a, T^c], T^b] = -4[\mathcal{T}^{cb}, T_b], \qquad (7.4.13)$$

where

$$\mathcal{T}^{ab} = \frac{1}{2}([T^a, T^c][T^b, T_c]) + (a \leftrightarrow b)) - \frac{1}{4}\eta^{ab}([T^c, T^d][T_c, T_d] - 2m^2 T^c T_c) = \mathcal{T}^{ba}$$
$$(7.4.14)$$

is the **matrix energy–momentum tensor**, for $a, b = 1, \ldots, D$. We have thus derived the matrix conservation law[8]

$$\boxed{[T_a, \mathcal{T}^{ab}] = 0.} \qquad (7.4.15)$$

This tensor plays an important role not only for the classical model (as assumed here) but also for the quantum theory; in particular, the one-loop interactions in the IKKT matrix model will be recognized as IIB supergravity interaction mediated by string modes coupled to the matrix energy–momentum tensor of the configuration. It is also interesting to observe that $\mathcal{T}^{ab}$ is traceless if $D = 4$ and $m^2 = 0$, as

$$\mathcal{T} = \mathcal{T}^{ab}\eta_{ab} = (4 - D)[T^a, T^b][T_a, T_b] + \frac{D}{2}m^2 T^a T_a. \qquad (7.4.16)$$

Ward identity

Since the matrix integration measure is invariant under gauge transformations, one expects that this conservation law holds also at the quantum level. However, the derivation cannot be repeated literally under the path integral, since the substitution $\Lambda \rightarrow T^c$ is "field-dependent." This suggests to consider the Schwinger–Dyson equation for a suitably adapted transformation. Thus, consider the following transformation under the path integral inspired by the classical derivation:

$$\delta T^a = i\{T^b, [T^a, \varepsilon_b]\} = -i\{[T^a, T^b], \varepsilon_b\} + i[T^a, \{T^b, \varepsilon_b\}], \qquad (7.4.17)$$

where ε^b is an arbitrary hermitian matrix. As elaborated in [180], this leads to the identity

$$\frac{1}{2}\delta S = \frac{i}{g^2}\mathrm{Tr}\varepsilon_b[T_a, \mathcal{T}^{ab}]. \qquad (7.4.18)$$

Moreover, it is not hard to see that this transformation preserves the matrix integration measure. Then one obtains the following Schwinger–Dyson equation or Ward identity (7.4.15)

$$\langle[T_a, \mathcal{T}^{ab}]\rangle = 0. \qquad (7.4.19)$$

---

[8] The generalization including fermions is discussed e.g. in [20, 160].

7.4.3 Branes, transversal fluctuations, and scalar fields

Now let T^a be some matrix configuration, which is considered as a background of the Yang–Mills matrix model (7.0.1). Such a background will typically describe an embedded quantum space(4.2.8)

$$T^a \sim t^a : \quad \mathcal{M} \hookrightarrow \mathbb{R}^D \tag{7.4.20}$$

considered as brane embedded in target space. In general, the dimension of $\mathcal{M}$ may differ from D; assume that dim $\mathcal{M} = 2n < D$. Then $\mathcal{M}$ can be viewed as a submanifold of $\mathbb{R}^D$, and it is plausible that transversal fluctuations of this submanifold can be interpreted as scalar fields on $\mathcal{M}$. To make this explicit, consider some point $\xi \in \mathcal{M}$, and assume – after a $SO(D)$ rotation if necessary – that the tangent space $T_\xi\mathcal{M} \cong \mathbb{R}^{2n} \subset \mathbb{R}^D$ is embedded along the first $2n$ coordinates. We can then introduce the following notation:

$$T^a = \begin{cases} T^\mu, & \mu = 1, \ldots, 2n \\ \phi^i, & i = 1, \ldots, D - 2n \end{cases} \tag{7.4.21}$$

so that the action (7.0.1) acquires the following form:

$$S[T^a] = S[T^\mu] + S[\phi^i],$$

$$S[\phi^i] = \frac{1}{g^2}\mathrm{Tr}\big(2[T^\mu, \phi^i][T_\mu, \phi^i] - 2m^2\phi^i\phi_i + [\phi^i, \phi^j][\phi_i, \phi_j]\big), \tag{7.4.22}$$

where Greek indices are contracted with $g_{\mu\nu} = \delta_{\mu\nu}$, or $g_{\mu\nu} = \eta_{\mu\nu}$ in Minkowski signature. Here $S[T^\mu]$ is the Yang–Mills matrix model in $d = 2n$ dimensions, and $S[\phi^i]$ is the action of $D - 2n$ scalar fields ϕ^i governed by the kinetic term

$$S_{\mathrm{kin}}[\phi] := \frac{2}{g^2}\mathrm{Tr}[T^\mu, \phi^i][T_\mu, \phi^i] \sim -\frac{2}{g^2}\int_\mathcal{M} \{t^\mu, \phi^i\}\{t_\mu, \phi^i\} \tag{7.4.23}$$

recalling that $[T^\mu, \phi^i] \sim i\{t^\mu, \phi^i\}$ in the semi-classical limit. The integral is defined via the symplectic volume form as in (4.2.37). Note that the hermitian matrices $\phi^i \in \mathrm{End}(\mathcal{H})$ are interpreted as real functions i.e. scalar fields on the quantum space $\mathcal{M}$, which are interacting via the quartic term $[\phi^i, \phi^j][\phi_i, \phi_j]$. This interaction will play an important role in the nonabelian case, while in the present abelian case it amounts to a four-derivative interaction at low energies arising from the Poisson brackets.

In the simplest case where $\mathcal{M} \cong \mathbb{R}^{2n}_\theta$ is a Moyal–Weyl quantum plane with $T^\mu = X^\mu$, the kinetic term takes the simple form

$$S_{\mathrm{kin}}[\phi] = -\frac{2}{g^2}\int_{\mathbb{R}^{2n}} \frac{d^{2n}x}{(2\pi L_{\mathrm{NC}}^2)^n} \gamma^{\mu\nu}\partial_\mu\phi^i\partial_\nu\phi^i, \qquad \gamma^{\mu\nu} = \theta^{\mu\mu'}\theta^{\nu\nu'}\eta_{\mu'\nu'}. \tag{7.4.24}$$

It is easy to see that the effective metric $\gamma_{\mu\nu}$ has the same signature as $\eta_{\mu\nu}$; however, their causal structures are incompatible. This strange feature will disappear on covariant quantum spaces.

The important point is that transversal fluctuations of quantum spaces in Yang–Mills matrix models can be interpreted as scalar fields on the brane. On the other hand, they also change the embedding $\mathcal{M} \subset \mathbb{R}^D$. This dual interpretation is a central feature of Yang–Mills matrix models, and underscores their geometric significance. Since the effective metric is universal for all fluctuations, it should be interpreted in terms of gravity.

7.4.4 Tangential fluctuations and gauge fields

Now we refine the discussion to include tangential fluctuations, setting $m^2 = 0$. For simplicity, we consider first the case where $\mathcal{M} \cong \mathbb{R}^{2n}_\theta$ is a Moyal–Weyl quantum plane, embedded along the first $2n$ coordinates in target space. This is realized by a background as in (7.4.21), with

$$\bar{T}^a = \begin{cases} \bar{X}^\mu, & \mu = 1, \ldots, 2n \\ 0, & i = 1, \ldots, D - 2n \end{cases} \tag{7.4.25}$$

with $[\bar{X}^\mu, \bar{X}^\nu] = i\theta^{\mu\nu}\mathbb{1}$. It is easy to see that this is a solution of the Yang–Mills matrix model (7.0.1) with $m^2 = 0$. Moreover, the full matrix algebra $\mathrm{End}(\mathcal{H})$ can be identified as algebra of functions on $\mathbb{R}^{2n}_\theta$. This observation is of utmost importance as we shall see.

Now consider fluctuations of that background. Such fluctuations can be parametrized as

$$T^a = \bar{T}^a + \mathcal{A}^a, \qquad \mathcal{A}^a = \begin{pmatrix} \mathcal{A}^\mu \\ \phi^i \end{pmatrix}, \tag{7.4.26}$$

where $\mathcal{A}^\mu$ and ϕ^i are "small" hermitian matrices. By the above observation they can be interpreted as smooth functions on $\mathbb{R}^{2n}_\theta$, i.e. $\mathcal{A}^\mu = \mathcal{A}^\mu(\bar{X}) \sim \mathcal{A}^\mu(x)$ and $\phi^i = \phi^i(\bar{X}) \sim \phi^i(x)$. This means that *all modes live on the brane*, and there is simply no space to describe modes propagating in the bulk, at least in the perturbative or weak coupling regime[9]. This innocent observation means that there is no need to compactify target space, in contrast to the orthodox approach to string theory.

As we have learned before, the action becomes most transparent in terms of

$$\mathcal{A}^\mu = \theta^{\mu\nu} A_\nu. \tag{7.4.27}$$

Using (3.1.64), one finds

$$\begin{aligned} [X^\mu, f] &= [\bar{X}^\mu + \mathcal{A}^\mu, f] = i\theta^{\mu\nu}(\partial_\nu f - i[A_\nu, f]) =: i\theta^{\mu\nu} D_\nu f, \\ [X^\mu, X^\nu] &= i\theta^{\mu\mu'}\theta^{\nu\nu'}(-\theta^{-1}_{\mu'\nu'} + F_{\mu'\nu'}), \end{aligned} \tag{7.4.28}$$

where $F_{\mu\nu} = \partial_\mu A_\nu - \partial_\nu A_\mu - i[A_\mu, A_\nu]$ is the $U(1)$ field strength on $\mathbb{R}^4_\theta$. The gauge symmetry $T^a \to U^{-1} T^a U$ or (7.4.2) of the matrix model acts on the fluctuations as

$$A_\mu \to U^{-1} A_\mu U + i U^{-1} \partial_\mu U, \qquad \phi^i \to U^{-1} \phi^i U. \tag{7.4.29}$$

This clearly has the structure of gauge transformations of Yang–Mills gauge fields coupled to scalar fields in the adjoint, which arise from the "covariant coordinates" $T^\mu = \bar{X}^\mu + \mathcal{A}^\mu$

---

[9] This is no longer justified for strong coupling and strong fluctuations, where a holographic description in terms of some bulk geometry may be more appropriate. We will stay away from this regime in the present book.

as in Section 6.6. Including the transversal scalar fields, the action (7.2.3) can now be written as

$$S = S[A] + S[\phi] = \frac{1}{g^2} \int\limits_{\mathbb{R}_\theta^{2n}} \frac{d^{2n}x}{(2\pi L_{NC}^2)^n} \Big(-\gamma^{\mu\mu'}\gamma^{\nu\nu'} F_{\mu\nu} F_{\mu'\nu'} - \gamma^{\mu\nu}\eta_{\mu\nu}$$

$$- 2\gamma^{\mu\nu} D_\mu\phi^i D_\nu\phi^j \delta_{ij} + [\phi^i, \phi^j][\phi^{i'}, \phi^{j'}]\delta_{ii'}\delta_{jj'} \Big) \qquad (7.4.30)$$

dropping surface terms and fermions, and assuming $2n \neq 2$. These formulas are exact (up to boundary terms) and recover the results of Section 6.6. Thus, on flat Moyal–Weyl backgrounds $\mathbb{R}_\theta^{2n}$, the matrix model can be viewed as a noncommutative gauge theory with effective flat metric $\gamma^{\mu\nu}$, coupled to $D - 2n$ scalar fields ϕ^i in the adjoint. Note that the dimensions are nonstandard, which can easily be rectified by absorbing suitable powers of L_{NC} in the metric and the scalar field. Then the Yang–Mills coupling constant is obtained as in (6.6.21).

Let us recapitulate this important result: The Yang–Mills matrix model for an infinite-dimensional (separable) Hilbert space $\mathcal{H}$ **is the same** as Yang–Mills gauge theory on $\mathbb{R}_\theta^{2n}$, including the scalar fields ϕ if $D > 2n$.

Fermions

It is straightforward to include fermions in this story. On the perturbed background $T^a = \bar{T}^a + \mathcal{A}^a$, the matrix action for the fermions take the form

$$\frac{1}{g^2}\text{Tr}\,\overline{\Psi}\Gamma_a[T^a, \Psi]) = \frac{1}{g^2} \int \frac{d^{2n}x}{(2\pi L_{NC}^2)^n} \left(\overline{\Psi}\gamma^\nu i(\partial_\nu - i[A_\nu, .])\Psi + \overline{\Psi}\Gamma_i[\phi^i, \Psi]\right), \quad (7.4.31)$$

where the effective gamma matrices

$$\gamma^\nu = \Gamma_\mu\theta^{\mu\nu}, \qquad \{\gamma^\mu, \gamma^\nu\} = 2\gamma^{\mu\nu} \qquad (7.4.32)$$

satisfy indeed the Clifford algebra corresponding to the metric $\gamma^{\mu\nu}$. The first term is clearly the standard gauge-invariant Dirac action on $\mathbb{R}_\theta^{2n}$, and the second term gives a Yukawa coupling of the scalar fields ϕ^i to the fermions. The decomposition of the D-dimensional spinor Ψ into $2n$-dimensional spinors on $\mathbb{R}^{2n}$ is straightforward, and will be given for the case of $\mathcal{N} = 4$ SYM in Section 11.1.2.

7.4.5 Stacks of branes and nonabelian Yang–Mills theory II

Given any irreducible solutions $T^a_{(i)}$ of the matrix model, we can consider *reducible* configurations consisting of several such blocks or branes:

$$T^a = \begin{pmatrix} T^a_{(1)} & & & \\ & T^a_{(2)} & & \\ & & \ddots & \\ & & & T^a_{(k)} \end{pmatrix}. \qquad (7.4.33)$$

This is clearly again a solution, which describes a collection or a "stack" of branes embedded in target space $\mathbb{R}^D$. For $m^2 = 0$, each brane may in fact be shifted by an arbitrary vector $T_{(i)}^a \to T_{(i)}^a + c_{(i)}^a \mathbb{1}$ in target space, due to the shift invariance of the matrix model.

It is natural to ask whether these branes are interacting with each other. At the classical level, the shift invariance implies that there is no interaction. We will see that quantum effects typically do induce an interaction between the branes, which is partially canceled in the maximally supersymmetric model.

A particularly interesting case arises if all these branes are identical. Then the configuration (7.4.33) can be written as follows:

$$T^a \otimes \mathbb{1}_k \quad \in \text{End}(\mathcal{H} \otimes \mathbb{C}^k). \tag{7.4.34}$$

We can again parametrize the most general fluctuations on this background as in (7.4.26),

$$T^a = \bar{T}^a + \mathcal{A}^a, \qquad \mathcal{A}^a = \begin{pmatrix} \mathcal{A}^\mu \\ \phi^i \end{pmatrix}, \tag{7.4.35}$$

but now the gauge fields $\mathcal{A}^\mu$ and the scalar fields ϕ^i take values in $\mathfrak{u}(k)$:

$$\mathcal{A}^\mu = \mathcal{A}_\alpha^\mu \lambda^\alpha, \qquad \phi^i = \phi_\alpha^i \lambda^\alpha, \tag{7.4.36}$$

where λ_α is a basis of the Lie algebra $\mathfrak{u}(k)$. On a stack of identical $\mathbb{R}_\theta^{2n}$ branes, we recover the $\mathfrak{u}(k)$-valued noncommutative Yang–Mills gauge theory as discussed in Section 6.6. The action including fermions takes precisely the same form as (7.4.30) and (7.4.31), with a trace over the internal $\mathfrak{u}(k)$.

It is worth pointing out that if $\mathcal{H}$ is infinite-dimensional, then the mathematical structure of the model is equivalent for all k, because $\text{End}(\mathcal{H} \otimes \mathbb{C}^k) \cong \text{End}(\mathcal{H})$ (6.6.22). Therefore, all NC Yang–Mills gauge theories with any gauge group $U(k)$ are equivalent, and realized on different vacua of one and the same theory, given by the Yang–Mills matrix model (7.2.3).

In the case of finite-dimensional Hilbert spaces $\mathcal{H}$, these basic features still apply in a modified sense. We recall from Section 7.3 that a nonvanishing mass is necessary to have any classical solutions. To obtain stable classical solutions, one should even add a cubic term such as in the ARS model (7.3.2). Then one finds indeed various vacua consisting of stacks of branes, and stacks of k identical branes lead again to a $\mathfrak{u}(k)$ noncommutative gauge theory. One may expect that similar effects arise even without such mass or cubic terms, if the matrices are sufficiently large and the branes are sufficiently long-lived.

In the same vein, one can find new nonperturbative solutions by adding one (or several) trivial 1×1 block-matrix $T_{(0)}^a = c^a$ (a "point brane") to the background (7.4.33). This turns out to describe a novel vortex-like solution of NC Yang–Mills theory on $\mathbb{R}_\theta^{2n}$ located at c^a, denoted as fluxons [144, 157]. These solutions exist even in the $U(1)$ case i.e. for single branes. They have no counterpart on commutative space, as their energy diverges as $\theta \to 0$. The above-mentioned equivalence can then be stated as $X^a \oplus c^a = S^\dagger X^a S$, where S is a partial isometry which is not unitary. NC Yang–Mills theory on $\mathbb{R}_\theta^{2n}$ also admits (anti-)self-dual instanton solutions which are analogous to the standard ones but have a minimum size, which lifts the standard singularity in the moduli space of small instantons. We refer to the excellent reviews [63, 195] for more details and references for this rich topic.

Perhaps the most distinctive feature of NC gauge theory is that the field strength transforms nontrivially as $\mathcal{F}^{ab} \to U^{-1}\mathcal{F}^{ab}U$ under noncommutative $U(1)$ gauge transformations, just like scalar fields. Recalling the discussion in Section 6.5, this means that $\mathcal{F}^{ab}$ transforms like a scalar field[10]

$$\delta_\Lambda \mathcal{F}^{ab} = \mathcal{L}_\xi \mathcal{F}^{ab} = \xi^\mu \partial_\mu \mathcal{F}^{ab} \tag{7.4.37}$$

under infinitesimal gauge transformations, which are diffeomorphisms generated by Hamiltonian vector fields ξ. This indicates that these theories – or rather their $U(1)$ sector – should be viewed as geometrical, and hence in some sense gravitational theories. In particular, there are no local gauge-invariant observables. This point of view will be elaborated in detail later. Finally, observe that the presence of a mass does not obstruct the gauge invariance of the model, unlike in commutative gauge theory. A mass term may stabilize certain brane geometries and gives a mass to the transversal scalar fields, but the tangential fluctuations always correspond to massless Yang–Mills gauge fields.

7.5 *Gauge fixing and perturbative quantization

To understand the stability of the background and the fluctuation spectrum, we need to elaborate the quadratic action for fluctuations around the background. Consider again the Yang–Mills matrix model (7.2.3) with mass term,

$$S[T] = \frac{1}{g^2}\text{Tr}\Big([T^a, T^b][T_a, T_b] - 2m^2 T^a T_a\Big). \tag{7.5.1}$$

Inserting fluctuations $T^a + \mathcal{A}^a$ of the background where $\mathcal{A}^a \in \text{End}(\mathcal{H})$ are arbitrary Hermitian matrices, this action can be expanded in $\mathcal{A}$ as

$$S[T + \mathcal{A}] = S[T] - \frac{4}{g^2}\text{Tr}\big(\mathcal{A}^a(\square + m^2)T_a\big) + S_2[\mathcal{A}] + O(\mathcal{A}^3). \tag{7.5.2}$$

The term linear in $\mathcal{A}$ vanishes provided T^a satisfies the equations of motion, and m^2 is chosen accordingly. The quadratic action governing the fluctuations is

$$S_2[\mathcal{A}] = \frac{2}{g^2}\text{Tr}\left(-\mathcal{A}_a(\mathcal{D}^2 + m^2)\mathcal{A}^a + \mathcal{G}(\mathcal{A})^2\right), \tag{7.5.3}$$

where

$$\mathcal{D}^2\mathcal{A} = (\square - 2\mathcal{I})\,\mathcal{A}, \qquad \mathcal{I}(\mathcal{A})^a = -[[T^a, T^b], \mathcal{A}_b] \tag{7.5.4}$$

is the vector Laplacian on $\mathcal{M}$, and $\mathcal{I}(\mathcal{A})$ is an $SO(D)$ intertwiner (5.2.101). Together with

$$\mathcal{G}(\mathcal{A}) = i[T^a, \mathcal{A}_a] \sim -\{t^a, \mathcal{A}_a\}, \tag{7.5.5}$$

gauge invariance is preserved. As usual this leads to flat directions, and a perturbative analysis requires gauge fixing. This can be achieved using the standard Faddeev–Popov

---

[10] On $\mathbb{R}^{2n}_\theta$, this reduces to $F_{\mu\nu} \to \xi^\sigma \partial_\sigma F_{\mu\nu}$, which is not the Lie derivative of a rank 2 tensor field. We will see in Section 9.1.3 how this is understood in terms of a frame.

procedure as worked out in the next paragraph, adding a suitable gauge-fixing term and imposing a gauge fixing condition, which removes the unphysical ghost mode in Minkowski signature. As a preparation, we show that the gauge-fixing condition $\mathcal{G}(\mathcal{A}) = 0$ can always be achieved:

Lemma 7.1 *For finite-dimensional matrices and for any nontrivial background $\bar{T}^a$, there is always a gauge such that $\mathcal{G}(\mathcal{A}) = i[\bar{T}^a, \mathcal{A}_a] = 0$.*

Proof Let $\mathcal{H} = \mathbb{C}^N$. For any fixed x, consider the real function $f(T) = \mathrm{Tr}(\bar{T}^a T_a)$ on a given $U(N)$ orbit. This orbit is compact, therefore the function $f(T)$ assumes its maximum at some $\bar{T}_a + \bar{\mathcal{A}}$. At this maximum, the variation vanishes, i.e.

$$0 = \delta_\Lambda f(\bar{T} + \bar{\mathcal{A}}) = i\mathrm{Tr}(\bar{T}^a[\bar{T}_a + \bar{\mathcal{A}}_a, \Lambda]) = i\mathrm{Tr}([\bar{T}^a, \bar{\mathcal{A}}_a]\Lambda) \qquad (7.5.6)$$

for any $\Lambda \in \mathfrak{su}(N)$, where $\delta_\Lambda T_a = i[T_a, \Lambda]$. This implies that

$$0 = [\bar{T}^a, \bar{\mathcal{A}}_a], \qquad (7.5.7)$$

which provides the desired gauge for $\bar{\mathcal{A}}_a$. $\square$

Faddeev–Popov or BRST gauge fixing

The quantization of the model is defined by integrating over the space of hermitian matrices T^a (7.2.1), (7.2.5). To be specific, we focus on the Euclidean case here; the generalization to the Minkowski case is straightforward by adopting our $i\varepsilon$ prescription. Rather than attempting a fully nonperturbative quantization[11], we are primarily interested in the quantum effective action around some given background $\overline{T}^a$. Splitting the matrices into background T^a and a fluctuating part $\mathcal{A}^a$ as in (7.4.1) (dropping the bar), the gauge symmetry acts as

$$\mathcal{A}^a \rightarrow \mathcal{A}_U^a := \mathcal{A}^a + U[T^a + \mathcal{A}^a, U^{-1}]. \qquad (7.5.8)$$

Then each point in the gauge orbit contributes to the partition function, and we are free to add an arbitrary weighting function to the action. To this end, consider the slightly generalized gauge fixing function

$$\mathcal{G}^\omega(\mathcal{A}) = i[T^a, \mathcal{A}_a] + \omega, \qquad (7.5.9)$$

where ω is an arbitrary hermitian $N \times N$ matrix. We can now apply the standard identity[12]

$$1 = \int\limits_{U(N)} dU \delta(\mathcal{G}^\omega(\mathcal{A}_U)) \det\left(\frac{\partial \mathcal{G}^\omega(\mathcal{A}_U)}{\partial U}\right) \qquad (7.5.10)$$

---

[11] At the nonperturbative level, all the different possible vacua of the matrix model corresponding to vastly different geometries would enter. This may be addressed using numerical simulations, see Section 11.2.

[12] Since we are only interested in the perturbative expansion, we ignore possible Gribov problems here.

using the parametrization $U = e^{i\Lambda}$ near $U = \mathbb{1}$, where the functional matrix of $\mathcal{G}^\omega(\mathcal{A})$ along the gauge orbit is

$$\frac{\partial \mathcal{G}^\omega(\mathcal{A}_U)}{\partial U} = \frac{\partial [T_a, [T^a + \mathcal{A}^a, \Lambda]]]}{\partial \Lambda} = [T_a, [T^a + \mathcal{A}^a, .]]. \tag{7.5.11}$$

This is a linear map acting on the space of hermitian $N \times N$ matrices, whose determinant can be written as

$$\det \left(\frac{\partial \mathcal{G}^\omega(\mathcal{A}_U)}{\partial U}\right) = \det \left([T_a, [T^a + \mathcal{A}^a, .]]\right) = \int dc d\bar{c} \exp \left[\mathrm{Tr}(-\bar{c}[T_a, [T^a + \mathcal{A}^a, c]])\right], \tag{7.5.12}$$

where $c, \bar{c}$ are Grassmann-valued matrices known as Faddeev–Popov ghosts. We can then follow the usual steps and rewrite the path integral in the form

$$\int dT e^{S[\mathcal{A}]} = \int d\mathcal{A} e^{S[\mathcal{A}]} \int dU \delta(\mathcal{G}^\omega(\mathcal{A}_U)) \det \left(\frac{\partial \mathcal{G}^\omega(\mathcal{A}_U)}{\partial U}\right)$$

$$= \int dU d\mathcal{A}_U e^{S[\mathcal{A}_U]} \delta(\mathcal{G}^\omega(\mathcal{A}_U)) \det \left(\frac{\partial \mathcal{G}^\omega(\mathcal{A}_U)}{\partial U}\right)$$

$$= \int dU \int d\mathcal{A} dc d\bar{c} \ \delta(\mathcal{G}^\omega(\mathcal{A})) e^{S[\mathcal{A}]} \exp \left[\mathrm{Tr}(-\bar{c}[T_a, [T^a + \mathcal{A}^a, c]])\right] \tag{7.5.13}$$

noting that $\int d\mathcal{A} = \int dT$, and using the gauge invariance of the action. Finally we can integrate over all ω with a Gaussian weight functions, which only changes the normalization of the partition function. Then

$$\int d\omega e^{-\frac{\omega^2}{2\xi}} dT e^{S[\mathcal{A}]} = \int dU \int d\mathcal{A} dc d\bar{c} \int d\omega e^{-\frac{\omega^2}{2\xi}} \delta(\mathcal{G}^\omega(\mathcal{A})) e^{S[\mathcal{A}]}$$

$$\cdot \exp \left[\mathrm{Tr}(-\bar{c}[T_a, [T^a + \mathcal{A}^a, c]])\right]$$

$$= \int d\mathcal{A} dc d\bar{c} e^{S[\mathcal{A}]} e^{-\frac{\mathcal{G}(\mathcal{A})^2}{2\xi}} \exp \left[\mathrm{Tr}(-\bar{c}[T_a, [T^a + \mathcal{A}^a, c]])\right], \tag{7.5.14}$$

where we have used the delta function to evaluate the integral over ω, and dropped all irrelevant normalization factors. Choosing $\xi = g^2/4$ such that the gauge-fixing term $S_{\mathrm{gf}} = -\mathcal{G}(\mathcal{A})^2/2\xi$ cancels the corresponding term in (7.5.3) we arrive at the quadratic action

$$S_2[\mathcal{A}] + S_{\mathrm{gf}} + S_{\mathrm{ghost}} = -\frac{2}{g^2} \mathrm{Tr} \left(\mathcal{A}_a (\mathcal{D}^2 + m^2) \mathcal{A}^a + 2\bar{c}[T_a, [T^a + \mathcal{A}^a, c]]\right). \tag{7.5.15}$$

The Faddeev–Popov ghosts will cancel the contribution from the unphysical negative modes of $\mathcal{A}^a$ in Minkowski signature. Adding also the higher-order terms in $\mathcal{A}$ in (7.5.2), this is the starting point for a perturbative analysis of the Yang–Mills matrix model. The physical Hilbert space of fluctuations $\mathcal{A}$ around the background T is then defined as usual in Yang–Mills theories as[13]

$$\mathcal{H}_{\mathrm{phys}} = \{\text{gauge-fixed on-shell modes}\} / _{\{\text{pure gauge modes}\}} \tag{7.5.16}$$

---

[13] Note that this is analogous to the symplectic reduction in Section 2.1.

at ghost number zero, which should be free of negative modes. This can also be recast in the BRST formalism, by adding the BRST-exact gauge fixing term

$$
\begin{aligned}
S_{\text{g.f.}} + S_{\text{ghost}} &= \frac{4}{g^2} \operatorname{Tr} \mathsf{s}\left(\bar{c}\, \mathcal{G}[\mathcal{A}] - \frac{1}{2}\bar{c}b \right) \\
&= \frac{4}{g^2} \operatorname{Tr}\left(b\, i[T^a, \mathcal{A}_a] - \frac{1}{2}b^2 - \bar{c}\, i[T^a, \mathsf{s}\mathcal{A}_a] \right)
\end{aligned}
\tag{7.5.17}
$$

in terms of the ghosts and antighosts c resp. $\bar{c}$ and the Nakanishi–Lautrup field b. Here the BRST transformations are defined as

$$
\begin{aligned}
\mathsf{s}\mathcal{A}_a &= i[T_a + \mathcal{A}_a, c], & \mathsf{s}\bar{c} &= b, \\
\mathsf{s}b &= 0, & \mathsf{s}c &= ic^2,
\end{aligned}
\tag{7.5.18}
$$

which satisfies

$$
\mathsf{s}^2 = 0. \tag{7.5.19}
$$

Eliminating b using its eom

$$
i[T^a, \mathcal{A}_a] = b, \tag{7.5.20}
$$

we arrive at

$$
S_{\text{g.f.}} + S_{\text{ghost}} = -\frac{2}{g^2} \operatorname{Tr}\left([T^a, \mathcal{A}_a][T^b, \mathcal{A}_b] - 2\bar{c}[T^a, [T_a + \mathcal{A}_a, c]] \right), \tag{7.5.21}
$$

thus recovering precisely the Faddeev–Popov action (7.5.15). The physical Hilbert space can then be characterized as

$$
\mathcal{H}_{\text{phys}} = \{\text{BRST-invariant on-shell modes}\} / \{\text{BRST exact modes}\} \tag{7.5.22}
$$

at ghost number zero. This is equivalent to (7.5.16), since the gauge fixing is recovered from the condition $0 \overset{!}{\sim} \mathsf{s}\bar{c} = b = i[T^a, \mathcal{A}_a]$ using (7.5.20), and pure gauges $i[T_a + \mathcal{A}_a, c] = \mathsf{s}\mathcal{A}_a$ are BRST exact. For finite-dimensional matrices, the matrix models can in principle be studied at the nonperturbative level without any gauge fixing, using e.g. Monte Carlo methods in the Euclidean case. Then the integral over the gauge orbit is finite and can be kept explicitly.

7.6 *Effective matrix action

We can define a quantum effective action for the matrix model along the lines of Section 6.1. The first step is to introduce sources $J_a \in \operatorname{End}(\mathcal{H})$ for each matrix T^a, and define the generating function

$$
Z[J_a] = \int DT^a\, e^{S[T] + \operatorname{Tr} T^a J_a} =: e^{W[J_a]}. \tag{7.6.1}
$$

Then the correlators can be obtained as

$$\langle T_{a_1} \ldots T_{a_n} \rangle = \frac{1}{Z[0]} \frac{\partial^n}{\partial J_{a_1} \ldots \partial J_{a_n}} Z[J] \Big|_{J=0}, \tag{7.6.2}$$

while $W[J]$ generates the connected correlators. By construction, $Z[J]$ and $W[J]$ are gauge invariant. We define the **classical matrix** in the presence of J via

$$T_{cl}^a := \langle T^a \rangle_J = \frac{\partial}{\partial J_a} W[J] \qquad \in \text{End}(\mathcal{H}) \tag{7.6.3}$$

or more explicitly

$$(T^a)_{cl} = \frac{\partial}{\partial J_a} W[J] = \frac{1}{Z[J]} \int DT \, T^a e^{S[T] + \text{Tr}(T^a J_a)}. \tag{7.6.4}$$

By construction, it inherits the standard gauge transformation law $T_{cl}^a \to U^{-1} T_{cl}^a U$. Then the **effective matrix action** is defined as Legendre transform of $W[J]$

$$\Gamma[T_{cl}^a] := W[J_a] - \text{Tr}(T_{cl}^a J_a) \Big|_{J=J(T_{cl})} \tag{7.6.5}$$

expressing J_a in terms of T_{cl}^a through (7.6.3). As usual, we conclude that the nontrivial vacua $T_{cl}^a = \langle T^a \rangle_{J=0}$ (7.3.1) are critical points of the effective action,

$$\boxed{\frac{\partial \Gamma[T_{cl}^a]}{\partial T_{cl}^a} = 0.} \tag{7.6.6}$$

Similar formulas hold in Minkowski signature, with the appropriate $i\varepsilon$ prescription. All these steps are perfectly well-defined for finite-dimensional matrices. Moreover, the effective matrix action $\Gamma[T^a]$ is gauge invariant, since all the steps in this procedure respect gauge invariance. At least for finite-dimensional matrices, one may therefore think of it as a generalized multi-trace matrix model, which is useful even at the perturbative level [36].

One-loop quantization

We recall from Section 6.1 that the one-loop contribution to the effective action is given by the Gaussian integral over the fluctuations $\mathcal{A}$ around some background T. In the Euclidean case, this is based on the basic formulas (A.0.1) and (A.0.2). Then

$$\int_{\text{1 loop}} d\mathcal{A} e^{S_2[\mathcal{A}] + S_{gf}} = e^{\Gamma_{\text{1 loop}}(T)}, \qquad \Gamma_{\text{1 loop}}(T) = -\frac{1}{2} \text{Tr} \ln(\mathcal{D}^2 + m^2) \tag{7.6.7}$$

where Tr is the trace over the space of modes $\mathcal{A}$, recalling our unconventional Euclidean sign convention (7.0.2). We will drop any normalization factors of the integral, which cancel in the correlation functions. In the Minkowski case, the oscillatory integrals are defined using the formula

$$\int_{\mathbb{R}} dx e^{i(-a+i\varepsilon)x^2} = \frac{\sqrt{\pi}}{\sqrt{-i(-a+i\varepsilon)}} = e^{i\pi/4} e^{-\frac{1}{2} \ln(a-i\varepsilon)} \tag{7.6.8}$$

for $a \in \mathbb{R}$ and $\varepsilon > 0$. Dropping the constant phase factor and generalized to the integral over hermitian $N \times N$ matrices, this gives

$$\int d\phi\, e^{i\phi(-\mathcal{K}-m^2+i\varepsilon)\phi} = \exp\left(-\frac{1}{2}\mathrm{Tr}\ln(\mathcal{K}+m^2-i\varepsilon)\right), \tag{7.6.9}$$

where typically $\mathcal{K} = \Box^2$ is the kinetic operator. For the Yang–Mills matrix model, we thus obtain

$$\int d\mathcal{A}\, e^{i(S_2[\mathcal{A}]+S_{gf})} = e^{i\Gamma_{1\,\mathrm{loop}}(T)}, \qquad \Gamma_{1\,\mathrm{loop}}(T) = \frac{i}{2}\mathrm{Tr}\ln(\mathcal{D}^2+m^2-i\varepsilon) \tag{7.6.10}$$

using the $i\varepsilon$ regularization (7.2.3). The i in the effective action will disappear in the evaluation of the trace, as elaborated in Section 12, cf. (7.6.23). Note also that the background T is not required to be a critical point of the classical action, and linear terms in $\mathcal{A}$ are always dropped in the evaluation of the effective action as they should be balanced by the external current, cf. [156]. Then the one-loop ghost term gives

$$\int dc\, d\bar{c}\, \exp\left(\mathrm{Tr}(\bar{c}[T_a,[T^a,c]])\right) = \exp\left(\mathrm{Tr}\ln(\Box-i\varepsilon)\right) \tag{7.6.11}$$

incorporating the appropriate regularization in Minkowski signature. It is sometimes useful to rewrite such expressions in terms of Schwinger parameters for the one-loop effective action, based on the identity

$$\int_0^\infty \frac{d\alpha}{\alpha}(e^{-aX}-e^{-aY}) = \log\frac{Y}{X}. \tag{7.6.12}$$

One can then write e.g.

$$\mathrm{Tr}\ln(\mathcal{D}^2+m^2) = -\mathrm{Tr}\int\limits_0^\infty \frac{d\alpha}{\alpha}\, e^{-\alpha(\mathcal{D}^2+m^2)}, \tag{7.6.13}$$

which has the form of a heat kernel as discussed in Section 12.1. In the infinite-dimensional case, such traces are typically divergent and need to be regularized by introducing an UV cutoff, e.g. by inserting a convergence factor $e^{-\frac{1}{\alpha L^2}}$. This is not necessary in the maximally supersymmetric case, which will be our main interest in the following.

7.6.1 Reducible versus irreducible backgrounds

We will mostly focus on irreducible matrix configurations in the following. This is justified, since reducible configurations have a smaller gauge orbit than irreducible ones (cf. the discussion in Section 7.3.1), and are therefore a null set. Nevertheless, it is interesting to consider reducible backgrounds (7.4.33)

$$T^a = \begin{pmatrix} T^a_{(1)} & & & \\ & T^a_{(2)} & & \\ & & \ddots & \\ & & & T^a_{(k)} \end{pmatrix} \tag{7.6.14}$$

interpreted as stacks of branes embedded in target space. If the constituent blocks are classical solutions, then so is the full configuration. Quantum effects will typically induce interactions between the branes, due to virtual off-diagonal modes which link the different branes. We can gain some insight into these interactions at one loop.

One-loop interaction between branes

To get some insight into the interactions induced at one loop, consider the case of two compact branes $\mathcal{M}_{(1,2)}$ defined in terms of a finite-dimensional matrix configurations $X^a_{(1,2)}$ displaced by a distance c. We assume for simplicity that the structure of the branes is identical, $X^a_{(1)} = X^a_{(2)} \equiv X^a$. Then

$$T^a = \begin{pmatrix} X^a & 0 \\ 0 & X^a + c^a \mathbb{1}_{(2)} \end{pmatrix}, \tag{7.6.15}$$

and

$$i[X^a, X^b] = \begin{pmatrix} \mathcal{F}^{ab} & 0 \\ 0 & \mathcal{F}^{ab} \end{pmatrix}. \tag{7.6.16}$$

We can then write the background as

$$T^a = X^a \otimes \mathbb{1}_2 + c^a \begin{pmatrix} 0 & 0 \\ 0 & \mathbb{1}_{(2)} \end{pmatrix}, \tag{7.6.17}$$

which can be viewed as $U(2)$ noncommutative gauge theory on $\mathcal{M}$ with a nontrivial vacuum, spontaneously breaking the gauge symmetry to $U(1) \times U(1)$ via the Higgs mechanism. The interactions of the branes are mediated by the off-diagonal modes. To understand how the effective action depends on the separation c in the Euclidean bosonic model, we need to evaluate

$$\Gamma_{1 \text{ loop}}[c] = -\frac{1}{2} \text{Tr} \ln(\mathcal{D}_T^2 + m^2). \tag{7.6.18}$$

$\mathcal{D}_T^2$ acts on off-diagonal block matrices as

$$\left[T^a, \begin{pmatrix} 0 & \phi_{(12)} \\ 0 & 0 \end{pmatrix} \right] = \begin{pmatrix} 0 & [X^a, \phi_{(12)}] - c^a \phi_{(12)} \\ 0 & 0 \end{pmatrix}$$

$$\mathcal{D}_T^2 \begin{pmatrix} 0 & \phi_{(12)} \\ 0 & 0 \end{pmatrix} = \begin{pmatrix} 0 & \mathcal{D}^2 \phi_{(12)} + c^a c_a \phi_{(12)} \\ 0 & 0 \end{pmatrix}, \tag{7.6.19}$$

where $\mathcal{D}^2$ is the operator (7.5.4) on a single brane X^a. Now we separate the trace over all modes into contributions from the diagonal and off-diagonal fluctuations:

$$\Gamma_{1 \text{ loop}} = \Gamma_{1 \text{ loop, diag}} + \Gamma_{1 \text{ loop, offdiag}}. \tag{7.6.20}$$

The diagonal contributions are the sum of the two contributions from the two branes, which is independent of the separation. The off-diagonal contribution can be expanded as follows:

$$\Gamma_{1 \text{ loop, offdiag}} = -\frac{1}{2} \text{Tr} \ln(\mathcal{D}^2 + m^2 + c^a c_a)$$

$$= -\frac{1}{2} \text{Tr} \ln(\mathcal{D}^2 + m^2) - \frac{1}{2} c^a c_a \text{Tr} \left(\frac{1}{\mathcal{D}^2 + m^2} \right) + O(c^4). \quad (7.6.21)$$

Therefore,

$$\Gamma_{1 \text{ loop}}[c] = \Gamma_{1 \text{ loop}}[0] - \frac{1}{2} c^a c_a \text{Tr} \left(\frac{1}{\mathcal{D}^2 + m^2} \right) + O(c^4). \quad (7.6.22)$$

This amounts to an attractive potential $V \sim c_a c^a$ in the Euclidean case, recalling our matrix model sign conventions (7.0.2), so that parallel branes attract each other.

In Minkowski signature, the conclusion is the same, but the evaluation is more tricky due to the singular $i\varepsilon$ structure. As discussed in detail in Section 12 and appendix D, one typically finds

$$\Gamma_{1 \text{ loop, M}} = \frac{i}{2} c^a c_a \text{Tr} \left(\frac{1}{\mathcal{D} + m^2 - i\varepsilon} \right) \sim -\frac{1}{2} c^a c_a \text{Tr} \delta(\mathcal{D} + m^2 - i\varepsilon) < 0, \quad (7.6.23)$$

using $\frac{1}{x - i\varepsilon} = P\frac{1}{x} + i\delta(x)$. Taking into account the standard decomposition of the action $S = T - V$ into kinetic and potential energy, this leads again to an attractive one-loop effective potential

$$V = -\Gamma_{1 \text{ loop, M}} \sim c_a c^a > 0 \quad (7.6.24)$$

between the branes. We will see that in supersymmetric models, this quadratic contribution to the potential cancels, leaving a weaker subleading interaction. If the matrix configuration is also compatible with SUSY (often denoted as Bogomol'nyi–Prasad–Sommerfield (BPS)), then the interaction cancels exactly. This is the case for flat parallel Moyal–Weyl quantum planes $\mathbb{R}^{2n}_\theta$; however, then the effective action for fluctuations on the branes is typically divergent, except for the maximally supersymmetric case as discussed in Chapter 11.

7.6.2 *Stabilization of covariant backgrounds

Now assume that the effective matrix action Γ or some approximation thereof – such as the one-loop effective action $\Gamma_{1 \text{ loop}}$ – is available. By construction, Γ is invariant under $U(\mathcal{H})$ gauge transformations, as well as under global $SO(D)$ rotations or some non-compact version thereof; in other words, these symmetries are anomaly free. Covariant quantum spaces such as S^2_N, H^4_n etc. respect part of this symmetry; typically, rotations are equivalent to gauge transformations of the corresponding matrix configurations. Carrying this over to the effective action, we will argue that quantum effects can stabilize covariant quantum spaces, mimicking the effect of mass terms or cubic terms in the matrix model as in the ARS model (7.3.2). This means that covariant quantum spaces are not just nice toy examples, but they are good candidates for dynamically stabilized backgrounds in the IKKT model.

The crucial property of covariant quantum spaces is that the underlying matrix configuration X^a are **vector operators**, i.e. they transform as (5.0.1)

$$U_g^{-1} X^a U_g = \Lambda^a_b(g) X^b, \quad (7.6.25)$$

where Λ_b^a is some linear action of the covariance group $G \subset SO(D)$. Now consider

$$\mathcal{X}_a := \left.\frac{\partial \Gamma[X]}{\partial X^a}\right|_X, \qquad \text{i.e. } (\mathcal{X}^a)_i^j = \left.\frac{\partial}{\partial (X^a)_j^i}\right|_X. \tag{7.6.26}$$

Due to the invariance of $\Gamma[X]$ under gauge transformations and $SO(D)$ rotations, it follows that $\mathcal{X}_a$ transforms also in the adjoint under gauge transformations and as vector under $SO(D)$ rotations. In particular if X^a is a vector operator, then so is $\mathcal{X}^a$, i.e.

$$U_g^{-1} \mathcal{X}^a U_g = \Lambda_b^a(g)\mathcal{X}^b. \tag{7.6.27}$$

Now we assume that the covariant quantum space under consideration has the property that the defining vector operators X^a in $\mathrm{End}(\mathcal{H})$ are uniquely determined (up to rescaling) by equivariance. This holds for S_n^2, S_N^4, H_n^4 from the decomposition of $\mathrm{End}(\mathcal{H})$ into irreps, cf. (3.3.17), (5.1.30). Assuming this uniqueness property, we can conclude that

$$\mathcal{X}_a = \alpha X_a \tag{7.6.28}$$

for some $\alpha \in \mathbb{R}$. This is very interesting: it means that to show that the equations of motion

$$\left.\frac{\partial \Gamma[X]}{\partial X^a}\right|_{rX} = \mathcal{X}_a \overset{!}{=} 0 \tag{7.6.29}$$

hold for the rescaled background rX^a for some suitable radius $r \in \mathbb{R}$, it suffices to show that the radial derivative vanishes. The appropriate normalization r can be determined simply by

$$\frac{d}{dr} V(r) = 0, \qquad V(r) := -\Gamma[rX]. \tag{7.6.30}$$

Therefore, all we need to do is compute the radial dependence of the effective potential $V(r)$ for the background under consideration, and show that it has a nontrivial minimum. That is relatively manageable due to the symmetry, and we will develop suitable methods to compute $V(r)$ at one loop. This may be used to justify the stability of fuzzy sphere solutions in matrix models even in the absence of flux terms, and it is easily applicable for supersymmetric models at least at one loop, cf. [183]. Of course, there might be other instabilities, but it is reasonable to expect that the radial mode is the leading instability; moreover, the argument could be extended to include other fluctuation modes.

This line of argument can be adapted to the case of covariant quantum spaces with a higher-dimensional space of vector operators, such as $\mathcal{M}^{3,1}$, possibly with fuzzy extra dimensions. For $\mathcal{M}^{3,1}$ there are two independent vector operators x^μ and t^μ, which requires some refinement of the argument. Combined with gauge invariance, one can then justify the existence of vacua of the structure (5.4.118) at the quantum level, which can be interpreted as (higher-spin extension of) FLRW space-time with generic scale function $a(t)$.

8 Fuzzy extra dimensions

The basic idea of Kaluza–Klein compactification is to reduce a higher-dimensional theory effectively to $3 + 1$ dimensions at low energies, by considering a target space of type $\mathcal{M}^{3,1} \times \mathcal{K}$ where $\mathcal{K}$ is a compact space with UV-scale harmonics. In string theory, this is commonly used to reduce $9 + 1$ dimensional target space to $3 + 1$ dimensions, albeit in a rather ad-hoc way, leading to a vast "landscape" of compactifications.

In this chapter, we will discuss a mechanism that seems formally similar but is in fact quite different: We consider a nonabelian (commutative or noncommutative) Yang–Mills theory on $\mathcal{M}^{3,1}$ with gauge group $U(N)$, which dynamically develops fuzzy extra dimensions via the Higgs effect in some nontrivial vacuum. The theory then behaves *in some intermediate regime* like a higher-dimensional gauge theory on

$$\mathcal{M}^{3,1} \times \mathcal{K}_N, \tag{8.0.1}$$

where $\mathcal{K}_N$ is some fuzzy space, with reduced rank of the residual gauge group. More precisely, it behaves like a higher-dimensional gauge theory compactified on $\mathcal{K}_N$ *below some energy scale* Λ_N, while for energies higher than Λ_N it behaves like a $3 + 1$ dimensional theory. The key point is that the internal space $\mathcal{K}_N$ – which may be a fuzzy sphere S_N, for example – has an intrinsic UV cutoff Λ_N, while its geometry disappears at higher energies. This is the typical feature of fuzzy spaces, as discussed previously. There is no limitation on the dimension of $\mathcal{K}_N$, except that it must be even. We will often drop the subscript N in the following.

8.1 Fuzzy extra dimensions in the Yang–Mills gauge theory

Consider a (commutative) Yang–Mills gauge theory on $\mathcal{M}^{3,1}$ with scalar fields ϕ_i in the adjoint of the gauge group[1] $SU(N)$, which may include fermions Ψ in the adjoint. We assume an action of the form

$$
\begin{aligned}
S_{\text{YM}} = \int d^4x \, \frac{1}{4g^2} \text{Tr}\Big(&- F^{\mu\nu} F_{\mu\nu} - 2D^\mu \phi^i D_\mu \phi_i + [\phi^i, \phi^j][\phi_i, \phi_j] - V_{\text{soft}}(\phi) \\
&+ \overline{\Psi} \gamma^\mu i D_\mu \Psi + \overline{\Psi} \Gamma^i [\phi_i, \Psi]\Big)
\end{aligned}
\tag{8.1.1}
$$

---

[1] Similar considerations apply to other gauge groups, and also e.g. for pure scalar field theories as long as the fields take values in some representation of a Lie algebra.

familiar from supersymmetric Yang–Mills gauge theory. Here $V_{\text{soft}}(\phi)$ could be quadratic and cubic terms in ϕ

$$V_{\text{soft}}(\phi) = f_{ijk}\phi_i\phi_j\phi_k + 2m^2\delta_{ij}\phi_i\phi_j, \tag{8.1.2}$$

where f_{ijk} is totally antisymmetric. In principle, there could also be more complicated terms which arise in the quantum effective potential, but we assume the present form for simplicity. The action is clearly invariant under gauge transformations

$$\phi_i(x) \to U^{-1}\phi_i(x)U \tag{8.1.3}$$

for $U(x) \in SU(N)$.

In the presence of such a nontrivial potential, the gauge theory typically admits several different vacua, where the vacuum expectation value (VEV) of the scalar fields is nontrivial. They arise from the (local and global) minima of[2]

$$V(\phi) = \frac{1}{4g^2}\big(- [\phi^i, \phi^j][\phi_i, \phi_j] + V_{\text{soft}}(\phi)\big). \tag{8.1.4}$$

We thus assume that the scalar fields acquire a nontrivial vacuum expectation value (VEV)

$$\langle\phi_i\rangle = m_{\mathcal{K}}\,\mathcal{K}_i, \qquad \mathcal{K}_i \in \mathfrak{su}(N), \tag{8.1.5}$$

which do not commute. Here $m_{\mathcal{K}} = m_{\mathcal{K}}(x)$ is a constant or a scalar field that sets the scale of ϕ, and $\mathcal{K}_i$ are dimensionless. For simplicity, we assume that the configuration is irreducible, i.e. the only matrix commuting with all $\mathcal{K}_i$ is the unit matrix. Then the gauge symmetry is broken spontaneously and completely. As discussed in Chapter 4, such a vacuum should be interpreted as a quantum space $\mathcal{K}$, with quantized space of functions

$$\text{End}(\mathcal{H}_{\mathcal{K}}) \cong \mathcal{C}(\mathcal{K}), \tag{8.1.6}$$

where $\mathcal{H}_{\mathcal{K}} \cong \mathbb{C}^N$. For example, in the case of three scalar fields with $V_{\text{soft}}(\phi) = \varepsilon_{ijk}\phi_i\phi_j\phi_k$, the minimum of the potential arises for $\mathcal{K}_i \sim \lambda_i$ being an $\mathfrak{su}(2)$ generator in some irreducible representation, and the ϕ_i define a fuzzy sphere S_N^2. Some other more interesting geometries will be discussed in Section 8.4.

In the next section, we will justify the claim that the effective geometry of the gauge theory arising on such a background is given by $\mathcal{M}^{3,1} \times \mathcal{K}$, by performing a mode analysis for the fluctuations and matching them with the Kaluza–Klein spectrum on $\mathcal{K}$.

8.2 Fluctuation modes for fuzzy extra dimensions

Define the "internal" matrix Laplacian $\Box_{\mathcal{K}}$ as

$$\Box_{\mathcal{K}} := m_{\mathcal{K}}^2\delta^{ij}[\mathcal{K}_i, [\mathcal{K}_j, .]]. \tag{8.2.1}$$

Diagonalizing $\Box_{\mathcal{K}}$ leads to discrete eigenmodes

$$\Box_{\mathcal{K}}\Upsilon_\Lambda = m_\Lambda^2\,\Upsilon_\Lambda, \qquad m_\Lambda^2 = m_{\mathcal{K}}^2\,\mu_\Lambda^2 \tag{8.2.2}$$

---

[2] Recall that the Lagrangian has the form $\mathcal{L} = T - V$ where T is the kinetic term and V the potential.

labeled by Λ. Based on the correspondence (8.1.6), these modes form a basis for the space of functions on $\mathcal{K}$ below some cutoff. We will see that from the field theory point of view on $\mathcal{M}^{3,1}$, the eigenmodes Υ_Λ play the role of Kaluza–Klein modes on the compact extra dimensions $\mathcal{K}$, with Kaluza–Klein masses given by μ_Λ^2. The scale of the KK masses is set by $m_\mathcal{K}^2$, while the

$$0 \leq \mu_\Lambda^2 \leq O(N^2) \tag{8.2.3}$$

are numerical values typically ranging from $O(1)$ to $O(N^2)$, which can be determined by group theory in simple cases. Moreover, the fluctuations of the scalar fields play the role of NC gauge fields on $\mathcal{K}$. To justify these claims, we expand the scalar potential around the vacuum

$$\phi_i = m_\mathcal{K}\big(\mathcal{K}_i + \mathcal{B}_i(x)\big) \tag{8.2.4}$$

so that $\mathcal{B}_i$ is dimensionless. According to (8.1.6), the $\mathcal{B}_i(x) \in \mathrm{End}(\mathcal{H}) \cong \mathcal{C}(\mathcal{K})$ can be viewed as functions on $\mathcal{K}$. Under a gauge transformation, they transform as

$$\mathcal{B}_i \to U\mathcal{B}_i U^{-1} + U[\mathcal{K}_i, U^{-1}]. \tag{8.2.5}$$

As discussed in Section 7.4.1, this means that the (tangential components of) $\mathcal{B}_i$ play the role of a gauge field on $\mathcal{K}$. Then

$$V[\phi] = V[\mathcal{K}] + V[\mathcal{B}] + \frac{m_\mathcal{K}^4}{g^2}\mathrm{Tr}\Big(\mathcal{B}^i\big(\Box_\mathcal{K} + \frac{m^2}{m_\mathcal{K}^2}\big)\mathcal{K}_i + \mathcal{K}^i\Box_\mathcal{B}\mathcal{B}_i + \frac{1}{2}\mathcal{B}^i\big(\mathcal{D}^2 + \frac{m^2}{m_\mathcal{K}^2}\big)\mathcal{B}_i$$
$$-\frac{1}{2}\mathcal{G}^2 + \frac{3}{4m_\mathcal{K}}\big(f_{ijk}\mathcal{B}_i\mathcal{K}_j\mathcal{K}_k + f_{ijk}\mathcal{B}_i\mathcal{B}_j\mathcal{K}_k\big)\Big), \tag{8.2.6}$$

where

$$\mathcal{G} = i[\phi_i, \mathcal{B}^i] \tag{8.2.7}$$

and

$$\mathcal{D}^2\mathcal{B}_i = \Box_K\mathcal{B}_i + 2[[\mathcal{K}^i, \mathcal{K}^j], \mathcal{B}_j]. \tag{8.2.8}$$

The linear term in $\mathcal{B}$ disappears assuming that the background satisfies

$$\big(\Box_\mathcal{K} + \frac{m^2}{m_\mathcal{K}^2}\big)\mathcal{K}_i + \frac{3}{4m_\mathcal{K}}f_{ijk}\mathcal{K}_j\mathcal{K}_k = 0. \tag{8.2.9}$$

For example, for $f_{ijk} = \epsilon_{ijk}$, these are the equations of motion for a fuzzy sphere S_N^2.

In the presence of such a nontrivial vacuum, the gauge bosons acquire a mass through the Higgs effect, which arises from the kinetic term of the scalar fields:

$$\int D^\mu\phi^i D_\mu\phi_i = \int (\partial^\mu\phi^i - i[A^\mu, \phi^i])(\partial_\mu\phi_i - i[A_\mu, \phi_i])$$
$$= m_\mathcal{K}^2\int \partial^\mu\mathcal{B}^i\partial_\mu\mathcal{B}_i - 2i\partial^\mu\mathcal{B}_i[A_\mu, \mathcal{K}_i + \mathcal{B}_i] - [A^\mu, \mathcal{K}^i + \mathcal{B}_i][A_\mu, \mathcal{K}_i + \mathcal{B}_i]$$
$$= m_\mathcal{K}^2\int \partial^\mu\mathcal{B}^i\partial_\mu\mathcal{B}_i + A^\mu\Box_\mathcal{K}A_\mu + S_{3,4}(A, B). \tag{8.2.10}$$

Here $S_{3,4}(A,B)$ are cubic and quartic terms in B and A, and we eliminated the mixed term

$$\int \partial^\mu \mathcal{B}_i[A_\mu, \mathcal{K}_i] = -\int A_\mu[\partial^\mu \mathcal{B}_i, \mathcal{K}_i] = 0 \tag{8.2.11}$$

by choosing a gauge

$$[\mathcal{K}^i, \mathcal{B}_i] = 0. \tag{8.2.12}$$

This is always possible due to Lemma (7.1). We can now obtain the masses of the nonabelian gauge modes from the quadratic action (8.2.10) by expanding the gauge fields A_μ into eigenmodes of $\Box_\mathcal{K}$

$$A_\mu(x) = \sum A_{\Lambda\mu}(x)\, \Upsilon_\Lambda. \tag{8.2.13}$$

Then the quadratic terms are diagonal, and we can read off the mass terms for the gauge modes

$$\int D^\mu \phi^i D_\mu \phi_i = m_\mathcal{K}^2 \int \sum \mu_\Lambda^2 A_\Lambda^\mu A_{\Lambda\mu} + \dots. \tag{8.2.14}$$

Therefore, the gauge fields acquire the same masses μ_Λ^2 as the scalar fields (8.2.2), which are interpreted again as Kaluza–Klein masses arising from the compact space $\mathcal{K}$. The fermionic modes also acquire analogous masses due to the Yukawa coupling $\overline{\Psi}\Gamma^i[\phi_i, \Psi]$, which can again be interpreted as Kaluza–Klein masses; the details depend on the type of fermions under consideration.

We conclude that the field theory **behaves like a higher-dimensional gauge theory** on $\mathcal{M}^{3,1} \times \mathcal{K}$ **below the UV-cutoff scale** of the fuzzy space $\mathcal{K} = \mathcal{K}_N$. This is nothing but the well-known Higgs effect in a geometric guise. At higher energies, the theory behaves like a four-dimensional Yang–Mills theory, which is well-defined as a quantum theory in contrast to genuine higher-dimensional gauge theories. Therefore, the present mechanism allows to define effectively higher-dimensional gauge theories in some range of energies. The same mechanism applies to the matrix model, where the entire geometry $\mathcal{M}^{3,1} \times \mathcal{K}$ emerges from the matrix model.

8.3 Fuzzy extra dimensions in matrix models

Now we return to the Yang–Mills matrix model of type (7.0.1), for matrices T^a, $a = 0, \dots, D$. To be specific, we consider a matrix configuration of the following type:

$$\begin{aligned} T^\mu &= X^\mu, & \mu &= 0, \dots, 3 \\ T^i &= m_\mathcal{K} \mathcal{K}^i, & i &= 4, \dots, D \end{aligned} \tag{8.3.1}$$

which act on $\mathcal{H} = \mathcal{H}_\mathcal{M} \otimes \mathcal{H}_\mathcal{K}$ and mutually commute[3]

$$[X^\mu, \mathcal{K}^i] = 0. \tag{8.3.2}$$

---

[3] More generally, $m_\mathcal{K}$ may be considered as a scalar field on $\mathcal{M}^{3,1}$, cf. Section 9.1.1.

The coupling constant g is absorbed in the matrices, which are thus dimensionless. The X^μ generate the quantum spacetime $\mathcal{M}^{3,1}$, while the $\mathcal{K}^i$ are dimensionless generators of the quantized space of functions $\text{End}(\mathcal{H}_\mathcal{K}) \cong \mathcal{C}(\mathcal{K})$ on some compact space $\mathcal{K}$, where $\mathcal{H}_\mathcal{K} \cong \mathbb{C}^N$. Such a configuration can be interpreted as quantum spacetime with fuzzy extra dimensions

$$\mathcal{M}^{3,1} \times \mathcal{K}. \tag{8.3.3}$$

At the classical level, such solutions arise e.g. in the presence of an extra cubic term in the action

$$S_{\text{soft}} = \text{Tr}(f_{ijk}T_iT_jT_k) \tag{8.3.4}$$

as in the ARS model (7.3.2). It is plausible that such backgrounds can be stabilized through quantum effects in the IKKT model without adding cubic terms, as indicated in Section 7.6.2 and in more detail in Chapter 12. Then the most general fluctuations

$$T^a \to T^a + \mathcal{A}^a, \qquad \mathcal{A}^a \in \text{End}(\mathcal{H}) \cong \mathcal{C}(\mathcal{M} \times \mathcal{K}) \tag{8.3.5}$$

can be expanded in KK harmonics on $\mathcal{K}$ as in the previous section, and the masses of the harmonics are determined by the spectrum of

$$\Box_\mathcal{K} = [T^i, [T_i, .]] \tag{8.3.6}$$

given by (8.2.2)

$$\Box_\mathcal{K} \Upsilon_\Lambda = m_\Lambda^2 \, \Upsilon_\Lambda, \qquad m_\Lambda^2 = m_\mathcal{K}^2 \, \mu_\Lambda^2. \tag{8.3.7}$$

This illustrates again the basic phenomenon of emergent quantum geometries in matrix models, which will play an essential role to recover both gravity and a nontrivial gauge theory of matter from the matrix model.

Some comments are in order. Although the product structure $\mathcal{M}^{3,1} \times \mathcal{K}$ may be reminiscent of string theory compactifications, there is a fundamental structural difference: in string theory compactifications, the geometry of target space itself is modified. Here, the matrix model background describes a brane of structure $\mathcal{M}^{3,1} \times \mathcal{K} \subset \mathbb{R}^{9,1}$, which is thought of as being embedded in the uncompactified target space. Therefore, the vast landscape of string theory compactifications is not encountered in such a scenario. In fact, "genuine" compactification of target space can be realized in matrix models [54, 197] e.g. via

$$U^{-1}T^iU = T^i + 2\pi R\mathbb{1}; \tag{8.3.8}$$

this will be discussed in Section 11.1.8. However, such backgrounds require an infinite-dimensional Hilbert space describing the transverse directions. This is a price we are not willing to pay, and quantum effects on such backgrounds would generically be divergent[4]

Finally, the mechanism of fuzzy extra dimensions should not be confused with attempts to attach a holographic higher-dimensional interpretation to the quantum effective action of some given model, as discussed in Section 11.1.7. The present mechanism is simply the

---

[4] There exist exceptional BPS backgrounds of this type that are protected from quantum corrections, but they do not allow any deformations and are hence of limited physical interest.

classical Higgs mechanism interpreted geometrically in terms of an effective gauge theory on $\mathcal{M} \times \mathcal{K}$, as appropriate in the weak coupling regime.

8.4 *Self-intersecting fuzzy extra dimensions and chiral fermions

The simplest explicit example of fuzzy extra dimensions is provided by a fuzzy sphere S_N^2 spanned by three transversal scalar fields in the presence of a cubic term (8.3.4) in the action. Here we briefly discuss a more interesting example of self-intersecting fuzzy extra dimensions, which involves six transversal scalar fields. This is precisely provided by the IKKT matrix model, or by $\mathcal{N} = 4$ SYM with appropriate cubic term in the potential.

We recall the construction of fuzzy $\mathbb{C}P^2$ in Section 3.3.2, which is given in terms of the 8 generators T_a, $a = 1, \ldots, 8$ of $\mathfrak{su}(3)$ in the representation $(N, 0)$, which satisfy

$$[T_a, T_b] = ic_{abc}T_c. \tag{8.4.1}$$

It will be useful to work with the following complex linear combinations or root generators

$$
\begin{aligned}
T_1^\pm &= \frac{1}{\sqrt{2}}(T_4 \pm iT_5), \\
T_2^\pm &= \frac{1}{\sqrt{2}}(T_6 \mp iT_7), \\
T_3^\pm &= \frac{1}{\sqrt{2}}(T_1 \mp iT_2)
\end{aligned}
\tag{8.4.2}
$$

while the Cartan generators T_3, T_8 commute. The weights of this representation (i.e. the eigenvalues of T_3 and T_8) are depicted in Figure 8.1. The geometry of the corresponding coadjoint orbit is best understood in terms of coherent states $|x\rangle = U_x|\Lambda\rangle$, which are given by the $SU(3)$ orbit of the highest weight state as discussed in Section 4.6.3. The highest weight state $|\Lambda\rangle$ satisfies

$$
\begin{aligned}
\langle \Lambda | T_\alpha^\pm | \Lambda \rangle &= 0 = \langle \Lambda | T_3 | \Lambda \rangle \\
\langle \Lambda | T_8 | \Lambda \rangle &= \frac{2N}{\sqrt{3}},
\end{aligned}
\tag{8.4.3}
$$

where $\alpha = 1, 2, 3$, so that $|x_0\rangle \equiv |\Lambda\rangle$ is located in the two-dimensional plane $\mathbb{R}_{3,8}^2 \subset \mathbb{R}^8 \cong \mathfrak{su}(3)$ corresponding to the Cartan subalgebra. The space $\mathcal{M} = SU(3)|\Lambda\rangle/_{U(1)}$ is four-dimensional since the stabilizer of Λ is $SU(2) \times U(1)$, and $T_{x_0}\mathcal{M}$ lies in the four-dimensional plane $\mathbb{R}_{4567}^4$ orthogonal to the Cartan subalgebra $\mathbb{R}_{3,8}^2$. Looking at the weight diagram, it is obvious that there are two analogous states $|\omega\Lambda\rangle$ and $|\omega^{-1}\Lambda\rangle$ which are obtained by a Weyl rotation ω with angle $2\pi/3$ in weight space. The corresponding coherent states $|\omega^\pm x_0\rangle$ are also located in the Cartan subalgebra since $\langle T_\alpha^\pm \rangle = 0$, and the corresponding tangent spaces of $\mathcal{M}$ lie in the four-dimensional planes $\mathbb{R}_{1267}^4$ and $\mathbb{R}_{1245}^4$, respectively.

Now consider the reduced matrix configuration T_i, $i = 1, 2, 4, 5, 6, 7$ corresponding to the root generators, after dropping the two commuting Cartan generators T_3 and T_8. Since

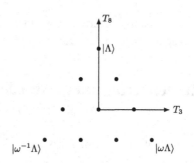

Figure 8.1 Weight lattice of the $(3, 0)$ irrep of $\mathfrak{su}(3)$, with highest weight state $|\Lambda\rangle$ and the two other extremal weight states $|\omega\Lambda\rangle$ and $|\omega^{-1}\Lambda\rangle$

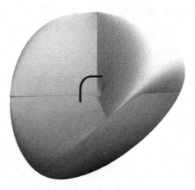

Figure 8.2 Three-dimensional section of squashed $\Pi\mathbb{C}P^2$, with zero mode connecting two sheets.

the root generators $T_\alpha^\pm$ of $\mathfrak{su}(3)$ satisfy simple commutation relations, it is easy to see that this matrix configuration is a solution of the Yang–Mills matrix model $S[T_i]$ for six matrices with cubic term (8.3.4) given by the reduced structure constants $c_{abc} \to f_{ijk}$ [173, 192]. The semi-classical geometry of this quantum space is clearly the projection

$$\mathbb{R}^8 \supset \mathbb{C}P^2 \xrightarrow{\Pi} \mathbb{R}^6 \tag{8.4.4}$$

along the Cartan subalgebra $\mathbb{R}^2_{3,8}$. This leads to a manifold or variety $\Pi\mathbb{C}P^2$ with a triple self-intersection at the origin, which we shall denote as squashed $\mathbb{C}P^2$. A projection of this variety to $\mathbb{R}^3$ is known as Roman surface, depicted in Figure 8.2. The self-intersecting structure of $\Pi\mathbb{C}P^2$ leads to several interesting features, most importantly to zero modes given by the string modes

$$\psi_{kl} = |\omega^k x_0\rangle\langle\omega^l x_0| \tag{8.4.5}$$

for $k \neq l \in \{\pm 1, 0\}$, linking the three sheets intersecting at the origin as indicated in Figure 8.2. These string zero modes are precisely the extremal weight states of $\text{End}(\mathcal{H})$, i.e. $[T_\alpha^+, \psi_{01}] = 0$ and similarly for their Weyl images.

Now consider fermions on such a background. Then the Dirac operator can be written as

$$\displaystyle\not{D}_{(6)}\Psi = \Gamma^i[T_i, \Psi] = \sum_{\alpha=1}^{3} \left(\gamma_\alpha[T_\alpha^+, \Psi] + \gamma_\alpha^\dagger[T_\alpha^-, \Psi] \right), \tag{8.4.6}$$

where

$$\gamma_1 = \frac{1}{\sqrt{2}}(\Gamma_4 - i\Gamma_5), \qquad \gamma_2 = \frac{1}{\sqrt{2}}(\Gamma_6 + i\Gamma_7), \qquad \gamma_3 = \frac{1}{\sqrt{2}}(\Gamma_1 - i\Gamma_2). \tag{8.4.7}$$

We hence obtain six fermionic zero modes linking the sheets at their intersection, with structure

$$\Psi_\Lambda = |\uparrow\uparrow\uparrow\rangle \otimes \psi_{01} \tag{8.4.8}$$

and similarly for their images $\Psi_{w\Lambda}$ under the six elements ω of the Weyl group of $\mathfrak{su}(3)$. Moreover, it follows easily that these states have well-defined chirality

$$\Gamma^{(6)}\Psi_{w\Lambda} = (-1)^{|w|}\Psi_{w\Lambda}, \tag{8.4.9}$$

where $\Gamma^{(6)} = \Gamma_1 \ldots \Gamma_6$ is the six-dimensional chirality operator. Using this squashed $\Pi\mathbb{C}P^2 \equiv \mathcal{K}$ as fuzzy extra dimensions $\mathcal{M}^{3,1} \times \mathcal{K}$ in the IKKT matrix model, the spinors satisfy the Majorana–Weyl condition in $9 + 1$ dimensions, so that these fermionic zero modes on $\mathcal{K}$ lead to three generations of chiral fermions on spacetime. Similar chiral zero modes arise from string modes linking point branes and six-dimensional variants of $\mathcal{K}$ [46, 173]. The chirality of these zero modes is correlated with the quantum numbers under the gauge group(s) arising on the branes or sheets. This leads to chiral gauge theories with rich low-energy structure, which can be in the ballpark of the (SUSY-extended) standard model, with analogous bosonic (string) modes playing the role of the Higgs fields. We refer to [173] for more details.

Geometry and dynamics in Yang–Mills matrix models

We have seen that the semi-classical matrix model on some given background configuration can be interpreted as a Poisson-sigma model (7.1.7) on a symplectic space $\mathcal{M}$ embedded in target space $\mathbb{R}^D$. This assumes that the background can be considered as (quantized) symplectic space, described by the algebra of functions $\mathcal{C}(\mathcal{M})$. In the present chapter, we will develop this geometrical interpretation further, and identify geometric structures familiar from gravity that arise naturally in the present framework. The considerations in this chapter will essentially go through for covariant quantum spaces in Chapter 10, provided we restrict ourselves to the effectively four-dimensional regime, where the $\mathfrak{hs}$ generators can be treated as constant.

9.1 Frame and tensor fields on generic branes

In this section, we will show how the basic objects $T^{\dot{a}}$ and[1]

$$\mathcal{F}^{\dot{a}\dot{b}} = i[T^{\dot{a}}, T^{\dot{b}}] \sim -\{T^{\dot{a}}, T^{\dot{b}}\} \tag{9.1.1}$$

(7.4.4) of the matrix model can be related to more standard geometric objects on a semi-classical background $\mathcal{M}$. Moreover, the gauge transformations (7.0.5) in the matrix model will be recognized as a subsector of diffeomorphisms on $\mathcal{M}$. This will allow to describe the effective gauge theory of geometry arising on $\mathcal{M}$ through the matrix model.

As a first step, we would like to find a geometric description of $T^{\dot{a}}$ and $\mathcal{F}^{\dot{a}\dot{b}}$ in terms of tensor fields. The key is to consider the *Hamiltonian vector fields* generated by the $T^{\dot{a}}$ and $\mathcal{F}^{\dot{a}\dot{b}}$:

$$E^{\dot{a}}[\phi] := \{T^{\dot{a}}, \phi\} \tag{9.1.2}$$

$$T^{\dot{a}\dot{b}}[\phi] := \{\mathcal{F}^{\dot{a}\dot{b}}, \phi\}, \tag{9.1.3}$$

for $\phi \in \mathcal{C}(\mathcal{M})$. These vector fields can be made more explicit in local coordinates x^μ on the $2n$-dimensional manifold $\mathcal{M}$: then $E^{\dot{a}} = E^{\dot{a}\mu}\partial_\mu$ and $T^{\dot{a}\dot{b}} = T^{\dot{a}\dot{b}\mu}\partial_\mu$, where

$$E^{\dot{a}\mu} := \{T^{\dot{a}}, x^\mu\}, \tag{9.1.4}$$

---

[1] The symbol $\mathcal{F}^{\dot{a}\dot{b}}$ will be reserved for commutators of matrix configuration in Yang–Mills matrix models, with indices transforming under the global $SO(D)$ or $SO(D-1, 1)$ symmetry. In contrast, the symbol $\theta^{\mu\nu} = \{x^\mu, x^\nu\}$ is used to indicate a Poisson tensor in local coordinates.

$$T^{\dot{a}\dot{b}\mu} := \{\mathcal{F}^{\dot{a}\dot{b}}, x^{\mu}\}. \tag{9.1.5}$$

Their significance will be clarified shortly. Note that the coupling g in the matrix model (7.0.1) is absorbed in the background $T^{\dot{a}}$, which is therefore dimensionless.

We should carefully distinguish the different types of indices: Greek indices $\mu, \nu = 1, \ldots, 2n$ will denote local coordinate indices on $\mathcal{M}$, which will play the role of tensor indices. Dotted Latin indices $\dot{a}, \dot{b} = 1, \ldots, D$ will indicate frame-like indices that are unaffected by a change of coordinates x^{μ}, but transform under the global $SO(D)$ or $SO(D-1,1)$ symmetry (7.0.6) of the matrix model. These frame-like indices will be raised and lowered with $\delta_{\dot{a}\dot{b}}$ or $\eta_{\dot{a}\dot{b}}$. In particular, $E^{\dot{a}\mu}$ provides D vector fields on $\mathcal{M}$, which will play a role of a (generalized[2]) *frame* on $\mathcal{M}$.

9.1.1 Effective metric and frame

Now we establish the interpretation of $E^{\dot{a}\mu}$ as frame. As in any field theory, the effective metric governing some field or fluctuation mode is encoded in the kinetic term of the action. To be specific, consider the case of transversal fluctuations around a matrix background $\mathcal{M}$; the case of parallel fluctuations is very similar apart from some technical details. The typical kinetic term in Yang–Mills matrix models has the structure as follows:[3]

$$S[\phi] = \mathrm{Tr}([T^{\dot{a}}, \phi][T_{\dot{a}}, \phi]) \sim -\frac{1}{(2\pi)^n} \int \Omega \, \{T^{\dot{a}}, \phi\}\{T_{\dot{a}}, \phi\}$$

$$= -\frac{1}{(2\pi)^n} \int \Omega \, E_{\dot{a}}{}^{\mu} E^{\dot{a}\nu} \partial_{\mu}\phi \partial_{\nu}\phi \tag{9.1.6}$$

in the semi-classical regime, using $\mathrm{End}(\mathcal{H}) \cong \mathcal{C}(\mathcal{M})$ and (9.1.4). Here

$$\Omega = \rho_M d^{2n}x \tag{9.1.7}$$

is the symplectic volume form. Assuming $\dim \mathcal{M} = 2n > 2$, this can be written in the familiar form

$$S[\phi] \sim -\int \frac{d^{2n}x}{(2\pi)^n} \rho_M \gamma^{\mu\nu} \partial_{\mu}\phi \partial_{\nu}\phi = -\int \frac{d^{2n}x}{(2\pi)^n} \sqrt{|G_{\mu\nu}|} G^{\mu\nu} \partial_{\mu}\phi \partial_{\nu}\phi. \tag{9.1.8}$$

We can read off the **effective metric** on $\mathcal{M}^{(2n)}$, which is given by

$$\boxed{G^{\mu\nu} = \rho^{-2} \gamma^{\mu\nu}, \qquad \gamma^{\mu\nu} = E_{\dot{a}}{}^{\mu} E^{\dot{a}\nu}.} \tag{9.1.9}$$

Here the **dilaton** ρ relates the symplectic density ρ_M to the Riemannian density $\sqrt{|G_{\mu\nu}|}$ via

$$L_{\mathrm{NC}}^{-2n} = \rho_M = \rho^{-2}\sqrt{|G_{\mu\nu}|}, \tag{9.1.10}$$

---

[2] Here "generalized" indicates that it need not be a basis; it may be over complete or even have smaller rank. Such generalized frames were also introduced by J. Madore in the noncommutative setting [132], albeit with extra requirements that we will not impose.

[3] Note that the coupling constant g in (7.4.23) has been absorbed in $T^{\dot{a}}$ and ϕ. The present analysis is analogous to Section 7.4.3 but more structured.

where L_{NC} is the scale of noncommutativity (4.2.27). Using $\sqrt{|G_{\mu\nu}|} = \rho^{2n}\sqrt{|\gamma_{\mu\nu}|}$, we obtain

$$\rho^{2n-2} = \frac{\rho_M}{\sqrt{|\gamma_{\mu\nu}|}}, \tag{9.1.11}$$

which is a dimensionless scalar function given by the ratio of two densities. At this point, it makes sense to assign the standard dimension mass to the scalar field ϕ, which is defined w.r.t. the background. Note that this discussion does not apply in two dimensions due to Weyl invariance, since it is not possible in that case to rewrite the kinetic term in standard Riemannian form.

At any given point on $\mathcal{M}^{2n}$, we can assume – possibly after a rotation in target space – that the brane is embedded locally along the first $2n$ components of target space via the semi-classical map

$$T^{\dot{a}} : \mathcal{M} \hookrightarrow \mathbb{R}^D. \tag{9.1.12}$$

Then the transversal components of the frame vanish, and the frame index can be restricted to $\dot{a} = 1, \ldots, 2n$. Furthermore,

$$\sqrt{|\gamma^{\mu\nu}|} = |\det E^{\dot{a}\mu}|, \tag{9.1.13}$$

so that (9.1.11) determines the dilaton as

$$\rho^{2n-2} = \rho_M \,|\det E^{\dot{a}\mu}|. \tag{9.1.14}$$

By construction, the effective metric $G_{\mu\nu}$ governs the physics of fluctuations on the brane. It corresponds essentially to the open string metric on the brane, cf. Section 6.7. We do not need to assume that the embedding of the brane in target space is nondegenerate. For example, the covariant quantum spaces considered in Sections 5 and 10 are realized via degenerate embeddings, which do not see the internal S^2 fiber. Then the same formulas still hold in local coordinates on the effective or reduced space(time) $\mathcal{M}$, and ρ_M is the symplectic density reduced over the fiber. All the geometric considerations in the following sections will then go through, provided the functions and tensors are understood as higher-spin $\mathfrak{hs}$ valued functions (and tensors) on $\mathcal{M}$. This will be made explicit in Section 10.

For symplectic branes $\mathcal{M}^{2n} \subset \mathbb{R}^D$ with nondegenerate embedding, one can locally choose $2n$ tangential components of the embedding

$$T^{\dot{\mu}} \sim x^{\mu}, \qquad \mu = 0, \ldots, 2n - 1 \tag{9.1.15}$$

denoted as local **Cartesian embedding coordinates** x^{μ} on $\mathcal{M}$. Then the frame is given by the local Poisson tensor

$$E^{\dot{\mu}\nu} = \{x^{\mu}, x^{\nu}\} = \theta^{\mu\nu} \tag{9.1.16}$$

and

$$|\det E^{\dot{\mu}\nu}| = |\det \theta^{\mu\nu}| = \rho_M^{-2} = L_{NC}^{4n}. \tag{9.1.17}$$

Then $\rho^{2n-2} = \rho_M^{-1}$, which together with (9.1.10) gives

$$\sqrt{|G_{\mu\nu}|} = \rho^{4-2n}. \tag{9.1.18}$$

In particular, the effective metric for four-dimensional branes is unimodular with

$$|G_{\mu\nu}| = 1, \qquad \rho^2 = L_{NC}^4. \tag{9.1.19}$$

However, this entails a strange assignment of dimensions that will not be adopted in the following; moreover this does not work for covariant quantum spacetime $\mathcal{M}^{3,1}$ where the background is given by momentum generators, cf. Section 10.2.1. Nevertheless, it is a hint that the physical notion of a cosmological constant will not apply in the framework of matrix models.

9.1.2 Frame and tensor fields

Assume that $\gamma^{\mu\nu}$ is nondegenerate on $\mathcal{M}$. We can then define a coframe by lowering the local index with $\gamma_{\mu\nu}$:

$$E^{\dot{a}}_{\ \mu} := E^{\dot{a}\nu}\gamma_{\mu\nu}. \tag{9.1.20}$$

It is often convenient to consider the coframe as a one-form on $\mathcal{M}$

$$\theta^{\dot{a}} := E^{\dot{a}}_{\ \mu}dx^{\mu} = E^{\dot{a}\mu}\gamma_{\mu\nu}dx^{\nu}. \tag{9.1.21}$$

Notice that this definition makes sense even if $D \neq \dim\mathcal{M}$, notably if $\mathcal{M} \subset \mathbb{R}^D$ is a submanifold (a brane) in target space. This is quite common in matrix models, and indicated by the name "generalized frame." In any case, we obtain

$$E^{\dot{a}\mu}E_{\dot{a}\nu} = \delta^{\mu}_{\nu}, \tag{9.1.22}$$

while

$$P_T^{\dot{a}\dot{b}} := E^{\dot{a}}_{\ \mu}E^{\dot{b}\mu} \tag{9.1.23}$$

is a projector on the tangent space,

$$P_T^{\dot{a}\dot{b}}P_{T\dot{b}}^{\ \ \dot{c}} = P^{\dot{a}\dot{c}}, \qquad P_{T\ \dot{b}}^{\ \dot{a}}E^{\dot{b}\nu} = E^{\dot{a}\nu}. \tag{9.1.24}$$

If $D = 2n = \dim\mathcal{M}$ and the frame is invertible, the coframe $E^{\dot{a}}_{\ \mu}$ is nothing but the inverse frame, with

$$P_T^{\dot{a}\dot{b}} = \eta^{\dot{a}\dot{b}} \qquad \text{if } D = \dim\mathcal{M}, \tag{9.1.25}$$

Now consider $T^{\dot{a}\dot{b}\mu}$. To understand its geometrical significance, we rewrite it using the Jacobi identity as follows:

$$\begin{aligned}
T^{\dot{a}\dot{b}\mu} = \{\mathcal{F}^{ab}, x^{\mu}\} &= -\{T^{\dot{a}}, \{T^{\dot{b}}, x^{\mu}\}\} + \{T^{\dot{b}}, \{T^{\dot{a}}, x^{\mu}\}\} \\
&= -\{T^{\dot{a}}, E^{\dot{b}\mu}\} + \{T^{\dot{b}}, E^{\dot{a}\mu}\} \\
&= -E^{\dot{a}\nu}\partial_{\nu}E^{\dot{b}\mu} + E^{\dot{b}\nu}\partial_{\nu}E^{\dot{a}\mu} = -[E^{\dot{a}}, E^{\dot{b}}]^{\mu}.
\end{aligned} \tag{9.1.26}$$

We replace the frame indices with coordinate indices by contracting with $E_{\dot{a}\sigma}E_{\dot{b}\kappa}$. Then

$$T_{\sigma\kappa}{}^{\mu} = -E_{\dot{b}\kappa}\,\partial_{\sigma}E^{\dot{b}\mu} + E_{\dot{a}\sigma}\,\partial_{\kappa}E^{\dot{a}\mu} = E^{\dot{a}\mu}(\partial_{\sigma}E_{\dot{a}\kappa} - \partial_{\kappa}E_{\dot{a}\sigma})$$
$$T_{\sigma\kappa}{}^{\dot{a}} = \partial_{\sigma}E^{\dot{a}}{}_{\kappa} - \partial_{\kappa}E^{\dot{a}}{}_{\sigma} \qquad\qquad (9.1.27)$$

using (9.1.22) in the last step. This is more transparent in the language of differential forms:

$$T^{\dot{a}} = \frac{1}{2}T_{\mu\nu}{}^{\dot{a}}dx^{\mu}dx^{\nu} = d\theta^{\dot{a}}. \qquad\qquad (9.1.28)$$

Therefore, the tangential components of $T^{\dot{a}\dot{b}\mu}$ encode the exterior derivative of the coframe, and in a sense it replaces the spin connection of the Cartan formalism. We will see that in the matrix model framework, the frame contains more physical information than in the Cartan formalism, and $T_{\sigma\kappa}{}^{\mu}$ can be interpreted as torsion of the Weitzenböck connection associated with the frame.

9.1.3 Gauge transformations and (symplectic) diffeomorphisms

By construction, $E^{\dot{a}\mu}$ and $T^{\dot{a}\dot{b}\mu}$ are rank 1 tensor fields on $\mathcal{M}$. As such, they satisfy the standard transformation laws under a change of coordinates $y^{\mu} = y^{\mu}(x)$:

$$E^{\dot{a}\mu}(y) = \{T^{\dot{a}}, y^{\mu}\} = \{T^{\dot{a}}, x^{\nu}\}\frac{\partial y^{\mu}}{\partial x^{\nu}} = E^{\dot{a}\nu}(x)\frac{\partial y^{\mu}}{\partial x^{\nu}}. \qquad (9.1.29)$$

Now we ask how they transform under a gauge transformation of the background

$$\delta_{\Lambda}T = \{\Lambda, T\} \qquad\qquad (9.1.30)$$

corresponding to the vector field $\xi = \{\Lambda, .\}$ (6.5.4). Then any Hamiltonian vector field $E = \{T, .\}$ on $\mathcal{M}$ transforms as

$$\begin{aligned}(\delta_{\Lambda}E)[\phi] &= \{\{\Lambda, T\}, \phi\} = \{\Lambda, \{T, \phi\}\} - \{T, \{\Lambda, \phi\}\}\\ &= \xi[E[\phi]] - E[\xi[\phi]]\\ &= [\xi, E][\phi] \end{aligned} \qquad (9.1.31)$$

for any $\phi \in \mathcal{C}(\mathcal{M})$, where $[\xi, E]$ is the Lie bracket of the vector fields ξ and E. Recalling the standard identity $\mathcal{L}_{\xi}E = [\xi, E]$, this is precisely the Lie derivative of the vector field E along ξ:

$$\delta_{\Lambda}E = \mathcal{L}_{\xi}E. \qquad\qquad (9.1.32)$$

Therefore, the frame transforms covariantly (i.e. as the Lie derivative) under the diffeomorphism generated by ξ. We also recall that the Lie derivative of the Poisson bivector field along any Hamiltonian vector field ξ on $\mathcal{M}$ vanishes $\mathcal{L}_{\xi}\{.,.\} = 0$ due to the Jacobi identity, so that δ_{Λ} is a symplectomorphism. The explicit transformation law of the vector field in local coordinates x^{μ} on $\mathcal{M}$ is obtained by setting $\phi = x^{\mu}$:

$$\delta_{\Lambda}E^{\mu} = \xi^{\rho}\partial_{\rho}E^{\mu} - E^{\rho}\partial_{\rho}\xi^{\mu} = \mathcal{L}_{\xi}E^{\mu}. \qquad (9.1.33)$$

This is the standard form of the Lie derivative of the vector field E^{μ} along the vector field ξ^{ν}.

These considerations extend immediately to higher-rank tensor fields as

$$\delta_\Lambda (E \otimes E') = (\mathcal{L}_\xi E) \otimes E' + E \otimes \mathcal{L}_\xi E' = \mathcal{L}_\xi (E \otimes E'). \tag{9.1.34}$$

In particular, the metric tensor $\gamma^{\mu\nu}$ transforms as follows:

$$\delta_\Lambda \gamma^{\mu\nu} \sim \mathcal{L}_\xi \gamma^{\mu\nu} = -\nabla^\mu_{(\gamma)} \xi^\nu - \nabla^\nu_{(\gamma)} \xi^\mu, \tag{9.1.35}$$

where the last form is a standard identity[4] for the Levi–Civita connection $\nabla_{(\gamma)}$ corresponding to $\gamma_{\mu\nu}$. Using the Leibniz rule

$$\mathcal{L}_\xi [\phi E] = \phi \mathcal{L}_\xi [E] + \mathcal{L}_\xi [\phi] E, \tag{9.1.36}$$

this generalizes to the effective metric $G^{\mu\nu} = \rho^{-2} \gamma^{\mu\nu}$ (9.1.9):

$$\delta_\Lambda G^{\mu\nu} \sim \mathcal{L}_\xi G^{\mu\nu} = -\nabla^\mu_{(G)} \xi^\nu - \nabla^\nu_{(G)} \xi^\mu, \tag{9.1.37}$$

where $\nabla_{(G)}$ is the Levi–Civita connection corresponding to $G_{\mu\nu}$. This is familiar from general relativity. As shown in Section 5.4.5, such gauge transformations can indeed realize the most general volume-preserving diffeomorphisms on the covariant spacetime $\mathcal{M}^{3,1}$, due to the internal space S^2.

Using these results, we can also define tensorial versions of the matrix field strength $\mathcal{F}^{\dot{a}\dot{b}}$ and the matrix-energy-momentum tensor $\mathcal{T}^{\dot{a}\dot{b}}$ (7.4.14) by contracting with the vielbein:

$$\mathcal{F}_{\mu\nu} := \mathcal{F}^{\dot{a}\dot{b}} E_{\dot{a}\mu} E_{\dot{b}\nu}, \tag{9.1.38}$$

$$\mathcal{T}_{\mu\nu} := \mathcal{T}^{\dot{a}\dot{b}} E_{\dot{a}\mu} E_{\dot{b}\nu}. \tag{9.1.39}$$

Since $\mathcal{T}^{\dot{a}\dot{b}}$ and $\mathcal{F}^{\dot{a}\dot{b}}$ transform as scalars $\mathcal{L}_\xi \mathcal{T}^{\dot{a}\dot{b}} = \{\Lambda, \mathcal{T}^{\dot{a}\dot{b}}\} = \xi^\mu \partial_\mu \mathcal{T}^{\dot{a}\dot{b}}$, the $\mathcal{F}_{\mu\nu}$ and $\mathcal{T}_{\mu\nu}$ transform as tensors under the diffeomorphisms generated by the Hamiltonian vector fields $\xi = \{\Lambda, .\}$:

$$\delta_\Lambda \mathcal{F}_{\mu\nu} = \mathcal{L}_\xi \mathcal{F}_{\mu\nu},$$

$$\delta_\xi \mathcal{T}_{\mu\nu} = \mathcal{L}_\xi \mathcal{T}_{\mu\nu}. \tag{9.1.40}$$

9.1.4 Noncommutative gauge fields and field strength

It is instructive to relate the geometric considerations in the previous paragraph to the noncommutative gauge theory point of view. Adding fluctuations $T^a = \bar{T}^a + \mathcal{A}^a$ to the background (dropping the dots) and rewriting the fluctuations $\mathcal{A}^a$ in the form

$$\mathcal{A}^a = E^{a\mu} A_\mu \tag{9.1.41}$$

generalizing (6.6.6), the deformed NC field strength takes the form

$$\begin{aligned}
\mathcal{F}^{ab} &= -\{\bar{T}^a + \mathcal{A}^a, \bar{T}^b + \mathcal{A}^b\} \\
&= \bar{\mathcal{F}}^{ab} - E^{a\mu} E^{b\nu} (\partial_\mu A_\nu - \partial_\nu A_\mu) - E^{a\mu} (\partial_\mu E^{b\nu}) A_\nu + E^{b\mu} (\partial_\mu E^{a\nu}) A_\nu - \{\mathcal{A}^a, \mathcal{A}^b\} \\
&= \bar{\mathcal{F}}^{ab} - E^{a\mu} E^{b\nu} F_{\mu\nu} + \{\bar{\mathcal{F}}^{ab}, x^\nu\} A_\nu + O(\mathcal{A}^2) \tag{9.1.42}
\end{aligned}$$

---

[4] This follows from (9.1.33) in Riemann normal coordinates at any given point $p \in \mathcal{M}$, noting that $\partial|_p \gamma^{\mu\nu} = 0$.

using (9.1.26), where $F_{\mu\nu} = \partial_\mu A_\nu - \partial_\nu A_\mu + \{A_\mu, A_\nu\}$. In particular, the tensorial form of the NC field strength (9.1.38) is obtained as

$$\mathcal{F}_{\mu\nu} = \bar{\theta}_{\mu\nu}^{-1} - F_{\mu\nu} + E_{a\mu}E_{b\nu}\{\bar{\mathcal{F}}^{ab}, x^\nu\}A_\nu + O(A^2), \tag{9.1.43}$$

which includes the contribution $\bar{\theta}_{\mu\nu}^{-1}$ of the background, cf. (9.1.56). For $\mathcal{A}$ arising from a gauge transformation of the background

$$\mathcal{A}_\Lambda^a = \delta_\Lambda T^a = \{\Lambda, T^a\} = -E^{a\mu}\partial_\mu\Lambda, \tag{9.1.44}$$

we have $A_\mu = -\partial_\mu\Lambda$, and (9.1.42) reduces to

$$\mathcal{F}^{ab} = \bar{\mathcal{F}}^{ab} - \{\bar{\mathcal{F}}^{ab}, x^\nu\}\partial_\nu\Lambda = \bar{\mathcal{F}}^{ab} + \{\Lambda, \bar{\mathcal{F}}^{ab}\}. \tag{9.1.45}$$

We recover precisely $\delta_\Lambda \mathcal{F}^{ab} = \mathcal{L}_\xi \mathcal{F}^{ab}$ (7.4.37) for $\xi = \{\Lambda, .\}$.

The general relation (9.1.42) can be written more transparently as

$$\boxed{\delta_\mathcal{A} \mathcal{F}^{ab} = \mathcal{L}_{\tilde{A}} \mathcal{F}^{ab} - E^{a\mu}E^{b\nu}F_{\mu\nu},} \tag{9.1.46}$$

where

$$\mathcal{L}_{\tilde{A}} \mathcal{F}^{ab} = \tilde{A}^\mu \partial_\mu \mathcal{F}^{ab} = \{\mathcal{F}^{ab}, x^\nu\}A_\nu, \qquad \tilde{A}^\mu = \theta^{\mu\nu}A_\nu. \tag{9.1.47}$$

This can be interpreted as a frame version of the infinitesimal Moser Lemma (1.1.17), with $\tilde{A}$ determining the flow of the symplectic structure. It states that the fluctuation $\mathcal{A}$ not only modifies the symplectic structure by F, but implies also a flow. Together with (9.1.41), this provides a relation between noncommutative gauge fields $\mathcal{A}^a$ and ordinary gauge fields A_μ in the semi-classical, linearized regime. That relation can be extended to finite maps defined by $\mathcal{A}^a$ and finite commutative gauge fields A_μ, where gauge transformations correspond to symplectic maps. By lifting this relation to the noncommutative case in the star product formulation, one obtains the Seiberg–Witten map [170]. This allows to rewrite a noncommutative Yang–Mills gauge theory in terms of a (complicated) commutative gauge theory for abelian or nonabelian gauge fields [110], and to separate $SU(n)$ gauge fields from the geometric trace-$U(1)$ gauge fields [178, 179, 180]. However, the matrix model formulation is much better suited to understand the quantization of NC gauge theory.

9.1.5 Geometric YM action on symplectic branes

Consider an irreducible brane given in terms of some matrix background T^a, which in the semi-classical regime defines a nondegenerate embedding

$$T^a : \quad \mathcal{M} \hookrightarrow \mathbb{R}^D \tag{9.1.48}$$

dropping the dots on the indices for better readability. Viewing the $T^a = T^a(y)$ as functions on $\mathcal{M}$ in local coordinates y^μ, we can compute their Poisson brackets as

$$\{T^a, T^b\} = \{T^a, y^\mu\}\partial_\mu T^b = E^{a\mu}\partial_\mu T^b$$
$$\{T^a, T^b\}\{T_a, T^c\} = \gamma^{\mu\nu}\partial_\mu T^b \partial_\nu T^c. \tag{9.1.49}$$

Now consider the embedding or induced metric on the brane $\mathcal{M} \subset \mathbb{R}^D$

$$\hat{g}_{\mu\nu} = \partial_\mu T^a \partial_\nu T_a, \tag{9.1.50}$$

which would be denoted as closed string metric in string theory language[5]. Writing the frame in terms of the Poisson tensor on $\mathcal{M}$ as

$$E^{a\mu} = \{T^a, y^\mu\} = -\theta^{\mu\sigma} \partial_\sigma T^a \tag{9.1.51}$$

or $\partial_\sigma T^a = -\theta^{-1}_{\sigma\mu} E^{a\mu}$, the induced metric can be written alternatively as

$$\hat{g}_{\mu\nu} = \partial_\mu T^a \partial_\nu T_a = \gamma^{\sigma\kappa} \theta^{-1}_{\sigma\mu} \theta^{-1}_{\kappa\nu}. \tag{9.1.52}$$

Then the bosonic action can be written in the semi-classical regime for $m^2 = 0$ as

$$S = \frac{1}{g^2} \text{Tr}[T^a, T^c][T_a, T_c] \sim -\frac{1}{g^2} \int_{\mathcal{M}} \frac{\Omega}{(2\pi)^n} \mathcal{L}, \tag{9.1.53}$$

where

$$\mathcal{L} = \{T^a, T^b\}\{T_a, T_b\} = \gamma^{\mu\nu} \hat{g}_{\mu\nu} = \gamma^{\mu\nu} \theta^{-1}_{\mu\rho} \theta^{-1}_{\nu\sigma} \gamma^{\rho\sigma} = \theta^{\mu\mu'} \theta^{\nu\nu'} \hat{g}_{\mu\nu} \hat{g}_{\mu'\nu'}. \tag{9.1.54}$$

This action governs the Poisson or symplectic structure and the embedding of $\mathcal{M} \subset \mathbb{R}^D$. Similarly, the geometric energy–momentum (e–m) tensor (7.4.14) can be written in the form

$$\mathcal{T}_{\mu\nu} = \mathcal{T}^{ab} E_{a\mu} E_{b\nu} = \hat{g}_{\mu\nu} - \frac{1}{4} \gamma_{\mu\nu} (\gamma^{\sigma\kappa} \hat{g}_{\sigma\kappa}), \tag{9.1.55}$$

which is recognized as the traceless part of the induced metric $\hat{g}_{\mu\nu}$. Finally, it follows that the abelian field strength tensor (9.1.38)

$$\mathcal{F}_{\mu\nu} = -\{T^a, T^b\} E_{a\mu} E_{b\nu} = -E^{a\mu} \partial_\mu T^b E_{b\nu} = \theta^{-1}_{\mu\nu} \tag{9.1.56}$$

is given by the symplectic form, which transforms trivially under the transformations (9.1.40)

$$\delta_\Lambda \mathcal{F}_{\mu\nu} = \mathcal{L}_\xi \mathcal{F}_{\mu\nu} = \mathcal{L}_\xi \theta^{-1}_{\mu\nu} = 0 \tag{9.1.57}$$

because $\xi^\mu = \{\Lambda, .\}$ is a symplectic vector field (1.1.27).

Fluctuations of the geometry

Now consider fluctuations of the brane governed by the semi-classical action (9.1.54). The degrees of freedom are given by the embedding of the brane as well as the symplectic form or the Poisson structure $\theta^{\mu\nu}$. More specifically, transversal brane fluctuations are encoded in $D - 2n$ transversal scalar fields, while the symplectic structure can be written locally as $\omega = dA$, which encodes $2n - 1$ degrees of freedom since $A = d\Lambda$ drops out. This matches with the $D - 1$ degrees of freedom of the matrix model on the background T^a modulo gauge transformations.

[5] $\hat{g}_{\mu\nu}$ should not be confused with the quantum metric $g_{\mu\nu}$ (4.2.19).

These degrees of freedom of the background can be parametrized as follows. Near any given point on $\mathcal{M}$ and after a suitable $SO(D-1,1)$ rotation, we can separate the matrix indices into $\mu = 0, \ldots, 2n-1$ tangential indices and $D-2n$ transversal indices i, such that $T^\mu =: X^\mu \sim x^\mu$ serves as Cartesian embedding coordinates (9.1.15) on $\mathcal{M}$. Then fluctuations of the symplectic form can be parametrized as

$$\delta\theta^{-1}_{\mu\nu} = \partial_\mu A_\nu - \partial_\nu A_\mu, \tag{9.1.58}$$

while the transversal fluctuations of the brane are simply captured by $D-2n$ scalar fields $\delta\phi^i$. Dropping an overall constant, the variation of the effective action (9.1.54) for $m^2 = 0$ is then

$$
\begin{aligned}
\delta S &= \int d^{2n}x \sqrt{|\theta^{-1}_{\mu\nu}|} \Big(\delta\mathcal{L}(x) + \frac{1}{2}\mathcal{L}(x)\,\theta^{\mu\nu}\delta\theta^{-1}_{\nu\mu} \Big) \\
&= \frac{1}{2} \int d^{2n}x \sqrt{|\theta^{-1}_{\mu\nu}|} \Big(\hat{g}_{\mu\nu}\theta^{\mu\mu'}\delta\theta^{\nu\nu'}\hat{g}_{\mu'\nu'} + \hat{g}_{\mu\nu}\theta^{\mu\mu'}\theta^{\nu\nu'}\delta\hat{g}_{\mu'\nu'} + \mathcal{L}(x)(\theta^{\mu\nu}\delta\theta^{-1}_{\nu\mu}) \Big) \\
&= \frac{1}{2} \int d^{2n}x \sqrt{|\theta^{-1}_{\mu\nu}|} \Big(\rho^4 G^{\eta\mu}\theta^{-1}_{\mu\nu}G^{\nu\rho}\delta\theta^{-1}_{\rho\eta} + \rho^2 G^{\mu\nu}\delta\hat{g}_{\mu\nu} + \mathcal{L}(x)(\theta^{\mu\nu}\delta\theta^{-1}_{\nu\mu}) \Big) \\
&= \int d^{2n}x \sqrt{|G|} \Big(\rho^2 G^{\eta\mu}G^{\nu\rho}\theta^{-1}_{\mu\nu}\partial_\rho A_\eta - \rho^{-2}\mathcal{L}\theta^{\rho\eta}\partial_\rho A_\eta + G^{\mu\nu}\partial_\mu\phi^i\partial_\nu\delta\phi^i\,\delta_{ij} \Big).
\end{aligned}
\tag{9.1.59}
$$

Recalling that $\partial_\mu A_\nu - \partial_\nu A_\mu = \nabla_\mu A_\nu - \nabla_\nu A_\mu$ where ∇ is the Levi–Civita connection associated with the effective metric G and using partial integration, we obtain

$$
\begin{aligned}
\delta S &= -\int d^{2n}x \sqrt{|G|}\, A_\eta \Big(G^{\eta\mu}\nabla^\nu(\rho^2\theta^{-1}_{\mu\nu}) - \nabla_\rho(\rho^{-2}\mathcal{L}\theta^{\rho\eta}) \Big) + \delta\phi^i\,\delta_{ij}\partial_\mu\Big(\sqrt{|G|}\,\partial^\mu\phi^j\Big) \\
&= -\int d^{2n}x \sqrt{|G|} \Big(A_\eta \Big(G^{\eta\mu}\nabla^\nu(\rho^2\theta^{-1}_{\mu\nu}) - \rho^{-2}\theta^{\rho\eta}\,\partial_\rho\mathcal{L} \Big) + \delta\phi^i\,\delta_{ij}\Box_G\phi^j \Big)
\end{aligned}
$$

recalling that $0 = \partial_\mu(\rho_M\,\theta^{\mu\nu}) = \sqrt{|G|}\,\nabla_\mu(\rho^{-2}\theta^{\mu\nu})$ (1.1.28) in the last step. This leads to the covariant equations of motion

$$
\begin{aligned}
\Box_G\phi^i &= 0 \\
\nabla^\mu(\rho^2\theta^{-1}_{\mu\nu}) &= \rho^{-2}G_{\rho\nu}\,\theta^{\rho\mu}\partial_\mu\mathcal{L}
\end{aligned}
\tag{9.1.60}
$$

which govern the geometry of semi-classical branes in the matrix model. The first equation of course applies to all components of the background, while the second equation governs the symplectic structure of the brane. The latter can also be obtained from the matrix conservation law (7.4.15) in the semi-classical regime [179], which suggests that it can be trusted at the quantum level, as long as the geometry is well-defined. These combined equations can be subsumed in the form

$$\Box_G T^a = 0 \tag{9.1.61}$$

for all a, which means that the embedding is harmonic w.r.t. the effective metric.

Reconstruction of symplectic branes

The results in the last paragraph can be used to address the following reconstruction problem: Can we realize some given symplectic manifold $(\mathcal{M}, \omega)$ as a solution of the matrix model in the semi-classical limit? According to our results, this amounts to finding some embedding functions $x^a : \mathcal{M} \hookrightarrow \mathbb{R}^D$ that are harmonic in the sense of (9.1.61). Restricting ourselves first to flat embeddings along some plane $\mathbb{R}^{2n}$, one needs to find harmonic coordinates x^μ in the sense

$$\Box_G x^\mu = 0 \tag{9.1.62}$$

or equivalently $\eta_{\nu\sigma}\{x^\nu, \{x^\sigma, x^\mu\}\} = 0$. Then $\mathcal{M}$ is recovered in terms of a solution x^μ of the semi-classical model, and the effective metric G can be computed. Moreover, we can quantize x^μ to obtain a set of matrices

$$X^\mu := \mathcal{Q}(x^\mu), \tag{9.1.63}$$

which will be approximate solutions of the matrix model. If we admit embedding in target space $\mathbb{R}^D$ with extrinsic curvature, there may be many different solutions. By exploiting such nontrivial embeddings for sufficiently high D, it is possible to recover any given metric $G^{\mu\nu}$ on $\mathcal{M}$, at least in some neighborhood.

Exercise 9.1.1   Show that for any four-dimensional symplectic brane, the tensor

$$J^\mu_\nu = \theta^{\mu\sigma} \hat{g}_{\sigma\nu} \tag{9.1.64}$$

satisfies a quartic characteristic equation

$$J^4 - \frac{1}{2} tr(J^2) J^2 \pm \rho^4 \mathbb{1} = 0, \tag{9.1.65}$$

where $+$ applies to Euclidean signature and $-$ to Minkowski signature.
Hint: transform $\theta^{\mu\nu}$ to block-diagonal form, and consider the eigenvalues of $\rho^{-2} J^2$.

9.1.6 Nonabelian gauge theory on curved symplectic branes

As discussed in section 7.4.5, nonabelian NC gauge theory arises from fluctuations on stacks of identical branes $\mathcal{M}$ in Yang–Mills matrix models. This generalizes to branes with arbitrary shape and structure, as long as they are irreducible quantum spaces. To see this, consider a stack of identical generic but nondegenerate branes realized by the matrix background $\bar{T}^a \otimes \mathbb{1}_k$, and add fluctuations

$$T^a = \bar{T}^a \otimes \mathbb{1}_k + \mathcal{A}^a_\alpha(y)\lambda^\alpha. \tag{9.1.66}$$

Since we do not want to change the geometry, the perturbations $\mathcal{A}^a$ take values in $\mathfrak{su}(k) \ni \lambda^\alpha$; the trace-$\mathfrak{u}(1)$ fluctuations would change the geometry as discussed above. A priori, all the $\mathcal{A}^a_\alpha(y)$ are scalar fields on $\mathcal{M}$, since the indices are global. Around any given point on $\mathcal{M}$, we can separate them into transversal and tangential modes using the tangential projector P^{ab}_T (9.1.23) and its orthogonal complement

$$P^{ab}_\perp := \eta^{ab} - P^{ab}_T. \tag{9.1.67}$$

Hence, for almost-commutative backgrounds, the background can be approximated locally by a stack of Moyal–Weyl quantum plane $\mathbb{R}^{2n}_\theta$ as in (7.4.25). Following the discussion in Section 7.4.5, we conclude that the tangential fluctuations describe $\mathfrak{su}(k)$-valued Yang–Mills gauge fields coupled to $D - 2n$ transversal $\mathfrak{su}(k)$-valued scalar fields ϕ^i, governed by the action (7.4.30) for $m^2 = 0$. Writing the $\mathfrak{u}(1)$ sector in terms of the effective metric as

$$\mathcal{L}_0 = \rho^2 G^{\mu\nu} \hat{g}_{\mu\nu} = \gamma^{\mu\nu} \hat{g}_{\mu\nu}, \tag{9.1.68}$$

the action becomes

$$S = \frac{1}{g^2} \int_{\mathcal{M}} \frac{d^{2n}y}{(2\pi)^n} \sqrt{|G|}\, \mathrm{tr}\Big(-\rho^{-2}\mathcal{L}_0 - \rho^2 G^{\mu\mu'} G^{\nu\nu'} F_{\mu\nu} F_{\mu'\nu'}$$
$$- 2G^{\mu\nu} D_\mu \phi^i D_\nu \phi^j \delta_{ij} + \rho^{-2}[\phi^i, \phi^j][\phi^{i'}, \phi^{j'}]\delta_{ii'}\delta_{jj'} \Big) + S_{\mathrm{CS}}. \tag{9.1.69}$$

Here F is the usual field strength of $\mathfrak{su}(k)$-valued gauge fields, and S_{CS} denotes an extra contribution that vanishes in the flat case $\mathbb{R}^{2n}_\theta$ and is usually dropped. This term arises from the $\mathcal{L}_{\tilde{A}}\mathcal{F}^{ab}$ contribution in (9.1.46), and can be obtained by carefully rewriting the noncommutative gauge fields $\mathcal{A}$ in terms of commutative gauge fields A using the second-order Seiberg–Witten map [178, 181]. An alternative derivation [179] is based on the matrix conservation law (7.4.15). One then obtains the following "would-be topological" Chern–Simons-like action

$$S_{\mathrm{CS}} = -\frac{1}{g^2} \int \frac{d^{2n}x}{(2\pi)^n} \sqrt{|\theta^{-1}_{\mu\nu}|}\, \mathrm{tr}\Big(F_{\mu'\nu'} F_{\mu\nu} \hat{\theta}^{\mu'\nu'} \theta^{\mu\nu} + 2F_{\mu'\mu} F_{\nu'\nu} \hat{\theta}^{\mu'\nu'} \theta^{\nu\mu}$$
$$- \frac{1}{2}\rho^{-2}\mathcal{L}_0\, \theta^{\mu\nu}\theta^{\mu'\nu'}\big(F_{\mu\nu}F_{\mu'\nu'} + 2F_{\mu\nu'}F_{\nu\mu'} \big) \Big)$$
$$\stackrel{n=2}{=} \frac{1}{2(2\pi)^n g^2} \int \mathcal{L}_0\, \mathrm{tr} F \wedge F, \tag{9.1.70}$$

where the last line holds for four-dimensional branes, and

$$\hat{\theta}^{\nu\eta} = G^{\nu\rho} g_{\rho\mu}(y)\theta^{\mu\eta} = -\rho^2\, G^{\nu\rho}\theta^{-1}_{\rho\mu} G^{\mu\eta} = -\hat{\theta}^{\eta\nu}. \tag{9.1.71}$$

It is appropriate here to discuss the dimensions. All the objects $\theta^{\mu\nu}, G_{\mu\nu}, \rho, \phi$ are defined in terms of the matrix variable T^a, which naturally has dimension length for the nondegenerate branes under consideration[6]; in particular, $\dim(g^2) = L^4$. If T^a describes a four-dimensional brane, then $\dim(\rho^2) = L^4$, and $G_{\mu\nu}$ is dimensionless. Then the dimensionless physical Yang–Mills coupling constant is given by (cf. (6.6.21))

$$\boxed{g^2_{\mathrm{YM}} = \frac{g^2}{\rho^2}.} \tag{9.1.72}$$

The same coupling constant governs the scalar fields upon rescaling

$$\phi^i = g\tilde{\phi}^i, \qquad \dim \tilde{\phi}^i = L^{-1}. \tag{9.1.73}$$

---

[6] On the covariant spacetime $\mathcal{M}^{3,1}$ considered in Sections 5.4 and 10, the matrices have dimension L^{-1}, and the assignments change accordingly. To avoid confusion, g will often be absorbed in the background.

In particular, it depends on the background whether the gauge theory governing the fluctuations is weakly or strongly coupled. On the covariant quantum space $\mathcal{M}^{3,1}$, the assignment of dimensions will be somewhat modified, but the conclusion is the same.

In principle, higher-order corrections in $\theta^{\mu\nu}$ to this semi-classical action could be obtained using Seiberg–Witten techniques, which allows to extend (9.1.40) to the noncommutative setting. This also allows to explicitly separate the $\mathfrak{su}(k)$ gauge fields from the geometric trace-$\mathfrak{u}(1)$ gauge fields. More details can be found in [178, 179, 180].

9.2 Weitzenböck connection, torsion, and Laplacian

A central part of any theory of gravity is a metric-compatible connection. The standard formulation of general relativity is based on the Levi–Civita connection, which is the unique torsion-free metric compatible connection. It turns out that in the present framework, it is more natural to work with another metric-compatible connection known as Weitzenböck connection. This is adapted to the preferred frame given by the matrix background as discussed before, and naturally encodes the matrix field strength that plays a central role in the matrix model.

The Weitzenböck connection[7] on $\mathcal{M}$ is defined such that the frame is covariantly constant (sometimes called parallel), $\nabla^{(W)}E_{\dot{a}} = 0$. More explicitly,

$$0 = \nabla_\nu^{(W)}E_{\dot{a}}{}^\mu = \partial_\nu E_{\dot{a}}{}^\mu + \Gamma_{\nu\rho}{}^\mu E_{\dot{a}}{}^\rho \tag{9.2.1}$$

or equivalently

$$0 = \nabla_{\dot{c}}^{(W)}E_{\dot{a}}{}^\mu = E_{\dot{c}}[E_{\dot{a}}{}^\mu] + \Gamma_{\dot{c}\nu}{}^\mu E_{\dot{a}}{}^\nu \tag{9.2.2}$$

where $\nabla_{\dot{c}} = \nabla_{E_{\dot{c}}}$. Note that there is no spin connection acting on the frame indices $\dot{a}$, which is sometimes indicated by denoting $E_{\dot{b}}$ as "inertial frame." If the frame is invertible, this is solved by the following connection coefficients

$$\Gamma_{\nu\sigma}{}^\mu = -E^{\dot{a}}{}_\sigma \partial_\nu E_{\dot{a}}{}^\mu = E_{\dot{a}}{}^\mu \partial_\nu E^{\dot{a}}{}_\sigma. \tag{9.2.3}$$

For any vector field V^μ on $\mathcal{M}$, the Weitzenböck covariant derivative is then

$$\nabla_\mu^{(W)}V^\nu = \partial_\mu V^\nu + \Gamma_{\mu\rho}{}^\nu V^\rho. \tag{9.2.4}$$

If the frame is not invertible, the Weitzenböck covariant derivative should be defined via

$$0 = \nabla_\nu^{(W)}E_{\dot{a}}{}^\mu := P_{\dot{a}}^{\dot{b}} \partial_\nu E_{\dot{b}}{}^\mu + \Gamma_{\nu\rho}{}^\mu E_{\dot{a}}{}^\rho. \tag{9.2.5}$$

Here $P_{\dot{a}}^{\dot{b}}$ is the tangential projector (9.1.23), which makes sure that $\nabla^{(W)}$ takes values in the tangent space. Then $\Gamma_{\nu\rho}{}^\mu$ is again given by (9.2.3), so that all the following considerations will essentially go through. For simplicity, this projector will be dropped for branes with trivial embedding such as the covariant spacetimes $\mathcal{M}^{3,1}$.

---

[7] The Weitzenböck connection is used in teleparallel formulation of gravity, which is essentially equivalent to Einstein gravity, see e.g. [2].

The Weitzenböck connection is always flat and metric compatible

$$R[X,Y]E_{\dot{a}} = \nabla_X^{(W)}\nabla_Y^{(W)}E_{\dot{a}} - \nabla_Y^{(W)}\nabla_X^{(W)}E_{\dot{a}} - \nabla_{[X,Y]}^{(W)}E_{\dot{a}} = 0$$

$$\nabla^{(W)}\gamma_{\mu\nu} = 0 \tag{9.2.6}$$

since the frame is parallel $\nabla^{(W)}E_{\dot{a}} = 0$. However, it typically has nonvanishing torsion,

$$T[X,Y] = \nabla_X^{(W)}Y - \nabla_Y^{(W)}X - [X,Y], \tag{9.2.7}$$

which is given by the antisymmetric part of the connection coefficients

$$T_{\mu\nu}{}^{\rho} = \Gamma_{\mu\nu}{}^{\rho} - \Gamma_{\nu\mu}{}^{\rho}$$

$$T_{\mu\nu}{}^{\dot{\alpha}} = T_{\mu\nu}{}^{\rho}E_{\rho}^{\dot{\alpha}} = \partial_\mu E^{\dot{\alpha}}{}_\nu - \partial_\nu E^{\dot{\alpha}}{}_\mu. \tag{9.2.8}$$

Comparing with (9.1.27) and rewriting the indices via the frame, this coincides with the tensor defined in the previous section in terms of Poisson brackets (9.1.5)

$$T_{\dot{\alpha}\dot{\beta}}{}^{\mu} = \{\mathcal{F}_{\dot{\alpha}\dot{\beta}}, x^\mu\}, \tag{9.2.9}$$

which is recognized as torsion of the Weitzenböck connection. Hence, the Weitzenböck connection provides a geometric interpretation of $\mathcal{F}_{\dot{\alpha}\dot{\beta}}$, which plays a fundamental role in the matrix model, and was interpreted as field strength from the NC gauge theory point of view in Section 7.4. This insight will lead to a more geometric understanding of the equations of motion of the matrix model.

Covariant conservation laws

Using these tools, we can recast the tangential components of the semi-classical matrix conservation law $\{T_b, \mathcal{T}^{ab}\} = 0$ (7.4.15) for the matrix energy–momentum tensor into a more geometric form. In the semi-classical regime, the conservation law for the tensors $\mathcal{T}_{\mu\nu}$ (9.1.39) can be written in terms of the Weitzenböck connection for the frame E as

$$\begin{aligned}
\nabla_\mu \mathcal{T}^{\mu\nu} &= \nabla_\mu(\mathcal{T}^{ab}E_a{}^\mu E_b{}^\nu)\\
&= E_b{}^\nu E_a{}^\mu \nabla_\mu(\mathcal{T}^{ab}) = E_b{}^\nu\{T_a, \mathcal{T}^{ab}\}\\
&= 0.
\end{aligned} \tag{9.2.10}$$

Dropping the (W) label on ∇. Similarly, the tangential components of the semi-classical matrix eom $\{T_b, \mathcal{F}^{ab}\} = -m^2 T^a$ can be written as

$$\nabla_\mu \mathcal{F}^{\mu\nu} = E_b{}^\nu\{T_a, \mathcal{F}^{ab}\} = -m^2 E_b{}^\nu T^b. \tag{9.2.11}$$

The rhs has no clear geometric meaning and will vanish in the cases of interest below, since $m^2 \approx 0$ should be considered as an IR regulator.

Relation with the Levi–Civita connection

Now consider the Levi–Civita connection $\nabla^{(\gamma)}$ for the metric $\gamma^{\mu\nu}$, which is obtained from the standard formula

$$\Gamma^{(\gamma)}{}_{\mu\nu}{}^{\sigma} = \frac{1}{2}\gamma^{\sigma\rho}\left(\partial_{\mu}\gamma_{\rho\nu} + \partial_{\nu}\gamma_{\rho\mu} - \partial_{\rho}\gamma_{\mu\nu}\right)$$
$$= \frac{1}{2}\left(\Gamma_{\mu}{}^{\sigma}{}_{\nu} + \Gamma_{\mu\nu}{}^{\sigma} + \Gamma_{\nu}{}^{\sigma}{}_{\mu} + \Gamma_{\nu\mu}{}^{\sigma} - \Gamma^{\sigma}{}_{\mu\nu} - \Gamma^{\sigma}{}_{\nu\mu}\right)$$
$$= \Gamma_{\mu\nu}{}^{\sigma} - K_{\mu\nu}{}^{\sigma}. \tag{9.2.12}$$

using $\nabla^{(W)}\gamma = 0$. Here $\Gamma_{\mu\nu}{}^{\sigma}$ are the Weitzenböck connection coefficients (9.2.3), and

$$K_{\mu\nu}{}^{\sigma} = \frac{1}{2}(T_{\mu\nu}{}^{\sigma} + T^{\sigma}{}_{\mu\nu} - T_{\nu}{}^{\sigma}{}_{\mu}) = -K_{\mu}{}^{\sigma}{}_{\nu} \tag{9.2.13}$$

is the **contorsion** of the Weitzenböck connection, which is antisymmetric in $\nu\sigma$. This provides the following relation between the Levi–Civita and the Weitzenböck connection:

$$\Gamma_{\mu\nu}{}^{\rho} = \Gamma^{(\gamma)}{}_{\mu\nu}{}^{\rho} + K_{\mu\nu}{}^{\rho}$$
$$\nabla_{\mu}^{(W)}V^{\nu} = \nabla_{\mu}^{(\gamma)}V^{\nu} + K_{\mu\rho}{}^{\nu}V^{\rho}. \tag{9.2.14}$$

Analogous formulas will apply to the effective frame discussed in Section 9.1.1.

Relation to the Cartan formalism and spin connection

Given the coframe $\theta^{\dot{b}} = E^{\dot{b}}{}_{\mu}dx^{\mu}$, one can introduce the standard spin connection one-form $\omega_{\dot{a}\dot{b}} = \omega_{\mu\dot{a}\dot{b}}dx^{\mu} = -\omega_{\dot{b}\dot{a}}$, which satisfies the first Cartan structure equations

$$d\theta^{\dot{a}} = -\omega^{\dot{a}}{}_{\dot{b}} \wedge \theta^{\dot{b}}. \tag{9.2.15}$$

Taking into account (9.1.28), this equation becomes

$$\frac{1}{2}(\partial_{\mu}E_{\dot{a}\nu} - \partial_{\nu}E_{\dot{a}\mu})dx^{\mu} \wedge dx^{\nu} = -\omega_{\mu\dot{a}}{}^{\dot{b}}E_{\dot{b}\nu}dx^{\mu} \wedge dx^{\nu} = \frac{1}{2}T_{\mu\nu\dot{a}}dx^{\mu} \wedge dx^{\nu} \tag{9.2.16}$$

so that

$$T_{\mu\nu\dot{a}} = -\omega_{\mu\dot{a}}{}^{\dot{b}}E_{\dot{b}\nu} + \omega_{\nu\dot{a}}{}^{\dot{b}}E_{\dot{b}\mu} = \omega_{\nu\dot{a}\mu} - \omega_{\mu\dot{a}\nu}. \tag{9.2.17}$$

Performing a cyclic permutation of the indices, we obtain

$$\mathcal{K}_{\mu\nu\dot{\beta}} = \omega_{\mu\nu\dot{\beta}} \tag{9.2.18}$$

in the frame E. This relates the spin connection ω in the Cartan formulation of Riemannian geometry to the Weitzenböck (con)torsion, and both encode the same information. In contrast to the Cartan formalism, the "teleparallel" framework is adapted to the preferred frame of the matrix model, which allows no local Lorentz transformations. This is not a problem, since gauge symmetries such as local Lorentz transformations are unphysical. The frame indices $\dot{a}, \dot{b} = 0, \ldots, 3$ only transform under the global $SO(3,1)$ symmetry of the matrix model. Hence, there is no point in introducing a spin connection. In the present context, the frame is a physical object, which contains more information than just the metric. This is reflected in the divergence constraint (9.2.21) for the frame, which is a consequence of the underlying Poisson structure.

9.2.1 Divergence constraint

The fact that the frame $E_{\dot{a}} = \{T^{\dot{a}}, .\}$ and the torsion $T^{\dot{a}\dot{b}} = \{\mathcal{F}^{\dot{a}\dot{b}}, .\}$ are Hamiltonian vector fields has important consequences. Since they preserve the symplectic volume form $\Omega = \rho_M d^{2n}x$, it follows that they satisfy the divergence constraint (1.1.30)

$$\partial_\mu(\rho_M E^{\dot{a}\mu}) = 0$$

$$\partial_\mu(\rho_M T^{\dot{a}\dot{b}\mu}) = 0 \qquad (9.2.19)$$

in local coordinates on $\mathcal{M}$. This can be rewritten in terms of the standard identity

$$\frac{1}{\sqrt{G}}\partial_\mu(\sqrt{G}V^\mu) = \nabla_\mu^{(G)}V^\mu \qquad (9.2.20)$$

in Riemannian geometry, where $\nabla^{(G)}$ denotes the Levi–Civita connection for the metric $G_{\mu\nu}$. Together with $\rho^{-2}\sqrt{G} = \rho_M$ (9.1.10), the divergence constraint takes the form

$$\nabla_\mu^{(G)}(\rho^{-2}E^{\dot{a}\mu}) = 0 = \partial_\mu(\rho_M E^{\dot{a}\mu}) \qquad (9.2.21)$$

and similarly for the torsion. These constraints also apply on covariant quantum spaces, but this requires some further discussion in Section 10.2.2. In particular, there are no local Lorentz transformations acting on the frame index, unlike in the Cartan formulation of general relativity. We will see in the next section that for any given classical metric $\gamma_{\mu\nu}$, there is always a frame $E_{\dot{a}\nu}$ which satisfies the divergence constraint.

As an example, we observe that the frame $E^{\dot{a}\mu} = \sinh(\eta)\eta^{\dot{a}\mu}$ of the cosmological $k = -1$ spacetime $\mathcal{M}^{3,1}$ discussed in Chapter 5 is the unique[8] $SO(3, 1)$–invariant frame of the form $E^{\dot{a}\mu} = f(\eta)\eta^{\dot{a}\mu}$ which is compatible with (9.2.21) for $\rho_M = 1/\sinh(\eta)$ (5.4.26). Generic FLRW frames of this type can be realized for suitable ρ_M.

Gauge transformations and divergence constraint

As a consistency check, we verify that the divergence constraint is preserved by the diffeomorphisms that arise from gauge transformations in the matrix model. We have seen in Section 9.1.3 that the frame transforms under a diffeomorphism as

$$\delta_\Lambda E_{\dot{a}}{}^\mu = \mathcal{L}_\xi E_{\dot{a}}{}^\mu = \xi^\nu \partial_\nu E_{\dot{a}}{}^\mu - E_{\dot{a}}{}^\nu \partial_\nu \xi^\mu, \qquad (9.2.22)$$

where $\xi^\mu = \{\Lambda, x^\mu\}$ satisfies again the divergence constraint

$$\partial_\mu(\rho_M \xi^\mu) = 0. \qquad (9.2.23)$$

Hence, the divergence constraint transforms as

$$\begin{aligned}
\delta_\Lambda \partial_\mu(\rho_M E_{\dot{a}}{}^\mu) &= \partial_\mu(\rho_M \xi^\nu \partial_\nu E_{\dot{a}}{}^\mu) - \partial_\mu(\rho_M E_{\dot{a}}{}^\nu \partial_\nu \xi^\mu) \\
&= \partial_\mu \partial_\nu(\rho_M \xi^\nu E_{\dot{a}}{}^\mu) - \partial_\mu \partial_\nu(\rho_M E_{\dot{a}}{}^\nu \xi^\mu) \\
&\quad - \partial_\mu(\partial_\nu(\rho_M \xi^\nu)E_{\dot{a}}{}^\mu) \\
&= -\partial_\mu(\partial_\nu(\rho_M \xi^\nu)E_{\dot{a}}{}^\mu). \qquad (9.2.24)
\end{aligned}$$

[8] However, there might be other solutions with inequivalent ρ_M leading to a different geometry.

Since the vector field $\xi^\nu = \{\Lambda, x^\mu\}$ is divergence-free, it follows that the divergence constraint of the frame $E^{\dot\alpha \mu}$ is preserved.

9.2.2 Effective frame

Since the effective metric is related to $\gamma_{\mu\nu}$ by a conformal factor, it is natural to define an effective frame as follows:

$$\mathcal{E}_{\dot\alpha}{}^\mu = \rho^{-1} E_{\dot\alpha}{}^\mu \tag{9.2.25}$$

such that

$$G^{\mu\nu} = \eta^{\dot\alpha\dot\beta} \mathcal{E}_{\dot\alpha}{}^\mu \mathcal{E}_{\dot\beta}{}^\nu. \tag{9.2.26}$$

This gives rise to an associated Weitzenböck connection $\tilde\nabla^{(W)}$ that is compatible with the effective metric,

$$\tilde\nabla^{(W)} \mathcal{E}_{\dot\alpha}{}^\mu = 0 = \tilde\nabla^{(W)} G^{\mu\nu}. \tag{9.2.27}$$

One must be very careful with the frame indices, as there are now two different frames in the game. It is hence safer to use the coordinate form. Then it is easy to show

$$\tilde\Gamma_{\nu\sigma}{}^\mu = \Gamma_{\nu\sigma}{}^\mu + \delta^\mu_\sigma \rho^{-1} \partial_\nu \rho. \tag{9.2.28}$$

For the covariant derivatives, this amounts to the simple rule

$$\tilde\nabla^{(W)}_\mu V^\sigma = \nabla^{(W)}_\mu V^\sigma + \rho^{-1} \partial_\mu \rho \, V^\sigma \tag{9.2.29}$$

and similarly for higher-rank tensors. Then the torsion tensor is

$$\mathcal{T}_{\mu\nu}{}^\sigma = \tilde\Gamma_{\mu\nu}{}^\sigma - \tilde\Gamma_{\nu\mu}{}^\sigma = T_{\mu\nu}{}^\sigma + \rho^{-1} \left(\delta^\sigma_\nu \partial_\mu \rho - \delta^\sigma_\mu \partial_\nu \rho \right), \tag{9.2.30}$$

and the effective contorsion is related to that of the basic frame as follows:

$$\mathcal{K}_{\mu\nu\sigma} = K_{\mu\nu\sigma} + \rho^{-1} \left(G_{\mu\nu} \partial_\sigma \rho - G_{\mu\sigma} \partial_\nu \rho \right) = -\mathcal{K}_{\mu\sigma\nu}. \tag{9.2.31}$$

Calligraphic fonts (or a tilde) indicate the effective frame. Nevertheless, it will be convenient to work with $E_{\dot\alpha}{}^\mu$ that is more fundamental in the matrix model, and insert the required factors of ρ as needed. One must then be very careful in rising and lowering the indices: the indices of the torsion $T_{\mu\nu}{}^\sigma$ are always raised and lowered with $\gamma_{\mu\nu}$. To avoid confusion, we will stick to the convention that the torsion tensor should always have two lower and one upper index.

9.2.3 Example: Fuzzy S^2_N

To illustrate the present formalism, consider the example of the fuzzy 2-sphere S^2_N; this is nontrivial in the sense that there are more frame indices than the dimension of $\mathcal{M} \cong S^2$. Recall that the fuzzy sphere is defined by the matrix configuration (3.3.13)

$$T^{\dot a} = c_N^{-2} J^a =: X^a, \qquad a = 1, 2, 3, \tag{9.2.32}$$

which defines the abstract quantum space $\mathcal{M} \cong S^2$. It is convenient to use two of the three Cartesian coordinates $y^\mu = x^\mu$, $\mu = 1, 2$ on target space $\mathbb{R}^3$ as local coordinates on the upper hemisphere. The frame is given by three vector fields

$$E^{\dot{a}\mu} = \{x^a, x^\mu\} = \frac{2}{N}\varepsilon^{a\mu b}x_b. \tag{9.2.33}$$

On the north pole ξ, this reduces to $E^{\dot{3}\mu}|_\xi = 0$ and $E^{\dot{\rho}\mu}|_\xi = \frac{2}{N}\varepsilon^{\rho\mu 3}$, which is tangential and given by the Poisson tensor. Since this frame is not invertible, the Weitzenböck connection is obtained using the definition (9.2.5), which boils down to the standard formula (9.2.3)

$$\Gamma_{v\sigma}{}^\mu = -E^{\dot{a}}{}_\sigma \partial_v E_{\dot{a}}{}^\mu = -\frac{2}{N}\varepsilon^{a\mu b}E_{\dot{a}\sigma}\partial_v x_b. \tag{9.2.34}$$

The connection coefficients vanish at the north pole:

$$\Gamma_{v\sigma}{}^\mu|_\xi = -\frac{2}{N}\varepsilon^{3\mu v'}\delta_{v'v}E_{\dot{3}\sigma} = 0 \tag{9.2.35}$$

so that the torsion tensor $T_{\mu v}{}^\sigma$ also vanishes. This is to be expected, since there is no rotational invariant rank three tensor on S^2. It is interesting to check consistency with (9.1.26):

$$T^{\dot{a}\dot{b}\mu}|_\xi = \{\mathcal{F}^{ab}, x^\mu\} = \varepsilon^{abc}\{x_c, x^\mu\} = \frac{2}{N}\varepsilon^{abc}\varepsilon^{c\mu 3}$$

$$= \frac{2}{N}(\delta^{a\mu}\delta^{b3} - \delta^{b\mu}\delta^{a3}). \tag{9.2.36}$$

Although this object has nonvanishing transversal components due to the extrinsic curvature, it vanishes upon contracting the indices with the frame. On higher-dimensional quantized coadjoint orbits, the torsion $T_{\mu v}{}^\sigma$ no longer vanishes, and encodes the structure constants of the underlying Lie algebra.

9.2.4 Divergence-free frames and metric

Now we give a heuristic argument that any given metric $G_{\mu v}$ can be realized in terms of a divergence-free frame. The first step is to determine the dilaton, which is obtained from (9.1.10) as

$$\rho^2 = \rho_M^{-1}\sqrt{|G|}. \tag{9.2.37}$$

The next step is to find some classical divergence-free frame $e^{\dot{a}\mu}$ such that

$$\rho^2 G^{\mu v} = \gamma^{\mu v} = \eta_{\dot{a}\dot{b}}e^{\dot{a}\mu}e^{\dot{b}v}. \tag{9.2.38}$$

Without the divergence constraint, there is six-dimensional $SO(3, 1)$ orbit of frames which achieve that (assuming $3 + 1$ dimensions to be specific). The constraints are easy to take into account in Cartesian coordinates x^μ: given any off-diagonal frame components $e^{\dot{a}\mu}$ with $\mu \neq a$, the diagonal components $e^{\dot{a}a}$ are determined by

$$\partial_a(\rho_M e^{\dot{a}a}) = -\sum_{\mu \neq \dot{a}} \partial_\mu(\rho_M e^{\dot{a}\mu}). \tag{9.2.39}$$

This can be viewed as set of four ordinary differential equations in x^a, which are solved by

$$e^{\dot{a}a} = -\rho_M^{-1} \int_{\xi^a}^{x^a} ds \sum_{\mu \neq a} \partial_\mu(\rho_M e^{\dot{a}\mu}) + e^{\dot{a}a}(\xi^a), \qquad (9.2.40)$$

where the value $e^{\dot{a}a}(\xi^a)$ at any given ξ^a can be chosen as desired. Hence, the divergence constraint can be solved by adjusting the diagonal frame elements, for arbitrary given 12 off-diagonal components $e^{\dot{a}\mu}$. This provides more than enough dof[9] to reproduce the 10 dof in $\gamma^{\mu\nu}$ (in Cartesian coordinates, say), leaving 2 undetermined dof. We conclude that for any $\gamma^{\mu\nu}$, there are always divergence-free frames $e^{\dot{a}\mu}$ that realize (9.2.38). The remaining two dof. can be thought of as encoded in the axionic vector field $\tilde{T}_\kappa \propto T^{\nu\sigma\mu}\varepsilon_{\nu\sigma\mu\kappa}$ (9.3.19), or in ρ and the density ρ_M (considered to be unrestricted). In particular, we can use this remaining freedom to restrict the axionic vector field $\tilde{T}_\kappa$ such that it can be written in terms of an axion $\tilde{\rho}$ as

$$\tilde{T}_\mu = \rho^{-2}\partial_\mu\tilde{\rho} \qquad (9.2.41)$$

for suitable ρ and ρ_M. This will indeed follow from the semi-classical equations of motion (9.3.21). However, then the effective metric $G_{\mu\nu}$ of the matrix model also determines the dilaton ρ, the axion $\tilde{\rho}$, and ρ_M, which means that the equivalence principle no longer applies. This is consistent with the reduced gauge invariance, as the diffeomorphisms are reduced to symplectomorphisms. It would be desirable to find an explicit formula for ρ, $\tilde{\rho}$, and ρ_M in terms of $G_{\mu\nu}$ assuming (9.2.41).

Furthermore, we will see in Sections 10.2.4 ff. that any divergence-free frame can be reconstructed in the form (9.1.4) on covariant quantum spacetime $\mathcal{M}^{3,1}$, and possible $\mathfrak{h}s$ components can be removed in local normal coordinates. In this sense, generic $3 + 1$ dimensional geometries can be implemented as covariant quantum spacetimes in the matrix model. Such a reconstruction is not possible in general on basic symplectic manifolds without the degrees of freedom arising from the internal S^2 or from transversal directions in target space.

9.2.5 *Contraction identities

Since the torsion of the Weitzenböck connection arises from a special class of frames which satisfy a divergence constraint, it also satisfies certain constraints. To understand them, we first observe the following relations:

Lemma 9.1 *The following identities hold for the Weitzenböck connection on $\mathcal{M}$:*

$$\Gamma_{\mu\sigma}{}^\mu = \rho_M^{-1}\partial_\sigma\rho_M \qquad (9.2.42)$$

$$\Gamma_{\sigma\mu}{}^\mu = -\partial_\sigma \ln(\sqrt{|\gamma^{\mu\nu}|}) \qquad (9.2.43)$$

$$\gamma^{\nu\rho}\Gamma_{\nu\rho}{}^\mu = \gamma^{\nu\rho}\Gamma^{(G)}{}_{\nu\rho}{}^\mu. \qquad (9.2.44)$$

---

[9] This could presumably be turned into a formal proof in some sufficiently small neighborhood of any given point ξ, by iteratively adjusting the off-diagonal elements to reproduce $\gamma^{\mu\nu}$ and the diagonal frame elements by (9.2.40), since the correction (9.2.40) for the diagonal frame elements vanish at ξ.

For the covariant quantum spaces $\mathcal{M}^{(3,1)}$, this holds for the derivatives ∂_μ defined in (5.4.29), which satisfy the identity $\partial_\mu(\rho_M \theta^{\mu\nu}) = 0$ (5.4.30).

Proof   The first identity follows from the divergence constraint (9.2.19) together with (9.2.3):

$$\Gamma_{\mu\sigma}{}^\mu = -E^{\dot{a}}{}_\sigma \partial_\mu E_{\dot{a}}{}^\mu = \rho_M^{-1} \partial_\mu \rho_M E^{\dot{a}}{}_\sigma E_{\dot{a}}{}^\mu = \rho_M^{-1} \partial_\mu \rho_M. \tag{9.2.45}$$

The second identity is also obtained from (9.2.3):

$$\begin{aligned}
\Gamma_{\sigma\mu}{}^\mu &= -E^{\dot{a}}{}_\mu \partial_\sigma E_{\dot{a}}{}^\mu = -E^{\dot{a}\rho} \gamma_{\rho\mu} \partial_\sigma E_{\dot{a}}{}^\mu \\
&= -\frac{1}{2}\gamma_{\rho\mu} \partial_\sigma \gamma^{\rho\mu} = -\partial_\sigma \ln(\sqrt{|\gamma^{\mu\nu}|})
\end{aligned} \tag{9.2.46}$$

using (9.1.20). Finally, (9.2.44) follows by contracting (9.2.3) with $\gamma^{\nu\sigma}$

$$\begin{aligned}
\gamma^{\nu\sigma} \Gamma_{\nu\sigma}{}^\mu &= -E^{\dot{a}\nu} \partial_\nu E_{\dot{a}}{}^\mu = -E^{\dot{a}\nu} \partial_\nu (E_{\dot{a}\sigma} \gamma^{\mu\sigma}) \\
&= \partial_\nu E^{\dot{a}\nu} E_{\dot{a}\sigma} \gamma^{\mu\sigma} - \partial_\nu \gamma^{\mu\nu} \\
&= -\Gamma_{\rho\sigma}{}^\rho \gamma^{\mu\sigma} - \partial_\nu \gamma^{\mu\nu} \\
&= -\rho_M^{-1} \partial_\sigma \rho_M \gamma^{\mu\sigma} - \partial_\nu \gamma^{\mu\nu} \\
&= -\rho^2 \frac{1}{\sqrt{|G_{\mu\nu}|}} \partial_\sigma \left(\sqrt{|G_{\mu\nu}|} G^{\mu\sigma}\right) = \gamma^{\mu\sigma} \Gamma^{(G)}{}_{\nu\rho}{}^\mu
\end{aligned} \tag{9.2.47}$$

using (9.2.42) and $\rho_M = \rho^{-2}\sqrt{|G_{\mu\nu}|}$ (9.1.10). $\square$

Contraction identity for the torsion

The identities in Lemma 9.1 lead to the following identities for the (con)torsion:

Lemma 9.2   *The (con)torsion associated with the frame $E_{\dot{\alpha}}{}^\mu$ on $\mathcal{M}^{(2n)}$ satisfies*

$$T_{\mu\sigma}{}^\mu = K_{\mu\sigma}{}^\mu = 2(n-1)\rho^{-1}\partial_\sigma \rho. \tag{9.2.48}$$

For the effective frame, this amounts to

$$\mathcal{T}_{\mu\sigma}{}^\mu = \mathcal{K}_{\mu\sigma}{}^\mu = -\rho^{-1}\partial_\sigma \rho. \tag{9.2.49}$$

Proof   Using Lemma 9.1, we can evaluate the contraction of the torsion

$$\begin{aligned}
T_{\mu\rho}{}^\mu &= \Gamma_{\mu\rho}{}^\mu - \Gamma_{\rho\mu}{}^\mu = \partial_\rho \left(\ln(\sqrt{|\gamma^{\mu\nu}|}) + \ln \rho_M \right) \\
&= \partial_\rho \ln(\rho_M \sqrt{|\gamma^{\mu\nu}|}) = \partial_\rho \ln(\rho^{2n-2})
\end{aligned} \tag{9.2.50}$$

using $\sqrt{|\gamma^{\mu\nu}|} = \rho^{2n-2}\rho_M^{-1}$ (9.1.11), and

$$K_{\mu\nu}{}^\mu = \frac{1}{2}(T_{\mu\nu}{}^\mu + T_\mu{}^\mu{}_\nu - T_\nu{}^\mu{}_\mu) = T_{\mu\nu}{}^\mu \tag{9.2.51}$$

using (9.2.13). For the effective frame, (9.2.28) gives

$$\tilde{\Gamma}_{\mu\sigma}{}^\mu = \Gamma_{\mu\sigma}{}^\mu + \rho^{-1}\partial_\sigma \rho, \qquad \tilde{\Gamma}_{\nu\mu}{}^\mu = \Gamma_{\nu\mu}{}^\mu + 2n\,\rho^{-1}\partial_\nu \rho \tag{9.2.52}$$

and therefore

$$\mathcal{T}_{\mu\sigma}{}^{\mu} = \tilde{\Gamma}_{\mu\sigma}{}^{\mu} - \tilde{\Gamma}_{\sigma\mu}{}^{\mu} = T_{\mu\sigma}{}^{\mu} - (2n-1)\rho^{-1}\partial_{\sigma}\rho = -\rho^{-1}\partial_{\sigma}\rho = \mathcal{K}_{\mu\sigma}{}^{\mu}. \quad (9.2.53)$$

$$\square$$

This contraction identity for the torsion can be seen to be equivalent to the divergence constraint for the frame [73].

9.2.6 Effective Laplacian

In the framework of Yang–Mills matrix models, fluctuations around some matrix background $T^{\dot{a}}$ are governed by the *matrix or Poisson Laplacian*

$$\square := [T^{\dot{a}}, [T_{\dot{a}}, .]] \sim -\{T^{\dot{a}}, \{T_{\dot{a}}, .\}\}: \quad \text{End}(\mathcal{H}) \to \text{End}(\mathcal{H}). \quad (9.2.54)$$

Here the indices are contracted with $\eta_{\dot{a}\dot{b}}$ or $\delta_{\dot{a}\dot{b}}$ in the Euclidean case. Not surprisingly, this operator is essentially the Laplacian associated with the effective metric:

Lemma 9.3 *The matrix or Poisson Laplacian $\square$ is related to the metric Laplacian $\square_G$ via*

$$\square\phi = -\{T^{\dot{a}}, \{T_{\dot{a}}, \phi\}\} = \rho^2\square_G\phi = -\frac{\rho^2}{\sqrt{|G_{\sigma\kappa}|}}\partial_{\mu}(\sqrt{|G_{\sigma\kappa}|}G^{\mu\nu}\partial_{\nu}\phi) \quad (9.2.55)$$

where G and ρ are defined via (9.1.9) and (9.1.10). This applies both in Euclidean and Minkowski signature, in dimensions larger than 2.

Proof The best way to show this relation is by considering the following bilinear form:

$$\langle\phi, \psi\rangle := \int \Omega\{T^{\dot{a}}, \phi^{\dagger}\}\{T_{\dot{a}}, \psi\} = \int \Omega\phi^{\dagger}\square\psi$$

$$\sim \int d^{2n}x\sqrt{|G|}G^{\mu\nu}\partial_{\mu}\phi\partial_{\nu}\psi = -\int d^{2n}x\sqrt{|G|}\phi\square_G\psi \quad (9.2.56)$$

for scalar fields ϕ, ψ, similar as in section 9.1.1 using partial integration. The claim then follows recalling that $\Omega = \rho^{-2}\sqrt{|G|}$ (9.1.10). This argument applies also to the covariant quantum spaces. Nevertheless, it is instructive to see how this works in a direct computation. Consider

$$\{T_{\dot{a}}, \{T^{\dot{a}}, \phi\}\} = E_{\dot{a}}{}^{\mu}\partial_{\mu}(E^{\dot{a}\nu}\partial_{\nu}\phi)$$

$$= \partial_{\mu}(E_{\dot{a}}{}^{\mu}E^{\dot{a}\nu}\partial_{\nu}\phi) - (\partial_{\mu}E_{\dot{a}}{}^{\mu})E^{\dot{a}\nu}\partial_{\nu}\phi$$

$$= \partial_{\mu}(\gamma^{\mu\nu}\partial_{\nu}\phi) + \Gamma_{\mu}{}^{\nu\mu}\partial_{\nu}\phi, \quad (9.2.57)$$

where $\Gamma_{\mu}{}^{\nu\mu}$ are Weitzenböck connection coefficients. The second term can be evaluated using (9.2.42), and we obtain

$$\{T_{\dot{a}}, \{T^{\dot{a}}, \phi\}\} = \partial_\mu(\gamma^{\mu\nu}\partial_\nu\phi) + \rho_M^{-1}\gamma^{\mu\nu}\partial_\mu\rho_M\partial_\nu\phi$$

$$= \rho_M^{-1}\partial_\mu(\rho_M\gamma^{\mu\nu}\partial_\nu\phi)$$

$$= \frac{\rho^2}{\sqrt{|G_{\mu\nu}|}}\partial_\mu(\sqrt{|G_{\mu\nu}|}G^{\mu\rho}\partial_\nu\phi).$$

□

9.3 Pre-gravity: Covariant equations of motion for frame and torsion

In this section, we will derive covariant equations of motion which govern the frame $E^{\dot{a}\mu}$ and the torsion $T_{\dot{a}\dot{b}}{}^\mu$ in the classical matrix model. These also apply to covariant quantum spaces, provided we restrict ourselves to the effectively four-dimensional regime where frame and torsion can be considered as a function of x, treating possible $\mathfrak{hs}$ generators as constant.

Consider the semi-classical equations of motion which results from the matrix eom (7.0.10) in vacuum

$$\{T^{\dot{a}}, \mathcal{F}_{\dot{a}\dot{b}}\} = -m^2 T_{\dot{b}}. \tag{9.3.1}$$

We obtain a tensorial equation for the torsion $T_{\dot{a}\dot{b}}{}^\mu = \{\mathcal{F}_{\dot{a}\dot{b}}, x^\mu\}$ in local coordinates x^μ by considering $\{., x^\mu\}$ of (9.3.1):

$$-m^2\{T_{\dot{b}}, x^\mu\} = \{\{T^{\dot{a}}, \mathcal{F}_{\dot{a}\dot{b}}\}, x^\mu\} = -\{\{\mathcal{F}_{\dot{a}\dot{b}}, x^\mu\}, T^{\dot{a}}\} - \{\{x^\mu, T^{\dot{a}}\}, \mathcal{F}_{\dot{a}\dot{b}}\}$$

$$= \{T^{\dot{a}}, T_{\dot{a}\dot{b}}{}^\mu\} + \{E^{\dot{a}\mu}, \mathcal{F}_{\dot{a}\dot{b}}\} \tag{9.3.2}$$

using the Jacobi identity. Using or assuming[10] that the frame and torsion $T_{\dot{a}\dot{b}}{}^\mu(x)$ and $E^{\dot{a}\mu}(x)$ are functions of x^μ, this can be written as

$$-m^2 E_{\dot{b}}{}^\mu = E^{\dot{a}\nu}\partial_\nu T_{\dot{a}\dot{b}}{}^\mu - T_{\dot{a}\dot{b}}{}^\nu\partial_\nu E^{\dot{a}\mu}. \tag{9.3.3}$$

The rhs can be recognizing as Lie bracket of the vector fields $E^{\dot{a}}$ and $T_{\dot{a}\dot{b}}$, so that

$$-m^2 E_{\dot{b}} = [E^{\dot{a}}, T_{\dot{a}\dot{b}}] = \mathcal{L}_{E^{\dot{a}}} T_{\dot{a}\dot{b}}. \tag{9.3.4}$$

In this form, the Poisson brackets are no longer explicit. Moreover, recalling that

$$T_{\dot{a}\dot{b}} = -[E_{\dot{a}}, E_{\dot{b}}] = -\mathcal{L}_{E_{\dot{a}}} E_{\dot{b}}, \tag{9.3.5}$$

the equation of motion can be written in the form

$$\left(\mathcal{L}_{E^{\dot{a}}}\mathcal{L}_{E_{\dot{a}}} - m^2\right)E_{\dot{b}} = 0. \tag{9.3.6}$$

This holds even for non-invertible frames.

---

[10] For covariant quantum spaces, this is justified in the effectively four-dimensional regime where possible $\mathfrak{hs}$ components can be treated as constant.

Equations of motion using the Weitzenböck connection

Now recall the Weitzenböck connection associated with the frame $E^{\dot{a}\mu}$. A covariant form of the eom is obtained by rewriting the second term in (9.3.2) as

$$\{E^{\dot{a}\mu}, \mathcal{F}_{\dot{a}\dot{b}}\} \sim -\partial_\nu E^{\dot{a}\mu} T_{\dot{a}\dot{b}}{}^\nu = \Gamma_{\nu\sigma}{}^\mu E^{\dot{a}\sigma} T_{\dot{a}\dot{b}}{}^\nu = (\Gamma_{\sigma\nu}{}^\mu + T_{\nu\sigma}{}^\mu)T^\sigma{}_{\dot{b}}{}^\nu. \tag{9.3.7}$$

The ΓT term on the rhs can be rewritten using

$$\nabla_\nu^{(W)} T^\nu{}_{\dot{b}}{}^\mu = \nabla_\nu^{(W)}(E_{\dot{a}}{}^\nu T^{\dot{a}}{}_{\dot{b}}{}^\mu) = E_{\dot{a}}{}^\nu \nabla_\nu^{(W)} T^{\dot{a}}{}_{\dot{b}}{}^\mu = E_{\dot{a}}{}^\nu (\partial_\nu T^{\dot{a}}{}_{\dot{b}}{}^\mu + \Gamma_{\nu\sigma}{}^\mu T^{\dot{a}}{}_{\dot{b}}{}^\sigma)$$
$$= \{T_{\dot{a}}, T^{\dot{a}}{}_{\dot{b}}{}^\mu\} + \Gamma_{\sigma\nu}{}^\mu T^\sigma{}_{\dot{b}}{}^\nu \tag{9.3.8}$$

since $\nabla^{(W)} E = 0$. Combining this, we obtain

$$-m^2 \{T_{\dot{b}}, y^\mu\} = \nabla_\nu^{(W)} T^\nu{}_{\dot{b}}{}^\mu + T_{\nu\sigma}{}^\mu T^\sigma{}_{\dot{b}}{}^\nu, \tag{9.3.9}$$

which amounts to

$$\nabla_\nu^{(W)} T^\nu{}_\rho{}^\mu + T_{\nu\sigma}{}^\mu T^\sigma{}_\rho{}^\nu = -m^2 \delta_\rho^\mu. \tag{9.3.10}$$

Upon lowering an index with $\gamma_{\mu\nu}$, we obtain

$$\boxed{\nabla_\nu^{(W)} T^\nu{}_{\rho\mu} + T_\nu{}^\sigma{}_\mu T_{\sigma\rho}{}^\nu = -m^2 \gamma_{\rho\mu}.} \tag{9.3.11}$$

This is a covariant equation of motion for the torsion, which allows to derive the equations for the frame and dilaton obtained in the following, cf. [13] or [73]. We shall take an independent approach, which does not rely on the Weitzenböck connection.

Equations of motion using the Levi–Civita connection

The eom (9.3.3) can be rewritten using the divergence constraints as follows:

$$-m^2 E_{\dot{b}}{}^\mu = E^{\dot{a}\nu}\partial_\nu T_{\dot{a}\dot{b}}{}^\mu - T_{\dot{a}\dot{b}}{}^\nu \partial_\nu E^{\dot{a}\mu}$$
$$= \rho_M^{-1}\partial_\nu\big(\rho_M(E^{\dot{a}\nu}T_{\dot{a}\dot{b}}{}^\mu - T^\mu{}_{\dot{b}}{}^\nu)\big)$$
$$= \rho_M^{-1}\partial_\nu\big(\rho_M(T^\nu{}_{\dot{b}}{}^\mu - T^\mu{}_{\dot{b}}{}^\nu)\big). \tag{9.3.12}$$

Contracting this equation with the inverse frame $E^{\dot{a}}{}_\mu$, we obtain

$$-m^2 P_T^{\dot{a}\dot{b}} = \rho_M^{-1}E^{\dot{a}}{}_\mu\partial_\nu\big(\rho_M(T^{\nu\dot{b}\mu} - T^{\mu\dot{b}\nu})\big)$$
$$= \rho_M^{-1}E^{\dot{a}}{}_\mu\partial_\nu\big(\rho_M T^{(AS)\nu\dot{b}\mu}\big) - \rho_M^{-1}E^{\dot{a}}{}_\mu\partial_\nu\big(\rho_M T^{\mu\nu\dot{b}}\big)$$
$$= \rho_M^{-1}E^{\dot{a}}{}_\mu\partial_\nu\big(\rho_M T^{(AS)\nu\dot{b}\mu}\big) - \rho_M^{-1}\partial_\nu\big(\rho_M E^{\dot{a}}{}_\mu T^{\mu\nu\dot{b}}\big) + \rho_M^{-1}(\partial_\nu E^{\dot{a}}{}_\mu)\rho_M T^{\mu\nu\dot{b}}$$
$$= \rho_M^{-1}E^{\dot{a}}{}_\mu\partial_\nu\big(\rho_M T^{(AS)\nu\dot{b}\mu}\big) - \rho_M^{-1}\partial_\nu\big(\rho_M T^{\dot{a}\nu\dot{b}}\big) + \frac{1}{2}T_{\nu\mu}{}^{\dot{a}}T^{\mu\nu\dot{b}}, \tag{9.3.13}$$

where

$$T^{(AS)}{}_{\dot{a}\dot{b}\dot{c}} = T_{\dot{a}\dot{b}\dot{c}} + T_{\dot{b}\dot{c}\dot{a}} + T_{\dot{c}\dot{a}\dot{b}} \tag{9.3.14}$$

is the totally antisymmetric component of the torsion.

The antisymmetric part of the torsion

The antisymmetric component of equation (9.3.13) gives

$$0 = E^{\dot{a}}{}_{\mu} \partial_\nu \left(\rho_M T^{(AS)\nu \dot{b}\mu} \right) - (\dot{a} \leftrightarrow \dot{b}) - \partial_\nu \left(\rho_M T^{\dot{a}\nu\dot{b}} - (\dot{a} \leftrightarrow \dot{b}) \right). \tag{9.3.15}$$

Recalling the divergence constraint $\partial_\nu \left(\rho_M T^{\dot{a}\dot{b}\nu} \right) = 0$ for the torsion, the last term can be written in terms of $T^{(AS)}$, and we obtain

$$\begin{aligned}
0 &= E^{\dot{a}}{}_{\mu} \partial_\nu \left(\rho_M T^{(AS)\nu \dot{b}\mu} \right) - (\dot{a} \leftrightarrow \dot{b}) - \partial_\nu \left(\rho_M T^{(AS)\dot{b}\dot{a}\nu} \right) \\
&= E^{\dot{a}}{}_{\mu} \partial_\nu \left(\rho_M T^{(AS)\nu\sigma\mu} E^{\dot{b}}{}_{\sigma} \right) - (\dot{a} \leftrightarrow \dot{b}) - \partial_\nu \left(\rho_M T^{(AS)\sigma\kappa\nu} E^{\dot{b}}{}_{\sigma} E^{\dot{a}}{}_{\kappa} \right) \\
&= 2 E^{\dot{a}}{}_{\mu} E^{\dot{b}}{}_{\sigma} \partial_\nu \left(\rho_M T^{(AS)\nu\sigma\mu} \right).
\end{aligned} \tag{9.3.16}$$

This provides a simple equation for the totally antisymmetric part of the torsion

$$\boxed{ \partial_\nu (\rho_M T^{(AS)\nu\sigma\mu}) = 0 = \nabla^{(G)}_\nu (\rho^{-2} T^{(AS)\nu\sigma\mu}). } \tag{9.3.17}$$

The second form is obtained using (9.1.10) and recalling the identity

$$\nabla^{(G)}_\nu A^{\nu\mu} = \frac{1}{\sqrt{G}} \partial_\nu \left(\sqrt{G} A^{\nu\mu} \right) \tag{9.3.18}$$

for antisymmetric $A^{\mu\nu}$, generalized to rank 3 totally AS tensors.

If $\mathcal{M}$ is four-dimensional, it is natural to express $T^{(AS)\nu}{}_{\rho\mu}$ in terms of a dual tensor $\tilde{T}_\sigma$,

$$T^{(AS)\nu}{}_{\rho\mu} =: -\sqrt{|G|} G^{\nu\nu'} \varepsilon_{\nu'\rho\mu\sigma} G^{\sigma\sigma'} \tilde{T}_{\sigma'}. \tag{9.3.19}$$

Then the equation of motion (9.3.17) for $T^{(AS)}$ becomes

$$0 = \nabla^{(G)}_\nu (\sqrt{|G|} G^{\nu\nu'} \varepsilon_{\nu'\rho\mu\sigma} \rho^2 G^{\sigma\sigma'} \tilde{T}_{\sigma'}) = \sqrt{|G|} G^{\sigma\sigma'} G^{\nu\nu'} \varepsilon_{\nu\rho\mu\sigma} \partial_{\nu'} (\rho^2 \tilde{T}_{\sigma'}). \tag{9.3.20}$$

This means that $\rho^2 \tilde{T}_\sigma$ is integrable, and can hence be written as

$$\rho^2 \tilde{T}_\mu = \partial_\mu \tilde{\rho} \tag{9.3.21}$$

for some function $\tilde{\rho}$. This function will be denoted as **axion**, for reasons that will become clear later. Hence the equations of motion for the torsion reduce the four degrees of freedom of $T^{(AS)}$ to a single scalar field $\tilde{\rho}$. Even though $\tilde{\rho}$ is not determined any further by (9.3.17), it does satisfy an equation of motion, which will be derived in the following.

These manipulations are more transparent using the language of differential forms, viewing $T^{(AS)\nu}{}_{\rho\mu}$ as the components of a three-form[11]

$$\begin{aligned}
T^{(AS)} &:= \frac{1}{3!} \gamma_{\nu\nu'} T^{(AS)\nu'}{}_{\rho\mu} dx^\nu \wedge dx^\rho \wedge dx^\mu = T_{\dot{a}} \wedge \theta^{\dot{a}} \\
&= \frac{1}{2} T_{\dot{a}\mu\nu} dx^\mu \wedge dx^\nu \wedge \theta^{\dot{a}} = \omega_{\dot{a}\dot{b}} \wedge \theta^{\dot{a}} \wedge \theta^{\dot{b}}
\end{aligned} \tag{9.3.22}$$

---

[11] Remember that the indices of $T^\nu{}_{\rho\mu}$ are always raised and lowered with $\gamma_{\mu\nu}$.

using (9.1.28) and the first Cartan structure equation (9.2.15). Then the vector field $\tilde{T}$ (9.3.19) can be extracted via

$$i_{\tilde{T}}\Omega := \theta \wedge d\theta = T^{(AS)}. \tag{9.3.23}$$

Moreover, the equation of motion (9.3.17) for $T^{(AS)}$ can be rewritten via the Hodge star with respect to G as

$$d * (\rho^4 T^{(AS)}) = 0. \tag{9.3.24}$$

Expressing the torsion form in terms of its Hodge dual

$$\rho^2 T^{(AS)} = *(\tilde{T}_\sigma dx^\sigma), \tag{9.3.25}$$

(9.3.24) reduces to

$$d(\rho^2 \tilde{T}) = 0, \qquad \text{therefore} \quad \rho^2 \tilde{T} = d\tilde{\rho} \tag{9.3.26}$$

consistent with (9.3.21).

Furthermore, the Bianchi identity $dT^{\dot{a}} = 0$ for the torsion two-form together with (9.3.22) implies

$$d * (\rho^{-2}\tilde{T}) = d(T^{\dot{a}} \wedge E_{\dot{a}}) = T^{\dot{a}} \wedge T_{\dot{a}}. \tag{9.3.27}$$

Expressing $\tilde{T}_\mu$ in terms of the axion $\tilde{\rho}$ using (9.3.21) results in

$$d * (\rho^{-4}d\tilde{\rho}) = T^{\dot{a}} \wedge T_{\dot{a}}, \tag{9.3.28}$$

which in local coordinates takes the explicit form

$$\boxed{\nabla^\mu_{(G)}(\rho^{-4}\partial_\mu\tilde{\rho}) = \frac{1}{4}\sqrt{|G|}^{-1}\varepsilon^{\nu\rho\mu\kappa}T_{\nu\rho}{}^{\dot{a}}T_{\kappa\mu\dot{a}}.} \tag{9.3.29}$$

Thus, $\tilde{\rho}$ is recognized as an axion-like field.

Equation of motion for the frame-valued torsion

Now we rewrite the eom (9.3.12) in terms of the antisymmetric torsion and using the identity (9.3.18):

$$-m^2 E_{\dot{b}}{}^\mu = \rho_M^{-1}\partial_\nu\left(\rho_M(T^\nu{}_{\dot{b}}{}^\mu - T^\mu{}_{\dot{b}}{}^\nu)\right)$$
$$= \rho^2\nabla^{(G)}_\nu\left(\rho^{-2}(T^\nu{}_{\dot{b}}{}^\mu - T^\mu{}_{\dot{b}}{}^\nu)\right) = \rho^2\nabla^{(G)}_\nu\left(\rho^{-2}(T^{(AS)\nu}{}_{\dot{b}}{}^\mu - T^{\mu\nu}{}_{\dot{b}})\right). \tag{9.3.30}$$

Lowering μ via $\gamma_{\mu\mu'} = \rho^{-2}G_{\mu\mu'}$ gives

$$-m^2 E_{\dot{b}\mu} = \nabla^{(G)\nu}\left(\rho^2(T_{\nu\mu\dot{b}} + T^{(AS)}{}_{\nu\dot{b}\mu})\right). \tag{9.3.31}$$

The last term can be written as

$$\nabla^{(G)\nu}\left(\rho^2 T^{(AS)}{}_{\nu\dot{b}\mu}\right) = \nabla^{(G)}_\nu(\rho^2 T^{(AS)\nu}{}_{\sigma\mu}\rho^{-2}E_{\dot{b}}{}^\sigma)$$
$$= \nabla^{(G)}_\nu(\rho^2 T^{(AS)\nu}{}_{\sigma\mu} G^{\sigma\kappa}E_{\dot{b}\kappa})$$

$$= T^{(AS)\nu\sigma}{}_\mu \nabla^{(G)}_\nu E_{\dot b\sigma}$$

$$= \frac{1}{2} T^{(AS)\nu\sigma}{}_\mu T_{\nu\sigma\dot b} \tag{9.3.32}$$

using the eom (9.3.17) and (9.1.27). Putting this together, the eom takes the simple form

$$\boxed{\nabla^{(G)\nu}\left(\rho^2 T_{\nu\mu}{}^{\dot a}\right) + \frac{1}{2} T^{(AS)\nu\sigma}{}_\mu T_{\nu\sigma}{}^{\dot a} = -m^2 E^{\dot a}{}_\mu.} \tag{9.3.33}$$

These equations for the frame are reminiscent of the Maxwell equations, amended by a nonlinear term involving the antisymmetric torsion. Expressing $T^{(AS)}$ in terms of the axion using (9.3.26), this can be written in terms of differential forms as

$$\boxed{d * (\rho^2 T^{\dot a}) = d\tilde\rho \wedge T^{\dot a} - m^2 * E^{\dot a}.} \tag{9.3.34}$$

Recalling that $T^{\dot a} = d\theta^{\dot a}$ (9.1.28), this equation states that the frame is essentially harmonic, up to corrections in terms of the axion and possibly a mass term.

Eom for the dilaton

Finally we can derive an equation of motion for the dilaton. This is most easily obtained by contracting the eom (9.3.10) with $\gamma^{\rho\mu}$ and recalling that $\nabla^{(W)}\gamma^{\rho\mu} = 0$. Using the contraction identities (9.2.48) and (9.2.14), this gives

$$-G^{\nu\nu'}\nabla^{(G)}_{\nu'}\rho^{-1}\partial_\nu\rho = -2m^2\rho^{-2} - \frac{1}{2} G^{\rho\mu} T_\nu{}^\sigma{}_\mu T_{\sigma\rho}{}^\nu \tag{9.3.35}$$

in four dimensions. This can be rewritten using (C.0.10) as

$$-\nabla^{(G)\mu}\left(\rho^{-1}\partial_\mu\rho\right) = -2\rho^{-2}m^2 + \frac{1}{4} T_\mu{}^\rho{}_\kappa T_{\nu\sigma}{}^\kappa G^{\mu\nu} + \frac{1}{2}\rho^{-4} G^{\mu\nu}\partial_\mu\tilde\rho\partial_\nu\tilde\rho. \tag{9.3.36}$$

9.3.1 Riemann tensor and torsion

Now we compute the Riemann curvature tensor for the Levi–Civita connection associated with the effective metric $G^{\mu\nu}$, starting from the definition in terms of Christoffel symbols

$$\mathcal{R}^{(G)}{}_{\mu\nu}{}^\lambda{}_\sigma = \partial_\mu\Gamma^{(G)}{}_{\nu\sigma}{}^\lambda - \partial_\nu\Gamma^{(G)}{}_{\mu\sigma}{}^\lambda + \Gamma^{(G)}{}_{\mu\rho}{}^\lambda\Gamma^{(G)}{}_{\nu\sigma}{}^\rho - \Gamma^{(G)}{}_{\nu\rho}{}^\lambda\Gamma^{(G)}{}_{\mu\sigma}{}^\rho. \tag{9.3.37}$$

We can evaluate this in Riemann normal coordinates w.r.t. $G_{\mu\nu}$ at any point $\xi \in \mathcal{M}$, where $\Gamma^{(G)}{}_{\nu\sigma}{}^\rho|_\xi = 0$. Using the relation (9.2.14) between the contorsion and the Christoffel symbols (generalized to the effective metric), this simplifies as

$$\mathcal{R}_{\mu\nu}{}^\lambda{}_\sigma = \partial_\mu(\tilde\Gamma_{\nu\sigma}{}^\lambda - \mathcal{K}_{\nu\sigma}{}^\lambda) - \partial_\nu(\tilde\Gamma_{\mu\sigma}{}^\lambda - \mathcal{K}_{\mu\sigma}{}^\lambda). \tag{9.3.38}$$

Now we use the fact that the curvature of the Weitzenböck connection vanishes,

$$0 = \partial_\mu\tilde\Gamma_{\nu\sigma}{}^\lambda - \partial_\nu\tilde\Gamma_{\mu\sigma}{}^\lambda + \tilde\Gamma_{\mu\rho}{}^\lambda\tilde\Gamma_{\nu\sigma}{}^\rho - \tilde\Gamma_{\nu\rho}{}^\lambda\tilde\Gamma_{\mu\sigma}{}^\rho. \tag{9.3.39}$$

Together with

$$\mathcal{K}_{\mu\nu}{}^{\sigma} = \tilde{\Gamma}_{\mu\nu}{}^{\sigma} \qquad \text{at } \xi \qquad (9.3.40)$$

in Riemann normal coordinates, we obtain the tensorial equations

$$\mathcal{R}_{\mu\nu}{}^{\lambda}{}_{\sigma} = -\nabla_{\mu}^{(G)}\mathcal{K}_{\nu\sigma}{}^{\lambda} + \nabla_{\nu}^{(G)}\mathcal{K}_{\mu\sigma}{}^{\lambda} - \mathcal{K}_{\mu\rho}{}^{\lambda}\mathcal{K}_{\nu\sigma}{}^{\rho} + \mathcal{K}_{\nu\rho}{}^{\lambda}\mathcal{K}_{\mu\sigma}{}^{\rho}$$
$$\mathcal{R}_{\nu\sigma} = -\nabla_{\mu}^{(G)}\mathcal{K}_{\nu\sigma}{}^{\mu} + \nabla_{\nu}^{(G)}\mathcal{K}_{\mu\sigma}{}^{\mu} - \mathcal{K}_{\mu\rho}{}^{\mu}\mathcal{K}_{\nu\sigma}{}^{\rho} + \mathcal{K}_{\nu\rho}{}^{\mu}\mathcal{K}_{\mu\sigma}{}^{\rho}. \qquad (9.3.41)$$

Hence, the Riemannian curvature can be expressed in terms of the (con)torsion of the Weitzenböck connection. In contrast to the more familiar form (9.3.37), it is no longer possible to eliminate the nonlinear terms in suitable coordinates, since $\mathcal{K}_{\nu\rho}{}^{\mu}$ is a tensor. The quadratic terms are negligible in the *linearized gravity regime* $\nabla\mathcal{K} \gg \mathcal{K}^2$, where the characteristic size L_{grav} of the local variations is much smaller than the length scale set by the background torsion.

9.3.2 Ricci scalar, torsion, and Einstein–Hilbert action

We have seen that the natural geometric objects in the matrix model framework are the frame and its associated Weitzenböck torsion. Using these results, we can express the Einstein–Hilbert action in terms of the torsion. To make this explicit, we write the Ricci scalar using the identity (9.3.41) as

$$
\begin{aligned}
\mathcal{R}^{(G)} &= 2\nabla_{(G)}^{\mu}\mathcal{K}_{\nu\mu}{}^{\nu} - \mathcal{K}_{\mu\rho}{}^{\mu}\mathcal{K}_{\nu\sigma}{}^{\rho}G^{\nu\sigma} + \mathcal{K}_{\nu\rho}{}^{\mu}\mathcal{K}_{\mu\sigma}{}^{\rho}G^{\nu\sigma} \\
&= \mathcal{K}_{\nu\rho}{}^{\mu}\mathcal{K}_{\mu\sigma}{}^{\rho}G^{\nu\sigma} - \mathcal{K}_{\mu\rho}{}^{\mu}\mathcal{K}_{\nu\sigma}{}^{\rho}G^{\nu\sigma} - 2\nabla_{(G)}^{\mu}(\rho^{-1}\partial_{\mu}\rho) \\
&= \mathcal{K}_{\nu\rho}{}^{\mu}\mathcal{K}_{\mu\sigma}{}^{\rho}G^{\nu\sigma} + 2\rho^{-2}G^{\mu\nu}\partial_{\mu}\rho\,\partial_{\nu}\rho - 2\nabla_{(G)}^{\mu}(\rho^{-1}\partial_{\mu}\rho) \\
&= -\frac{1}{4}T^{\mu}{}_{\sigma\rho}T_{\mu\sigma'}{}^{\rho}G^{\sigma\sigma'} - \frac{1}{2}T^{\mu}{}_{\sigma\rho}T_{\mu}{}^{\rho}{}_{\sigma'}G^{\sigma\sigma'} + 2\rho^{-2}G^{\mu\nu}\partial_{\mu}\rho\,\partial_{\nu}\rho - 2\nabla_{(G)}^{\mu}(\rho^{-1}\partial_{\mu}\rho)
\end{aligned}
$$
$$(9.3.42)$$

using the contraction relation (9.2.49), (9.2.31), and the identity (C.0.2) in the last step. No equations of motion were used. The second term can be related to the first using the identity (C.0.10) via the totally antisymmetric torsion, which gives

$$\mathcal{R} = -\frac{1}{2}T^{\mu}{}_{\sigma\rho}T_{\mu\sigma'}{}^{\rho}G^{\sigma\sigma'} - \frac{1}{2}\tilde{T}_{\nu}\tilde{T}_{\mu}G^{\mu\nu} + 2\rho^{-2}G^{\mu\nu}\partial_{\mu}\rho\,\partial_{\nu}\rho - 2\nabla_{(G)}^{\mu}(\rho^{-1}\partial_{\mu}\rho). \qquad (9.3.43)$$

Integration with some coupling function $G_N(x)$, this gives the identity

$$
\begin{aligned}
\int d^4x \frac{\sqrt{|G|}}{G_N}\mathcal{R} = \int d^4x \frac{\sqrt{|G|}}{G_N}\Big(& -\frac{1}{2}T^{\mu}{}_{\sigma\rho}T_{\mu\sigma'}{}^{\rho}G^{\sigma\sigma'} - \frac{1}{2}\tilde{T}_{\nu}\tilde{T}_{\mu}G^{\mu\nu} + 2\rho^{-2}\partial_{\mu}\rho\,\partial^{\mu}\rho \\
& + 2\rho^{-1}\partial_{\mu}\rho\,G_N\partial^{\mu}G_N^{-1}\Big)
\end{aligned}
$$
$$(9.3.44)$$

which will play an important role in the following.

9.3.3 On-shell Ricci tensor and effective energy–momentum tensor

Now we impose the classical eom. (9.3.11) for the torsion obtained from the matrix model action. The expression (9.3.41) for the Ricci tensor can then be reduced to the following form (see [187] for details):

$$
\mathcal{R}_{\nu\mu} = -\frac{1}{2}(T_\rho{}^\delta{}_\mu T_{\nu\delta}{}^\rho + T_\rho{}^\delta{}_\nu T_{\mu\delta}{}^\rho) - K_\delta{}^\rho{}_\nu K_\rho{}^\delta{}_\mu + 2\rho^{-2}\partial_\nu\rho\partial_\mu\rho
$$
$$
- G_{\nu\mu}\left(\rho^{-2}m^2 + \frac{1}{2}T_\nu{}^\sigma{}_\delta T_{\sigma\rho}{}^\nu G^{\delta\rho}\right). \tag{9.3.45}
$$

The first three terms can be rewritten using (C.0.1) and (C.0.9) as

$$
-\frac{1}{2}T_{\mu\sigma}{}^\rho T_\rho{}^\sigma{}_\nu - \frac{1}{2}T_{\nu\sigma}{}^\rho T_\rho{}^\sigma{}_\mu - K_\sigma{}^\rho{}_\mu K_\sigma{}^\sigma{}_\nu = \frac{1}{4}T^{(AS)\sigma}{}_{\rho\mu} T^{(AS)}{}_\sigma{}^\rho{}_\nu - T_\mu{}^{\sigma\rho} T_{\nu\sigma\rho}. \tag{9.3.46}
$$

Therefore, the on-shell Ricci tensor (9.3.45) can be written as

$$
\mathcal{R}_{\nu\mu} = \frac{1}{4}T^{(AS)\sigma}{}_{\rho\mu} T^{(AS)}{}_\sigma{}^\rho{}_\nu - T_{\mu\sigma}{}^\rho T_\nu{}^\sigma{}_\rho + 2\rho^{-2}\partial_\nu\rho\partial_\mu\rho
$$
$$
- \frac{1}{4}G_{\nu\mu}\left(4\rho^{-2}m^2 - T^\sigma{}_{\nu\delta} T_\sigma{}^\nu{}_\rho G^{\delta\rho} + \frac{1}{3}T^{(AS)\sigma}{}_{\rho\mu} T^{(AS)}{}_\sigma{}^\rho{}_\nu G^{\mu\nu}\right) \tag{9.3.47}
$$

and

$$
\mathcal{R} = -\frac{1}{12}T^{(AS)\sigma}{}_{\rho\mu} T^{(AS)}{}_\sigma{}^\rho{}_\nu G^{\mu\nu} + 2\rho^{-2}\partial_\mu\rho\partial^\mu\rho - 4\rho^{-2}m^2. \tag{9.3.48}
$$

Note that these relations hold only in the classical matrix model, without quantum corrections.

The fact that (9.3.47) is quadratic in T and $\partial\rho$ is very interesting from the gravity point of view: Consider linearized fluctuations $T^{\dot{a}} \to T^{\dot{a}} + \mathcal{A}^{\dot{a}}$ around a flat (or almost-flat) background, with (almost-) vanishing torsion and constant dilaton ρ. Such fluctuations give rise to linearized metric fluctuations $\delta G_{\mu\nu} = G_{\mu\nu} + h_{\mu\nu}[\mathcal{A}]$, so that the linearized contribution to the Ricci tensor vanishes or is suppressed by the background torsion $T_{\mu\sigma}{}^\rho$ or $\partial\rho$. Therefore,

> linearized on-shell metric fluctuations around a flat background are Ricci-flat

$$\tag{9.3.49}$$

as in GR, assuming that m^2 is negligible. However, at the nonlinear level, the classical solutions of the matrix model are in general not Ricci-flat.

Effective energy–momentum tensor

To gain some insights into the on-shell formula (9.3.47) for $\mathcal{R}_{\mu\nu}$, we collect the different contributions to the Einstein tensor in the classical model

$$\mathcal{G}_{\mu\nu} = \mathcal{R}_{\mu\nu} - \frac{1}{2}G_{\mu\nu}\mathcal{R}$$

$$= \frac{1}{4}T^{(AS)}{}_{\sigma}{}^{\rho}{}_{\mu}T^{(AS)\sigma}{}_{\rho\nu} - T_{\mu\sigma}{}^{\rho}T_{\nu}{}^{\sigma}{}_{\rho} + 2\rho^{-2}\partial_{\nu}\rho\partial_{\mu}\rho$$

$$+ \frac{1}{4}G_{\nu\mu}\left(4\rho^{-2}m^2 + T^{\sigma}{}_{\nu\delta}T_{\sigma}{}^{\nu}{}_{\rho}G^{\delta\rho} - \frac{1}{6}T^{(AS)}{}_{\sigma}{}^{\rho}{}_{\mu}T^{(AS)\sigma}{}_{\rho\nu}G^{\mu\nu} - 4\rho^{-2}\partial^{\mu}\rho\partial_{\mu}\rho\right)$$

$$=: \mathbf{T}_{\mu\nu}[E^{\dot\alpha}] + \mathbf{T}_{\mu\nu}[\rho] + \mathbf{T}^{(AS)}_{\mu\nu}[T] + \rho^{-2}m^2G_{\mu\nu}. \tag{9.3.50}$$

In the last line, we *define* the effective e–m tensors of the various contributions. In this sense, the effective e–m tensor of the dilaton is

$$\mathbf{T}_{\mu\nu}[\rho] = 2\rho^{-2}\left(\partial_{\mu}\rho\partial_{\nu}\rho - \frac{1}{2}G_{\mu\nu}\partial^{\sigma}\rho\partial_{\sigma}\rho\right)$$

$$= 2\left(\partial_{\mu}\sigma\partial_{\nu}\sigma - \frac{1}{2}G_{\mu\nu}\partial^{\kappa}\sigma\partial_{\kappa}\sigma\right), \qquad \rho = e^{-\sigma}. \tag{9.3.51}$$

For the contribution of $T^{(AS)}$, we can use (C.0.7) and (C.0.8), so that

$$\mathbf{T}^{(AS)}_{\mu\nu}[\tilde{T}] = \frac{1}{4}\left(T^{(AS)\sigma}{}_{\rho\mu}T^{(AS)}{}_{\sigma}{}^{\rho}{}_{\nu} - \frac{1}{6}G_{\mu\nu}(T^{(AS)\sigma}{}_{\rho\kappa}T^{(AS)}{}_{\sigma}{}^{\rho}{}_{\lambda}G^{\kappa\lambda})\right)$$

$$= \frac{1}{2}\left(\tilde{T}_{\mu}\tilde{T}_{\nu} - \frac{1}{2}G_{\mu\nu}(\tilde{T}^{\rho}\tilde{T}_{\rho})\right). \tag{9.3.52}$$

Using the eom. (9.3.21) for $\tilde{T}_{\mu}$, this reduces to the e–m tensor of the axion,

$$\mathbf{T}^{(AS)}_{\mu\nu}[\tilde{T}] = \frac{1}{2}\rho^{-4}\left(\partial_{\mu}\tilde\rho\partial_{\nu}\tilde\rho - \frac{1}{2}G_{\mu\nu}(G^{\sigma\sigma'}\partial_{\sigma}\tilde\rho\partial_{\sigma'}\tilde\rho)\right) =: \mathbf{T}_{\mu\nu}[\tilde\rho]. \tag{9.3.53}$$

Finally, the e–m tensor of the frame is

$$\mathbf{T}_{\mu\nu}[E] = -T_{\mu\sigma}{}^{\rho}T_{\nu}{}^{\sigma}{}_{\rho} + \frac{1}{4}G_{\mu\nu}T^{\sigma}{}_{\nu\delta}T_{\sigma}{}^{\nu}{}_{\rho}G^{\delta\rho}$$

$$= -\rho^2\left(T_{\mu\sigma}{}^{\dot\alpha}T_{\nu\rho\dot\alpha}G^{\rho\sigma} - \frac{1}{4}G_{\mu\nu}(T_{\rho\kappa}{}^{\dot\alpha}T_{\sigma\kappa'\dot\alpha}G^{\rho\sigma}G^{\kappa\kappa'})\right). \tag{9.3.54}$$

The latter has the structure of a *negative* Maxwell-like energy-momentum tensor

$$\mathbf{T}_{\mu\nu}[A] = F_{\mu\rho}F_{\nu\sigma}G^{\rho\sigma} - \frac{1}{4}G_{\mu\nu}(F_{\rho\sigma}F^{\rho\sigma}), \qquad F = dA \tag{9.3.55}$$

summed over each frame component $E^{\dot\alpha}$, with opposite sign for spacelike and timelike components. These effective energy–momentum tensors should not be confused with the matrix energy–momentum tensor (7.4.14).

9.3.4 Example: Spherically symmetric solutions

To illustrate these results, we shall discuss the simplest spherically symmetric static solution of the geometric equations of motion, which can be considered as local perturbations of an asymptotically flat $3 + 1$ dimensional background. We impose the asymptotic form of the frame

$$E_{\dot\alpha}{}^{\mu} \to \delta_{\dot\alpha}^{\mu} \qquad \text{as} \quad r \to \infty \tag{9.3.56}$$

in Cartesian coordinates x^μ, and adopt the notation

$$t = x^0, \qquad r^2 = x^i x^j \delta_{ij}. \tag{9.3.57}$$

The asymptotic frame is $SO(3)$-invariant if both the frame index $\dot\alpha$ as well as x^μ are transformed as vectors, which is natural due to the global $SO(3)$ symmetry of the matrix model. We shall impose this symmetry also for the perturbed frame. Then the most general spherically symmetric ansatz for the frame $E^{\dot\alpha}{}_\mu$ is

$$\begin{aligned}
E^0{}_0 &= A, \\
E^i{}_0 &= Ex^i, \\
E^0{}_i &= Dx^i, \\
E^i{}_j &= Fx^i x^j + \delta^i{}_j B + S\epsilon_{ijm}x^m,
\end{aligned} \tag{9.3.58}$$

where A, B, D, E, F, and S are assumed to be functions of r only, restricting ourselves to static configurations. D and F can be eliminated using a change of coordinates $t \to f(r)+t$ and $x^i \to g(r)x^i$, which is understood from now on. In terms of differential forms $E^{\dot\alpha} = E^{\dot\alpha}{}_\mu dx^\mu$, the (co)frame is then

$$E^0 = Adt, \qquad E^i = Bdx^i + Ex^i dt + S\epsilon_{ijm}x^m dx^j \tag{9.3.59}$$

and the associated torsion two-form $T^{\dot\alpha} = dE^{\dot\alpha}$ is obtained as

$$\begin{aligned}
T^0 &= dE^0 = A'dr \wedge dt, \\
T^i &= dE^i = B'dr \wedge dx^i + Edx^i \wedge dt + x^i E' dr \wedge dt + S\epsilon_{ijm}dx^m \wedge dx^j \\
&\quad + S'\epsilon_{ijm}x^m dr \wedge dx^j.
\end{aligned} \tag{9.3.60}$$

Clearly the totally antisymmetric part $T^{(AS)}$ of the torsion vanishes if $S = 0$.

All spherically symmetric asymptotically flat solutions with this structure were obtained in [13], including contributions from dilaton and axion. They reproduce the linearized Schwarzschild geometry for large r but deviate from it at the nonlinear level. Here, we only discuss the simplest solution, which is obtained for $S = E = 0$. Then the dilaton ρ can be computed e.g. from (9.2.48), which for the spacelike components $\mu = k$ gives

$$-\frac{2}{\rho}\partial_k\rho = A^{-1}\partial_k A + 2B^{-1}\partial_k B \tag{9.3.61}$$

while the timelike component $\mu = 0$ is trivial. Hence, ρ is determined by

$$AB^2\rho^2 = c \tag{9.3.62}$$

for some constant c, which we will set to $c = 1$ in the following. This gives

$$\sqrt{|G|}\,\rho^{-2} = B. \tag{9.3.63}$$

It is straightforward to check that these relations also imply the divergence constraint for the frame. Now we use the equations of motion (9.3.34) for the torsion, which using the previous assumptions as well as $m = 0$ reduces to

$$d * (\rho^2 T^{\dot\alpha}) = 0. \tag{9.3.64}$$

The spacelike components $\dot{\alpha} = 1, 2, 3$ imply that B is constant, hence $B = 1$ using (9.3.56). Then the timelike component for $\dot{\alpha} = 0$ implies

$$\frac{d}{dr}\left(r^2(A^{-1})'\right) = 0, \tag{9.3.65}$$

which – taking into account the asymptotic form (9.3.56) – has the solution

$$A^{-1} = 1 + \frac{2M}{r} \tag{9.3.66}$$

for some constant M. This leads to the following metric and dilaton:

$$ds_G^2 = G_{\mu\nu}dx^\mu dx^\nu = -\frac{1}{(1 + \frac{2M}{r})}dt^2 + \left(1 + \frac{2M}{r}\right)(dr^2 + r^2 d\Omega^2),$$

$$\rho^2 = 1 + \frac{2M}{r}, \tag{9.3.67}$$

which reproduces the linearized Schwarzschild metric, but deviates from the full Schwarzschild metric at the nonlinear level. However, the geometry will be modified through quantum effects, which lead to the induced Einstein–Hilbert action at one loop as discussed in Chapter 12.

Discussion

The presence of a nonvanishing torsion in the present framework leads immediately to the question: What is the physical significance of $T^{\mu\nu\sigma}$?

In the present matrix model framework, the fundamental object is the matrix configuration, which can be viewed as a potential for the frame. Hence, the frame plays a more fundamental role than in GR, where it is just a local choice of gauge. The torsion is the Lie bracket of this frame, which is a very natural object. An explicit relation between the torsion and the (Levi–Civita) spin connection was given in (9.2.17); however, the torsion is a tensor here which encodes physically significant information, rather than some gauge-dependent auxiliary object. As discussed in Section 9.2.4, the frame encodes not only the metric but also an axionic vector field (via the torsion), which may be traded for a dilaton, an axion and ρ_M.

Moreover, the fact that the on-shell Ricci tensor (9.3.47) is quadratic in the torsion implies that the vacuum geometries in the "pre-gravity" theory defined by the classical matrix model are Ricci-flat as long as the torsion is negligible, but differ significantly from GR in the regime where the torsion is significant. This is also seen in the explicit solutions (9.3.67), and will be corroborated in the following chapters.

For the cosmological background $\mathcal{M}^{3,1}$ considered in Section 10.1, we will see that the torsion tensor reduces to its pure trace component, which in turn encodes the timelike vector field defining the cosmic FLRW background. Hence, the torsion does not constitute a new geometric structure on that background.

Finally, it is appropriate to point out an alternative interpretation of the matrix model [92], where the matrices are interpreted as operators $T_a \cong \nabla_a$ acting on bundles over some space(time). This also leads to equations that formally look like (9.3.11). The appealing feature is that one can then formally derive the vacuum Einstein equations; however, the space of modes in $\mathrm{End}(\mathcal{H})$ is then vastly bigger, and the absence of ghosts is doubtful. This is somewhat analogous to pre-quantization in the context of geometric quantization, where differential operators with desired algebraic properties are constructed on a space that is far too big and contains pathological sectors.

10 Higher-spin gauge theory on quantum spacetime

Our results concerning the frame and the effective geometry suggests that physical spacetime could arise as quantized brane $\mathcal{M}^{3,1}$ within a Yang–Mills matrix model. However, there are several potential obstacles for obtaining a realistic gravitational theory in this way:

- The Poisson structure corresponds to an antisymmetric tensor field $\theta^{\mu\nu}$ that breaks Lorentz invariance. This breaking is not manifest because $\theta^{\mu\nu}$ is hidden in the frame at leading order, but subleading contributions may be problematic. Quantum corrections might also depend on $\theta^{\mu\nu}$, which however is mitigated in the maximally supersymmetric IKKT model.
- Related to the same issue, it is not obvious how to obtain rotationally invariant solution of the classical matrix model, since the Poisson structure typically violates this symmetry.
- In principle, general embeddings of branes $\mathcal{M}^{3,1} \hookrightarrow \mathbb{R}^{9,1}$ provide enough dof to describe the most general $3 + 1$ dimensional metric, at least locally. However, the geometric dynamics of the brane will be governed not only by an induced gravity action as discussed in Section 12 but also by the bare action, which prefers harmonic embeddings. It is not clear if this can conspire to give a physically viable theory of gravity.
- The gauge transformations in Yang–Mills matrix models can be interpreted as symplectomorphisms on quantum spacetime. This gauge group is significantly smaller than the usual diffeomorphism invariance in $3 + 1$ dimensions.

To resolve these issues, we will consider a more sophisticated class of backgrounds, given by matrix configurations $T^{\dot\alpha}$ with $\dot\alpha = 0, \ldots, 3$ which describe **covariant quantum spaces** as discussed in Section 5.4. On these backgrounds, the mentioned problems are avoided, at the expense of introducing higher-spin modes.

The covariant quantum spaces under consideration are quantizations of the six-dimensional symplectic space $\mathbb{C}P^{2,1}$, which is an equivariant S^2 bundle over $3 + 1$ dimensional spacetime, with local structure (5.2.36)

$$\mathbb{C}P^{2,1} = \mathcal{M}^{3,1} \,\bar{\times}\, S^2. \tag{10.0.1}$$

Here, $\bar{\times}$ indicates the equivariant bundle structure, which means that the bundle and its projection are compatible with the action of the space-like isometry group $SO(3,1)$. This structure coincides with the covariant spacetime discussed in Section 5.4, but the frame and metric will be more general in the following.

10.1 Cosmic FLRW spacetime $\mathcal{M}^{3,1}$

To set the stage, we recall the cosmological spacetime described in Section 5.4, which arises from the matrix background

$$T^{\dot{\alpha}} = \frac{1}{R} M^{\dot{\alpha}4}. \tag{10.1.1}$$

We have seen (5.4.6) that this is a solution of the Yang–Mills matrix model (7.0.1) for

$$m^2 = -\frac{3}{R^2}, \tag{10.1.2}$$

which is understood from now on. The $T^{\dot{\alpha}}$ generate the algebra $\mathrm{End}(\mathcal{H}_n) \sim \mathcal{C}$, which is interpreted as quantized algebra of functions on the symplectic space $\mathbb{C}P^{2,1}$, which is an S^2 bundle over the cosmological spacetime $\mathcal{M}^{3,1}$. This is a covariant quantum space, in the sense that $T^{\dot{\alpha}}$ transforms like (5.0.1) under $SO(3, 1)$. In the semi-classical regime, we can choose

$$x^{\mu} \sim X^{\mu} = rM^{\mu 5} \tag{10.1.3}$$

as Cartesian coordinates on $\mathcal{M}^{3,1}$, while

$$t^{\mu} \sim T^{\mu} = \frac{1}{R} M^{\mu 4} \tag{10.1.4}$$

with $x_{\mu} t^{\mu} = 0$ will be used to parametrize the spacelike S^2 fiber. This coincides with the undeformed cosmic background (10.1.1), but the latter will be deformed in the following. Both x^{μ} and t^{μ} transform as vectors under the global $SO(3, 1)$.

Frame, metric, and torsion

Following the general strategy discussed in Section 9.1.1, we can extract the effective metric on $\mathcal{M}^{3,1}$ from the kinetic term for scalar fields:

$$
\begin{aligned}
S[\phi] = \mathrm{Tr}[T^{\dot{\alpha}}, \phi][T_{\dot{\alpha}}, \phi] &\sim -\frac{1}{(2\pi)^n} \int_{\mathbb{C}P^{2,1}} \Omega \, \{T^{\dot{\alpha}}, \phi\}\{T_{\dot{\alpha}}, \phi\} \\
&= -\int_{\mathcal{M}^{3,1}} d^4 x \, \rho_M(x) \, \gamma^{\mu\nu} \partial_\mu \phi \partial_\nu \phi \\
&= -\int_{\mathcal{M}^{3,1}} d^4 x \, \sqrt{|G_{\mu\nu}|} \, G^{\mu\nu} \partial_\mu \phi \partial_\nu \phi.
\end{aligned} \tag{10.1.5}
$$

The coupling constant g in the matrix model is again absorbed in the background $T^{\dot{\alpha}}$, which is therefore dimensionless. This determines a frame $E^{\dot{\alpha}} = \{T^{\dot{\alpha}}, \cdot\}$ and a metric $\gamma^{\mu\nu}$, which for the undeformed background are given by

$$E^{\dot\alpha} = E^{\dot\alpha\mu}\partial_\mu, \qquad E^{\dot\alpha\mu} = \eta^{\dot\alpha\mu}\sinh(\eta),$$

$$\gamma^{\mu\nu} = \eta_{\dot\alpha\dot\beta}E^{\dot\alpha\mu}E^{\dot\beta\nu} = \sinh^2(\eta)\eta^{\mu\nu} \tag{10.1.6}$$

in Cartesian coordinates, so that the effective metric on $\mathcal{M}^{3,1}$ is

$$G_{\mu\nu} = \rho^2\gamma_{\mu\nu} = \sinh(\eta)\eta_{\mu\nu},$$

$$\rho^2 = \rho_M\sqrt{|\gamma^{\mu\nu}|} = \sinh^3(\eta) = \rho_M^{-1}\sqrt{|G_{\mu\nu}|}. \tag{10.1.7}$$

Here, $\rho_M \sim \sinh(\eta)^{-1}$ is the effective density (5.4.26) on the $3+1$ dimensional spacetime $\mathcal{M}^{3,1}$, which arises by reducing the symplectic volume form Ω over the fiber. Accordingly, the projection $[.]_0$ to the spin 0 sector $\mathcal{C}^0$ in the action is understood. As elaborated in Section 5.4.2, this metric defines a $SO(3,1)$-invariant FLRW metric on $\mathcal{M}^{3,1}$ which is conformal to the induced metric $\eta_{\mu\nu}$; in particular, the causal structures coincide. Note that the ratio of the Riemannian and symplectic volume grows with the cosmic expansion. The Weitzenböck torsion is most easily computed from (9.1.26) using

$$\mathcal{F}^{\dot\alpha\dot\beta} = -\{T^{\dot\alpha}, T^{\dot\beta}\} = -\frac{1}{R^2}M^{\dot\alpha\dot\beta}, \tag{10.1.8}$$

which gives

$$T^{\dot\alpha\dot\beta\mu} = \{\mathcal{F}^{\dot\alpha\dot\beta}, x^\mu\} = -\frac{1}{R^2}\left(\eta^{\dot\alpha\mu}x^{\dot\beta} - \eta^{\dot\beta\mu}x^{\dot\alpha}\right) \tag{10.1.9}$$

in Cartesian coordinates. In tensorial form, this is

$$T_{\nu\sigma}{}^\mu = -\frac{1}{R^2\rho^2}\left(\delta^\mu_\nu\tau_\sigma - \delta^\mu_\sigma\tau_\nu\right), \tag{10.1.10}$$

where

$$\tau_\mu = G_{\mu\nu}\tau^\nu = G_{\mu\nu}x^\nu = \sinh(\eta)\eta_{\mu\nu}x^\nu \tag{10.1.11}$$

is the global timelike FLRW vector field along the cosmic expansion.

10.2 Deformed spacetime $\mathcal{M}^{3,1}$ and higher spin

10.2.1 Frame and metric

Now we want to admit generic perturbations or deformations of this cosmological spacetime, which is considered as a model for our universe. This is achieved by replacing the matrix background $\bar T^{\dot\alpha}$ by some generic deformation as in (7.4.1),

$$T^{\dot\alpha} = \bar T^{\dot\alpha} + \mathcal{A}^{\dot\alpha} \qquad \in \mathrm{End}(\mathcal{H}_n) \sim \mathcal{C}. \tag{10.2.1}$$

We will *not* restrict ourselves to the linearized theory around $\bar T^{\dot\alpha}$, but consider $T^{\dot\alpha}$ as new background, which in the semi-classical regime amounts to four deformed functions on $\mathbb{C}P^{2,1}$. This deformed background defines a frame

$$E^{\dot\alpha} = \{T^{\dot\alpha}, .\}. \tag{10.2.2}$$

We can still use the undeformed Cartesian coordinates[1] x^μ and t^μ to describe spacetime $\mathcal{M}^{3,1}$ and the fiber S^2 of $\mathbb{C}P^{2,1}$; then the symplectic form and hence $\rho_M \sim \sinh(\eta)^{-1}$ are undeformed. Then the geometrical considerations of Section 9 essentially go through, provided we restrict ourselves to the effectively four-dimensional regime (5.4.76) where the derivatives along the internal S^2 fiber can be neglected (i.e. assuming that only modes with low spin are relevant). We can then consider

$$E^{\dot\alpha} = E^{\dot\alpha\mu}(x,t)\partial_\mu \tag{10.2.3}$$

as $\mathfrak{hs}$-valued frame acting on $\mathcal{M}^{3,1}$ with associated $\mathfrak{hs}$-valued Weitzenböck connection and torsion, treating the t^μ generators as constant. In this sense, the frame defines a $\mathfrak{hs}$-valued $3 + 1$ dimensional effective metric

$$G_{\mu\nu} = \rho^2 \gamma_{\mu\nu} \tag{10.2.4}$$

on $\mathcal{M}^{3,1}$, where

$$\gamma^{\mu\nu} = E_{\dot\alpha}{}^\mu E^{\dot\alpha\nu}$$
$$\rho^2 = \rho_M^{-1}\sqrt{|G_{\mu\nu}|} = \rho_M \det E^{\dot\alpha\nu} \tag{10.2.5}$$

cf. (9.1.10), (9.1.14). We will see that at any given point, there is always a choice of local generalized Riemann normal coordinates $\tilde{y}^\mu$ where all higher-spin components of the (effective) frame and metric vanish. This provides the basis for a physical interpretation in terms of a $3 + 1$ dimensional spacetime. The coupling constant g in the matrix model is again absorbed in the background, so that the matrices $T^{\dot\alpha}$ and the dilaton ρ are dimensionless.

10.2.2 Divergence constraint on covariant quantum spaces

The derivation of the divergence constraint in Section 9.2.1 does not apply directly to the case of covariant quantum spaces. Nevertheless, the same constraint holds. This can be seen by recalling the identities $\eth_a\{Y,x^a\} = 0$ (5.2.53) and $x_a\{Y,x^a\} = 0$ for any $Y \in \mathcal{C}$, which imply that $V^a = \{Y,x^a\}$ is a divergence-free tangential vector field on H_n^4. As shown in Lemma 5.8, this implies that the vector field $V^\mu = \{Y,x^\mu\}$ on $\mathcal{M}^{3,1}$ is conserved in the sense

$$\partial_\mu(\rho_M V^\mu) = 0 \tag{10.2.6}$$

for any $\mathfrak{hs}$-valued Y, where ∂_μ is the derivation defined in (5.4.29). Alternatively, this follows from (5.4.30).

---

[1] The x^μ and t^μ can be viewed as "Darboux-like" coordinates on the bundle, in terms of which the Poisson structure has the canonical form (5.4.9).

10.2.3 Higher-spin gauge trafos and volume-preserving diffeos

Now consider the gauge transformations $T^{\dot\alpha} \to UT^{\dot\alpha}U^{-1}$ of the matrix model, or their infinitesimal versions

$$T^{\dot\alpha} \to -i[\Lambda, T^{\dot\alpha}], \qquad \Lambda = \Lambda^\dagger \in \mathrm{End}(\mathcal{H}) \tag{10.2.7}$$

for $U = e^{i\Lambda}$. For the $3+1$ dimensional cosmological background $\mathcal{M}^{3,1}$ (10.1.1), they can be interpreted as higher-spin gauge transformations, which include volume-preserving diffeos on spacetime. To see this, we parametrize the generator $\Lambda \in \mathcal{C}^s \subset \mathrm{End}(\mathcal{H})$ as follows:

$$\Lambda = \Lambda_{\mu_1 \ldots \mu_s}(x) t^{\mu_1} \ldots t^{\mu_s} \qquad \in \mathcal{C}^s \tag{10.2.8}$$

in terms of higher-spin fields on $\mathcal{M}^{3,1}$, as discussed in Section 5.4.3. For example, consider the spin 1 gauge transformations generated by

$$\Lambda = \Lambda_\mu(x) t^\mu \qquad \in \mathcal{C}^1. \tag{10.2.9}$$

Functions $\phi \in \mathcal{C}^0$ transform under these spin 1 gauge trafos as

$$\delta_\Lambda \phi = \{\Lambda, \phi\} = \mathcal{L}_\xi \phi = \xi^\mu \partial_\mu \phi, \qquad \xi^\mu = \{\Lambda, x^\mu\} \tag{10.2.10}$$

which is a diffeomorphism[2] generated by a vector field ξ^μ. However, this vector field typically contains also some spin 2 components

$$\xi^\mu = \xi^\mu_{(0)} + \xi^\mu_{(2)} \qquad \in \mathcal{C}^0 \oplus \mathcal{C}^2 \tag{10.2.11}$$

determined by $\xi^\mu_{(2)} = \mathcal{D}^{++}(\xi^\mu_{(0)})$. Both components satisfy the divergence constraint (10.2.6) $\partial_\mu(\rho_M \xi^\mu) = 0$, which can be written covariantly as in Section 9.2.1

$$\nabla^{(G)}_\mu(\rho^{-2}\xi^\mu) = 0, \tag{10.2.12}$$

where $\nabla^{(G)}$ is the Levi–Civita covariant derivative associated with the effective metric (10.2.4). In this sense, ξ^μ corresponds to a volume-preserving diffeo. Recalling the results of Section 9.1.3, it follows that these spin 1 gauge transformations induce the usual gauge transformations of the metric

$$\delta_\Lambda G_{\mu\nu} = \nabla^{(G)}_\mu \xi_\nu + \nabla^{(G)}_\nu \xi_\mu = \mathcal{L}_\xi G_{\mu\nu}. \tag{10.2.13}$$

Hence, the spin 1 gauge transformations can be interpreted as (modified) volume-preserving diffeomorphisms on $\mathcal{M}^{3,1}$. More generally, (10.2.10) provides a geometric interpretation of higher-spin gauge transformations: They are simply Hamiltonian vector fields on $\mathbb{C}P^{2,1}$, which generate the symplectomorphism group. Since these preserve the symplectic volume on $\mathbb{C}P^{2,1}$, the resulting $\mathfrak{hs}$-valued diffeos on $\mathcal{M}^{3,1}$ are volume-preserving in the sense of (10.2.12).

---

[2] The restriction to time-like Λ_μ is reflected by the divergence constraint.

10.2.4 Geometric backgrounds and frame reconstruction

For covariant quantum spaces, the frame is in general higher-spin valued, and so is the metric. To describe real physics, we should be able to recover classical geometries in terms of backgrounds that contain no significant higher-spin contributions. This leads to the following problem: For any given classical divergence-free frame $e_{\dot{\alpha}}{}^{\mu} \in \mathcal{C}^0$, we would like to find generators $T_{\dot{\alpha}}$ such that the (generally $\mathfrak{hs}$-valued) reconstructed frame

$$E_{\dot{\alpha}}{}^{\mu} = \{T_{\dot{\alpha}}, x^{\mu}\} \qquad (10.2.14)$$

reproduces the classical frame through its spin 0 component $\mathcal{C}^0$:

$$[E_{\dot{\alpha}}{}^{\mu}]_0 = [\{T_{\dot{\alpha}}, x^{\mu}\}]_0 = e_{\dot{\alpha}}{}^{\mu}. \qquad (10.2.15)$$

This problem of **frame reconstruction** is of central importance, since gravity requires to realize generic geometries in the matrix model framework. A solution of this problem was given in Sections 5.4.6 as follows:

$$T_{\dot{\alpha}} = \mathcal{D}^{+}(e_{\dot{\alpha}}) = -3(\Box_H - 4r^2)^{-1}\big(\{e_{\dot{\alpha}}{}^{\mu}, x_{\mu}\} + \frac{1}{x_4}\{x_4, x_{\mu}e_{\dot{\alpha}}{}^{\mu}\}\big), \qquad \in \mathcal{C}^1 \quad (10.2.16)$$

where

$$E_{\dot{\alpha}}{}^{\mu} = \{T_{\dot{\alpha}}, x^{\mu}\} = e_{\dot{\alpha}}{}^{\mu} + \mathcal{D}^{++}(e_{\dot{\alpha}}{}^{\mu}) \in \mathcal{C}^0 \oplus \mathcal{C}^2 \qquad (10.2.17)$$

contains also $\mathcal{C}^2$ contributions as in (5.4.111). The component in $\mathcal{C}^2$ vanishes only for very special "pure backgrounds."

Pure backgrounds

Consider the class of **pure backgrounds** $T_{\dot{\alpha}} \in \mathcal{C}^1$, which have the property that the frame is a pure function,

$$E_{\dot{\alpha}}{}^{\mu} = \{T_{\dot{\alpha}}, x^{\mu}\} \qquad \in \mathcal{C}^0. \qquad (10.2.18)$$

According to Lemma 5.7, all such backgrounds are some linear combination of $\mathfrak{so}(4, 1)$ generators

$$T_{\dot{\alpha}} = e_{\dot{\alpha};bc}\theta^{bc} \qquad \in \mathcal{C}^1. \qquad (10.2.19)$$

This leads to the $\mathcal{C}^0$-valued frame and torsion

$$E_{\dot{\alpha}}{}^{\mu} = e_{\dot{\alpha};bc}\{\theta^{bc}, x^{\mu}\} = -r^2 e_{\dot{\alpha};bc}(\eta^{b\mu}x^c - \eta^{c\mu}x^b),$$
$$T_{\dot{\alpha}\dot{\beta}}{}^{\mu} = -\{\{T_{\dot{\alpha}}, T_{\dot{\beta}}\}, x^{\mu}\} = r^2 c_{\dot{\alpha}\dot{\beta};ab}(\eta^{a\mu}x^b - \eta^{b\mu}x^a), \qquad (10.2.20)$$

where $\{T_{\dot{\alpha}}, T_{\dot{\beta}}\} = c_{\dot{\alpha}\dot{\beta};ab}\theta^{ab}$. These configurations comprise the cosmic background solution (10.1.1), which is recovered for

$$T_{\dot{\alpha}} = t_{\dot{\alpha}} = \frac{1}{2R}(\delta_c^4 \eta_{\dot{\alpha}b} - \delta_b^4 \eta_{\dot{\alpha}c})\theta^{bc}. \qquad (10.2.21)$$

In particular, any classical frame of the form

$$e_{\dot{\alpha}}{}^{\mu} = \sinh(\eta)\tilde{e}_{\dot{\alpha}}^{\mu} \qquad (10.2.22)$$

with constant (!) $\tilde{e}_{\dot{\alpha}}^{\mu} \in \mathbb{R}$ is reproduced by the pure background

$$T_{\dot{\alpha}} = \tilde{e}_{\dot{\alpha}}^{\mu} t_{\mu}. \qquad (10.2.23)$$

More generally, all such frames $E_{\dot{\alpha}}{}^{\mu}$ (10.2.20) are in the kernel of the map $\mathcal{D}^{++}$ (5.2.121) that links the components in $\mathcal{C}^0$ and $\mathcal{C}^2$, and they are closed under $SO(4,1)$ gauge transformations. They provide clean configurations which can be used as a starting point for a perturbative approach around any point $\xi \in \mathcal{M}^{3,1}$.

Generic backgrounds and local linearization

In general, the reconstructed frame $E^{\dot{\alpha}} = \{T^{\dot{\alpha}}, .\}$ will contain components in $\mathcal{C}^2$, which are obtained from the $\mathcal{C}^0$ components via $\mathcal{D}^{++}$. However, we have seen that any given classical frame $e^{\dot{\alpha}}$ can be reproduced at some fixed point p by a pure frame, which is in the kernel of $\mathcal{D}^{++}$. Writing the classical frame in the form[3]

$$e_{\dot{\alpha}}^{\mu} = \bar{e}_{\dot{\alpha}}^{\mu} + \delta e_{\dot{\alpha}}^{\mu}, \qquad (10.2.24)$$

where $\bar{e}_{\dot{\alpha}}^{\mu}$ is pure (hence essentially constant) and $\delta e_{\dot{\alpha}}^{\mu}$ vanishes at p, the reconstructed frame is given by

$$E_{\dot{\alpha}}^{\mu} = \bar{e}_{\dot{\alpha}}^{\mu} + \delta E_{\dot{\alpha}}^{\mu}, \qquad \delta E_{\dot{\alpha}}^{\mu} = (1 + \mathcal{D}^{++})\delta e_{\dot{\alpha}}^{\mu} \qquad (10.2.25)$$

with corresponding metric

$$\gamma^{\mu\nu} = \bar{e}_{\dot{\alpha}}^{\mu} \bar{e}^{\dot{\alpha}\nu} + \bar{e}_{\dot{\alpha}}^{\mu} \delta E^{\dot{\alpha}\nu} + \delta E_{\dot{\alpha}}^{\mu} \bar{e}^{\dot{\alpha}\nu} + O(\delta E^2)$$

$$[\gamma^{\mu\nu}]_0 = e_{\dot{\alpha}}^{\mu} e^{\dot{\alpha}\nu} + O(\delta E^2). \qquad (10.2.26)$$

Since the ħs contribution will drop out at the linearized level in the action, this justifies the present reconstruction procedure for the frame and the metric, in the linearized regime. More specifically, the $\mathcal{C}^2$ component of $\delta E_{\dot{\alpha}}^{\mu}$ has typically the same size as the classical perturbation $\delta e_{\dot{\alpha}}^{\mu}$ of the frame, since $\mathcal{D}^{++}$ is norm-preserving (5.2.122). Moreover, since the torsion $T \sim e^{-1}\partial e$ is set by the curvature scale $\mathcal{R} \sim T^2$, the frame is essentially constant in a neighborhood which is small compared to the curvature scale of gravity. Hence, in the linearized regime, the frame is well-approximated by its classical spin 0 component $e_{\dot{\alpha}}^{\mu}$, and the metric is well approximated by its classical component $[G_{\mu\nu}]_0$; this will be clarified further in Section 10.2.5.

Now consider the torsion and curvature, which are given by derivatives of the frame. In the linearized regime, the torsion of the reconstructed frame satisfies

$$T_{\dot{\alpha}\dot{\beta}}{}^{\mu} \approx (1 + \mathcal{D}^{++})\left[T_{\dot{\alpha}\dot{\beta}}{}^{\mu}\right]_0 \qquad (10.2.27)$$

---

[3] Note that this can be achieved using a volume-preserving diffeo, i.e. using a gauge transformation.

to a good approximation due to (5.2.123), and similarly for the contorsion $\mathcal{K}$. It is thus no longer evident that the $\mathfrak{hs}$ component is smaller than the classical $\mathcal{C}^0$ contribution. Similarly, the linearized Riemannian curvature tensor can be written using(9.3.41) as

$$\mathcal{R}_{\mu\nu}{}^{\lambda}{}_{\sigma} \approx -\partial_{\mu}\mathcal{K}_{\nu\sigma}{}^{\lambda} + \partial_{\nu}\mathcal{K}_{\mu\sigma}{}^{\lambda} \approx (1 + \mathcal{D}^{++})[\mathcal{R}_{\mu\nu}{}^{\lambda}{}_{\sigma}]_0$$
$$\mathcal{R}_{\nu\sigma} \approx -\partial_{\mu}\mathcal{K}_{\nu\sigma}{}^{\mu} + \partial_{\nu}\mathcal{K}_{\mu\sigma}{}^{\mu} \approx (1 + \mathcal{D}^{++})[\mathcal{R}_{\nu\sigma}]_0 \qquad (10.2.28)$$

dropping the quadratic contributions in the (con)torsion, assuming that the torsion of the background is negligible. Therefore, the $\mathfrak{hs}$ components of the linearized Einstein–Hilbert term $\int[\mathcal{R}]_0$ drop out. Moreover, the isometry property (5.2.122) implies that the quadratic contribution of a gravitational action of the form (9.3.44) is essentially equivalent to the classical $\mathcal{C}^0$ contribution. Hence, the geometrical tensors can be considered as effectively classical in the linearized regime.

We can summarize these considerations as follows: The perturbations of the frame lead to classical and $\mathcal{C}^2$-valued contributions to the frame, torsion, and curvature of comparable size, but the $\mathcal{C}^2$ components can be neglected in the linearized regime.

In the nonlinear regime, the $\mathfrak{hs}$ contributions do contribute, so that the reconstruction procedure needs to be refined. However, as discussed in Section 10.2.5, one can always choose local normal coordinates such that all $\mathfrak{hs}$ components of the frame vanish at some given point $p \in \mathcal{M}^{3,1}$. In this sense, the metric always reduces to a classical $3 + 1$ dimensional metric, which governs the local physics near p. One can then use a suitably modified reconstruction procedure around p in the weak gravity regime, where the curvature scale is much smaller than all other physical scales. Then the previous statements apply locally, and the global geometry can be recovered by combining the locally linearized patches.

Another possibility is to refine the reconstruction procedure by adding suitable $\mathfrak{hs}$ corrections to the generators $T_{\dot{\alpha}}$, such that the $\mathfrak{hs}$ components of the frame vanish in a larger region. This is illustrated for spherically symmetric geometries in [73]. The best approach to recover general classical geometries remains to be clarified.

Example: Newtonian frame

To illustrate the reconstruction of classical geometries and the extra $\mathfrak{hs}$ components, consider the following classical static (comoving) frame

$$e^{\dot{0}0} = -\sinh(\eta)(1 + 2V(\vec{x})) = -\sinh(\eta) + \delta e^{\dot{0}0}, \qquad e^{\dot{k}l} = \sinh(\eta)\delta^{\dot{k}l} \qquad (10.2.29)$$

in Cartesian coordinates x^{μ}, with all off-diagonal components vanishing. This can be seen as a perturbation of the pure background $\bar{e}^{\dot{\alpha}}$ through V, which decays $V(\vec{x}) \to 0$ as $|\vec{x}| \to \infty$. For simplicity, we assume that ρ_M is locally constant, so that the divergence constraint is (approximately) satisfied. The corresponding metric and dilaton are

$$\gamma^{\mu\nu} = \sinh^2(\eta)\text{diag}(-(1 + 2V)^2, 1, 1, 1), \qquad \rho^2 = \frac{\rho_M}{\sqrt{|\gamma_{\mu\nu}|}} = \sinh^3(\eta)(1 + 2V)$$

$$G^{\mu\nu} = \sinh^{-1}(\eta)(1 + 2V)^{-1}\text{diag}(-(1 + 2V)^2, 1, 1, 1). \qquad (10.2.30)$$

Dropping the (locally constant) factor $\sinh^{-1}(\eta) = \rho_M$, this corresponds to a Newtonian potential $V(x)$. We can reconstruct this frame via $E^{\dot{\alpha}} = \{T^{\dot{\alpha}}, .\}$ using the map $\mathcal{D}^+$ as discussed before. The only nontrivial component is $\delta e^{\dot{0}} = -2V\partial_0 \equiv V^\mu \partial_\mu$, for which we find

$$\{x^\mu, V_\mu\} = \{x^0, 2V\} = 2\theta^{0i}\partial_i V \approx 2r^2 R t^i \partial_i V$$

$$x^\mu V_\mu = 2Vx^0$$

$$\{x^4, \frac{1}{x^4}x^\mu V_\mu\} \sim 2\{x^4, V\} = 2r^2 R t^\mu \partial_\mu V \tag{10.2.31}$$

using (5.4.34). Therefore, (5.4.110) gives

$$\delta T^{\dot{0}} = \mathcal{D}^+(\delta e^{\dot{0}}) = -12(\Box_H - 4r^2)^{-1}(t^i \partial_i V) \approx -12\Box_H^{-1}(t^i \partial_i V) \tag{10.2.32}$$

for static V in the effectively 4D regime. Note that for rotationally invariant $V = V(\vec{x}^2)$, this simplifies as $t^i \partial_i V = 2t^i x_i V' = -2t^0 x_0 V'$ due to (5.4.7c).

In practice, it is easier to reconstruct the frame by making an ansatz

$$T^{\dot{0}} = t^{\dot{0}}(1 + w(x)). \tag{10.2.33}$$

Then

$$E^{\dot{0}\mu} = \{t^{\dot{0}}(1 + w), x^\mu\} = \sinh(\eta)\eta^{\dot{0}\mu}(1 + w) - t^0\theta^{\mu i}\partial_i w, \tag{10.2.34}$$

where the second term contains the higher-spin contributions in $\mathcal{C}^2$ as discussed before. Clearly $E^{\dot{0}0} \gg E^{\dot{0}i}$ at late times $\eta \to \infty$, hence it suffices to consider $E^{\dot{0}0}$. Then

$$E^{\dot{0}0} = -\sinh(\eta)(1 + w) - t^0\theta^{0i}\partial_i w$$

$$= -\sinh(\eta)(1 + w) - r^2 R t^0 t^i \partial_i w \quad \in \mathcal{C}^0 \oplus \mathcal{C}^2, \tag{10.2.35}$$

with classical component

$$[E^{\dot{0}0}]_0 \approx -\sinh(\eta)\left(1 + w - \frac{1}{3}x^i \partial_i w\right) \tag{10.2.36}$$

using (5.4.20). Comparing with (10.2.29), we obtain

$$(1 - \frac{1}{3}x^i \partial_i)w \overset{!}{=} 2V(x) \tag{10.2.37}$$

with the solution

$$w(x) = 6\int_1^\infty \frac{dt}{t^4}V(tx). \tag{10.2.38}$$

This means that $w(x)$ is of similar magnitude as $V(x)$. We can thus estimate the $\mathcal{C}^2$ component

$$[E^{\dot{0}0}]_2 \approx |r^2 R t^0 t^i \partial_i w| \approx \sinh(\eta)|\vec{x}|\partial w \tag{10.2.39}$$

using $r|t_0| \sim \frac{|\vec{x}|}{R}$, which indicates that the $\mathcal{C}^0$ and $\mathcal{C}^2$ components of the derivative contributions to δE around the nonderivative (pure) background $\bar{e}^{\dot{0}0} = -\sinh(\eta)$ have comparable magnitude. Since the $\mathcal{C}^2$ contributions enter in the action only at second order, they can be neglected in the linearized approximation. However, they cannot be neglected for the torsion. We conclude that the derivative contributions lead to classical and $\mathcal{C}^2$-valued contributions to the curvature and torsion of comparable size.

For example, consider a Newtonian potential $V = -\frac{MG}{|\vec{x}|}$ centered at $\vec{x} = 0$. Then

$$w(x) = -\frac{3}{2}\frac{MG}{|\vec{x}|},\tag{10.2.40}$$

and we obtain

$$E^{\dot{0}0} \approx -\sinh(\eta)\left(1 - \frac{3}{2}\frac{MG}{|\vec{x}|} - \frac{3}{2}r^2R^2t_0^2\frac{MG}{|\vec{x}|^3}\right).\tag{10.2.41}$$

This has indeed the desired classical component

$$[E^{\dot{0}0}]_0 = -\sinh(\eta)\left(1 - \frac{3}{2}\frac{MG}{|\vec{x}|} - \frac{1}{2}R^2\kappa^{00}\frac{MG}{|\vec{x}|^3}\right)$$

$$= -\sinh(\eta)\left(1 - 2\frac{MG}{|\vec{x}|}\right) = e^{\dot{0}0},\tag{10.2.42}$$

noting that $\kappa^{00} = \frac{|\vec{x}|^2}{R^2}$. The $\mathcal{C}^2$ component

$$[E^{\dot{0}0}]_2 \approx \frac{3}{2}\sinh(\eta)\,r^2R^2[t_0^2]_2\frac{MG}{|\vec{x}|^3}\tag{10.2.43}$$

has the same magnitude as $\delta e^{\dot{0}0}$ as $rR|t_0| \sim |\vec{x}|$.

Local reconstruction

In practice, reconstructing some given geometry using the map $\mathcal{D}^+$ is difficult to carry out explicitly. As in the previous example, it is more convenient to make an ansatz for the matrix background near some reference point p as follows:

$$T^{\dot{\alpha}} = \tilde{e}^{\dot{\alpha}\mu}t_\mu + \tilde{e}_\rho^{\dot{\alpha}\mu}y^\rho t_\mu + \frac{1}{2}\tilde{e}_{\rho\sigma}^{\dot{\alpha}\mu}y^\rho y^\sigma t_\mu + \ldots,\tag{10.2.44}$$

where

$$y^\mu = x^\mu - p^\mu\tag{10.2.45}$$

are Cartesian coordinates centered at p, and the coefficients $\tilde{e}^{\dot{\alpha}\mu}$, $\tilde{e}_\rho^{\dot{\alpha}\mu}$ etc. are constant. This leads to the frame

$$E^{\dot{\alpha}\mu} = \{T^{\dot{\alpha}}, x^\mu\} = \sinh(\eta)\tilde{e}^{\dot{\alpha}\mu} + \tilde{e}_\rho^{\dot{\alpha}\nu}t_\nu\theta^{\rho\mu} + \left(\sinh(\eta)\tilde{e}_\sigma^{\dot{\alpha}\mu} + \tilde{e}_{\rho\sigma}^{\dot{\alpha}\nu}t_\nu\theta^{\rho\mu}\right)y^\sigma + O(y^2).$$

$$\tag{10.2.46}$$

Dropping $\tilde{e}_{\rho\sigma}^{\dot{\alpha}\nu}$, we can reconstruct any $\mathcal{C}^0$ component of the frame to linear order in y by a suitable choice of $\tilde{e}_\sigma^{\dot{\alpha}\mu}$ and $\tilde{e}^{\dot{\alpha}\mu}$ in

$$E^{\dot{\alpha}\mu} = \{T^{\dot{\alpha}}, x^\mu\} = \sinh(\eta)\tilde{e}^{\dot{\alpha}\mu} + \tilde{e}_\rho^{\dot{\alpha}\nu}t_\nu\theta^{\rho\mu} + \sinh(\eta)\tilde{e}_\sigma^{\dot{\alpha}\mu}y^\sigma.\tag{10.2.47}$$

This entails some $\mathcal{C}^2$ component of the frame, in accord with the general discussion. That component is characterized by the derivative $\sinh(\eta)\tilde{e}_\sigma^{\dot{\alpha}\mu} = \partial_\sigma E^{\dot{\alpha}\mu}|_p$ of the frame, which in turn characterizes the torsion. Therefore, we expect that the scale of the C^2 component of the frame is set by the torsion. In fact, we can locally eliminate this C^2 component of the frame at p – and more generally any $\mathfrak{hs}$ component – by choosing suitably adapted local normal coordinates.

10.2.5 Local normal coordinates

This section establishes the crucial fact that any given $\mathfrak{hs}$-valued matrix background defines locally some ordinary $3+1$ dimensional geometry. To see this, we observe that the frame $E_{\dot\alpha}{}^\mu = \{T_{\dot\alpha}, x^\mu\}$ and all derived objects such as the torsion are tensors, considering the frame indices $\dot\alpha$ as fixed. This means that they transform as $\mathfrak{hs}$-valued tensors under coordinate redefinitions

$$\tilde{y}^\mu = \phi^\mu(x), \tag{10.2.48}$$

i.e.

$$\tilde{E}_{\dot\alpha}{}^\mu = \{T_{\dot\alpha}, \tilde{y}^\mu\} = \frac{\partial\phi^\mu}{\partial x^\nu} E_{\dot\alpha}{}^\nu \tag{10.2.49}$$

assuming that $\phi^\mu(x)$ depends only on x^μ. However, the general framework laid out so far does not depend on the specific choice of base and fiber coordinates x^μ and t^ν. We can take advantage of this freedom to cast any $\mathfrak{hs}$-valued frame locally into some standard form in terms of suitably adapted $\mathfrak{hs}$-valued coordinates

$$\tilde{y}^\mu = \phi^\mu(x, t). \tag{10.2.50}$$

Then the higher-spin components of the frame can always be eliminated locally, in analogy to Riemann normal coordinates in GR:

Lemma 10.1 *Let $E_{\dot\alpha} = \{T_{\dot\alpha}, .\}$ be any $\mathfrak{hs}$-valued frame. Consider some point $p \in \mathcal{M}^{3,1}$, and assume that $E_{\dot\alpha}{}^\mu|_p$ is invertible. Choose shifted coordinates $y^\mu = x^\mu - p^\mu$ which are centered at p, i.e.*

$$y^\mu|_p = 0. \tag{10.2.51}$$

Then we can construct local coordinates $\tilde{y}^\mu$ around p of the form

$$\tilde{y}^\mu = \phi^\mu_\sigma(t)y^\sigma + \phi^\mu{}_{\alpha\dot\beta}(t)y^\alpha y^{\dot\beta} \tag{10.2.52}$$

such that the frame $\tilde{E}_{\dot\alpha}{}^\mu = \{T_{\dot\alpha}, \tilde{y}^\mu\}$ satisfies

$$\tilde{E}_{\dot\alpha}{}^\mu\big|_p = \delta^\mu_{\dot\alpha}, \qquad \gamma^{\mu\nu}\big|_p = \eta^{\mu\nu}$$
$$\partial_\sigma \gamma^{\mu\nu}\big|_p = 0 \tag{10.2.53}$$

at $p \in \mathcal{M}^{3,1}$.

In other words, all $\mathfrak{hs}$ components of the frame and the metric can be absorbed locally by this change of coordinates, corresponding to a local inertial system.

Proof   Consider first the linear coordinate transformation

$$\tilde{y}^{\mu} := \phi^{\mu}_{\sigma} y^{\sigma}, \tag{10.2.54}$$

where $\phi^{\mu}_{\sigma} = \phi^{\mu}_{\sigma}(t)$ are independent of y but may depend on the fiber. Then

$$\tilde{E}_{\dot{\alpha}}{}^{\mu}\big|_{p} = \{T_{\dot{\alpha}}, \tilde{y}^{\mu}\}\big|_{p} = \{T_{\dot{\alpha}}, \phi^{\mu}_{\sigma} y^{\sigma}\}\big|_{p} = \left(\phi^{\mu}_{\sigma}\{T_{\dot{\alpha}}, y^{\sigma}\}\right)\big|_{p}$$
$$= \phi^{\mu}_{\sigma} E_{\dot{\alpha}}{}^{\sigma}\big|_{p} = \delta^{\mu}_{\dot{\alpha}} \tag{10.2.55}$$

since $y(p) = 0$, for

$$\phi^{\mu}_{\sigma} := \left(E_{\sigma}{}^{\mu}\big|_{p}\right)^{-1}. \tag{10.2.56}$$

Thus, all $\mathfrak{hs}$ components can be eliminated at any given point provided $E_{\dot{\alpha}}{}^{\mu}\big|_{p}$ is invertible.

Next, we refine these coordinates and impose a normal form also on the derivatives of the frame. One might hope to impose $\partial E_{\dot{\alpha}}{}^{\mu}\big|_{p} = 0$, but this would be asking too much: the torsion is a tensor characterized by the derivatives $\partial E_{\dot{\alpha}}{}^{\mu}$, which cannot be removed by a change of coordinates. All we can ask for is that the derivatives of the metric $\partial \gamma^{\mu\nu}\big|_{p} = 0$ vanish at p, as in Riemann normal coordinates. Assuming $E_{\dot{\alpha}}{}^{\mu}\big|_{p} = \delta^{\mu}_{\dot{\alpha}}$ as before, this amounts to

$$0 = \eta^{\dot{\alpha}\dot{\nu}}\partial_{\sigma}E_{\dot{\alpha}}{}^{\mu}\big|_{p} + \eta^{\mu\dot{\beta}}\partial_{\sigma}E_{\dot{\beta}}{}^{\nu}\big|_{p}. \tag{10.2.57}$$

To achieve this, consider local coordinates centered at p of the form

$$\tilde{y}^{\mu} := y^{\mu} + \phi^{\mu}_{\alpha\beta} y^{\alpha} y^{\beta}, \tag{10.2.58}$$

where $\phi^{\mu}_{\alpha\beta}$ is a constant (possibly $\mathfrak{hs}$-valued) tensor which is symmetric in $\alpha\beta$. Then

$$\frac{\partial \tilde{y}^{\mu}}{\partial y^{\nu}} = \delta^{\mu}_{\nu} + 2\phi^{\mu}_{\nu\beta} y^{\beta} \tag{10.2.59}$$

so that

$$\partial_{\nu}\tilde{E}_{\dot{\alpha}}{}^{\mu}\big|_{p} = \partial_{\nu}\left((\delta^{\mu}_{\sigma} + 2\phi^{\mu}_{\sigma\beta} y^{\beta}) E_{\dot{\alpha}}{}^{\sigma}\right)\big|_{p}$$
$$= 2\phi^{\mu}_{\sigma\nu} E_{\dot{\alpha}}{}^{\sigma}\big|_{p} + (\delta^{\mu}_{\sigma} + 2\phi^{\mu}_{\sigma\beta} y^{\beta})\partial_{\nu} E_{\dot{\alpha}}{}^{\sigma}\big|_{p}$$
$$= 2\phi^{\mu}_{\dot{\alpha}\nu} + \partial_{\nu}E_{\dot{\alpha}}{}^{\mu}\big|_{p}. \tag{10.2.60}$$

Thus, (10.2.57) holds for $\tilde{E}$ if

$$0 = \eta^{\dot{\alpha}\nu}(\partial_{\sigma}E_{\dot{\alpha}}{}^{\mu}\big|_{p} + 2\phi^{\mu}_{\dot{\alpha}\sigma}) + \eta^{\dot{\alpha}\mu}(\partial_{\sigma}E_{\dot{\alpha}}{}^{\nu}\big|_{p} + 2\phi^{\nu}_{\dot{\alpha}\sigma}) \tag{10.2.61}$$

or

$$\eta^{\alpha\nu}\phi^{\mu}_{\alpha\sigma} + \eta^{\alpha\mu}\phi^{\nu}_{\alpha\sigma} = e^{\mu\nu}{}_{\sigma}, \tag{10.2.62}$$

where

$$e^{\mu\nu}{}_{\sigma} := -\frac{1}{2}\left(\eta^{\dot{\alpha}\nu}(\partial_{\sigma}E_{\dot{\alpha}}{}^{\mu})\big|_{p} + (\mu \leftrightarrow \nu)\right) = e^{\nu\mu}{}_{\sigma}. \tag{10.2.63}$$

This is solved by

$$\phi^\nu{}_{\rho\sigma} = \frac{1}{2}\left(e_\rho{}^\nu{}_\sigma + e_\sigma{}^\nu{}_\rho - e_{\sigma\rho}{}^\nu\right), \qquad (10.2.64)$$

which is manifestly symmetric in $\rho\sigma$, and all indices are raised and lowered with $\eta^{\mu\nu}$. $\quad\square$

In these $\tilde{y}^\mu$ coordinates, the matrix d'Alembertian takes the simple form $\square = \eta^{\mu\nu}\partial_\mu\partial_\nu$ at p, because the derivatives of $\gamma^{\mu\nu}$ vanish at p. In particular, the metric and the local propagation is always four-dimensional. Moreover, the Poisson brackets $\{T_{\dot\alpha}, \tilde{y}^\mu\}$ have locally the same structure as for the cosmic background $\mathcal{M}^{3,1}$. Hence, the basic features of the higher-spin gauge theory discussed in Section 10.3 can be carried over locally to a generic background[4]. The algebra of functions is redefined accordingly[5] as space of functions in $\tilde{y}^\mu$

$$\tilde{\mathcal{C}}^0 = \{\phi(\tilde{y})\}; \qquad (10.2.65)$$

the tilde will typically be dropped. All functions can then be written as

$$\phi(\tilde{y}^\mu, t) = \sum_{s,m}\phi_{sm}(\tilde{y})Y^{sm}(t), \quad \in \bigoplus_s \tilde{\mathcal{C}}^s \qquad (10.2.66)$$

where the $Y^{sm}(t)$ generate the spin s module $\tilde{\mathcal{C}}^s$ in the vicinity of p. However, the constraints (5.4.7c) take a different form in these coordinates.

An analogous normal form can be achieved for the effective metric, but not for both metrics simultaneously:

Effective metric and higher-spin manifold

The results of the previous section can be extended immediately to the effective metric, if we use the effective frame $\mathcal{E}_{\dot\alpha}{}^\mu$ (9.2.25) instead of $E_{\dot\alpha}{}^\mu$ in Lemma 10.1. We can then find local coordinates around $p \in \mathcal{M}^{3,1}$ such that

$$\mathcal{E}_{\dot\alpha}{}^\mu\big|_p = \delta_{\dot\alpha}^\mu, \qquad G_{\mu\nu}\big|_p = \eta_{\mu\nu}$$

$$\partial_\sigma G_{\mu\nu}\big|_p = 0. \qquad (10.2.67)$$

These correspond to local "free-falling elevator" coordinates, or generalized $\mathfrak{hs}$- valued Riemann normal coordinates. In this sense, the local physics is always governed by the standard $3+1$ dimensional Minkowski metric $G_{\mu\nu} = \eta_{\mu\nu}$, even in the presence of a strong gravitational field. All results of the previous chapter go through in a sufficiently small domain, as long as the divergence constraint is applicable in the normal coordinates at the length scale under consideration. In particular, linearized vacuum geometries are expected to be Ricci-flat (9.3.49), as long as the dilaton is approximately constant. Note that if the frame has significant $\mathfrak{hs}$ components, the effective metric can only be identified in local normal coordinates where the $\mathfrak{hs}$ components vanish.

---

[4] Note that the explicit form of the coordinate algebra was used only indirectly in the divergence constraint, which will apply at least locally also in the new coordinates.

[5] Note that such a $\mathfrak{hs}$ modification of x^μ amounts to a redefinition of the bundle projection.

The existence of local normal coordinates suggests that a background in the presence of strong gravity should be described globally as a *higher-spin manifold*, which is locally described via local normal coordinates in terms of $3 + 1$ dimensional tangent spaces equipped with a metric, patched together with $\mathfrak{hs}$-valued transition functions. The details of such a concept remain to be elaborated.

Locally undeformed $\mathfrak{hs}$-valued coordinates

Finally, we note that at the linearized level, the local elimination of the $\mathfrak{hs}$-valued frame components $E_{\dot\alpha}{}^{\mu} = \bar{e}_{\dot\alpha}{}^{\mu} + \delta E_{\dot\alpha}{}^{\mu}$ with $\delta E_{\dot\alpha}{}^{\mu} \in \mathfrak{hs}$ can be achieved with $\mathfrak{hs}$-valued coordinates whose C^0 part is undeformed, i.e.

$$\tilde{y}^{\mu} := y^{\mu} + O(\delta Ey), \tag{10.2.68}$$

where $O(\delta Ey)$ vanishes at the coordinate origin. Here, $y^{\mu} = x^{\mu} - p^{\mu}$ are shifted global Cartesian coordinates around the reference point p, and the $\mathfrak{hs}$ components vanish at p. Indeed, for the local coordinates $\tilde{y}^{\mu} = \phi^{\mu}_{\nu} y^{\nu}$ with

$$\phi^{\mu}_{\sigma} := \bar{e}_{\dot\alpha}{}^{\mu} E_{\sigma}{}^{\mu}\big|_{p}^{-1} = \delta^{\mu}_{\sigma} + O(\delta E) \tag{10.2.69}$$

(which implies (10.2.68)) the frame reduces to (10.2.55)

$$\tilde{E}_{\dot\alpha}{}^{\mu}\big|_{p} = \phi^{\mu}_{\sigma} E_{\alpha}{}^{\sigma}\big|_{p} = \bar{e}_{\dot\alpha}{}^{\mu}. \tag{10.2.70}$$

This means that the effective geometry near p can be extracted simply by projecting to the C^0 component of the frame, in the linearized regime. This applies in particular to some neighborhood in local normal coordinates, so that the global geometry can be obtained as usual by patching together local $\mathfrak{hs}$-valued coordinates.

Example: Newtonian background

To illustrate the local normal coordinates, consider again the Newtonian background (10.2.33)

$$T^{\dot 0} = t^{0}(1 + w(x)), \quad T^{\dot k} = t^{k}, \tag{10.2.71}$$

where $w(x)$ is static, with frame

$$E^{\dot 0\mu} = \sinh(\eta)\eta^{\dot 0\mu}(1 + w) - t^{0}\theta^{\mu i}\partial_{i}w, \quad E^{\dot k\mu} = \sinh(\eta)\eta^{\dot k\mu}. \tag{10.2.72}$$

At late times, it suffices to keep the diagonal terms with $E^{\dot 0 0}$ given by (10.2.35). Now consider some reference point p, and let $y^{\mu} = x^{\mu} - p^{\mu}$ be Cartesian coordinates centered at p. We wish to determine new local coordinates

$$\tilde{y}^{\mu} = \phi^{\mu}_{\nu} y^{\nu} \tag{10.2.73}$$

such that the frame reduces to $\tilde{E}^{\dot\alpha\mu}\big|_{p} = \eta^{\dot\alpha\mu}$. According to (10.2.55), this amounts to

$$\phi^{\mu}_{\sigma} E^{\dot\alpha\sigma} = \eta^{\dot\alpha\mu}, \tag{10.2.74}$$

which is solved by a diagonal matrix ϕ_σ^μ with nontrivial entries

$$\phi_0^0 = \frac{1}{\sinh\eta(1+w) + r^2 R t^0 t^i \partial_i w|_p}, \qquad \phi_i^i = \frac{1}{\sinh\eta}. \tag{10.2.75}$$

As observed before, the ℏs contributions could even be eliminated locally with $\mathcal{C}^2$-valued Cartesian coordinates in the linearized regime.

Local gauge transformations versus normal coordinates

Not all of the changes of coordinates considered so far arise from gauge transformations. This means that the formulation of physics is not necessarily the same in all local normal coordinates. A natural question is then what normal form of the metric and frame can be achieved using gauge transformations. We restrict ourselves again to transformations that preserve the point $p \in \mathcal{M}^{3,1}$ under consideration, i.e.

$$\mathcal{L}_\xi x^\mu = \{\Lambda, x^\mu\} = \xi^\mu \qquad \text{vanishes at } p \in \mathcal{M}^{3,1}. \tag{10.2.76}$$

Then (9.1.33)

$$\delta_\Lambda E^\mu = \xi^\rho \partial_\rho E^\mu - E^\rho \partial_\rho \xi^\mu = -E^\rho \partial_\rho \xi^\mu \tag{10.2.77}$$

at p, assuming the effectively four-dimensional regime, where derivatives along the internal S^2 can be neglected. Since spin 1 gauge transformations correspond to volume-preserving vector fields, we can always achieve $[\mathcal{E}_{\dot\alpha}{}^\mu]_0 = c\delta_{\dot\alpha}^\mu$ at any point p using such a gauge transformation. However, the ℏs components of the frame at p cannot be gauged away in general, and a precise characterization of a suitable normal form remains to be established.

10.3 Linearized ℏs gauge theory and fluctuation modes

Now we will discuss generic properties of the ℏs gauge theory defined by the semi-classical matrix model on $\mathcal{M}^{3,1}$, and elaborate the higher-spin valued fluctuation modes in the linearized regime. As expected, the linearized metric fluctuations will turn out to be Ricci-flat. The discussion should generalize to more general pure backgrounds – such as the $k = 0$ geometry in Section 5.5 – with some modifications.

As discussed in Section 7.5, the gauge-invariant quadratic action 7.5.3 governing the fluctuations $T^\alpha + \mathcal{A}^\alpha$ on the background is replaced upon gauge-fixing by (7.5.15)

$$S_2[\mathcal{A}] + S_{\text{gf}} = -\frac{2}{g^2} \int \frac{\Omega}{(2\pi)^3} \mathcal{A}_\alpha (\mathcal{D}^2 + m^2) \mathcal{A}^\alpha. \tag{10.3.1}$$

Here, $\mathcal{D}^2 = \Box - 2\mathcal{I}$ is the vector d'Alembertian with

$$\mathcal{I}(A)^\alpha := \{\{t^\alpha, t^\beta\}, \mathcal{A}_\beta\} = -\frac{1}{r^2 R^2}\{\theta^{\alpha\beta}, A_b\} = -\frac{1}{r^2 R^2}\tilde{\mathcal{I}}(A)^\alpha \tag{10.3.2}$$

(cf. (5.4.90)) and the ghost and fermionic terms are dropped. Our first task is therefore to find a complete set of eigenmodes $\mathcal{A}$ of $\mathcal{D}^2$ on the background T^α. We work in the semi-classical regime for simplicity, although all steps generalize to the noncommutative setting.

10.3.1 Diagonalization of $\mathcal{D}^2$ and fluctuation spectrum

The most general fluctuation modes can be expanded into higher-spin sectors according to (5.4.60), (5.4.61)

$$\mathcal{A}^\alpha = A^\alpha(x) + A^\alpha_\mu(x)\, t^\mu + A^\alpha_{\mu\nu}(x)\, t^\mu t^\nu + \cdots \ \in\ \mathcal{C}^0 \oplus \mathcal{C}^1 \oplus \mathcal{C}^2 \oplus \cdots. \tag{10.3.3}$$

The goal of this section is to organize these into eigenmodes of $\mathcal{D}^2$, taking advantage of the underlying $\mathfrak{so}(4,2)$ structure. On flat spacetime, the Poincare group would suggest an organization into plane waves with extra helicity structure. On $\mathcal{M}^{3,1}$, the isometry group $SO(3,1)$ is only six-dimensional and should not be confused with Lorentz symmetry, since there is no (local or global) boost invariance. Accordingly, the $\mathcal{C}^s$ sectors in (10.3.3) decompose into the subsectors $\mathcal{C}^{(s,k)}$ as discussed in Section 5.4.4. In particular, the would-be massive spin s modes decompose into submodes that carry the degrees of freedom of massless spin $s - k$ modes for $k = 0, \ldots, s$. However, these submodes are linked by the extra structure provided by $\mathfrak{so}(4,2)$, notably the $D^\pm$ maps (5.4.83).

An explicit construction of the eigenmodes of $\mathcal{D}^2$ can be obtained as follows [174, 188]. The most obvious eigenmodes are the pure gauge modes

$$\mathcal{A}^{(g)}_\alpha[\phi^{(s)}] = \{t_\alpha, \phi^{(s)}\} \quad \in \mathcal{C}^s, \tag{10.3.4}$$

which are by construction zero modes of the gauge-invariant action (7.5.3), provided the mass parameter is given by $m^2 = -\frac{3}{R^2}$ such that $\mathcal{M}^{3,1}$ is a classical solution. It is then easy to verify either using gauge invariance or by direct verification that

$$\mathcal{D}^2 \mathcal{A}^{(g)}_\alpha[\phi] = \mathcal{A}^{(g)}_\alpha\Big[\big(\Box + \frac{3}{R^2}\big)\phi\Big]. \tag{10.3.5}$$

This means that $\mathcal{A}^{(g)}_\alpha[\phi]$ is an eigenmode of $\mathcal{D}^2$ if $\Box\phi = \lambda\phi$, which provides a tower of $\mathfrak{hs}$-valued eigenmodes determined by $\phi^{(s)} \in \mathcal{C}^s$. Since the most general fluctuations $\mathcal{A} \in \mathbb{R}^4 \otimes \mathcal{C}$ comprise four times the degrees of freedom of $\mathcal{C}$, one should expect three further towers of $\mathfrak{hs}$-valued eigenmodes. They can be obtained by acting with $D^\pm$ on $\mathcal{A}^{(g)}_\alpha$, observing that the intertwiner relations (5.4.89) together with $[D^\pm, \mathcal{I}] = 0$ imply

$$\mathcal{D}^2 D^+ \mathcal{A}^{(s)} = D^+(\mathcal{D}^2 + \frac{2s+2}{R^2})\mathcal{A}^{(s)},$$

$$\mathcal{D}^2 D^- \mathcal{A}^{(s)} = D^-(\mathcal{D}^2 - \frac{2s}{R^2})\mathcal{A}^{(s)}, \qquad \mathcal{A}^{(s)} \in \mathcal{C}^{(s)}. \tag{10.3.6}$$

Two of these modes are obtained explicitly from

$$D^\pm \mathcal{A}^{(g)}_\alpha[\phi^{(s)}] = \mathcal{A}^{(g)}_\alpha[D^\pm \phi^{(s)}] + \frac{1}{R}\mathcal{A}^{(\pm)}_\alpha[\phi^{(s)}] \tag{10.3.7}$$

using the Jacobi identity, where

$$\mathcal{A}^{(\pm)}_\alpha[\phi^{(s)}] = \{x_\alpha, \phi^{(s)}\}\big|_{s\pm1} \equiv \{x_\alpha, \phi^{(s)}\}_\pm \quad \in \mathcal{C}^{s\pm1} \tag{10.3.8}$$

are the modes defined in (5.2.99) restricted to $\alpha = 0, \ldots, 3$. Using (10.3.6), we obtain

$$\mathcal{D}^2 \mathcal{A}_\alpha^{(+)}[\phi^{(s)}] = \mathcal{A}_\alpha^{(+)}\Big[\big(\Box + \frac{2s+5}{R^2}\big)\phi^{(s)}\Big], \qquad (10.3.9)$$

$$\mathcal{D}^2 \mathcal{A}_\alpha^{(-)}[\phi^{(s)}] = \mathcal{A}_\alpha^{(-)}\Big[\big(\Box + \frac{-2s+3}{R^2}\big)\phi^{(s)}\Big]. \qquad (10.3.10)$$

Hence, these are eigenmodes of $\mathcal{D}^2$ if $\Box\phi = \lambda\phi$. Acting on $\mathcal{A}^{(g)}$ with higher powers of $(D^+)^k$ or $(D^+)^k$ does not lead to independent new modes. However, we do get a fourth tower of eigenmodes by acting with D^+D^-, or equivalently D^-D^+. We choose

$$\mathcal{A}_\alpha^{(n)}[\phi^{(s)}] := D^+ \mathcal{A}_\alpha^{(-)}[\phi^{(s)}], \qquad \in \mathcal{C}^s \qquad (10.3.11)$$

which due to (10.3.6) satisfies

$$\mathcal{D}^2 \mathcal{A}_\alpha^{(n)}[\phi^{(s)}] = \mathcal{A}_\alpha^{(n)}\Big[\big(\Box + \frac{3}{R^2}\big)\phi^{(s)}\Big]. \qquad (10.3.12)$$

This provides the following list of eigenmodes of $\mathcal{D}^2$ in $\mathbb{R}^4 \otimes \mathcal{C}^s$

$$\{\mathcal{A}_\alpha^{(g)}[\phi^{(s)}], \ \mathcal{A}_\alpha^{(+)}[\phi^{(s-1)}], \ \mathcal{A}_\alpha^{(-)}[\phi^{(s+1)}], \ \mathcal{A}_\alpha^{(n)}[\phi^{(s)}]\}. \qquad (10.3.13)$$

They can be organized more coherently upon inserting $D^\pm$ using (5.4.89). Then for any eigenmode of $\Box\phi^{(s)} = \lambda\phi^{(s)}$, we obtain four towers of *regular eigenmodes* of $\mathcal{D}^2$

$$\tilde{\mathcal{A}}^{(i)}[\phi] = \begin{pmatrix} \mathcal{A}^{(+)}[D^-\phi] \\ \mathcal{A}^{(-)}[D^+\phi] \\ \mathcal{A}^{(n)}[\phi] \\ r^2 R \mathcal{A}^{(g)}[\phi] \end{pmatrix}, \qquad i,j \in \{+,-,n,g\} \qquad (10.3.14)$$

for $\phi = \phi^{(s)}$, with the same eigenvalue

$$\boxed{\begin{aligned} \mathcal{D}^2 \tilde{\mathcal{A}}^{(+)}[\phi] &= \big(\lambda + \frac{3}{R^2}\big)\tilde{\mathcal{A}}^{(+)}[\phi] \\[4pt] \mathcal{D}^2 \tilde{\mathcal{A}}^{(-)}[\phi] &= \big(\lambda + \frac{3}{R^2}\big)\tilde{\mathcal{A}}^{(-)}[\phi] \\[4pt] \mathcal{D}^2 \tilde{\mathcal{A}}^{(g)}[\phi] &= \big(\lambda + \frac{3}{R^2}\big)\tilde{\mathcal{A}}^{(g)}[\phi] \\[4pt] \mathcal{D}^2 \tilde{\mathcal{A}}^{(n)}[\phi] &= \big(\lambda + \frac{3}{R^2}\big)\tilde{\mathcal{A}}^{(n)}[\phi] \end{aligned}} \qquad (10.3.15)$$

as well as the $\mathcal{A}^{(-)}[\phi^{(s,0)}]$ mode which is not included in (10.3.14). In particular, we obtain the following *on-shell modes* $(\mathcal{D}^2 - \frac{3}{R^2})\mathcal{A} = 0$

$$\{\tilde{\mathcal{A}}^{(+)}[\phi^{(s)}], \ \tilde{\mathcal{A}}^{(-)}[\phi^{(s)}], \ \tilde{\mathcal{A}}^{(g)}[\phi^{(s)}], \ \tilde{\mathcal{A}}^{(n)}[\phi^{(s)}]\} \qquad \text{for} \quad \Box\phi^{(s)} = 0$$

$$\mathcal{A}^{(-)}[\phi^{(s,0)}] \qquad \text{for} \quad \big(\Box - \frac{2s}{R^2}\big)\phi^{(s,0)} = 0. \qquad (10.3.16)$$

To establish independence of these modes, we need to distinguish them using some extra observable. Since $\mathcal{I}$ is related to the total $SO(3,1)$ Casimir (5.4.91) and commutes with $\mathcal{D}^2$, one can find a basis of common eigenvectors of $\mathcal{D}^2$ and $\mathcal{I}$ in $\mathbb{R}^4 \otimes \mathcal{C}$. This can be achieved with some effort by evaluating $\mathcal{I}$ on the modes using (5.4.93), and similar expressions for $\mathcal{A}^{(g,n)}$. The details of this computation can be found in [188]; here we shall simply quote the main results.

Generically, it turns out that for admissible $\phi^{(s)} \in \mathcal{C}^s$ (i.e. square-integrable principal series irreps of $SO(4,1)$), these four eigenmodes $\tilde{\mathcal{A}}^{(g,\pm,n)}[\phi^{(s)}]$ of $\mathcal{D}^2$ can be decomposed into four eigenmodes $\{v_\pm, v'_\pm\}$ of $\mathcal{I}$ with distinct eigenvalues and are therefore linearly independent. However, this depends on the $SO(3,1)$ sub-structure $\phi^{(s,k)} \in \mathcal{C}^{(s,k)}$, and there are some exceptions as follows:

- For $k = 0$, the eigenmodes $v_\pm$ are replaced by a single mode v_0, since $\tilde{\mathcal{A}}^{(+)}[\phi^{(s,0)}] = \mathcal{A}^{(+)}[D^-\phi^{(s,0)}] \equiv 0$. This is supplemented by the extra mode $v_0^- := \mathcal{A}^{(+)}[\phi^{(s,0)}]$ as indicated in (10.3.16). This yields again four independent eigenmodes $\{v_0, v_0^-, v'_\pm\}$.
- For $s = 0$, there are only two independent (scalar) modes $\mathcal{A}^{(-)}[D^+\phi^{(0)}]$ and $\mathcal{A}^{(g)}[\phi^{(0)}]$.
- For $k = s$, the modes $v_\pm$ coincide and are replaced by the modes v_{extra}. This leaves three independent modes, all of which are essentially scalars since $\phi^{(s,s)} = (D^+)^s \phi^{(0)}$.

10.3.2 Inner product, Hilbert space, and no ghosts

Having identified the eigenmodes of $\mathcal{D}^2$, we can compute their inner products

$$\langle \mathcal{A}, \mathcal{A}' \rangle = \text{Tr}(\mathcal{A}_\alpha{}^\dagger \mathcal{A}'^\alpha) = \int \frac{\Omega}{(2\pi)^3} \mathcal{A}_\alpha{}^* \mathcal{A}'^\alpha \qquad (10.3.17)$$

defined by contracting the indices of $\mathcal{A}$ with $\eta^{\alpha\beta}$, and integrating using the $SO(4,2)$ invariant measure (5.4.25) inherited from $\mathbb{C}P^{2,1}$, which is well-defined for admissible modes[6]. Notice that this inner product governs precisely the quadratic action (10.3.1), and it is not positive definite in general due to the $\eta^{\alpha\beta}$. However, it turns out to be positive definite for the physical gauge-fixed on-shell modes, thus defining the **physical Hilbert space**

$$\mathcal{H}_{\text{phys}} = \{\text{gauge-fixed on-shell modes}\}/_{\{\text{pure gauge modes}\}}. \qquad (10.3.18)$$

This no-ghost statement is crucial for a meaningful physical gauge theory.

To understand how the physical picture of time evolution arises in the present framework, we note that the on-shell relation $\Box \phi = 0$ determines the Casimir $C^2[\mathfrak{so}(4,1)]$ via (5.4.3) for a given $SO(3,1)$ mode, corresponding to some irreducible tensor field configuration on a spacelike H^3. This means that the "state" at any given time-slice H^3 completely determines the time evolution (up to forward or backward direction), which is governed by the effective metric $G_{\mu\nu}$ on $\mathcal{M}^{3,1}$. In other words, the physical Hilbert space $\mathcal{H}_{\text{phys}}$ is determined as usual by its restriction to any spacelike slice H^3. Since all these results are based on the $\mathfrak{so}(4,2)$ Lie algebra relations, they generalize immediately to the fully

---

[6] It is important to observe that the modes are integrable over the entire $\mathcal{M}^{3,1}$, rather than just the spacelike H^3. The reason is that admissible modes are principal series unitary irreps of $SO(4,2)$ with $C^2[\mathfrak{so}(4,2)] > 9/2$, and correspond to square-integrable tensor fields on H^4 or $\mathbb{C}P^{2,1}$ [188]. This allows to use invariance relations such as $(D^-)^\dagger = -D^+$, which is important for the explicit evaluation of the inner products of the $\mathcal{A}^{(i)}$. Although semi-classically one could define an inner product based on H^3, this would not make sense in the fully NC case.

noncommutative matrix model setting. Hence, the time evolution is captured by $SO(4,1)$ group theory, even though $\mathcal{M}^{3,1}$ admits only spacelike $SO(3,1)$ isometries, and even though there is no commutative time. In this sense, the concept of time evolution and the counting of degrees of freedom applies also in the noncommutative case, leading to a picture that corresponds exactly to the usual setup in field theory. This would be completely obscured in a star product approach, where the higher derivatives might falsely suggest a much more complicated Cauchy problem.

In the remainder of this chapter, we will discuss the no-ghost statement and the explicit form of the physical Hilbert space in some detail. This can be established following [188] by computing the inner product matrix

$$\mathcal{G}^{(i,j)} = \left\langle \tilde{\mathcal{A}}^{(i)}[\phi], \tilde{\mathcal{A}}^{(j)}[\phi'] \right\rangle, \qquad i \in \{\pm, n, g\}. \tag{10.3.19}$$

The four eigenmodes $\{v_\pm, v'_\pm\}$ discussed before are found to be mutually orthogonal, as they must be. The norm of the $v_\pm$ can be computed explicitly, and is found to be positive for admissible $\phi^{(s,k)}$. The signature of $\mathcal{G}^{(i,j)}$ can be determined via $\det \mathcal{G}^{(i,j)}$, and is found to be $(+,+,+,-)$ for $s \neq k \neq 0$, both for on-shell and off-shell modes. For the above-mentioned exceptional cases, the following is are found:

- For $k = 0$, the three independent modes $\tilde{\mathcal{A}}^{(-,n,g)}$ have an inner product matrix with signature $(+,+,-)$, supplemented by the positive definite orthogonal mode $\mathcal{A}^{(-)}[\phi^{(s,0)}]$.
- For $s = 0$, there are only two independent modes $\tilde{\mathcal{A}}^{(-,g)}$, which have an inner product matrix with signature $(+,-)$.
- For $k = s$, the three independent modes $\tilde{\mathcal{A}}^{(\pm,g)}$ have inner product matrix with signature $(+,+,-)$, while $\tilde{\mathcal{A}}^{(n)}$ turns out to be a linear combination of them, cf. (10.3.27).

Completeness

Now we want to understand whether the set of modes discussed so far is complete, i.e. if they span the space of all fluctuations $\mathcal{A}$. This can be addressed by counting the number of degrees of freedom (dof), i.e. real scalar fields on $\mathcal{M}^{3,1}$, at each sector $\mathcal{A}_\alpha \in \mathcal{C}^s \otimes \mathbb{R}^4$. We discuss separately the following cases:

1) $\mathcal{A}_\alpha \in \mathcal{C}^0 \otimes \mathbb{R}^4$

This sector contains four dof. Among these modes, only the spin 1 mode $\mathcal{A}^{(-)}[\phi^{(1)}]$ and the spin 0 modes $\mathcal{A}^{(g)}[\phi^{(0)}]$ are in $\mathcal{C}^0 \otimes \mathbb{R}^4$, while $\mathcal{A}^{(n)}[\phi^{(0)}]$ vanishes. The inner product considerations imply that these modes are independent. Clearly $\mathcal{A}_\alpha^{(-)}[\phi^{(1)}]$ i.e. $\phi^{(1)}$ encodes the most general spacelike vector field on $\mathcal{M}^{3,1}$, which amounts to three degrees of freedom. Together with the spin 0 mode $\mathcal{A}^{(g)}[\phi^{(0)}]$ we obtain four dof, which is precisely the content of $\mathcal{C}^0 \otimes \mathbb{R}^4$, and that list of modes is complete.

2) $\mathcal{A}_\alpha \in \mathcal{C}^s \otimes \mathbb{R}^4$

This sector contains $4(2s+1)$ dof. It is convenient to ignore the (s,k) substructure of the $\mathcal{C}^s$ here. Among our modes, $\mathcal{A}^{(-)}[\phi^{(s+1)}]$, $\mathcal{A}^{(n)}[\phi^{(s)}]$, $\mathcal{A}^{(g)}[\phi^{(s)}]$ and $\mathcal{A}^{(+)}[\phi^{(s-1)}]$ are in

$C^s \otimes \mathbb{R}^4$. If they were all independent, this would provide all the $(2s + 3) + 2(2s + 1) + (2s - 1) = 4(2s + 1)$ dof. The previous results show that this is indeed the case *except* for the scalar sector $k = s$, which provides only three rather than four modes since $\mathcal{A}^{(n)}[\phi^{(s,s)}]$ is dependent. Therefore, there must be *one exceptional scalar dof* for each $s \geq 1$,

$$\mathcal{A}^{(ex,s)} \in C^s \otimes \mathbb{R}^4, \qquad s \geq 1. \tag{10.3.20}$$

Since none of the regular scalar modes $\tilde{\mathcal{A}}^{(i)}$ is null, we can choose $\mathcal{A}^{(ex,s)}$ to be orthogonal to all $\tilde{\mathcal{A}}^{(i)}$. Due to the explicit form of the $\mathcal{A}^{(i)}$, this implies that the $\mathcal{A}^{(ex,s)}$ can be chosen as follows:

$$\{t^\alpha, \mathcal{A}^{(ex,s)}_\alpha\} = 0 = \{x^\alpha, \mathcal{A}^{(ex,s)}_\alpha\}, \qquad \mathcal{A}^{(ex,s)} = (D^+)^{s-1} \mathcal{A}^{(ex,1)}. \tag{10.3.21}$$

Orthogonality implies that this sector is respected by $\mathcal{D}^2$, and the physical constraint is satisfied. $\mathcal{A}^{(ex,1)}$ can be expressed by including an ansatz of the form

$$\mathcal{A}^{(ex,1)}_\alpha[\phi] = t_\alpha D^+ \phi + \tilde{\mathcal{A}}^{(g)}_\alpha[D\tilde{\phi}] + \tilde{\mathcal{A}}^{(+)}_\alpha[D\phi_+] + \tilde{\mathcal{A}}^{(-)}_\alpha[D\phi_-] \tag{10.3.22}$$

but its explicit form is not known. Further details are discussed in [188]. Taking these exceptional modes into account, we have recovered all $4(2s + 1)$ dof in $C^s \otimes \mathbb{R}^4$, so that the list of modes is complete. Together with the results obtained so far, we obtain:

Theorem 10.2 *The space of all modes $\mathcal{A} \in C \otimes \mathbb{R}^4$ is spanned by the $\tilde{\mathcal{A}}^{(i)}[\phi^{(s)}]$ modes (10.3.14) along with the $\mathcal{A}^{(-)}[\phi^{(s,0)}]$ for all $s \geq 0$ and the exceptional modes $\mathcal{A}^{(ex,s)}$ for $s \geq 1$. A basis is obtained by dropping $\tilde{\mathcal{A}}^{(n)}[\phi^{(s,s)}]$ and $\tilde{\mathcal{A}}^{(+)}[\phi^{(s,0)}]$.*

In particular, the C^1-valued fluctuations of the background arise from the 12 independent modes $\mathcal{A}^{(-)}[\phi^{(2)}]$, $\mathcal{A}^{(n,g)}[\phi^{(1)}]$, $\mathcal{A}^{(+)}[\phi^{(0)}]$ and $\mathcal{A}^{(ex,1)}$ which lead to $5 + 3 + 3 + 1 = 12$ degrees of freedom, which provides precisely the dof of a divergence-free frame.

A simpler way to count degrees of freedom is to use the expansion (10.3.3) in terms of four spacelike tensor fields $\mathcal{A}^{\dot\alpha}_{\mu_1 \ldots \mu_s}$, which provides $4(2s + 1)$ dof. However, this is not useful to diagonalize $\mathcal{D}^2$, and it hides the group-theoretical origin from $\mathfrak{so}(4,1) \subset \mathfrak{so}(4,2)$. From a representation theory point of view, we have elaborated the decomposition of

$$C \otimes \mathbb{R}^4 = \oplus (\ldots) \tag{10.3.23}$$

into $SO(3,1)$ irreps. It is natural to expect that each irrep in C arises with multiplicity 4 on the rhs, and we have seen that this holds indeed for the generic, regular modes.

Physical Hilbert space and no ghost

Now we are in a position to determine the Hilbert space of physical modes for the action (10.3.1), taking into account the gauge fixing condition

$$\{t^\alpha, \mathcal{A}_\alpha\} = 0. \tag{10.3.24}$$

We start with the observation that an (admissible, i.e. integrable) fluctuation mode $\mathcal{A}$ satisfies the gauge-fixing condition if and only if it is orthogonal to all pure gauge modes,

$$\langle \mathcal{A}^{(g)}, \mathcal{A} \rangle = 0. \tag{10.3.25}$$

In particular, all on-shell pure gauge modes $\mathcal{A}^{(g)}$ (10.3.16) satisfy (10.3.24) and therefore null. Now consider a generic on-shell mode $\tilde{\mathcal{A}}^{(i)}[\phi]$, $i \in \{+ - ng\}$ determined by some $\phi \in \mathcal{C}^{(s,k)}$ with $\Box \phi = 0$ and $s > k > 0$. Since that four-dimensional space of modes has signature $(+++-)$ as shown before and $\mathcal{A}^{(g)}$ is null, the gauge-fixing constraint (10.3.25) leads to a three-dimensional subspace with signature $(+ + 0)$, which contains $\mathcal{A}^{(g)}$. Then the physical Hilbert space defined as in (10.3.18) comprises two modes with positive norm. This establishes the generic part of

Theorem 10.3 *The space $\mathcal{H}_{phys}$ (10.3.18) of admissible solutions of $\left(\mathcal{D}^2 - \frac{3}{R^2}\right)\mathcal{A} = 0$ which are gauge-fixed $\{t^\alpha, \mathcal{A}_\alpha\} = 0$ modulo pure gauge modes inherits a positive-definite inner product, and forms a Hilbert space.*

Proof The same argument works for the on-shell modes $\tilde{\mathcal{A}}^{(i)}[\phi] \in \mathcal{C}^s$ with $\phi \in \mathcal{C}^{(s,0)}$. For $s \neq 0$ there are two physical modes. One is given by a linear combination of the $\tilde{\mathcal{A}}^{(i)}[\phi]$, $i \in \{-ng\}$ which has signature $(+ + -)$ before gauge fixing. In addition, there is an extra on-shell physical mode $\mathcal{A}^{(-)}[\phi^{(s,0)}] \in \mathcal{C}^{s-1}$ for $\left(\Box - \frac{2s}{R^2}\right)\phi^{(s,0)} = 0$ (10.3.16). For $s = 0$, the same argument shows that no physical mode arises from the $\tilde{\mathcal{A}}^{(i)} \in \mathcal{C}^0$ with $i \in \{-g\}$ which has signature $(+-)$ before gauge fixing. Similarly, one physical mode arises from the scalar on-shell modes $\tilde{\mathcal{A}}^{(i)}[\phi] \in \mathcal{C}^s$, $i \in \{+ - g\}$ with $\phi \in \mathcal{C}^{(s,s)}$. Finally, the exceptional modes $\mathcal{A}^{(ex,s)}$ (10.3.21) are physical, and their norm is positive because the $2s + 1$ dof in $\mathcal{C}^s \otimes \mathbb{R}^4$ with negative norm are already accounted for by the regular modes as shown before. Together with the completeness Theorem 10.2, the statement follows. Recall also that the admissibility condition implies square-integrability as discussed before. $\quad\square$

Since the inner product (10.3.17) for vector modes is precisely realized in the quadratic action (10.3.1), the present theorem is tantamount to the statement that the quadratic action is free of ghosts, i.e. there are no physical modes with negative norm. Although this result was established only at the semi-classical level, the derivation would essentially go through in the noncommutative case using the $\mathfrak{so}(4, 2)$-covariant quantization map $\mathcal{Q}$, with minor adaptions due to the cutoff. Hence, the theorem is expected to hold also in the noncommutative case.

An explicit description for the lowest spin modes in $\mathcal{H}_{phys}$ is as follows.

The physical Yang–Mills modes $\mathcal{A}_\alpha \in \mathcal{C}^0$

As explained before, the off-shell modes $\mathcal{A}_\alpha \in \mathcal{C}^0$ comprise the spin 1 mode $\mathcal{A}^{(-)}[\phi^{(1)}]$ and the spin 0 modes $\mathcal{A}^{(g)}[\phi^{(0)}]$. Among these, only the spin 1 modes $\mathcal{A}^{(-)}[\phi^{(1,0)}]$ are physical, and

$$\mathcal{H}_{phys} \cap \mathcal{C}^0 = \{\mathcal{A}^{(-)}[\phi] \text{ for } \phi \in \mathcal{C}^{(1,0)}, \ \left(\Box - \frac{2}{R^2}\right)\phi = 0\}. \tag{10.3.26}$$

These modes satisfy $\partial^\alpha \mathcal{A}_\alpha = 0 = x^\alpha \mathcal{A}_\alpha$ due to (5.4.88), and describe the two physical degrees of freedom of a massless Yang–Mills (or Maxwell) vector field which should be treated along the lines of Section 9.1.4.

The physical gravity modes $\mathcal{A}_\alpha \in \mathcal{C}^1$

They arise from the $4 \cdot 3 = 12$ off-shell modes $\mathcal{A}^{(-)}[\phi^{(2)}]$, $\mathcal{A}^{(n)}[\phi^{(1)}]$, $\mathcal{A}^{(g)}[\phi^{(1)}]$ and $\mathcal{A}^{(ex,1)}$ modulo the relation

$$2r^2 R \mathcal{A}^{(g)}[\phi^{(1,1)}] + 5\mathcal{A}^{(-)}[D^+\phi^{(1,1)}] + 2\mathcal{A}^{(+)}[D^-\phi^{(1,1)}] - 6\mathcal{A}_\alpha^{(n)}[\phi^{(1,1)}] = 0. \quad (10.3.27)$$

Among these, all $\mathcal{A}^{(-)}[\phi^{(2)}]$ are physical due to (5.4.88), and so is $\mathcal{A}^{(ex,1)}$, whose on-shell condition is not known explicitly. By combining the previous results, one finds [188]

$$\mathcal{H}_{\text{phys}} \cap \mathcal{C}^1 = \{\mathcal{A}^{(-)}[\phi] \text{ for } \phi \in \mathcal{C}^{(2,*)}, \ (\Box - \frac{4}{R^2})\phi = 0\}$$

$$\cup \ \{\mathcal{A}^{(ex,1)}; \ (\mathcal{D}^2 - \frac{3}{R^2})\mathcal{A}^{(ex,1)} = 0\}. \quad (10.3.28)$$

These 5+1 modes satisfy $\{t^\alpha, \mathcal{A}_\alpha\} = 0$, and $x^\alpha \mathcal{A}_\alpha^{(-)}[\phi^{(2,0)}] = 0$. They describe deformations of the background t^α, and therefore govern the linearized gravity sector on $\mathcal{M}^{3,1}$. They should be considered as "would-be massive" modes, since $\frac{3}{R^2}$ corresponds to a tiny cosmic mass scale, while the two modes in $\mathcal{A}_\alpha^{(-)}[\phi^{(2,0)}]$ should be viewed as massless. In particular, there are more propagating dof than in GR.

The physical higher-spin modes $\mathcal{A}_\alpha \in \mathcal{C}^s$.

One can similarly determine the physical modes $\mathcal{A}_\alpha \in \mathcal{C}^s$ with $s \geq 2$ among all $4(2s+1)$ modes. The most obvious physical modes are the "massless" $\mathcal{A}^{(-)}[\phi^{(s+1,0)}]$ modes for $(\Box - \frac{2(s+1)}{R^2})\phi^{(s)} = 0$. The remaining modes are more cumbersome to work out, and the details can be found in [188]; one finds another mode parametrized by $\Box \phi^{(s,0)} = 0$, as well as two series of modes given by linear combinations of the $\tilde{\mathcal{A}}^{(i)}[\phi^{(s,k)}]$ modes with $\Box \phi^{(s,k)} = 0$ for $k \neq 0$, with $\tilde{\mathcal{A}}^{(n)}[\phi^{(s,s)}]$ replaced by $\mathcal{A}^{(ex,s)} = (D^+)^{s-1}\mathcal{A}^{(ex,1)}$. These amount to $2(2s+1)$ degrees of freedom of two "would-be massive" spin s modes, decomposed into irreducible spin $s - k$ modes of the spacelike $SO(3,1)$. Some of these modes might combine into massive modes in the interacting theory, which remains to be clarified.

10.3.3 Linearized frame, metric, and Schwarzschild solution

Now consider the linearized fluctuations of the frame due to the modes under consideration:

$$E_{\dot\alpha}^\mu = \{T_{\dot\alpha}, x^\mu\} = \bar{e}_{\dot\alpha}^\mu + \{\mathcal{A}_{\dot\alpha}, x^\mu\} \equiv \bar{e}_{\dot\alpha}^\mu + \delta E_{\dot\alpha}^\mu \quad (10.3.29)$$

cf. (10.2.25), where $\bar{e}_{\dot\alpha}^\mu = \sinh(\eta)\delta_{\dot\alpha}^\mu$. This leads to the linearized metric fluctuation

$$\gamma^{\mu\nu} = \bar{\gamma}^{\mu\nu} + \delta_A \gamma^{\mu\nu} + O(\mathcal{A}^2),$$

$$\delta_A \gamma^{\mu\nu} = \sinh(\eta)\{\mathcal{A}^\mu, x^\nu\}_0 + (\mu \leftrightarrow \nu). \quad (10.3.30)$$

Note that only the projection $\{\mathcal{A}^\mu, x^\nu\}_0$ to $\mathcal{C}^0$ is relevant in the kinetic action for fields on $\mathcal{M}^{3,1}$, and only $\mathcal{A} \in \mathcal{C}^1$ can contribute to $[\delta_A \gamma^{\mu\nu}]_0$; therefore we restrict ourselves to

$\mathcal{A} \in \mathcal{C}^1$ henceforth. To evaluate this explicitly, it is convenient to consider the following rescaled graviton mode:

$$h^{\mu\nu}[\mathcal{A}] := \{\mathcal{A}^\mu, x^\nu\}_0 + (\mu \leftrightarrow \nu), \qquad h[\mathcal{A}] = 2\{\mathcal{A}^\mu, x_\mu\}_0. \tag{10.3.31}$$

Including the conformal factor in (10.2.4), this leads to the effective metric fluctuation

$$\delta G^{\mu\nu} = \beta^2 \big(h^{\mu\nu} - \frac{1}{2}\eta^{\mu\nu} h\big). \tag{10.3.32}$$

The conformal factor $\beta = \sinh(\eta)^{-1}$ reflects the cosmic expansion, and drops out upon lowering the indices with $\bar{G}_{\mu\nu}$. Then the fluctuations of the effective background effective metric take the form

$$(G_{\mu\nu} - \delta G_{\mu\nu})\mathrm{d}x^\mu \mathrm{d}x^\nu = -\mathrm{d}t^2 + a^2(t)\mathrm{d}\Sigma^2 - (h_{\mu\nu} - \frac{1}{2}\eta_{\mu\nu} h)\mathrm{d}x^\mu \mathrm{d}x^\nu, \tag{10.3.33}$$

where the cosmic scale factor and the comoving time are given by (5.4.44), and the indices of the Cartesian coordinates in

$$h_{\mu\nu}[\mathcal{A}] = \{\mathcal{A}_\mu, x_\nu\}_0 + (\mu \leftrightarrow \nu) \tag{10.3.34}$$

are lowered with $\eta_{\mu\nu}$. In this form, the contributions of the various on- and off-shell modes $\mathcal{A}^{(i)}$ are easy to evaluate, and are properly related to the cosmic background geometry.

Propagating modes

The physical $\mathcal{A}^{(-)}[\phi^{(2,0)}]$ modes with $(\Box - \frac{4}{R^2})\phi = 0$ describe massless gravitons with two propagating degrees of freedom. They are transversal and traceless in the following sense

$$\{t^\mu, h_{\mu\nu}\} = 0 = x^\mu h_{\mu\nu} = h \tag{10.3.35}$$

due to (5.4.88) and $\{t_\mu, \phi^{(2)}\}_0 = 0$. The physical $\mathcal{A}^{(-)}[\phi^{(2,k)}]$ modes for $k = 1, 2$ describe vector-like propagating modes, which satisfy the following gauge condition [186]:

$$\partial_\mu(\beta h^{\mu\nu}) = 0. \tag{10.3.36}$$

Those could be viewed as "would-be massive" graviton modes. Using that constraint, it is easy to see that the linearized Ricci tensor vanishes up to the cosmic curvature scale,

$$2R_{(\text{lin})}^{\mu\nu}[G + \delta G] = 0 + O\big(\frac{\partial G^{\mu\nu}}{x_4}\big), \tag{10.3.37}$$

consistent with the general result (9.3.49). This is not in conflict with standard results of GR, since the vector-like modes $\mathcal{A}^{(-)}[\phi^{(2,k)}]$ for $k = 1, 2$ are "almost" diffeomorphisms of the form $\nabla_\mu \xi_\nu + (\mu \leftrightarrow \nu)$. In particular, they almost decouple from matter.

Quasi-static solutions

It is interesting to identify solutions which correspond to the linearized Schwarzschild solution. More generally, any static linearized solution of GR is recovered from the physical mode $\mathcal{A}^{(-)}[\phi^{(2,2)}] = \mathcal{A}^{(-)}[D^+D\phi]$ for $\phi \in \mathcal{C}^0$ satisfying the quasi-static condition (5.4.46)

$$\tau\phi = a(t)\frac{\partial}{\partial t}\phi = -2\phi. \tag{10.3.38}$$

This means "almost-static" on a cosmic time scale, and for late times reduces using $a(t) \sim \frac{3}{2}t$ (5.4.53) to the mild time dependence

$$\phi(t) \sim t^{-4/3}. \tag{10.3.39}$$

Then the on-shell condition $\left(\Box + \frac{2}{R^2}\right)\phi = 0$ reduces using (5.4.97) to

$$\Delta^{(3)}\phi = 0. \tag{10.3.40}$$

The resulting metric perturbation is

$$\delta G_{\mu\nu}\, dx^\mu dx^\nu \stackrel{\tau \to -2}{=} \phi'(dt^2 + a(t)^2 d\Sigma^2) \tag{10.3.41}$$

after subtracting a (large but unphysical) pure gauge contribution and rescaling ϕ as

$$\phi' = \lim_{\tau \to -2} \beta(\tau + 2)\phi. \tag{10.3.42}$$

As shown in [186], this reproduces essentially the linearized (quasi-) static metric perturbation of the cosmic background $\mathcal{M}^{3,1}$ in GR. In particular, we recover the linearization of the spherically symmetric Schwarzschild–McVittie solution on a FLRW background, consistent with (9.3.67), modified by a slow decay with the cosmic expansion and an exponential suppression at cosmic distance scales.

10.4 Pre-gravity and matter

We have seen that all essential structures of a gravitational theory arise naturally on covariant quantum spaces in matrix models, including a higher-spin generalization of diffeo invariance and a frame, with sufficiently many degrees of freedom for gravity. Moreover, the linearized classical equations of motion arising from the matrix model (i.e. without loop corrections) describe Ricci-flat metric perturbations around the background.

Nevertheless, this theory differs significantly from GR, as the matrix model action has *two derivative less* than the Einstein–Hilbert action. Indeed, the linearized perturbations of the frame $\delta E \sim \partial \mathcal{A}$ (10.3.29) are derivatives[7] of the perturbations $\mathcal{A}$, which should be viewed as *potential* or generator for the frame. Hence, the Yang–Mills type matrix model action for $\mathcal{A}$

$$S \sim \mathrm{Tr}[T,T][T,T] \sim \int \partial\mathcal{A}\partial\mathcal{A} \sim \int \delta E\delta E \tag{10.4.1}$$

---

[7] Note that according to the discussion in Section 5.4.3, the bracket $\{., x^\mu\}$ acts predominantly as a derivation of $\mathcal{A}$ along x, while the bracket with the t generators is subleading except for extreme IR contributions.

amounts to a *potential* for the perturbations δE of the frame, while the linearized Einstein–Hilbert action governs the derivatives of the frame

$$S_{\text{E-H}} \sim \int \partial(\delta E)\partial(\delta E). \tag{10.4.2}$$

This fundamental difference of the matrix model action and the Einstein–Hilbert action means that the usual derivation of the Einstein equations in the presence of matter fails. If the frame was the fundamental dof, then adding matter to (10.4.1) would merely lead to some local deformation of the frame. However, since $\mathcal{A}$ is the fundamental degree of freedom, matter will couple to the geometrical equations of motion equations through higher derivatives, leading roughly to the structure

$$\delta R_{\mu\nu} \sim \partial\partial T_{\mu\nu}. \tag{10.4.3}$$

This means that the *classical* matrix model should be viewed as a sort of **pre-gravity** theory: It is a consistent theory of dynamical geometry whose vacuum solutions share important features of GR, but the effect of matter on the geometry is different, and much more local. These remarks also apply to the basic four-dimensional (non-covariant) quantum spaces discussed in Chapter 9.

However, this seemingly bad news is actually very good: The pre-gravity theory defined by the matrix model has the great advantage that it is well suited for quantization, precisely because it has less derivatives than the E-H action, and it is not governed by a dimensionful coupling constant. In contrast, general relativity is a good effective (classical) theory, but not well suited for quantization. The crucial point is that the Einstein–Hilbert action will arise from quantum effects, i.e. in the quantum effective action. This is expected on general grounds as pointed out long time ago by Sakharov [163], and it will be derived explicitly in Sections 12 and 12.4. This suggests to understand gravity as a quantum effect on quantum spacetime, which is a perfectly reasonable and appealing scenario.

Part IV

Matrix theory and gravity

Matrix theory: Maximally supersymmetric matrix models

In the class of Yang–Mills matrix models, there is one model that is singled out by maximal supersymmetry, and thus uniquely well-behaved upon quantization. This model is known as IKKT or IIB matrix model [102], which was first proposed as a constructive definition of IIB string theory. Having a preferred model should not be seen as a restriction, but as a welcome selection principle in physics. Our aim will be to explore this distinguished model, and to understand if and how (near-) realistic physics can arise within that model.

There is a related model known as BFSS model [21, 57], which is also distinguished by maximal supersymmetry. This model involves a classical time, which seems less natural for the present approach. The BFSS model and its relation to the IKKT model will be discussed briefly in Chapter 13.

11.1 The IKKT or IIB matrix model at one loop

The IKKT or IIB matrix model [102] is a particular Yang–Mills matrix model in $9 + 1$ dimensions, defined by the action

$$S[T, \Psi] = \frac{1}{g^2} \text{Tr}\big([T^a, T^b][T_a, T_b] + \overline{\Psi}\Gamma_a[T^a, \Psi]\big), \tag{11.1.1}$$

where $T^a \in \text{End}(\mathcal{H})$, $a = 0, \ldots, 9$ are Hermitian matrices, and Ψ are Majorana–Weyl spinors of $SO(9, 1)$ whose entries are (Grassmann-valued) matrices. Indices are contracted with the $SO(9, 1)$-invariant tensor η^{ab}. This model admits a manifest $SO(9, 1)$ symmetry acting as

$$T^a \to \Lambda(g)^a_b T^b, \quad \Psi_\alpha \to \tilde{\pi}(g)^\beta_\alpha \Psi_\beta, \tag{11.1.2}$$

where $\tilde{\pi}(g)$ denotes the spinorial representation of the universal covering group $\widetilde{SO}(9, 1)$ acting on the fermions Ψ, and $\Lambda(g)$ denotes the vector representation. Since T^a enters only via commutators, there is also a shift symmetry[1]

$$T^a \to T^a + c^a \mathbb{1}, \quad c^a \in \mathbb{R}. \tag{11.1.3}$$

---

[1] Even though this shift affects only the trace-$U(1)$ part of the model that drops out from the commutators, this "center" symmetry is important for the emergence of a macroscopic spacetime.

The fundamental gauge symmetry of the matrix model extends to fermions as follows:

$$T^a \to U^{-1} T^a U, \qquad \Psi \to U^{-1} \Psi U \tag{11.1.4}$$

for $U \in U(\mathcal{H})$. This is considered as a gauge symmetry, i.e. configurations related by such transformations are considered as equivalent.

The distinctive property of the IKKT model among all Yang–Mills matrix models is *maximal supersymmetry*. This is best understood by observing that the model is nothing but the dimensional reduction of $U(N)$ $\mathcal{N} = 1$ Super-Yang–Mills theory on $\mathbb{R}^{9,1}$ to zero dimensions, which is defined by the action

$$S_{\text{YM}} = \int d^{10} y \, \frac{1}{g^2} \text{Tr}\Big(-F^{ab} F_{ab} + \bar{\Psi} \gamma^a i D_a \Psi \Big). \tag{11.1.5}$$

Keeping only constant fields leads to (11.1.1), which thereby inherits the following SUSY transformations (7.0.9):

$$\delta \Psi = i[T^a, T^b] \Sigma^{(\psi)}_{ab} \epsilon$$
$$\delta T^a = i \bar{\epsilon} \Gamma^a \Psi, \tag{11.1.6}$$

where $\Sigma^{(\psi)}_{ab} = \frac{i}{4}[\Gamma_a, \Gamma_b]$, and ϵ is a (Grassmann-) matrix-valued spinor acting as SUSY generator. In addition, there is another (almost trivial) supersymmetry given by

$$\delta \Psi = \xi$$
$$\delta T^a = 0, \tag{11.1.7}$$

where ξ is another Grassmann-valued spinor. Taken together, these transformations provide the analog of the $\mathcal{N} = 2$ SUSY of type IIB string theory. Since 10-dimensional field theory without gravity does not admit $\mathcal{N} = 2$ SUSY, one may expect that the model contains gravity.

There are several ways to motivate the IKKT model. It was introduced in [102] as a quantization of the Green–Schwarz worldsheet action for IIB superstring theory in $\mathbb{R}^{9,1}$, which in Schild gauge takes the form

$$S_{\text{Schild}} = \int d^2 \sigma \sqrt{g} \left(-\{T_a, T_b\}\{T^a, T^b\} - i\bar{\psi} \Gamma_a \{T^a, \psi\} \right). \tag{11.1.8}$$

This can be viewed as a classical version of (11.1.1). The matrix model should in fact be considered as a second quantization of this action, since block-matrix configurations correspond to independent worldsheets. A link with light-cone string field theory was established in [12, 75], based on matrix versions of Wilson loops. Moreover, the interaction between branes in IIB supergravity is reproduced by the quantized matrix model, as discussed in Section 11.1.7. Last but not least, the IKKT model is rooted in previous work on the large N reduction of lattice gauge theory and the (twisted) Eguchi–Kawai model [68, 80]. The underlying common theme is the emergence of a macroscopic gauge theory from basic matrix models.

However, there is no need to appeal to string theory and exotic dimensions. We will see that the IKKT model can be viewed as nonperturbative formulation of $\mathcal{N} = 4$ noncommutative Super-Yang–Mills on $\mathbb{R}^4_\theta$. This is not only the most distinguished NC field theory in four dimensions, it is essentially the only model of NC field theory without pathological UV/IR mixing in four dimensions (cf. Section 6.4), such that the quantized theory leads to (almost-) local physics.

11.1.1 Fermions

The most important property of the IKKT model is the suppression of quantum effects due to maximal supersymmetry, which provides a crucial selection principle. This is closely related to the statement that among all possible string theories, critical (super)string theory in 10 dimensions is distinguished because it is free from anomalies at the quantum level, notably the conformal anomaly in the worldsheet formulation.

Supersymmetry requires that for each bosonic degree of freedom, there is a corresponding fermionic degree of freedom. As is well-known, Dirac spinors in even dimensions D contain $2^{D/2}$ degrees of freedom, and chiral or Weyl spinors contain $2^{D/2-1}$ dof. In dimensions 2 and 10, it is possible to impose one further constraint, namely invariance under charge conjugation, leading to Majorana–Weyl (MW) fermions. Hence, in 10 dimensions, Weyl spinors have 16 dof, and MW fermions have 8 dof. These eight dof correspond precisely to the $10 - 2$ physical bosonic degrees of freedom of a Yang–Mills matrix model in 10 dimensions because it is a gauge theory, and gauge theories in D dimensions have $D - 2$ physical degrees of freedom (one is removed by the gauge fixing condition, and another one by quotienting out pure gauge modes). Therefore, a supersymmetric Yang–Mills matrix model in 10 dimensions requires the fermions to be Majorana–Weyl, and this is the maximal possible number of dimensions. There are more possibilities in lower dimensions with less overall supersymmetry, but this entails insufficient cancellation of UV/IR mixing.

Furthermore, it is well-known that the MW condition in $D = 10$ is only consistent in Minkowski signature, with $SO(9, 1)$-invariant metric η^{ab}. Therefore, the IKKT model is only defined in Minkowski signature, which is quite remarkable. A Euclidean version may be defined as a suitable continuation of the Minkowski case after integrating out the fermions.

Minkowski case

For Dirac fermions, the (Grassmann) functional integral can be evaluated as

$$\int d\bar{\Psi}d\Psi \, e^{i\mathrm{Tr}\bar{\Psi}\Gamma_a[T^a,\Psi]} = \det(i\Gamma_0\slashed{D}) = \sqrt{\det \slashed{D}^2}. \tag{11.1.9}$$

However, the fermions in the IKKT model are (matrix-valued) Majorana–Weyl spinors of $SO(1, 9)$, which means that (cf. [102, 146])

$$\Psi_C = \mathcal{C}\overline{\Psi}^t = \mathcal{C}\Gamma_0^t\Psi^\star = \Psi, \tag{11.1.10}$$

where $\Gamma_a^t = \mathcal{C}\Gamma_a\mathcal{C}^{-1}$ (the transpose refers only to the spinor indices). Hence, in a MW basis, the spinor entries are Hermitian matrices. Moreover in $9 + 1$ dimensions, $\mathcal{C} = \mathcal{C}^{-1t}$ anticommutes with the chirality operator $\Gamma = i\Gamma_0 \dots \Gamma_9$, so that the symmetric matrix

$$\tilde{\Gamma}_a = C\Gamma_a \tag{11.1.11}$$

is well-defined for Weyl spinors. Using $\overline{\Psi} = \Psi^t C^{-1t}$, the fermionic action can be written as

$$\mathrm{Tr}\overline{\Psi}\Gamma_a[T^a, \Psi] = \mathrm{Tr}\Psi^t C^{-1t}\Gamma_a[T^a, \Psi] = \mathrm{Tr}\Psi_\alpha \tilde{\Gamma}_a^{\alpha\beta}[T^a, \Psi_\beta], \tag{11.1.12}$$

where Ψ are 16-dimensional Weyl spinors, or equivalently

$$\mathrm{Tr}\overline{\Psi}\Gamma_a[T^a, \Psi] = \mathrm{Tr}\Psi^t C \slashed{D} P_+ \Psi, \qquad P_\pm = \frac{1}{2}(1 \pm \Gamma), \tag{11.1.13}$$

for 32-component Dirac spinors. Then the Grassmann integral over the MW spinors yields

$$e^{i\Gamma^\Psi[T]} := \int d\Psi e^{i\mathrm{Tr}\overline{\Psi}\Gamma_a[T^a, \Psi]} = \mathrm{Pfaff}(\tilde{\Gamma}_a^{\alpha\beta}[T^a, .])$$

$$= \pm\sqrt{\det \tilde{\Gamma}_a^{\alpha\beta}[T^a, .]} = \pm\sqrt{\det(C\slashed{D}_+)}, \tag{11.1.14}$$

where $C\slashed{D}_+$ denotes $C\slashed{D}$ restricted the positive chirality spinors. Taking the square gives

$$e^{2i\Gamma^\Psi[T]} = \det(\slashed{D}_+) = \sqrt{\det \slashed{D}_+^2} = e^{\frac{1}{2}\mathrm{Tr}\ln \slashed{D}_+^2}. \tag{11.1.15}$$

Together with

$$\slashed{D}^2 = \square - \Sigma_{ab}^{(\psi)}[\mathcal{F}^{ab}, .], \tag{11.1.16}$$

where $\Sigma_{ab}^{(\psi)}$ denotes the $\mathfrak{so}(9,1)$ representation (11.1.29) on Weyl spinors and $\mathcal{F}^{ab} = i[T^a, T^b]$ (9.1.1), this leads to the fermionic contribution to the effective action

$$\Gamma^\Psi[T] = -\frac{i}{4}\mathrm{Tr}\big(\ln(\square - \Sigma_{ab}^{(\psi)}[\mathcal{F}^{ab}, .])\big). \tag{11.1.17}$$

Euclidean case

Since the Pfaffian makes sense for any antisymmetric matrix, one can use (11.1.14) to define a Wick-rotated fermionic induced action $\Gamma_E^\psi[X]$ also in the Euclidean case (at least for finite-dimensional matrices), by replacing $\Gamma_0 \to i\Gamma_{10}$. However, then $\tilde{\Gamma}_a^{\alpha\beta}[X^a,]$ is in general not a Hermitian operator for chiral $SO(10)$ spinors, and the effective action has both real and imaginary contributions. The real part of the action can be extracted from

$$\det\big((C\slashed{D})^\dagger C\slashed{D}_+\big) = \det(\slashed{D}_+^2) = e^{-2\,\mathrm{Re}(\Gamma_E^\psi[X])}, \tag{11.1.18}$$

which is real because $\slashed{D}^2$ is a Hermitian operator. Therefore, we can write

$$\Gamma_E^\psi[X] = \Gamma_E^{\psi,\mathrm{real}}[X] + i\Gamma_E^{\psi,\mathrm{imag}}[X]$$

$$= -\frac{1}{4}\mathrm{Tr}\log(\slashed{D}_+^2) + i\Gamma_{WZ}. \tag{11.1.19}$$

The imaginary part is known as Wess–Zumino contribution, which incorporates the chiral anomaly contribution due to the integration over fermions, cf. [204].

11.1.2 $\mathcal{N} = 4$ noncommutative super-Yang–Mills

In this section, we explain why the IKKT model is nothing but the nonperturbative formulation of $\mathcal{N} = 4$ noncommutative Super-Yang–Mills on $\mathbb{R}_\theta^4$. This is seen along the lines of Section 7.4.3, by expanding the model around the brane solution or background $\mathbb{R}_\theta^4$ and identifying the bosonic and fermionic field content of $\mathcal{N} = 4$ SYM from the IKKT model [11]. The result is fairly obvious, since ordinary $\mathcal{N} = 4$ SYM is the dimensional reduction of $\mathcal{N} = 1$ SYM on $\mathbb{R}^{10}$ to $\mathbb{R}^4$, while the IKKT model is the reduction of $\mathcal{N} = 1$ SYM on $\mathbb{R}^{10}$ to zero dimensions.

To make this explicit, we separate the 10 bosonic matrices T^a of the IKKT model as in (7.4.21)

$$T^a = \begin{cases} T^\mu, & \mu = 0, \dots, 3 \\ \phi^i, & i = 4, \dots, 9. \end{cases} \tag{11.1.20}$$

Now we write

$$T^\mu = X^\mu \mathbb{1}_k + \theta^{\mu\nu} A_\nu, \qquad D_\mu = -i\theta_{\mu\nu}^{-1}[T^\nu, .] = \partial_\nu - i[A_\nu, .], \tag{11.1.21}$$

where X^μ generate the Moyal–Weyl quantum plane $\mathbb{R}_\theta^4$. Then the action takes the form (cf. (7.4.30), (9.1.69))

$$S = \int \frac{d^4x}{(2\pi)^2} \frac{1}{g^2} \text{Tr}\left(-F^{\mu\nu}F_{\mu\nu} - 2D^\mu\phi^i D_\mu\phi_i + [\phi^i, \phi^j][\phi_i, \phi_j] \right)$$
$$+ \text{Tr}\left(\bar{\Psi}\gamma^\mu iD_\mu\Psi + \bar{\Psi}\Gamma^i[\phi_i, \Psi] \right), \tag{11.1.22}$$

which is precisely noncommutative $\mathcal{N} = 4$ $U(k)$ SYM on $\mathbb{R}_\theta^4$. Here $F_{\mu\nu}$ is the field strength, $D_\mu = \partial_\mu - i[A_\mu, \cdot]$ the gauge covariant derivative, ϕ^i are six scalar fields, and the MW spinor Ψ of $SO(9, 1)$ reduces to four Weyl or Majorana spinors of $SO(3, 1)$. All fields take values in $\mathfrak{u}(k)$ and transform in the adjoint of the $U(k)$ gauge symmetry. The global $SO(6)_R$ symmetry is manifest. As in Section 7.4.5, we observe that any rank k of the noncommutative gauge group is equivalent at the nonperturbative (matrix model) level.

11.1.3 One-loop effective action in Minkowski signature

The unique property of the IKKT matrix model becomes manifest in the one-loop effective action, which will be discussed in this section. In the physical case of Minkowski signature, the quantization of the model is defined in terms of the $i\varepsilon$ prescription discussed in Section 7.2. Thus, the bosonic part of the model is modified as follows:

$$S_\varepsilon[T] = \frac{1}{g^2} \text{Tr}\left([T^a, T^b][T_a, T_b] + i\varepsilon \sum_a (T^a)^2 \right), \tag{11.1.23}$$

where $a, b = 0, \dots, 9$, and the $i\varepsilon$ term does *not* involve $\eta^{ab} = (-1, 1, \dots, 1)$. Recalling the discussion of the kinetic term e.g. in (7.4.24) or (9.1.8), this has the structure $S = \int \mathcal{L}$

where $\mathcal{L} = T - V$ is a field-theoretic Lagrangian in Minkowski signature, with kinetic term T and potential V. Moreover, the integral

$$Z_{\varepsilon} = \int dT e^{iS_{\varepsilon}[T]} \tag{11.1.24}$$

over the matrices is absolutely convergent, and boils down at the perturbative level to the following "Feynman regularized" kinetic quadratic form[2]

$$i\mathcal{A}_a\left(-\eta^{ab}\Box + i\varepsilon\delta^{ab}\right)\mathcal{A}_b \tag{11.1.25}$$

for the matrix fluctuations $\mathcal{A}^a$. In the following, we will restrict ourselves to the one-loop contribution to the effective action, which arises from Gaussian integration around some given matrix background T^a as discussed in Section 7.6, defined in Minkowski signature through the $i\varepsilon$ prescription. Taking into account the fermions and the ghosts in the IKKT model, the one-loop effective action is defined by

$$\int_{1 \text{ loop}} dT d\Psi d\bar{c} dc \, e^{iS[T,\Psi,c]} = e^{i(S_0[T]+\Gamma_{1 \text{ loop}}[T])} = e^{i\Gamma_{\text{eff}}[T]}. \tag{11.1.26}$$

Using (11.1.17), the Gaussian integral formula (7.6.10) for

$$\mathcal{D}^2 = \Box - 2\mathcal{I} = \Box - \Sigma_{ab}^{(\mathcal{A})}[\mathcal{F}^{ab}, .] \tag{11.1.27}$$

and similarly for the ghosts, one obtains the following formula [35, 47, 102]:

$$\begin{aligned}
\Gamma_{1\text{loop}}[T] &= \frac{i}{2}\text{Tr}\left(\log(\Box-i\varepsilon-\Sigma_{ab}^{(\mathcal{A})}[\mathcal{F}^{ab}, .]) - \frac{1}{2}\log(\Box-i\varepsilon-\Sigma_{ab}^{(\psi)}[\mathcal{F}^{ab}, .]) - 2\log(\Box-i\varepsilon)\right) \\
&= \frac{i}{2}\text{Tr}\left(\sum_{n>0}\frac{1}{n}\left(-((\Box-i\varepsilon)^{-1}\Sigma_{ab}^{(\mathcal{A})}[\mathcal{F}^{ab}, .])^n + \frac{1}{2}((\Box-i\varepsilon)^{-1}\Sigma_{ab}^{(\psi)}[\mathcal{F}^{ab}, .])^n\right)\right) \\
&= \frac{i}{2}\text{Tr}\left(-\frac{1}{4}((\Box-i\varepsilon)^{-1}\Sigma_{ab}^{(\mathcal{A})}[\mathcal{F}^{ab}, .])^4 + \frac{1}{8}((\Box-i\varepsilon)^{-1}\Sigma_{ab}^{(\psi)}[\mathcal{F}^{ab}, .])^4\right. \\
&\quad\left. + \mathcal{O}(\Box^{-1}[\mathcal{F}^{ab}, .])^5\right)
\end{aligned} \tag{11.1.28}$$

using the results in Appendix B.4 in the last line, which is specific for the IKKT model. Here

$$\begin{aligned}
(\Sigma_{ab}^{(\psi)})_{\beta}^{\alpha} &= \frac{i}{4}[\Gamma_a, \Gamma_b]_{\beta}^{\alpha} \\
(\Sigma_{ab}^{(\mathcal{A})})_d^c &= i(\delta_b^c\delta_{ad} - \delta_a^c\delta_{bd})
\end{aligned} \tag{11.1.29}$$

are $SO(9,1)$ generators acting on the spinor or vector representation, respectively. The $2\log\Box$ term arises from the ghost contribution. The crucial point is that the leading terms in this expansion cancel due to maximal supersymmetry, so that the first nonvanishing term

---

[2] In the presence of a mass term, this would amount to the standard Feynman replacement $m^2 \to m^2 - i\varepsilon$; note also that the timelike fluctuation modes are taken care of by the ghost term in (11.1.28).

in this expansion arises for $n = 4$. This leading fourth order term is given by the following expression:

$$\Gamma_{1\text{loop};4}[T] = \frac{i}{8}\text{Tr}\left(-((\Box - i\varepsilon)^{-1}\Sigma_{ab}^{(\mathcal{A})}[\mathcal{F}^{ab}, .])^4 + \frac{1}{2}((\Box - i\varepsilon)^{-1}\Sigma_{ab}^{(\psi)}[\mathcal{F}^{ab}, .])^4 \right)$$

$$= \frac{i}{4}\text{Tr}\left((\Box - i\varepsilon)^{-1}[\mathcal{F}^{a_1 b_1}, \ldots (\Box - i\varepsilon)^{-1}[\mathcal{F}^{a_4 b_4}, .]]]] \right)$$

$$\left(-4\eta_{b_1 a_2}\eta_{b_2 a_3}\eta_{b_3 a_4}\eta_{b_4 a_1} - 4\eta_{b_1 a_2}\eta_{b_2 a_4}\eta_{b_4 a_3}\eta_{b_3 a_1} - 4\eta_{b_1 a_3}\eta_{b_3 a_2}\eta_{b_2 a_4}\eta_{b_4 a_1} \right.$$
$$\left. + \eta_{b_1 a_2}\eta_{b_2 a_1}\eta_{b_3 a_4}\eta_{b_4 a_3} + \eta_{b_1 a_3}\eta_{b_3 a_1}\eta_{b_2 a_4}\eta_{b_4 a_2} + \eta_{b_1 a_4}\eta_{b_4 a_1}\eta_{b_2 a_3}\eta_{b_3 a_2} \right).$$
$$(11.1.30)$$

This vanishes for constant fluxes $\mathcal{F} = const\mathbb{1}$, which holds in particular for Moyal–Weyl quantum planes. More generally, the interaction between parallel $\mathbb{R}_\theta^{2n}$ branes with identical constant field strength $\mathcal{F}^{ab}$ vanishes at one loop. This is a reflection of the supersymmetry (often denoted as BPS property) of such backgrounds, as familiar from string theory. If the branes are not parallel or have different field strengths, then there is a residual interaction, which turns out to be consistent with the interaction of the corresponding branes in IIB supergravity. This interaction is much weaker than in the non-supersymmetric case discussed in Section 7.6.1.

Note also that the coupling constant g drops out from $\Gamma_{1\text{loop}}$, so that the bare action S_0 dominates for small g. Furthermore, we observe the following scaling relations:

$$S_0[cT, c^{\frac{3}{2}}\Psi] = c^4 S_0[T, \Psi]$$
$$\Gamma_{1\text{loop}}[cT] = \Gamma_{1\text{loop}}[T] \qquad (11.1.31)$$

for $c \in \mathbb{R}$. In particular if $\Gamma_{1\text{loop}}$ leads to a negative contribution to the effective potential, the bare action is negligible for small c, but will prevent c from getting too large. This is important for the stabilization of nontrivial backgrounds.

Evaluation of the trace using string modes

For backgrounds given by a quantized symplectic space $\mathcal{M}$, the trace over Hermitian matrices in End($\mathcal{H}$) can be evaluated explicitly using the basic formula (6.2.77)

$$\text{Tr}_{\text{End}(\mathcal{H})}\mathcal{O} = \int_{\mathcal{M} \times \mathcal{M}} \frac{\Omega_x \Omega_y}{(2\pi)^m} \left(\begin{matrix} x \\ y \end{matrix} \middle| \mathcal{O} \middle| \begin{matrix} x \\ y \end{matrix} \right). \qquad (11.1.32)$$

Here $\left| \begin{matrix} x \\ y \end{matrix} \right) = |x\rangle\langle y| \in \text{End}(\mathcal{H})$ are string modes linking $x, y \in \mathcal{M}$, which are built out of coherent states $|x\rangle$ on the underlying[3] quantized symplectic space $\mathcal{M}$. The string modes are very useful for loop computations, because they have good localization properties in *both* position and momentum as discussed in Section 6.2. In particular,

---

[3] This formula is exact for the invariant symplectic form on quantized homogeneous spaces such as $\mathbb{C}P^3$ or $\mathbb{C}P^{2,1}$. It is also applicable for generic deformations thereof, because the underlying symplectic space is rigid, so that the undeformed symplectic form and coherent states can be used even for a deformed matrix configuration.

$$(\Box - i\varepsilon)^{-1} \Big|_y^x\Big) \sim \frac{1}{|x - y|^2 + 2\Delta^2 - i\varepsilon} \Big|_y^x\Big)$$

$$(\Box - i\varepsilon)^{-1}[\mathcal{F}^{ab}, .] \Big|_y^x\Big) \sim \frac{1}{|x - y|^2 + 2\Delta^2 - i\varepsilon} \delta\mathcal{F}^{ab}(x, y) \Big|_y^x\Big)$$

$$\delta\mathcal{F}^{ab}(x, y) = \mathcal{F}^{ab}(x) - \mathcal{F}^{ab}(y), \tag{11.1.33}$$

where $|x - y|$ is the distance in target space $(\mathbb{R}^{9,1}, \eta)$, and $2\Delta^2 \sim L_{\text{NC}}^2$ is the uncertainty scale on $\mathcal{M}$. We can therefore evaluate the one-loop integral (11.1.30) approximately as follows:

$$\Gamma_{1\text{loop};4}[T] \sim \frac{i}{4} \int_{\mathcal{M} \times \mathcal{M}} \frac{\Omega_x \Omega_y}{(2\pi)^m} \frac{\delta\mathcal{F}^{a_1 b_1}(x, y)\delta\mathcal{F}^{a_2 b_2}(x, y)\delta\mathcal{F}^{a_3 b_3}(x, y)\delta\mathcal{F}^{a_4 b_4}(x, y)}{(|x - y|^2 + 2\Delta^2 - i\varepsilon)^4}$$

$$3\Big(-4\eta_{b_1 a_2}\eta_{b_2 a_3}\eta_{b_3 a_4}\eta_{b_4 a_1} + \eta_{b_1 a_2}\eta_{b_2 a_1}\eta_{b_3 a_4}\eta_{b_4 a_3}\Big)$$

$$= \frac{3i}{4} \int_{\mathcal{M} \times \mathcal{M}} \frac{\Omega_x \Omega_y}{(2\pi)^m} \frac{V_4[\delta\mathcal{F}(x, y)]}{(|x - y|^2 + 2\Delta^2 - i\varepsilon)^4}, \tag{11.1.34}$$

where

$$V_4[\delta\mathcal{F}] = -4\text{tr}(\delta\mathcal{F}^4) + (\text{tr}\delta\mathcal{F}^2)^2, \tag{11.1.35}$$

where tr denotes the trace over the matrix indices ab, which are contracted with the target space metric η_{ab}. The prefactor i is an artifact as $\Gamma_{1\text{loop};4}[T]$ picks up an extra i upon evaluating the $i\varepsilon$ regularization, which basically gives

$$\int_{\mathcal{M} \times \mathcal{M}} \frac{\Omega_x \Omega_y}{(2\pi)^m} \frac{V_4[\delta\mathcal{F}(x, y)]}{(|x - y|^2 + 2\Delta^2 - i\varepsilon)^4} \sim i \int_{\mathcal{M} \times \mathcal{M}} \frac{\Omega_x \Omega_y}{(2\pi)^m} V_4[\delta\mathcal{F}(x, y)]\delta\big((|x - y|^2 + 2\Delta^2)^4\big)$$

$$\tag{11.1.36}$$

upon dropping the contribution from the principal value, cf. (7.6.23). Note that $\frac{1}{(|x-y|^2+2\Delta^2-i\varepsilon)^4}$ is essentially the massless $9 + 1$ dimensional Feynman propagator in position space, for distances larger than the scale of noncommutativity. Here the maximal SUSY cancellations are crucial, which lead to a residual interaction that decays like $|x - y|^{-8}$ as appropriate for a massless field in $9 + 1$ dimensions, modified at short distances by the NC cutoff Δ^2. In fact, we will see in Section 11.1.7 that this one-loop induced action can be interpreted in terms of IIB supergravity in $\mathbb{R}^{9,1}$. However, it arises only as effective interaction of objects on the brane $\mathcal{M} \subset \mathbb{R}^{9,1}$, while all the physical dof live on $\mathcal{M}$. From the point of view of $3 + 1$ dimensional branes $\mathcal{M}$, the r^{-8} behavior thus amounts to a weak, short-distance interaction, which is a small correction to the long range interactions including gravity mediated by massless fields in $3 + 1$ dimensions.

It is remarkable that the double integral in (11.1.34) is already the explicit bi-local *one-loop effective action in position space* on $\mathcal{M}$, just like (6.3.13) is the nonlocal action of scalar field theory on fuzzy spaces. In contrast to the non-supersymmetric case, the nonlocality is significantly weaker, and "almost-local" from a $3 + 1$ dimensional point of view. This method of computing loops is very remarkable and has no analog in ordinary QFT, where no analog of string modes exists. We will see in Chapter 12 that this action includes the Einstein–Hilbert action for the effective metric on $\mathcal{M}^{3,1}$, but only in the presence of fuzzy extra dimensions, i.e. for a background brane with the structure $\mathcal{M}^{3,1} \times \mathcal{K} \subset \mathbb{R}^{9,1}$. Without $\mathcal{K}$, it yields a higher-derivative action which is quite distinct from ordinary gravity.

11.1.4 The one-loop effective action for the Euclidean IKKT model

Now consider the Euclidean IKKT model, where the bosonic action is negative definite in our conventions. Therefore, the one-loop effective action is defined through (cf. Section 7.6)

$$e^{\Gamma_{\text{eff}}[T]} = \int_{\text{1 loop}} dT d\Psi dc \, e^{S_0[T,\Psi,c]} \tag{11.1.37}$$

and can be evaluated as before using (7.6.7), with the fermionic contribution (11.1.19) arising from the continuation of the Pfaffian as already discussed. We will drop the Wess–Zumino contribution for simplicity[4]. The full one-loop effective action is then

$$\Gamma_{\text{eff}}[T] = S_0[T] + \Gamma_{1\text{loop}}[T] \tag{11.1.38}$$

which leads to the same form as in (11.1.28) but without the prefactor i:

$$\Gamma_{1\text{loop}}[T] = -\frac{1}{2}\text{Tr}\Big(\log(\Box - \Sigma_{ab}^{(\mathcal{A})}[\mathcal{F}^{ab},.]) - \frac{1}{2}\log(\Box - \Sigma_{ab}^{(\psi)}[\mathcal{F}^{ab},.]) - 2\log(\Box) \Big)$$
$$= -\frac{1}{2}\text{Tr}\Big(-\frac{1}{4}(\Box^{-1}\Sigma_{ab}^{(\mathcal{A})}[\mathcal{F}^{ab},.])^4 + \frac{1}{8}(\Box^{-1}\Sigma_{ab}^{(\psi)}[\mathcal{F}^{ab},.])^4 + \mathcal{O}(\Box^{-1}[\mathcal{F}^{ab},.])^5 \Big) \tag{11.1.39}$$

with $a, b = 1, \ldots, 10$. The crucial point is again that the leading terms in this expansion cancel due to maximal supersymmetry, so that the first nonvanishing term in this expansion is $n = 4$.

The trace over $\text{End}(\mathcal{H})$ can again be evaluated using string modes. This gives

$$\Gamma_{1\text{loop};4}[T] = -\frac{1}{4} \int_{\mathcal{M}\times\mathcal{M}} \frac{\Omega_x \Omega_y}{(2\pi)^m} \frac{\delta\mathcal{F}^{a_1 b_1}(x,y)\delta\mathcal{F}^{a_2 b_2}(x,y)\delta\mathcal{F}^{a_3 b_3}(x,y)\delta\mathcal{F}^{a_4 b_4}(x,y)}{(|x-y|^2 + 2\Delta^2)^4}$$
$$3\Big(-4\delta_{b_1 a_2}\delta_{b_2 a_3}\delta_{b_3 a_4}\delta_{b_4 a_1} + \delta_{b_1 a_2}\delta_{b_2 a_1}\delta_{b_3 a_4}\delta_{b_4 a_3} \Big)$$
$$= -\frac{3}{4} \int_{\mathcal{M}\times\mathcal{M}} \frac{\Omega_x \Omega_y}{(2\pi)^m} \frac{V_4[\delta\mathcal{F}(x,y)]}{(|x-y|^2 + 2\Delta^2)^4}, \tag{11.1.40}$$

where $V_4[\delta\mathcal{F}]$ has the same form as in (11.1.35), with indices contracted by δ_{ab} instead of η_{ab}. We observe again that the interaction between parallel $\mathbb{R}_\theta^{2n}$ branes with identical field strength vanishes, as $\delta F_{ab} = 0$.

The significance of V_4 becomes more transparent using the following observation [203] pertinent to Euclidean signature: If $\delta\mathcal{F}^{ab}$ has rank ≤ 4, then

---

[4] The WZ contribution vanishes on backgrounds which do not span the full 10-dimensional target space.

$$-V_4[\delta\mathcal{F}] = 4\text{tr}(\delta\mathcal{F}\delta\mathcal{F}\delta\mathcal{F}\delta\mathcal{F}) - (\text{tr}\delta\mathcal{F}\delta\mathcal{F})^2$$
$$= 4(\delta\mathcal{F}_+^{ab}\delta\mathcal{F}_{+ba})(\delta\mathcal{F}_-^{cd}\delta\mathcal{F}_{-dc}), \qquad \delta\mathcal{F}_\pm = \delta\mathcal{F} \pm \star\delta\mathcal{F}$$
$$\geq 0, \tag{11.1.41}$$

where $\star$ denotes the four-dimensional Hodge star with respect to $\delta_{\mu\nu}$. This describes an *attractive* interaction, which vanishes only in the (anti-) self-dual case $\delta\mathcal{F} = \pm \star \delta\mathcal{F}$. It strongly suggests that branes may form bound states through quantum effects, as well as a preference for branes with dimensions ≤ 4. Even though this observation requires Euclidean signature, it does apply in Minkowski signature to backgrounds with fuzzy extra dimensions $\mathcal{M}^{3,1} \times \mathcal{K} \subset \mathbb{R}^{9,1}$, which will be crucial in Chapter 12.

11.1.5 Selection and stabilization of branes at one loop

As soon as the brane or background configuration $\mathcal{M}$ is deformed and no longer BPS, the one-loop effective potential is nontrivial. This provides a selection mechanism for the branes and their dimension, which strongly suggests that four-dimensional branes are preferred.

Brane dimension and UV finiteness

We recall the discussion on UV/IR mixing in NC field theory in Section 6.2.3, where the would-be UV divergences in loop integrals were seen to transmute to IR divergences or strong nonlocality in the effective action. In the present maximally supersymmetric cases, this conclusion is changed, since the one-loop contribution starts with the fourth power of the propagator $\frac{1}{(\square^2)^4}$, which translates to $\frac{1}{(|x-y|^2+2\Delta^2)^4}$ in the string basis. The significance of the trace or integral then depends strongly on the dimension of $\mathcal{M}$. If the dimension is four or less, then the integral $\int \Omega_y \frac{1}{(|x-y|^2+2\Delta^2)^4}$ is convergent and *almost-local*, i.e. confined to a region of size Δ around x in position space. This is the position space version of the statement that $\text{Tr}\frac{1}{(\square^2)^4}$ is UV finite. Then the effective action reduces to an almost-local contribution of $V_4[\delta\mathcal{F}]$, which is perfectly well defined. This is expected to generalize to higher loops, because the underlying NC $\mathcal{N} = 4$ SYM theory is UV finite [91, 107] just like its commutative counterpart. Hence, the resulting physics will be almost-local. Moreover, the attractive potential (11.1.41) in the Euclidean one-loop effective action suggests that such branes can be stable.

For higher-dimensional branes, the conclusion is completely different. For six-dimensional branes $\mathcal{M}^6$, the one-loop effective action is still convergent and almost-local in the sense just discussed; however, higher loop configurations are expected to be divergent or nonlocal, because the classical six-dimensional theory is UV divergent and not renormalizable. Therefore, branes of dimension 6 and higher are pathological, i.e. unstable at the quantum level. This observation strongly suggests[5] that *quantum effects select the dimension of spacetime to be 4* at weak coupling. Note that this does not exclude backgrounds of the structure $\mathcal{M}^{3,1} \times \mathcal{K}_n$ where the compact fuzzy space $\mathcal{K}_n$ supports only

---

[5] Of course finiteness also holds for two-dimensional branes, but they are not expected to be favored by quantum effects.

a finite number of dof; such backgrounds will be important in Chapter 12. For the same reason, covariant quantum spaces such as $\mathcal{M}^{3,1}$ are not excluded.

In the Euclidean case – and for branes with Euclidean embedding – we can furthermore exploit the observation (11.1.41) that $V_4[\delta\mathcal{F}] < 0$ for four-dimensional branes $\mathcal{K}^4$, which thus acquire a binding energy. This suggests that such fuzzy extra dimensions might be stabilized by quantum effects, with a preference for four-dimensional $\mathcal{K}$. However, this ignores the interactions between $\mathcal{K}$ and $\mathcal{M}^{3,1}$, which will be taken into account in Section 12.4.2.

Intersecting branes

A particularly intriguing situation arises for (self)intersecting branes, which lead to chiral zero modes realized by fermionic string modes connecting the two branes at their intersection [46]. Being chiral these zero modes are expected to contribute significantly to the one-loop effective action, because they break supersymmetry. Such configurations $\mathcal{K}$ are particularly interesting for fuzzy extra dimensions as discussed in Section 8.4, and deserve to be studied in detail.

The mini-landscape

Due to the existence of a vast number of different (approximate) solutions or saddle points of the matrix model, we face the following selection problem: which – if any – of the many solutions are preferred? This could be viewed as a baby-version of the landscape problem in string theory. Nevertheless, there is a crucial difference: In the present framework, the stability and the longevity of each such "vacuum" is a well-defined problem, which can in principle be studied and answered in terms of the quantum effective action of the matrix model. This is not the case in the traditional approach of string theory, where the landscape is much worse: it arises from different 10-dimensional compactifications of target space, which are very hard to even compare, lacking a suitable background-independent formulation of string theory. The present mini-landscape arises only as different branes embedded in one and the same uncompactified target space $\mathbb{R}^{9,1}$. Therefore, the problem is in fact a blessing, because it offers a mechanism which could in principle explain why the physics as we see it is preferred among other possible scenarios.

This problem has been under investigation at the nonperturbative level via numerical simulations of the matrix model, as discussed in Section 11.2. However, this turns out to be very challenging, and no conclusive picture has emerged so far. It may also be that there is no unique answer, and different possible worlds or universes might be consistent, i.e. sufficiently long-lived. If that is the case, we would be in a similar situation as in traditional string theory. However, the space of consistent vacua would be far smaller than in string theory, so that the matrix model is certainly more predictive.

11.1.6 Flux terms and other soft SUSY breaking terms

The generation of fuzzy extra dimensions in directions transversal to the spacetime brane will play an essential role to get interesting physics. One way to stabilize such fuzzy extra dimensions is by adding an explicit cubic "flux" term to the action as in Section 8.3, replacing the bosonic action by

$$S[T] = \frac{1}{g^2} \text{Tr}\big([T^a, T^b][T_a, T_b] - 2m^2 T^a T_a + f_{abc}[T^a, T^b] T^c\big). \tag{11.1.42}$$

Here, f_{abc} is totally antisymmetric, and typically related to structure constants of some Lie algebra. This is perfectly fine at the classical level; however, the extra terms break supersymmetry, and thereby reintroduce the problem of UV/IR mixing (even though the breaking is "soft").

To see this, we compute again the one-loop effective action, including the extra terms in the action. We assume that f_{abc} is nonvanishing only if all indices are transversal, i.e. $a, b, c = 4, \ldots, 9$. For example, $f_{abc} \sim \varepsilon_{abc}$ for $a, b, c = 7, 8, 9$ would lead to the formation of a fuzzy sphere. Expanding this action around a background T^a leads upon gauge fixing to

$$S[T + \mathcal{A}] = S[T] + \frac{1}{g^2} \text{Tr}\big((-4\Box T_c + 3f_{abc}[T^a, T^b])\mathcal{A}^c\big) + S_{2,\text{gf}}[\mathcal{A}] + O(\mathcal{A}^3), \tag{11.1.43}$$

where the linear term vanishes on the background solution, and

$$S_{2,\text{gf}}[\mathcal{A}] = \frac{1}{g^2} \text{Tr}\left(-2\mathcal{A}_a(\mathcal{D}^2 + m^2)\mathcal{A}^a + 3f_{abc}[T^a, \mathcal{A}^b]\mathcal{A}^c - \bar{c}[T_a, [T^a, c]]\right) \tag{11.1.44}$$

includes the ghosts. It is convenient to rewrite the $[T, \mathcal{A}]\mathcal{A}$ term as follows:

$$3\text{Tr}f_{abc}[T^a, \mathcal{A}^b]\mathcal{A}^c = 2\text{Tr}\mathcal{A}\Gamma_a^{(\mathcal{A})}[T^a, .]\mathcal{A}, \tag{11.1.45}$$

where

$$(\Gamma_a^{(\mathcal{A})})_{bc} = \frac{3}{2} f_{abc} \tag{11.1.46}$$

acts on the space of bosonic fluctuations $\mathcal{A}$. Then the (Euclidean, to be specific) one-loop effective action can be written as follows:

$$\begin{aligned}
\Gamma_{1\text{loop}}[T] &= -\frac{1}{2}\text{Tr}\Big(\log(\Box + m^2 - \Sigma_{ab}^{(\mathcal{A})}[\mathcal{F}^{ab}, .] - \Gamma_a^{(\mathcal{A})}[T^a, .]) \\
&\quad - \frac{1}{2}\log(\Box - \Sigma_{ab}^{(\psi)}[\mathcal{F}^{ab}, .]) - 2\log\Box\Big) \\
&= -\frac{1}{2}\text{Tr}\Big(\sum_{n>0}\frac{1}{n}\big(-(\Box^{-1}(\Sigma_{ab}^{(\mathcal{A})}[\mathcal{F}^{ab}, .] + \Gamma_a^{(\mathcal{A})}[T^a, .] - m^2))\big)^n \\
&\quad + \frac{1}{2}(\Box^{-1}\Sigma_{ab}^{(\psi)}[\mathcal{F}^{ab}, .])^n\Big)\Big) \\
&= \Gamma_{1\text{loop}}^{\text{IKKT}}[T] - \frac{m^2}{2}\text{Tr}\Box^{-1} + \frac{\alpha}{2}\text{Tr}(\Box^{-1}[T^a, .]\Box^{-1}[T^b, .]\delta_{ab}^{\mathcal{K}}) + \cdots \tag{11.1.47}
\end{aligned}$$

assuming

$$\begin{aligned}
\text{tr}(\Gamma_a^{(\mathcal{A})}) &= 0 \\
\text{tr}(\Gamma_a^{(\mathcal{A})}\Gamma_b^{(\mathcal{A})}) &= \alpha\,\delta_{ab}^{\mathcal{K}} \\
\text{tr}(\Gamma_a^{(\mathcal{A})}\Sigma_{bc}^{(\mathcal{A})}) &= 0. \tag{11.1.48}
\end{aligned}$$

Here $\delta_{ab}^{\mathcal{K}}$ is a delta-like tensor for the transversal indices attached to $\mathcal{K}$. In particular, the term $\mathrm{Tr}(\square^{-1}\Gamma_a^{(\mathcal{A})}[T^a, .]) = 0$ vanishes. Assuming moreover $[\square, [T_a, .]] = 0$ for the transversal indices a, the extra contribution from the flux term can be written as

$$\mathrm{Tr}(\square^{-1}[T^a, .]\square^{-1}[T^b, .]\delta_{ab}^{\mathcal{K}}) = \mathrm{Tr}(\square^{-2}\square_{\mathcal{K}}) \qquad (11.1.49)$$

in terms of the matrix Laplacian for $\mathcal{K}$

$$\square_{\mathcal{K}} := \delta_{ab}^{\mathcal{K}}[T^a, [T^b, .]]. \qquad (11.1.50)$$

To understand the meaning of the extra $\mathrm{Tr}\square^{-1}$ and $\mathrm{Tr}(\square^{-2}\square_{\mathcal{K}})$ terms, we recall the discussion on UV/IR maxing in Section 6.2.3 where such traces were recognized as nonlocal IR-divergent integral interpreted as nonlocal interaction. Since $\square = \square_T$ depends on the background, these terms provide a nonlocal contribution to the geometrical dynamics. More divergent terms are expected at higher loops, since these models are no longer UV (and therefore IR) finite. This underscores the special status of the IKKT model: any deformation of its unique form induces a significant and probably unacceptable impact on the quantum effective action.

In the case of non-supersymmetric matrix models, the same issue arises, but much more severely. Taking the Euclidean case to be specific, the one-loop contribution (11.1.39) becomes

$$\Gamma_{1\text{loop, bosonic}}[T] = -\frac{1}{2}\mathrm{Tr}\left(\log(\square - \Sigma_{ab}^{(\mathcal{A})}[\mathcal{F}^{ab}, .])\right)$$

$$= -\frac{1}{2}\left(\mathrm{Tr}\log\square - \frac{5}{11}\mathrm{Tr}(\square^{-1}[\mathcal{F}^{ab}, .]\square^{-1}[\mathcal{F}_{ab}, .]) + \cdots\right) \qquad (11.1.51)$$

using $\mathrm{tr}\Sigma_{ab}^{(\mathcal{A})} = 0$. The leading term can be written in the form

$$\mathrm{Tr}\log\square \sim \int_{\mathcal{M}} d^4x\, d^4y\, \ln\left(|x - y|^2\right) \qquad (11.1.52)$$

on a four-dimensional quantum space $\mathcal{M}^4$, using the string representation of the trace. This is divergent on non-compact spaces, and completely nonlocal even on compact spaces. It cannot be dismissed as an irrelevant constant, since it does depend nontrivially and nonlocally on the fluctuations of $\mathcal{M}$, and it cannot be absorbed by suitable counterterms as in renormalizable QFT. The second term in (11.1.51) is similar to (11.1.47), and can be written in the form

$$\mathrm{Tr}\left(\square^{-1}[\mathcal{F}^{ab}, .]\square^{-1}[\mathcal{F}^{ab}, .]\right) \sim \int d^4x\, d^4y \frac{(\mathcal{F}^{ab}(x) - \mathcal{F}^{ab}(y))^2}{(x - y)^4}. \qquad (11.1.53)$$

This is a nonlocal interaction term[6], which cannot be consistently truncated as a local effective action. Hence, only the IKKT model has a chance to produce a sufficiently local effective action in four dimensions.

---

[6] An $\frac{1}{(x-y)^4}$ interaction in four dimensions would lead to a divergent contribution for uniformly distributed background objects, and can therefore not be considered as local. This is in contrast to the $\frac{1}{(x-y)^8}$ interactions arising in the maximally supersymmetric model.

Mass term and broken shift invariance

Now consider the contribution of a mass term $\mathrm{Tr}(m^2 T_a T^a)$ in the IKKT model (11.1.47), in the absence of any extra flux terms. Somewhat surprisingly, it also leads to a nonlocal contribution

$$\mathrm{Tr}\square^{-1} \sim \int_{\mathcal{M}} d^4x d^4y \frac{1}{|x-y|^2} \tag{11.1.54}$$

to the effective action in position space, which is somewhat milder than in the non-SUSY case but still non-local and divergent in four dimensions. This suggests that m^2 should not be considered as physical in the large N limit. It may nevertheless be useful as a regularization parameter for finite N, provided it scales as $m^2 \to 0$ for $N \to \infty$ in some appropriate way.

It is also interesting to observe that the mass term breaks the shift invariance (7.0.7). Nevertheless, the following scaling relations hold:

$$S_0[cT, c^{\frac{3}{2}}\Psi; cm] = c^4 S_0[T, \Psi; m]$$
$$\Gamma_{1\mathrm{loop}}[cT, cm] = \Gamma_{1\mathrm{loop}}[T, m]. \tag{11.1.55}$$

It follows that

$$\Gamma_{1\mathrm{loop}}[T, m] = \Gamma_{1\mathrm{loop}}\left[\frac{T}{m}\right] \tag{11.1.56}$$

is a function of the dimensionless matrix variables $\frac{T}{m}$. Moreover, the trace-$U(1)$ sector $T^a \sim \mathbb{1}$ of the model does not couple to the trace-free part of matrices. This implies that no "hard shift-breaking terms" such as $\mathrm{Tr}(T_a T_b T^{ab})$ arise in the effective action, because these would entail a nontrivial coupling with the trace-$U(1)$ sector. Therefore, the quantum effective action can be written in terms of $\mathcal{F}^{ab}$ and $\mathrm{Tr}(T^a)\mathrm{Tr}(T_a)$.

11.1.7 UV/IR mixing, supergravity, string theory, and $\mathcal{N} = 4$ SYM

Given the results obtained for the one-loop effective action, we are in a position to make a more detailed comparison with string theory. To see the relation, we recall that the effective potential (11.1.35) can be written using $\delta\mathcal{F}(x,y) = \mathcal{F}_x - \mathcal{F}_y$ as

$$\begin{aligned}
V_4[\delta\mathcal{F}(x,y)] &= -4\mathrm{tr}\delta\mathcal{F}(x,y)^4 + (\mathrm{tr}\delta\mathcal{F}(x,y)^2)^2 \\
&= -4\mathrm{tr}(\tilde{T}_x^2) + (\mathrm{tr}\tilde{T}_x)^2 + (x \leftrightarrow y) \\
&\quad + 4\big(4\mathrm{tr}(\mathcal{F}_x\mathcal{F}_x\mathcal{F}_x\mathcal{F}_y) - \mathrm{tr}(\mathcal{F}_x\mathcal{F}_y)\,\mathrm{tr}(\mathcal{F}_x\mathcal{F}_x) + (x \leftrightarrow y)\big) \\
&\quad - 16\mathrm{tr}(\tilde{T}_x\tilde{T}_y) + 2\mathrm{tr}\tilde{T}_x\mathrm{tr}\tilde{T}_y \\
&\quad - 8\mathrm{tr}(\mathcal{F}_x\mathcal{F}_y\mathcal{F}_x\mathcal{F}_y) + 4(\mathrm{tr}(\mathcal{F}_x\mathcal{F}_y))^2.
\end{aligned} \tag{11.1.57}$$

Here, tr denotes the trace over the matrix indices ab, and

$$\tilde{T}^{ab}[\mathcal{F}] = \mathcal{F}^{ac}\mathcal{F}^b_{\ c} \tag{11.1.58}$$

is the matrix energy–momentum tensor (7.4.14) without the trace term. To proceed, we define

$$D(x - y) := \frac{1}{(|x - y|^2 + \Delta^2 - i\varepsilon)^4} \sim \frac{1}{|x - y|^8}, \qquad (11.1.59)$$

which for distances $|x - y|^2 \gg \Delta^2$ coincides with the massless Feynman propagator in $9 + 1$ dimensions. Then the effective interaction induced at one loop can be written as

$$\Gamma_{1\text{loop};4}[T] \sim \frac{3i}{4} \int_{\mathcal{M} \times \mathcal{M}} \frac{\Omega_x \Omega_y}{(2\pi)^m} \Big(2V_4(\mathcal{F}(x))\, D(x - y)$$

$$+ 16\big(\mathcal{F}^{ae}(x)\mathcal{F}_{ef}(x)\mathcal{F}^{fb}(x) + \tfrac{1}{4}\mathcal{F}^{ab}(x)\mathcal{F}^{ef}(x)\mathcal{F}_{ef}(x)\big)D^{(AS)}_{ab;cd}(x,y)\,\mathcal{F}^{dc}(y)$$

$$- 8\tilde{T}^{ab}(x)D^{(S)}_{ab;cd}(x,y)\,\tilde{T}^{cd}(y)$$

$$+ 4\mathcal{F}^{ab}(x)\mathcal{F}^{ef}(x)D^{(AS)}_{abef;cdgh}(x,y)\,\mathcal{F}^{cd}(y)\mathcal{F}^{gh}(y)\Big), \qquad (11.1.60)$$

and similarly in Euclidean signature replacing $\frac{3i}{4} \to -\frac{3}{4}$, where

$$D^{(S)}_{ab;cd}(x,y) = \big(\eta_{ac}\eta_{bd} + \eta_{ad}\eta_{bc} - \tfrac{1}{4}\eta_{ab}\eta_{cd}\big)D(x - y)$$

$$D^{(AS)}_{abef;cdgh}(x,y) = \big(\eta_{ac}\eta_{bd}\eta_{eg}\eta_{fh} + \eta_{ac}\eta_{bh}\eta_{ed}\eta_{fg} - \eta_{ac}\eta_{bg}\eta_{ed}\eta_{fh}\big)D(x - y)$$

$$D^{(AS)}_{ab;cd}(x,y) = \big(\eta_{ac}\eta_{bd} - \eta_{ad}\eta_{bc}\big)D(x - y). \qquad (11.1.61)$$

For $|x - y|^2 \gg \Delta^2$, the third line in (11.1.60) can be interpreted in terms of a graviton exchange in target space, and $D^{(S)}_{ab;cd}$ is indeed the graviton propagator on $\mathbb{R}^{9,1}$ in de Donder gauge. The last line can be attributed to the exchange of a rank 4 antisymmetric tensor with propagator $D^{(AS)}_{abef;cdgh}$, and the first line can be interpreted in terms of a dilaton exchange [120, 200] coupling the background density Ω_y to

$$V_4(\mathcal{F}) = -4\tilde{T}^{ab}(x)\tilde{T}_{ab}(x) + (\tilde{T}^{ab}(x)\eta_{ab})^2. \qquad (11.1.62)$$

Finally, the second line can be interpreted as exchange of an antisymmetric rank 2 tensor field B_{ab}, which couples to branes [200]. Hence, all terms are recognized as interaction mediated by the exchange of massless fields in IIB supergravity in target space $\mathbb{R}^{9,1}$, coupled to a brane or background carrying the field strength $\mathcal{F}^{ab}$.

This interpretation becomes more transparent for a matrix background consisting of two distinct irreducible quantum spaces $\mathcal{M}_A \oplus \mathcal{M}_B$ as in (7.4.33), corresponding to two branes embedded in target space $\mathbb{R}^{9,1}$. In string theory, these are described as solitonic objects in IIB supergravity, which interact by exchanging massless supergravity modes (i.e. the graviton, dilaton, and the antisymmetric two-form and four-form fields). It was indeed found in [47, 48, 64, 102, 150, 198, 120] that the interaction in the matrix model matches with supergravity at least in the long-distance limit. While these computations were originally limited to highly symmetric branes, the present techniques based on string modes apply to rather generic branes, as long as they admit (quasi)coherent states. To see this, we decompose the field strength in the background $\mathcal{M}_A \oplus \mathcal{M}_B$ as

$$\mathcal{F}(x) = \mathcal{F}_A(x) + \mathcal{F}_B(x), \qquad (11.1.63)$$

where $\mathcal{F}_A$ and $\mathcal{F}_B$ are assumed to have nonoverlapping support in target space. Plugging this into (11.1.60), the effective action decomposes as

$$\Gamma_{1\text{loop};4}[\mathcal{F}_A, \mathcal{F}_B] = \Gamma_{1\text{loop};4}[\mathcal{F}_A] + \Gamma_{1\text{loop};4}[\mathcal{F}_B] + \Gamma_{1\text{loop};4}^{\text{int}}[\mathcal{F}_A, \mathcal{F}_B], \qquad (11.1.64)$$

where the interaction terms acquire the familiar form

$$\Gamma_{1\text{loop};4}^{\text{int}}[\mathcal{F}_A, \mathcal{F}_B] = \frac{3i}{2} \int\limits_{\mathcal{M}_A \times \mathcal{M}_B} \frac{\Omega_x \Omega_y}{(2\pi)^m} \Big(-8\tilde{T}_A^{ab}(x) D_{ab;cd}^{(S)}(x,y) \tilde{T}_B^{cd}(y)$$

$$+ 16\big(\mathcal{F}_A^{ae}(x)\mathcal{F}_{ef}^A(x)\mathcal{F}_A^{fb}(x) + \frac{1}{4}\mathcal{F}_A^{ab}(x)\mathcal{F}_A^{ef}(x)\mathcal{F}_{Aef}(x)\big)$$

$$\cdot D_{ab;cd}^{(AS)}(x,y) \, \mathcal{F}_B^{dc}(y)$$

$$+ 4\mathcal{F}_A^{ab}(x)\mathcal{F}_A^{ef}(x) D_{abef;cdgh}^{(AS)}(x,y) \, \mathcal{F}_B^{cd}(y)\mathcal{F}_B^{gh}(y)\Big). \qquad (11.1.65)$$

The factor 2 arises from the two possible associations $x \leftrightarrow A, y \leftrightarrow B$ and vice versa. This can clearly be interpreted as interactions due to the exchange of IIB supergravity modes between the A and B branes as explained before. In fact, the computation is very close to the origin of IIB supergravity interactions in string theory, which are also mediated by massless strings stretched between the branes. Moreover, this derivation generalizes easily to the nonabelian case modeled by stacks of coincident branes, by simply amending (11.1.60) with an internal trace. By comparing this form of the effective action in target space with the interpretation in terms of $\mathcal{N} = 4$ SYM (11.1.22), we obtain directly a "holographic" relation between $\mathcal{N} = 4$ SYM and linearized IIB supergravity on $\mathbb{R}^{9,1}$. This will be generalized in the following to the "near-horizon limit" $AdS^5 \times S^5$ for a stack of branes, which has come to fame as AdS/CFT correspondence [134]. However, it is evident that the propagating degrees of freedom are confined to the brane(s) $\mathcal{M}^{3,1}$ here, at least in the weak-coupling regime.

It is important to stress that this correspondence works only between critical (super) string theory in $9 + 1$ dimensions and the maximally supersymmetric IKKT matrix model. Indeed, IIB supergravity in target space is obtained in string theory as a consistency condition to avoid the conformal anomaly, while in matrix models the quantization is pathological except in the maximally SUSY IIB model. In contrast, both noncritical string theory and generic matrix models would lead to pathologies.

Stacks of branes, $\mathcal{N} = 4$ SYM, and holography

It is interesting to elaborate the one-loop effective action from the point of view of $\mathcal{N} = 4$ SYM (11.1.22) with nontrivial but slowly varying scalar fields ϕ^i and gauge fields A_μ on $3 + 1$ dimensional branes $\mathcal{M}^{3,1} \subset \mathbb{R}^{9,1}$. The $\mathfrak{u}(1)$ contribution from a single brane vanishes in the flat limit, and will be understood in terms of gravity in the next chapter. For a stack of n branes, the contribution from the nonabelian $\mathfrak{su}(n)$ modes does survive in that limit, and can be extracted from the previous results. For simplicity we assume that $[\Box_\mathcal{K}, \mathcal{F}_{ab}] = 0$ for the background under consideration, where $\Box_\mathcal{K} = [\phi^i, [\phi_i, .]]$ is the matrix Laplacian from the transversal sector as in Section 8.3. The integral over the propagators is convergent and almost-local, and can be evaluated in (local) Cartesian embedding coordinates (9.1.19) using $\Omega = d^4 x \rho_M$ and $\rho^{-2} = L_{\text{NC}}^{-4} = \rho_M$ as

$$\int \frac{d^4 y}{(|x - y|^2 + \Box_{\mathcal{K}} + 2\Delta^2 - i\varepsilon)^4} = \frac{i\pi^2}{6} \frac{1}{(\Box_{\mathcal{K}} + 2\Delta^2)^2}. \tag{11.1.66}$$

Such integrals will be discussed in more detail in Section 12.3 from a phase space point of view. Then the leading nonabelian contribution to the one-loop effective action takes the form

$$\Gamma_{1\text{loop};4}[T] = \frac{3i}{4} \int\limits_{\mathcal{M} \times \mathcal{M}} \frac{\Omega_x \Omega_y}{(2\pi)^4} \text{Tr}_n \left(\frac{-4\text{tr}([\mathcal{F}, .]^4) + (\text{tr}[\mathcal{F}, .]^2)^2}{(|x - y|^2 + \Box_{\mathcal{K}} + 2\Delta^2 - i\varepsilon)^4} \right)$$

$$\approx -\frac{1}{128\pi^2} \int\limits_{\mathcal{M}} d^4 x \, \rho^{-4} \text{Tr}_n \left(\frac{-4\text{tr}([\mathcal{F}, .]^4) + (\text{tr}[\mathcal{F}, .]^2)^2}{(\Box_{\mathcal{K}} + 2\Delta^2)^2} \right) \tag{11.1.67}$$

for slowly varying fields, where Tr_n indicates the trace over $\mathfrak{su}(n)$.

Now consider a specific background consisting of $n - 1$ coincident $\mathbb{R}^{3,1}_\theta$ branes and a single "test-brane" with transversal separation d, which is realized in $\mathcal{N} = 4$ language by nonabelian scalar fields Φ^i assuming a nontrivial VEV

$$\Phi^i = \phi^i(x)\lambda, \qquad \lambda = \text{diag}(1 - n, 1, \ldots, 1) \tag{11.1.68}$$

for some vector $\phi^i \in \mathbb{R}^6$ in the transversal directions. The gauge symmetry on such a background is broken spontaneously to $SU(n - 1) \times U(1)$, where λ is the generator of the extra $U(1)$. The internal matrix Laplacian becomes

$$\Box_{\mathcal{K}} = (\phi^i \phi_i) [\lambda, [\lambda, .]], \tag{11.1.69}$$

which has only a single nonvanishing eigenvalue $n^2(\phi^i \phi_i)$ with multiplicity $2(N - 1)$. Hence, these $2(n - 1)$ modes acquire a mass $\Box_{\mathcal{K}} = n^2(\phi^i \phi_i)$ in the loop. We are interested in the one-loop effective action for the "Coulomb branch," i.e. the unbroken (massless) $U(1)$ sector which consists of the scalar and gauge fields taking values proportional to λ. We therefore consider a background of the form

$$\mathcal{F}^{ab} = \mathcal{F}_0^{ab} \mathbb{1}_n + \mathcal{F}_1^{ab} \lambda, \tag{11.1.70}$$

where

$$\mathcal{F}_0^{ab} = \begin{pmatrix} -\theta^{\mu\nu} & 0 \\ 0 & 0 \end{pmatrix}, \qquad \mathcal{F}_1^{ab} = \begin{pmatrix} -\theta^{\mu\mu'}\theta^{\nu\nu'} F_{\mu'\nu'} & \theta^{\mu\mu'} D_{\mu'}\phi^i \\ -\theta^{\nu\nu'} D_{\nu'}\phi^j & 0 \end{pmatrix} \tag{11.1.71}$$

(cf. (7.4.28)) describes the background branes $\mathcal{M} \cong \mathbb{R}^{3,1}_\theta$ and the massless $U(1)$ sector, respectively, noting that $[\Phi_i, \Phi_j] = 0$; here, we decomposed $a = (\mu, i)$ into spacetime and transversal indices. Then Tr_n can be evaluated noting that $[\mathcal{F}, .] = \mathcal{F}_1[\lambda, .]$, which gives

$$\Gamma_{U(1)} = -\frac{n-1}{4(4\pi)^2} \int_{\mathcal{M}} \frac{d^4x}{(\phi^i\phi_i)^2} \rho^{-4} \left(-4(\mathcal{F}_1\eta\mathcal{F}_1\eta\mathcal{F}_1\eta\mathcal{F}_1\eta) + (\mathcal{F}_1\eta\mathcal{F}_1\eta)^2 \right)$$

$$= -\frac{n-1}{4(4\pi)^2} \int_{\mathcal{M}} \frac{d^4x}{(\phi^i\phi_i)^2} \rho^4 \left(-4F_{\mu\nu}F^{\nu\eta}F_{\eta\rho}F^{\rho\mu} + (F_{\mu\nu}F^{\mu\nu} + 2\rho^{-2}D_\mu\phi_i D^\mu\phi^i)^2 \right.$$

$$\left. + 16\rho^{-2}D^\mu\phi_i D_\nu\phi^i F^{\nu\eta}F_{\eta\mu} - 8\rho^{-4}D_\mu\phi_i D_\nu\phi^i D^\mu\phi_j D^\nu\phi^j \right) \qquad (11.1.72)$$

using

$$\rho^{-4}(\mathcal{F}_1\eta\mathcal{F}_1\eta) = -F_{\mu\nu}F^{\mu\nu} - 2\rho^{-2}D_\mu\phi_i D^\mu\phi^i,$$

$$\rho^{-8}(\mathcal{F}_1\eta\mathcal{F}_1\eta\mathcal{F}_1\eta\mathcal{F}_1\eta) = F_{\mu\nu}F^{\nu\eta}F_{\eta\rho}F^{\rho\mu} - 4\rho^{-2}D^\mu\phi_i D_\nu\phi^i F^{\nu\eta}F_{\eta\mu}$$

$$+ 2\rho^{-4}D_\mu\phi_i D_\nu\phi^i D^\mu\phi_j D^\nu\phi^j. \qquad (11.1.73)$$

Here the spacetime indices are contracted with $G_{\mu\nu}$, and ϕ^i has dimension length. For $D_\mu\phi^i = 0$, we recover the well-known $\frac{F^4}{\phi^4}$ term in the low-energy effective action for the Coulomb branch of $\mathcal{N} = 4$ SYM [28]. The present $SO(9,1)$ invariant setup provides also the contributions from the scalar fields. Now we combine this with the bare matrix model action for the present nonabelian background, which can be written as (9.1.69)

$$S_0[X] = \frac{1}{g^2}\mathrm{Tr}([X^a, X^b][X_a, X_b])$$

$$\sim -\frac{n(n-1)}{(2\pi)^2 g^2} \int_{\mathcal{M}} d^4x \left(2D_\mu\phi^i D^\mu\phi_i + \rho^2 F_{\mu\nu}F^{\mu\nu} - \frac{1}{2}\mathcal{L}_0 F \wedge F \right), \qquad (11.1.74)$$

where $\mathcal{L}_0 = \gamma^{\mu\nu}\eta_{\mu\nu}$ and the factor $n(n-1)$ arises from $\mathrm{Tr}_n\lambda^2$. Remarkably, the effective action $S_0 + \Gamma_{U(1)}$ for the Coulomb branch – which should be viewed as effective action for the single test brane in the background of the stack of $n-1$ coincident $\mathbb{R}^{3,1}_\theta$ branes – is consistent with the expansion of the following Dirac–Born–Infeld (DBI) action on $\mathcal{M}$ [203]

$$S_{\mathrm{DBI}} = T_3 \int_{\mathcal{M}} d^4x \left(\sqrt{|\det(\mathcal{G}_{\mu\nu} + \rho F_{\mu\nu})|} - \frac{|\phi^2|^2}{Q}\sqrt{|\mathcal{G}_{\mu\nu}|} \right) = S_0 + \Gamma_{U(1)} + O(\partial)^6$$

$$(11.1.75)$$

up to higher derivatives supplemented by the Chern–Simons term $\sim \int \eta F \wedge F$, for

$$\mathcal{G}_{\mu\nu} = \frac{|\phi^2|}{\sqrt{Q}} G_{\mu\nu} + \frac{\sqrt{Q}}{|\phi^2|} D_\mu\phi^i D_\nu\phi_i,$$

$$Q = \frac{\rho^2 g^2}{2n}, \qquad T_3 = \frac{n(n-1)}{\pi^2 g^2} \qquad (11.1.76)$$

(recall that $\rho \sim L_{NC}^2$ and $g = g_{YM}\rho^2$ and ϕ has dimension length here[7]). Indeed, using

$$\det(\mathcal{G}_{\mu\nu} + \rho F_{\mu\nu}) = \det(\mathcal{G}_{\mu\nu})\left(1 + \frac{1}{2}\rho^2 F_{\mu\nu}F_{\mu'\nu'}\mathcal{G}^{\mu\mu'}\mathcal{G}^{\nu\nu'} + \rho^4 \det(F_{\mu\nu})\right),$$

$$\det(F_{\mu\nu}) = \frac{1}{8}(F_{\mu\nu}F^{\mu\nu})^2 - \frac{1}{4}F_{\mu\nu}F^{\nu\eta}F_{\eta\rho}F^{\rho\mu},$$

$$\sqrt{\det(\mathcal{G}_{\mu\nu})} = \frac{|\phi^2|^2}{Q} + \frac{1}{2}D_\mu\phi^i D^\mu\phi_i$$

$$+ \frac{1}{8}\frac{Q}{|\phi^2|^2}\left((D_\mu\phi^i D^\mu\phi_i)^2 - 2D_\mu\phi^i D^\nu\phi_i D_\nu\phi^j D^\mu\phi_j\right) + O(\partial A, \partial\phi)^6$$

$$\tag{11.1.77}$$

and recalling that $|G_{\mu\nu}| = 1$ (9.1.19), one finds

$$\sqrt{\det(\mathcal{G}_{\mu\nu} + \rho F_{\mu\nu})} = \sqrt{\det\mathcal{G}_{\mu\nu}}\left(1 + \frac{1}{4}\rho^2 F_{\mu\nu}F^{\mu\nu} - \frac{1}{2}\frac{Q}{|\phi^2|^2}\rho^2 F_{\mu\nu}F^{\mu\nu'}D^\nu\phi^i D_{\nu'}\phi_i\right.$$

$$\left. + \frac{1}{32}\rho^4\left((F_{\mu\nu}F^{\mu\nu})^2 - 4F_{\mu\nu}F^{\nu\eta}F_{\eta\rho}F^{\rho\mu}\right)\right) + O(A, \phi)^6.$$

$$\tag{11.1.78}$$

This reproduces our results for $S_0 + \Gamma_{U(1)}$ to the given order. We can understand the geometrical meaning of S_{DBI} recalling that the single D3 brane $\mathcal{M}$ is modeled by the unbroken $U(1)$ component, whose displacement in the transversal $\mathbb{R}^6$ is given by ϕ^i. Then the action (11.1.75) is recognized as the DBI action on a brane with metric $\mathcal{G}_{\mu\nu}$, which is the pullback to $\mathcal{M}$ of the $9 + 1$ dimensional $AdS^5 \times S^5$ target space metric

$$ds^2 = \frac{|\phi^2|}{\sqrt{Q}}dx^\mu dx_\mu + \frac{\sqrt{Q}}{|\phi^2|}(d\phi^2 + \phi^2 d\Omega^5). \tag{11.1.79}$$

We have thus shown that the (one-loop) effective action of the IKKT matrix model on $\mathbb{R}_\theta^{3,1}$ – which is nothing but noncommutative $\mathcal{N} = 4$ SYM – can be interpreted in terms of an effective $AdS^5 \times S^5$ metric in target space. This is the basic statement of the AdS/CFT correspondence [134]. The present result is in fact more general, since the effective four-dimensional brane metric $G_{\mu\nu}$ can be quite generic and need not be flat.

This $AdS^5 \times S^5$ metric, in turn, arises as a "near-horizon" limit $\frac{Q}{|\phi^2|^2} \gg 1$ (i.e. for small transversal distance $|\phi|$) from the "black p-brane" target space geometry with metric

$$ds^2 = \frac{1}{\sqrt{H}}dx^\mu dx_\mu + \sqrt{H}(d\phi^2 + \phi^2 d\Omega^5) \rightarrow \frac{|\phi^2|}{\sqrt{Q}}dx^\mu dx_\mu + \frac{\sqrt{Q}}{|\phi^2|}(d\phi^2 + \phi^2 d\Omega^5),$$

$$H = 1 + \frac{Q}{|\phi^2|^2} \quad \text{for} \quad \frac{Q}{|\phi^2|^2} \gg 1, \tag{11.1.80}$$

which is a solution of IIB supergravity [134]. Having established that the one-loop effective action is consistent with linearized IIB supergravity in target space, it is plausible that the effective target space geometry of the full theory in the present background is given by (11.1.80).

---

[7] To compare with the supergravity, one should take into account the present nonstandard normalization (11.1.68) and the dimensions pointed out before. For a detailed discussion from the point of view of supergravity in the presence of a B-field, see [4, 31, 135].

These computations demonstrate that quantum effects in the IKKT model modify the effective metric seen by test branes in target space $\mathbb{R}^{9,1}$. This metric is found to be $AdS^5 \times S^5$ in a background given by a stack of $\mathbb{R}^{3,1}_\theta$ branes as in IIB supergravity, thus establishing the link with string theory. The mechanism is consistent with the familiar picture of holography and the AdS/CFT correspondence for $\mathcal{N} = 4$ SYM, and the matrix model framework allows a transparent derivation in terms of string modes.

The relation with IIB supergravity in target space at the quantum level can also be confirmed through scattering processes in target space for states generated by suitable vertex operators, with similar structure similar as in string theory [120, 121]. For example, $V_{ab}(k) = \mathrm{Str}(e^{ikT} \tilde{T}_{ab})$ using (11.1.58) corresponds[8] to the graviton vertex operator, which nicely matches with (11.1.60). They describe processes from the point of view of 10-dimensional target space rather than on the branes.

However, in the weak coupling regime, the holographic target space metric is merely a nice interpretation of the effective action on the brane $\mathcal{M}^{3,1}$. All low-energy excitations are confined to the brane(s) at weak coupling, and there is nothing propagating off the brane in target space. This is so because all fluctuations are given by matrices in $\mathrm{End}(\mathcal{H})$, whose low-energy sector comprises only functions *on* the brane. Therefore, the physically relevant metric for fluctuations is the effective metric $G_{\mu\nu}$ as described in the previous chapters, and the effective gravity is determined by the dynamics of this metric. That metric is $3 + 1$ dimensional on $\mathcal{M}^{3,1}$ branes, hence there is no need to compactify target space or to worry about the associated landscape. We only need to understand the dynamics of the brane geometry, which will be studied in the next chapter including quantum effects at the one loop. We will see that for branes of the structure $\mathcal{M}^{3,1} \times \mathcal{K} \subset \mathbb{R}^{9,1}$, the emergent or induced gravity on $\mathcal{M}^{3,1}$ is indeed governed by an induced Einstein–Hilbert action, amended by extra terms.

11.1.8 Compactification of target space

In the string theory literature, compactifications of the IKKT model on S^1 are defined [54, 197] by imposing the following type of constraints on the matrices

$$T^0 + R\mathbb{1} = U^{-1}T^0 U,$$
$$[U, T^a] = 0, \qquad a = 1, \ldots, 9 \qquad (11.1.81)$$

cf. (8.3.8)). This can be interpreted as compactification of the timelike direction on $S^1 \cong \mathbb{R}/\sim$ by identifying $t^0 \sim t^0 + R$; spacelike matrices can be compactified similarly. The irreducible solutions of these constraints can be written as

$$T^0 = -iR\frac{d}{dt} + A(t), \qquad T^a = X^a(t),$$
$$U = e^{it} \qquad (11.1.82)$$

---

[8] Here Str denotes the trace of symmetrized operators, similar as in (11.1.30).

where $A(t)$ and $X^a(t)$ are matrix-valued function on S^1 realized as

$$A(t) = \sum_n A_n U^n = A(t)$$

$$T^a = \sum_n T_n^a U^n = X^a(t), \qquad a = 1, \ldots, 9 \qquad (11.1.83)$$

with Hilbert space $\mathcal{H} = \mathcal{H}_0 \otimes L^2(S^1)$ given by the space of $\mathcal{H}_0$ – valued functions on S^1. A similar procedure applies to fermions, where usually antiperiodic boundary conditions are imposed. Observing

$$[T^0, T^a(t)] = -iD_t T^a(t), \qquad D_t = \frac{d}{dt} + i[A, .] \qquad (11.1.84)$$

for $a = 1, \ldots, 9$, the IKKT model takes the following form:

$$S = \frac{1}{g^2} \int dt \mathrm{Tr}\left(\frac{1}{2}(D_t T^a)(D_t T_a) + \frac{1}{4}[T^a, T^b][T_a, T_b] - \frac{i}{2}\Psi^T D_t \Psi + \frac{1}{2}\Psi^T \Gamma_a[T^a, \Psi]\right).$$
$$(11.1.85)$$

This is a model for nine bosonic as well as fermionic matrices on a timelike circle S^1, which will be recognized in Chapter 13 as BFSS model (13.2.1). Note that the present compactification prescription is distinct from simply using $T^0 = -iR\frac{d}{dt} + A$ as a background in the IKKT model, due to the extra constraints (11.1.81) on the T^a. Without the constraint, the matrices would also depend on $p = i\frac{d}{dt}$, and generate the space of functions on a fuzzy cylinder generated by U and p. Hence, in the semi-classical regime, the matrices would be matrix-valued functions on a four-dimensional cylinder, with degenerate kinetic term and metric along the S^1 direction only, as discussed in Section 7.3.1.

Compactifying similarly d of the matrix directions leads to a d-dimensional matrix-valued field theory on T^d. Moreover, the classical tori can be replaced by noncommutative tori, defined by relations of the type

$$U_i U_j = q_{ij} U_j U_i. \qquad (11.1.86)$$

This leads to rich and interesting structures, and stringy features such as T-duality can be recovered at the classical level [54].

However, such a compactification of target space necessarily introduces infinitely many degrees of freedom per unit volume. Then the matrix model no longer provides a nonperturbative framework for a quantum theory, and the one-loop effective action would generically be divergent in $3 + 1$ dimensions[9]. In fact, the action (11.1.85) is obtained only after dropping an infinite factor from the trace on $L^2(S^1)$. This is in contrast to the fuzzy compact extra dimensions $\mathcal{K}_N$ discussed in Chapters 8 and 12, where the effective dimensions are reduced through spontaneous symmetry breaking on suitable vacua with finitely many dof per volume, preserving the nonperturbative framework of matrix models. We will therefore not consider such compactifications any further.

[9] Except possibly for special SUSY-preserving backgrounds; this has not been studied in detail.

11.2 Numerical and nonperturbative aspects

The basic hypothesis underlying our approach to matrix models is the existence of a vacuum condensate given by some almost-commutative matrix configuration, leading to spontaneous symmetry breaking (SSB) of the global and gauge symmetry of the model. This is analogous to a nontrivial vacuum and the associated Higgs effect in a nonabelian QFT.

Of course, the fundamental virtue of matrix models is that they provide a non-perturbative framework for quantization, by integrating over the space of matrices. There has indeed been extensive work on numerical simulations of Yang–Mills matrix models, aiming to establish this hypothesis and related statements at the nonperturbative level. We shall briefly discuss some of the results pertinent to the IIB model, and refer to [9] for more details and references.

11.2.1 Euclidean model

In the Euclidean case, the quantization of Yang–Mills matrix models is well-defined without any regularization. This is evident in the presence of a mass term, but not completely obvious for $m^2 = 0$ due to the flat directions reflected by the shift symmetry (7.0.7); nevertheless the integrals are well-defined [15, 16, 127]. Without fermions, the models can be simulated using conventional Monte-Carlo simulation techniques. In the presence of fermions, the Pfaffian (11.1.14) is complex due to the Wess–Zumino contribution (11.1.19). This means that a severe sign problem arises in the numerical simulations, due to the oscillatory nature of the integral. Then conventional Monte-Carlo simulations are no longer adequate, and more sophisticated methods such as the complex Langevin method are required.

The most basic question to be addressed by such numerical simulations is whether or not SSB of the global $SO(10)$ symmetry occurs, corresponding to some nontrivial vacuum with dimension less than 10. Such an SSB can be detected numerically by measuring the following order parameter:

$$S^{ab} = \langle \mathrm{Tr}(T^a T^b) \rangle, \tag{11.2.1}$$

where $\langle . \rangle$ denotes the averaging in the matrix integral, i.e. the Euclidean analog of (7.2.5). If the eigenvalues of this matrix separate into d large and $10 - d$ small ones, this indicates that SSB towards a $SO(d)$–invariant vacuum occurs. Numerical investigations of these questions support the following picture [5, 6, 7, 8]: in the purely bosonic 10-dimensional YM model, no such SSB occurs. However, SSB does occur in the presence of fermions, precisely due to the non-trivial phase of the Pfaffian. It appears that the $SO(10)$ symmetry is spontaneously broken to $SO(3)$, consistent with results obtained using a Gaussian approximation method [148]. The specific nature of such a nontrivial vacuum has not yet been determined in the numerical simulations of the 10-dimensional model.

11.2.2 Lorentzian model

As discussed in Section 7.2, the Lorentzian IKKT model needs to be regularized for the quantization to be well-defined. Such a regularization is provided by the $i\varepsilon$ term in (7.2.3), which seems essentially equivalent to the method used in numerical simulations[10] in [8, 147] based on the complex Langevin method. The relation with the Euclidean model is then no longer manifest, and the vacua of the Lorentzian model need to be established independently. To interpret such simulations, one may diagonalize the timelike matrix $T^0 = \mathrm{diag}(\lambda_1, \ldots, \lambda_N)$ with increasing eigenvalues $\lambda_1 < \cdots < \lambda_N$, as in (7.2.11). In this basis, the remaining matrices are found to be band-diagonal, i.e. dominated by matrix elements close to the diagonal. This means that the dominant matrix configurations are almost-commutative as discussed in Chapter 4, which strongly supports the adequacy of the present framework. Moreover, one can devise observables which are designed to measure the evolution of the size of such a vacuum in numerical simulations. Even though the numerical results are not yet fully conclusive at the time of writing, there are clear hints for SSB and the emergence of a spacetime with reduced dimension.

Given sufficiently reliable results of the simulations of the IKKT model, the resulting matrix configurations can and should of course be "measured" with the methods presented in Section 4, cf. [167], which should allow to determine their detailed structure.

Due to the complex nature of the path integral, different vacua may contribute with different phase factors, thus obscuring the identification of a vacuum structure. It is conceivable that vacua with block-matrix structure as in (7.4.33) are important, leading to a multiverse-type scenario [14]. Other interesting proposals include the use of the master equation to analyze the IKKT model [122].

---

[10] A different regularization and a simplified treatment of the sign problem in [118] led to early results indicating an expanding $3+1$ dimensional vacuum. However, the simplified treatment was later recognized as inadequate.

12 Gravity as a quantum effect on quantum spacetime

We consider the IKKT model for a matrix configuration corresponding to a quantum space $\mathcal{M}$. The aim of this chapter is to understand the geometrical meaning of the one-loop effective action arising on $\mathcal{M}$, and its interpretation in terms of gravity. For simple $3 + 1$ dimensional quantized symplectic spaces $\mathcal{M}$, this will lead to a higher-derivative geometric action. We will then add fuzzy extra dimensions corresponding to a background $\mathcal{M} \times \mathcal{K}$ as in Section 8.3, and show that this leads to an effective Einstein–Hilbert-type action, expressed in terms of torsion. Finally in Section 12.4, we consider the case of a covariant quantum space $\mathcal{M}^{3,1}$, taking into account the contributions from the internal fiber S_n^2. The maximal supersymmetry of the IKKT model is crucial for the cancellation of lower-order terms, and hence for obtaining gravity.

12.1 Heat kernel and induced gravity on commutative spaces

It is well-known that quantum effects in field theory covariantly coupled to a metric lead to gravity-type terms in the effective action, including notably the Einstein–Hilbert term, as well as a cosmological constant. This "induced gravity" mechanism was pointed out first by Sakharov in 1967, and subsequently developed further from various points of view[1]. However, it typically leads to a huge cosmological constant, which constitutes the notorious cosmological constant problem. We will see that a better-behaved variant of this mechanism arises within the IKKT model, which will play a crucial role to understand gravity. Most importantly, the cosmological constant problem does not arise in the present framework.

The basic idea is best understood in terms of the heat kernel expansion[2], and its generalization to the present framework. Consider

$$H(\alpha) := \mathrm{Tr} e^{-\alpha \Delta} = \sum_{n \geq 0} \alpha^{\frac{n-d}{2}} S_n, \qquad (12.1.1)$$

where d is the dimension of $\mathcal{M}$, and Δ is the Laplacian on $\mathcal{M}$ with metric $g_{\mu\nu}$; more generally it could be any operator which is bounded from below. $H(\alpha)$ basically counts the number of eigenmodes of Δ below the UV cutoff scale $\alpha^{-2} =: \Lambda^2$. If Δ was bounded (such

---

[1] It can be seen as a precursor of the spectral action principle [52].

[2] The name "heat kernel" arises from the Euclidean case, where $e^{-\alpha\Delta}$ is the fundamental solution of the heat equation.

as on compact fuzzy spaces), then $H(\alpha)$ would be analytic in α. On ordinary manifolds where Δ is unbounded from above, $H(\alpha)$ can only be an asymptotic expansion, i.e. the series (12.1.1) is formal and diverges as $\alpha \to 0$. The S_n are local functionals known as Seeley–de Witt coefficients

$$S_n = \int d^d x \sqrt{|g|}\, \mathcal{L}_n, \qquad (12.1.2)$$

which respect all (gauge and global) symmetries of $H(\alpha)$. Typically, only terms with even n are nonvanishing. The leading terms are as follows: ·

$$S_0 = \frac{1}{16\pi^2} \int d^4 x \sqrt{|g|}$$
$$S_2 = \frac{1}{16\pi^2} \int d^4 x \sqrt{|g|}\, \frac{1}{6}\mathcal{R}[g] \qquad (12.1.3)$$

etc., starting with the cosmological constant term for $n = 0$, and the Einstein–Hilbert action for $n = 2$. The S_n are *local* functionals of the background metric and its derivatives up to order n. Their explicit computation is similar to perturbative quantum field theory computations in the presence of some UV cutoff Λ [77, 209]. In other words, the heat kernel expansion provides the derivative expansion of the (generally nonlocal) "effective action" $\operatorname{Tr}\exp(-\alpha\Delta)$. Once $H(\alpha)$ is known, the analogous expansion for observables of the type $\operatorname{Tr}\big(f(\Delta)e^{-\alpha\Delta}\big)$ can be obtained using a Laplace transformation.

The leading term S_0 was obtained in Section 6.2.3 using the trace formula (6.2.66). It arises generally as leading term in the one-loop effective action and typically induces a UV-divergent cosmological constant $\sim \Lambda^4$ in four dimensions, which is finite only in the presence of some UV cutoff Λ given e.g. by the SUSY breaking scale. This is the infamous cosmological constant problem, which suggests that there is something wrong with the conventional understanding of gravity. The Einstein–Hilbert term is subleading, but is also UV divergent unless there is a suitable cutoff. We will see that these problems disappear in the present matrix model framework, where the Einstein–Hilbert term will arise directly as the leading contribution, and can be computed using the formalism of string states.

Heat kernel in the noncommutative setting

Now consider the case of a UV-finite theory on a compact noncommutative space, described e.g. by a matrix model with finite-dimensional Hilbert space $\mathcal{H}$. It is then natural to define the following analog of the heat kernel in terms of the matrix Laplacian,

$$H(\alpha) = \operatorname{Tr} e^{-\alpha\Box} \qquad (12.1.4)$$

which is analytic in α. The leading term in this series is

$$H(0) = \operatorname{Tr}\mathbb{1} = \dim(\mathcal{H}) \sim \int_{\mathcal{M}} \frac{\Omega}{(2\pi)^n} = \operatorname{Vol}_\Omega \mathcal{M}, \qquad (12.1.5)$$

which is finite for compact $\mathcal{M}$, and has no dynamical significance. This already suggests that the cosmological constant problem should not arise in such a setting. However, the relation with the commutative case is quite subtle.

We have seen in Section 6.2 that the UV sector of noncommutative spaces is governed by string modes, which are highly nonlocal. Whenever the trace over modes in $\text{End}(\mathcal{H})$ is dominated by a UV regime, these will dominate, and invalidate the classical results. Therefore, the heat kernel should approximate the commutative one for large $\alpha > \Lambda_{\text{NC}}^{-2}$, where only IR modes contribute; in that regime, one can use the semi-classical relation $\Box \sim \rho^2 \Box_G$. In contrast, for small $\alpha \to 0$ the highly noncommutative UV regime dominates, leading to a totally nonlocal heat kernel. Hence, the classical behavior of the heat kernel should be recovered in the IR regime $\alpha > \Lambda_{\text{NC}}^2$, rather than the UV regime $\alpha \to 0$ where nonlocal effects will dominate. In this sense the Seeley–de Witt coefficients can actually be recovered in the IR regime of noncommutative field theory [36], but the full effective action is typically totally different from the commutative one.

We conclude once again that a reasonable gravity theory on noncommutative spaces can only be expected from a UV-finite theory, hence from the IKKT model. Rather than focusing on the heat kernel $H(\alpha)$, it is then more appropriate to compute the one-loop effective action directly, and extract the gravitational sector. We will see that this leads indeed to an induced Einstein–Hilbert term similar to S_2 in (12.1.3). However, the maximal SUSY of the IKKT model implies a very specific and richer scenario where the Einstein–Hilbert-like action arises only in the presence of fuzzy extra dimensions, which lead to spontaneous breaking of SUSY.

12.2 Higher-derivative local action on simple branes

We start with the simplest possible noncommutative background

$$T^\alpha = X^\alpha \sim x^\alpha : \quad \mathcal{M} \hookrightarrow \mathbb{R}^{9,1} \tag{12.2.1}$$

for $\alpha = 0, \ldots, 3$ describing a quantized four-dimensional symplectic space embedded in target space, with $T^a = 0$ for $a = 4, \ldots, 9$. Then

$$i[X^\alpha, X^\beta] = \mathcal{F}^{\alpha\beta} \tag{12.2.2}$$

reduces in the semi-classical regime to the Poisson tensor on $\mathcal{M}$, which may be x – dependent. Now consider the one-loop effective action $\Gamma_{1\text{loop};4}[T]$ (11.1.34) on this background evaluated using the integral over string modes, given by the quartic term (11.1.35)

$$V_4[\delta \mathcal{F}] = -4\text{tr}\delta \mathcal{F}^4 + (\text{tr}\delta \mathcal{F}^2)^2$$
$$= -4\left(\delta\mathcal{F}^{\alpha\beta}\delta\mathcal{F}_{\beta\gamma}\delta\mathcal{F}^{\gamma\delta}\delta\mathcal{F}_{\delta\alpha}\right) + \left(\delta\mathcal{F}^{\alpha\beta}\delta\mathcal{F}_{\alpha\beta}\right)\left(\delta\mathcal{F}^{\gamma\delta}\delta\mathcal{F}_{\gamma\delta}\right). \tag{12.2.3}$$

Here indices are contracted with $\eta_{\alpha\beta}$, and (11.1.33)

$$\delta\mathcal{F}^{\alpha\beta} = \mathcal{F}^{\alpha\beta}(x) - \mathcal{F}^{\alpha\beta}(y). \tag{12.2.4}$$

This action takes the form

$$\Gamma_{1\,\text{loop},4}[\mathcal{M}] = \frac{3i}{4}\text{Tr}\Big(\frac{V_4}{(\Box - i\varepsilon)^4}\Big) = \frac{3i}{4(2\pi)^4}\int\limits_{\mathcal{M}\times\mathcal{M}} \Omega_x\Omega_y\,\frac{O(\mathcal{F}(x) - \mathcal{F}(y))^4}{((x-y)^2 - i\varepsilon)^4} \qquad (12.2.5)$$

representing the contribution of string modes linking x to y. This nonlocal interaction is well-defined, but clearly quite different from four-dimensional gravity. We can elaborate the local limit using the trace formula (6.2.85) in terms of the semi-classical wave-packets $\psi_k(y) = \psi_{k,y}^{(L)}$. Then the numerator leads to fourth order derivative terms such as

$$\{\mathcal{F}^{\alpha\beta},\{\mathcal{F}_{\alpha\beta},\{\mathcal{F}^{\gamma\delta},\{\mathcal{F}_{\gamma\delta},\psi_k\}\}\}\} \sim \{\mathcal{F}^{\alpha\beta},y^\mu\}\{\mathcal{F}_{\alpha\beta},y^\nu\}\{\mathcal{F}^{\gamma\delta},y^\rho\}\{\mathcal{F}_{\gamma\delta},y^\sigma\}\partial_\mu\partial_\nu\partial_\rho\partial_\sigma\psi_k$$

$$= T^{\alpha\beta\mu}T_{\alpha\beta}{}^\nu T^{\gamma\delta\rho}T_{\gamma\delta}{}^\sigma\partial_\mu\partial_\nu\partial_\rho\partial_\sigma\psi_k$$

$$\approx T^{\alpha\beta\mu}T_{\alpha\beta}{}^\nu T^{\gamma\delta\rho}T_{\gamma\delta}{}^\sigma k_\mu k_\nu k_\rho k_\sigma\psi_k \qquad (12.2.6)$$

assuming that $\mathcal{F}^{\alpha\beta}$ is slowly varying. Here we recall the relation with the torsion tensor[3] (9.1.26)

$$\{\mathcal{F}^{\alpha\beta},y^\mu\} = T^{\alpha\beta\mu} \qquad (12.2.7)$$

in local coordinates. Similarly, the full four–derivative contribution of (12.2.3) takes the form

$$V_{4,\mathcal{M}}[\delta\mathcal{F}] \approx \Big(T^{\alpha\beta\mu}T_{\alpha\beta}{}^\nu T^{\gamma\delta\rho}T_{\gamma\delta}{}^\sigma - 4T^{\alpha\beta\mu}T_{\beta\gamma}{}^\nu T^{\gamma\delta\rho}T_{\delta\alpha}{}^\sigma\Big)k_\mu k_\nu k_\rho k_\sigma\psi_{k;y}$$

$$=: V_4(T)^{\mu\nu\rho\sigma}k_\mu k_\nu k_\rho k_\sigma\psi_{k;y}. \qquad (12.2.8)$$

Therefore, the local contribution to the effective action is

$$\Gamma_{1\,\text{loop},4}[\mathcal{M}] = \frac{3i}{4(2\pi)^4}\int\limits_{\mathcal{M}} d^4x\int d^4k\,\frac{V_4(T)^{\mu\nu\rho\sigma}k_\mu k_\nu k_\rho k_\sigma}{(k\cdot k - i\varepsilon)^4}, \qquad (12.2.9)$$

where

$$k\cdot k \equiv k_\mu k_\nu\gamma^{\mu\nu} = \rho^2 G^{\mu\nu}k_\mu k_\nu \qquad (12.2.10)$$

involves the effective metric (6.2.38), recalling $\Box \sim \rho^2\Box_G = \Box_\gamma$ in local coordinates with $\partial\gamma = 0$ (cf. (6.2.37)). Since it involves a contraction of four torsion tensors, this amounts to a higher-derivative contribution to the local geometric action.

At face value, the integral over k is logarithmically divergent both in the IR and the UV, but both are artifacts of the present expansion. In the IR, the k integration is effectively cut off at the background curvature scale, where (12.2.6) is no longer valid. This can be effectively taken into account by adding a suitable mass term to the propagator. In the UV, the loop integral (12.2.5) is regular due to the decay $\frac{1}{((x-y)^2 - i\varepsilon)^4}$ for the long string modes. Hence, the present result gives a subleading higher-derivative contribution to the gravitational action, which will be dropped. We also recall that it can be interpreted as IIB supergravity interaction on $\mathcal{M}$ due to the exchange of $9+1$ dimensional gravitons

---

[3] We assume that $\mathcal{M}$ is a four-dimensional space with Poisson tensor $\{x^\alpha, x^\beta\} = \theta^{\alpha\beta} = -\mathcal{F}^{\alpha\beta}$. In the case of covariant quantum spaces, some complications arise due to the internal S^2 sphere, which will be considered later.

(and other modes) coupling to the matrix energy–momentum tensors $\mathcal{T}_{\alpha\beta}$, which is certainly well-defined and decays like $(x - y)^{-8}$.

12.3 Einstein–Hilbert action from fuzzy extra dimensions

Now consider a background with structure $\mathcal{M}^{3,1} \times \mathcal{K}$, where $\mathcal{M}^{3,1}$ plays the role of spacetime and $\mathcal{K}$ are fuzzy extra dimensions. Then the one-loop effective action in the IKKT model contains a term that can be interpreted either as supergravity interaction between $\mathcal{K}$ and $\mathcal{M}$, or as Einstein–Hilbert action on $\mathcal{M}$. The effective Newton constant on $\mathcal{M}$ is determined by the scale of $\mathcal{K}$.

As discussed in Section 8.3, such backgrounds are realized by matrix configurations where the first $3 + 1$ matrices

$$T^{\alpha} \sim x^{\alpha} : \quad \mathcal{M} \hookrightarrow \mathbb{R}^{9,1}, \qquad \alpha = 0, \ldots, 3 \tag{12.3.1}$$

describe a quantized four-dimensional symplectic space $\mathcal{M}$ as before, and the remaining six matrices

$$T^{i} \sim m_{\mathcal{K}} \mathcal{K}^{i} : \quad \mathcal{K} \hookrightarrow \mathbb{R}^{9,1} \qquad i = 4, \ldots, 9 \tag{12.3.2}$$

describe $\mathcal{K}$. Hence, $\mathcal{M}$ is embedded along the first four coordinates in target space, and $\mathcal{K} \subset \mathbb{R}^{9,1}$ along the six transversal directions. The precise form of $\mathcal{K}$ is not important in the following. The Hilbert space is assumed to factorize as $\mathcal{H} = \mathcal{H}_{\mathcal{M}} \otimes \mathcal{H}_{\mathcal{K}}$, where $\mathcal{H}_{\mathcal{K}} \cong \mathbb{C}^n$ corresponds to the compact quantum space $\mathcal{K}$. Then the trace factorizes accordingly as

$$\text{Tr}_{\text{End}(\mathcal{H})} = \text{Tr}_{\mathcal{M}} \times \text{Tr}_{\mathcal{K}} \tag{12.3.3}$$

using the short notation

$$\text{Tr}_{\mathcal{M}} := \text{Tr}_{\text{End}(\mathcal{H}_{\mathcal{M}})}, \qquad \text{Tr}_{\mathcal{K}} := \text{Tr}_{\text{End}(\mathcal{H}_{\mathcal{K}})}. \tag{12.3.4}$$

The matrix d'Alembertian then decomposes as

$$\Box = [T^{\alpha}, [T_{\alpha}, .]] + [T^{i}, [T_{i}, .]] = \Box_{\mathcal{M}} + \Box_{\mathcal{K}}. \tag{12.3.5}$$

Let $\Upsilon_{\Lambda} \in \text{End}(\mathcal{H}_{\mathcal{K}})$ be the eigenmodes of

$$\Box_{\mathcal{K}} \Upsilon_{\Lambda} = m_{\Lambda}^{2} \Upsilon_{\Lambda}, \qquad m_{\Lambda}^{2} = m_{\mathcal{K}}^{2} \mu_{\Lambda}^{2} \tag{12.3.6}$$

with dimensionless μ_{Λ}^{2} as in (8.3.7). We then expand the most general $\phi \in \text{End}(\mathcal{H})$ into modes

$$\phi = \sum_{\Lambda} \phi_{\Lambda}, \qquad \phi_{\Lambda} = \phi_{\Lambda}(x) \Upsilon_{\Lambda} \tag{12.3.7}$$

which satisfy

$$\Box \phi_{\Lambda} = (\Box_{\mathcal{M}} + m_{\Lambda}^{2}) \phi_{\Lambda}. \tag{12.3.8}$$

Therefore, the ϕ_Λ mode acquires a mass m_Λ^2 on $\mathcal{M}$, just like in standard Kaluza–Klein compactification. Then the trace over $\text{End}(\mathcal{H}_\mathcal{K})$ can be evaluated directly by summing over the ON basis Υ_Λ, and the trace over $\text{End}(\mathcal{H}_\mathcal{M})$ is evaluated by integrating over the string modes as before. Alternatively, it is also possible to evaluate $\text{Tr}_\mathcal{K}$ in terms of string modes on $\mathcal{K}$, observing that the string modes on $\mathcal{M} \times \mathcal{K}$ factorize into string modes on $\mathcal{M}$ and $\mathcal{K}$ as in Section 6.2.6.

The key for the following consideration is the decomposition of the potential (12.2.3) into contributions from $\mathcal{M}$, from $\mathcal{K}$, and from mixed contributions:

$$
\begin{aligned}
V_4[\delta\mathcal{F}] =& \left(\delta\mathcal{F}^{\alpha\beta}\delta\mathcal{F}_{\alpha\beta}\right)\left(\delta\mathcal{F}^{\gamma\delta}\delta\mathcal{F}_{\gamma\delta}\right) - 4\left(\delta\mathcal{F}^{\alpha\beta}\delta\mathcal{F}_{\beta\gamma}\delta\mathcal{F}^{\gamma\delta}\delta\mathcal{F}_{\delta\alpha}\right) \\
&+ \left(\delta\mathcal{F}^{ij}\delta\mathcal{F}_{ij}\right)\left(\delta\mathcal{F}^{kl}\delta\mathcal{F}_{kl}\right) - 4\left(\delta\mathcal{F}^{ij}\delta\mathcal{F}_{jk}\delta\mathcal{F}^{kl}\delta\mathcal{F}_{li}\right) \\
&+ 2\left(\delta\mathcal{F}^{\alpha\beta}\delta\mathcal{F}_{\alpha\beta}\right)\left(\delta\mathcal{F}^{ij}\delta\mathcal{F}_{ij}\right).
\end{aligned}
\tag{12.3.9}
$$

There are no other terms, since the mixed commutators $[T^\alpha, \mathcal{K}^i] = 0$ vanish due to the product structure of the background. The contractions are understood with $\eta_{\alpha\beta}$ or δ_{ij}.

$\mathcal{K}$ contribution and effective potential

Consider the contribution of the $\mathcal{K}$ terms

$$
V_{4,\mathcal{K}} = \left(\delta\mathcal{F}^{ij}\delta\mathcal{F}_{ij}\right)\left(\delta\mathcal{F}^{kl}\delta\mathcal{F}_{kl}\right) - 4\left(\delta\mathcal{F}^{ij}\delta\mathcal{F}_{jk}\delta\mathcal{F}^{kl}\delta\mathcal{F}_{li}\right).
\tag{12.3.10}
$$

Going back to (11.1.30) and summing over the modes (12.3.7), $\delta\mathcal{F}^{ij}$ indicates the term

$$
\delta\mathcal{F}^{ij} = [\mathcal{F}^{ij}, \phi_\Lambda(x)\Upsilon_\Lambda] = \phi_\Lambda(x)[\mathcal{F}^{ij}, \Upsilon_\Lambda]
\tag{12.3.11}
$$

so that $V_{4,\mathcal{K}}$ is a fourth order derivation acting on $\text{End}(\mathcal{H}_\mathcal{K})$. For simplicity, we assume that Υ_Λ is a common eigenvector of both $\Box_\mathcal{K}$ and $V_{4,\mathcal{K}}$; if not, the following consideration can be adapted, as discussed later. Then $V_{4,\mathcal{K}}$ reduces to

$$
\left([\mathcal{F}^{ij}, [\mathcal{F}_{ij}, [\mathcal{F}^{kl}, [\mathcal{F}_{kl}, .]]]] - 4[\mathcal{F}^{ij}, [\mathcal{F}_{jk}, [\mathcal{F}^{kl}, [\mathcal{F}_{li}, .]]]]\right)\Upsilon_\Lambda =: m_\mathcal{K}^8 V_{4,\Lambda}\Upsilon_\Lambda,
\tag{12.3.12}
$$

where $m_\mathcal{K}$ is the scale parameter of $\mathcal{K}$. Thus, the trace reduces to

$$
\text{Tr}\left(\frac{V_{4,\mathcal{K}}}{(\Box - i\varepsilon)^4}\right) = m_\mathcal{K}^8 \sum_\Lambda V_{4,\Lambda} \text{Tr}_\mathcal{M}\left(\frac{1}{(\Box_\mathcal{M} + m_\Lambda^2 - i\varepsilon)^4}\right).
\tag{12.3.13}
$$

The trace over $\text{End}(\mathcal{M})$ can be evaluated in the string basis as before,

$$
\text{Tr}_\mathcal{M}\left(\frac{1}{(\Box_\mathcal{M} + m_\Lambda^2 - i\varepsilon)^4}\right) \approx \frac{1}{(2\pi)^4} \int_{\mathcal{M}\times\mathcal{M}} \Omega_x\Omega_y \frac{1}{((x-y)^2 + m_\Lambda^2 - i\varepsilon)^4}.
\tag{12.3.14}
$$

The y integration is a convergent short-range integral for $\dim \mathcal{M} = 4$, so that the discussion for UV-convergent traces on quantum spaces in Section 6.2.3 applies. We can thus evaluate it as

$$\mathrm{Tr}_{\mathcal{M}}\left(\frac{1}{(\Box_{\mathcal{M}} + m_\Lambda^2 - i\varepsilon)^4}\right) \approx \frac{1}{(2\pi)^4} \int_{\mathcal{M}} d^4x \int d^4k \langle \Psi_{k,x}^{(L)}, \frac{1}{(\Box_{\mathcal{M}} + m_\Lambda^2 - i\varepsilon)^4} \Psi_{k,x}^{(L)} \rangle$$

$$\approx \frac{1}{(2\pi)^4} \int_{\mathcal{M}} d^4x \sqrt{G} \int \frac{d^4k}{\sqrt{G}} \frac{1}{(k \cdot k + m_\Lambda^2 - i\varepsilon)^4} \qquad (12.3.15)$$

using (6.2.85), and assuming the normalization $\langle \Psi_{k,x}^{(L)}, \Psi_{k,x}^{(L)} \rangle = 1$. This boils down to

$$\kappa_{(4)} := \int \frac{d^4k}{\sqrt{G}} \frac{1}{(k \cdot k + m^2 - i\varepsilon)^4} = i\frac{\pi^2}{6m^4} \rho^{-4} \qquad (12.3.16)$$

using (D.0.3), where $k \cdot k \equiv \rho^2 G^{\mu\nu} k_\mu k_\nu$. Thus, the contribution of $\mathcal{K}$ to the effective action is

$$\Gamma_{1\,\mathrm{loop}}^{\mathcal{K}} = \frac{3i}{4} \mathrm{Tr}\left(\frac{V_{4,\mathcal{K}}}{(\Box - i\varepsilon)^4}\right) = -\frac{\pi^2}{8(2\pi)^4} \int_{\mathcal{M}} d^4x \sqrt{G}\, \rho^{-4} m_{\mathcal{K}}^4 \left(\sum_\Lambda \frac{1}{\mu_\Lambda^4} V_{4,\Lambda}\right)$$

$$= -\frac{1}{8(4\pi)^2} \int_{\mathcal{M}} \Omega\, \rho^{-2} m_{\mathcal{K}}^4 \left(\sum_\Lambda \frac{1}{\mu_\Lambda^4} V_{4,\Lambda}\right), \qquad (12.3.17)$$

where

$$d^4x \rho_M = \Omega = \sqrt{G}\, \rho^{-2} \qquad (12.3.18)$$

is the symplectic volume form on $\mathcal{M}^{3,1}$, and μ_Λ^2 is the dimensionless KK mass (8.3.7). This term contributes to the effective potential for the $\mathcal{K}^i$ that form the extra dimension. Note that the integral measure is given by the symplectic volume form $\Omega = d^4x \rho_M$ rather than the Riemannian one, which will have important implications for the cosmological constant problem.

Mixed contributions and induced gravity action

The most interesting contribution comes from the mixed term in (12.3.9)

$$V_{4,\mathrm{mix}} = 2\left(\delta\mathcal{F}^{\alpha\beta}\delta\mathcal{F}_{\alpha\beta}\right)\left(\delta\mathcal{F}^{ij}\delta\mathcal{F}_{ij}\right). \qquad (12.3.19)$$

As before, we expand all modes into product states of the form $\phi_\Lambda = \phi_\Lambda(x)\Upsilon_\Lambda$; then the first factor acts only on $\phi_\Lambda(x)$, and the second only on Υ_Λ. We assume again that Υ_Λ is a common eigenvector of both $\Box_{\mathcal{K}}$ and the second order derivation $\left(\delta\mathcal{F}^{ij}\delta\mathcal{F}_{ij}\right)$ acting on $\mathrm{End}(\mathcal{H}_{\mathcal{K}})$. Then

$$\left(\delta\mathcal{F}^{ij}\delta\mathcal{F}_{ij}\right)\Upsilon_\Lambda = [\mathcal{F}^{ij}, [\mathcal{F}_{ij}, \Upsilon_\Lambda]] = m_{\mathcal{K}}^4 C_\Lambda^2 \Upsilon_\Lambda, \qquad C_\Lambda^2 > 0 \qquad (12.3.20)$$

where $C_\Lambda^2 > 0$ depends on the structure of $\mathcal{K}$, and it is positive since the corresponding target space metric is Euclidean.

Next, we need to evaluate $\left(\delta\mathcal{F}^{\alpha\beta}\delta\mathcal{F}_{\alpha\beta}\right)$ acting on $\mathrm{End}(\mathcal{H}_{\mathcal{M}})$. Using the string representation, the trace $\mathrm{Tr}_{\mathcal{M}}\left(\frac{V_{4,\mathrm{mix}}}{\Box^4}\right)$ leads again to a nonlocal integral $\int_{\mathcal{M}\times\mathcal{M}} \Omega_x \Omega_y$ as in (12.3.14), which is convergent for large $|x - y|$ and hence almost-local. We compute it again using the trace formula (6.2.85) in terms of the wave packets $\Psi_{k,y}^{(L)} \sim \psi_{k,y}^{(L)}(x)$

with size L, which we choose somewhat shorter than the typical scale of the background geometry. For the numerator, we recall the formula (6.2.76)

$$v[\psi_{k;y}] = v^\mu \partial_\mu \psi_{k;y} \approx ik_\mu v^\mu \psi_{k;y}, \qquad k > \frac{1}{L}. \tag{12.3.21}$$

This allows to evaluate the differential operator $\delta\mathcal{F}^{\alpha\beta}\delta\mathcal{F}^{\alpha\beta} \sim -\{\mathcal{F}^{\alpha\beta}, \{\mathcal{F}^{\alpha\beta}, .\}\}$ acting on the wave packets $\psi_{k;y}$ as

$$\{\mathcal{F}^{\alpha\beta}, \psi_{k;y}\} \approx ik_\mu\{\mathcal{F}^{\alpha\beta}, x^\mu\}\psi_{k;y} = ik_\mu T^{\alpha\beta\mu}\psi_{k;y}$$

$$\{\mathcal{F}^{\alpha\beta}, \{\mathcal{F}_{\alpha\beta}, \psi_{k;y}\}\} \approx ik_\mu\{\mathcal{F}^{\alpha\beta}, T_{\alpha\beta}{}^\mu\psi_{k;y}\}$$

$$= ik_\mu\{\mathcal{F}^{\alpha\beta}, T_{\alpha\beta}{}^\mu\}\psi_{k;y} + ik_\mu T^{\alpha\beta\mu}\{\mathcal{F}_{\alpha\beta}, \psi_{k;y}\}$$

$$= ik_\mu\{\mathcal{F}^{\alpha\beta}, T_{\alpha\beta}{}^\mu\}\psi_{k;y} - T^{\alpha\beta\mu}\{\mathcal{F}_{\alpha\beta}, x^\nu\}k_\mu k_\nu \psi_{k;y}$$

$$\approx -T^{\alpha\beta\mu}T_{\alpha\beta}{}^\nu k_\mu k_\nu \psi_{k;y}, \qquad k > \frac{1}{L}. \tag{12.3.22}$$

Here, we can neglect the first term assuming that the curvature scale is smaller than k^2. Putting this together, $V_{4,\text{mix}}$ acting on $\psi_{k;y}\Upsilon_\Lambda$ is approximately diagonal, and reduces to

$$V_{4\text{mix}}[\psi_{k;y}\Upsilon_\Lambda] = 2m_\mathcal{K}^4 C_\Lambda^2 T^{\alpha\beta\mu}T_{\alpha\beta}{}^\nu k_\mu k_\nu \psi_{k;y}\Upsilon_\Lambda. \tag{12.3.23}$$

Hence, the full trace reduces using (6.2.85) to the following local integral:

$$\text{Tr}\left(\frac{V_{4,\text{mix}}}{(\Box - i\varepsilon)^4}\right) = \frac{2m_\mathcal{K}^4}{(2\pi)^4}\int_\mathcal{M} d^4x\sqrt{G}\, T^{\alpha\beta\mu}T_{\alpha\beta}{}^\nu \sum_\Lambda C_\Lambda^2 \int \frac{d^4k}{\sqrt{G}}\frac{k_\mu k_\nu}{(k\cdot k + m_\Lambda^2 - i\varepsilon)^4}$$

$$= \frac{2m_\mathcal{K}^4}{(2\pi)^4}\int_\mathcal{M} d^4x\sqrt{G}\left(\sum_\Lambda C_\Lambda^2 \tilde{\kappa}_{(4)}(m_\Lambda)\right)T^{\alpha\beta\mu}T_{\alpha\beta}{}^\nu G_{\mu\nu}. \tag{12.3.24}$$

As before, $k\cdot k = \gamma^{\mu\nu}k_\mu k_\nu$ is defined via the effective metric without dilaton, and

$$\tilde{\kappa}_{(4)}G_{\mu\nu} = \int \frac{d^4k}{\sqrt{G}}\frac{k_\mu k_\nu}{(k\cdot k + m^2 - i\varepsilon)^4} \tag{12.3.25}$$

where $\tilde{\kappa}_{(4)} = \tilde{\kappa}_{(4)}(m)$ can be computed by taking the trace

$$4\rho^2\tilde{\kappa}_{(4)} + m^2\kappa_{(4)} = \int \frac{d^4k}{\sqrt{G}}\frac{k\cdot k + m^2}{(k\cdot k + m^2 - i\varepsilon)^4}$$

$$= \int \frac{d^4k}{\sqrt{G}}\frac{1}{(k\cdot k + m^2 - i\varepsilon)^3} = \kappa_{(3)} \tag{12.3.26}$$

in terms of $\kappa_{(3)}$ and $\kappa_{(4)}$, which are computed in (D.0.1). This gives

$$4\rho^2\tilde{\kappa}_{(4)}(m) = \frac{i\pi^2}{3m^2}\rho^{-4}. \tag{12.3.27}$$

Now recall that the mass scale of the KK modes is set by $m_\mathcal{K}$ through (8.3.7). Moreover, recalling the conventions in Section 9.1.1 that the indices of the torsion tensor are covariantized in terms of the frame $E^{\dot\alpha}_\mu$ corresponding to the metric $\gamma^{\mu\nu} = \rho^2 G^{\mu\nu}$, we have

$$\rho^{-2} T^{\dot\alpha\dot\beta\mu} T_{\dot\alpha\dot\beta}{}^\nu G_{\mu\nu} = T_{\rho\sigma}{}^\mu T_{\rho'\sigma'}{}^\nu \gamma^{\rho\rho'} \gamma^{\sigma\sigma'} \gamma_{\mu\nu} = T^\rho{}_{\sigma\mu} T_\rho{}^\sigma{}_\nu \gamma^{\mu\nu} = \rho^2 T^\rho{}_{\sigma\mu} T_\rho{}^\sigma{}_\nu G^{\mu\nu}.$$

$$(12.3.28)$$

Collecting these results, we obtain

$$\Gamma^{\mathcal{K}-\mathcal{M}}_{1\,\text{loop}} = \frac{3i}{4} \text{Tr}\left(\frac{V_{4,\text{mix}}}{(\square - i\varepsilon)^4}\right) = -\frac{1}{(2\pi)^4} \int_{\mathcal{M}} d^4x \sqrt{G}\, \rho^{-2} m^2_{\mathcal{K}} c^2_{\mathcal{K}}\, T^\rho{}_{\sigma\mu} T_\rho{}^\sigma{}_\nu G^{\mu\nu}$$

$$= -\frac{1}{2} \int_{\mathcal{M}} d^4x\, \frac{\sqrt{G}}{16\pi G_N}\, T^\rho{}_{\sigma\mu} T_\rho{}^\sigma{}_\nu G^{\mu\nu}. \qquad (12.3.29)$$

Here,

$$c^2_{\mathcal{K}} = \frac{\pi^2}{8} \sum_\Lambda \frac{C^2_\Lambda}{\mu^2_\Lambda} > 0 \qquad (12.3.30)$$

is finite due to the fuzzy nature of $\mathcal{K}$, and

$$\frac{1}{G_N} = \frac{2c^2_{\mathcal{K}}}{\pi^3} \rho^{-2} m^2_{\mathcal{K}} \qquad (12.3.31)$$

plays the role of the Newton constant i.e. the Planck scale, which is set by the compactification scale via $\rho^{-2} m^2_{\mathcal{K}}$. This can be rewritten using the identity (9.3.44) as

$$\boxed{\Gamma^{\mathcal{K}-\mathcal{M}}_{1\,\text{loop}} = \int d^4x\, \frac{\sqrt{|G|}}{16\pi G_N} \left(\mathcal{R} + \frac{1}{2}\tilde{T}_\nu \tilde{T}_\mu G^{\mu\nu} - 2\rho^{-2}\partial_\mu\rho\partial^\mu\rho + 2\rho^{-1}\partial_\mu\rho\, G^{-1}_N \partial^\mu G_N\right).}$$

$$(12.3.32)$$

Here, $\tilde{T}_\mu$ encodes the totally antisymmetric part of the torsion (9.3.19), which reduces to the axion using the on-shell relation $\tilde{T}_\mu = \rho^{-2}\partial_\mu\tilde{\rho}$ (9.3.21). Hence, for $\tilde{T}_\mu = 0 = \partial\rho$ we recover the Einstein–Hilbert action $\int G^{-1}_N \mathcal{R} + \mathcal{L}_{\text{matter}}$ coupled to matter. This leads to (modified) Einstein equations with the appropriate sign, since our sign convention entails $\frac{\delta S_{\text{matter}}}{\delta G^{\mu\nu}} = -T_{\mu\nu}$.

It is gratifying that the Planck scale is related to the Kaluza–Klein scale for the fuzzy extra dimensions $\mathcal{K}$, and hence to the effective UV cutoff on $\mathcal{K}$. Note that G^{-1}_N inherits the appropriate dimension mass^2 from $m^2_{\mathcal{K}}$, as discussed in Section 8.3. The presence of the dilaton may seem reminiscent of Brans–Dicke theory; however, ρ is not an independent field here due to the divergence constraint, as discussed in Section 9.2.4. Note that the effective Newton constant G_N is determined dynamically in terms of the dilaton and $m^2_{\mathcal{K}}$. The stabilization of the latter will be discussed further in Section 12.4.4.

Before we proceed, let us recapitulate this important result: the one-loop effective action for $\mathcal{M}^{3,1} \times \mathcal{K}$ leads to a gravitational action on $\mathcal{M}^{3,1}$ including the Einstein–Hilbert term. On the other hand, this induced action can also be interpreted as IIB supergravity interaction between $\mathcal{K}$ and $\mathcal{M}$ in target space $\mathbb{R}^{9,1}$. This term can hence be considered as

almost-local IIB interaction. In other words, *the local* $3 + 1$ *dimensional gravity* **action** arises from the *nonlocal IIB supergravity* **interaction** in $9 + 1$ dimensions. Since all fluctuations propagate on the spacetime brane $\mathcal{M}$ in the weak coupling regime, there is no need to compactify target space (unlike in string theory), and no landscape problem arises.

$\mathcal{M}$ contributions revisited

Finally, we reconsider the contribution of the terms

$$V_{4,\mathcal{M}} = \left(\delta\mathcal{F}^{\alpha\beta}\delta\mathcal{F}^{\alpha\beta}\right)\left(\delta\mathcal{F}^{\gamma\delta}\delta\mathcal{F}^{\gamma\delta}\right) - 4\left(\delta\mathcal{F}^{\alpha\beta}\delta\mathcal{F}^{\beta\gamma}\delta\mathcal{F}^{\gamma\delta}\delta\mathcal{F}^{\delta\alpha}\right) \tag{12.3.33}$$

arising from $\mathcal{M}$, in the presence of $\mathcal{K}$. Summing over the modes (12.3.7), we can evaluate the trace over $\text{End}(\mathcal{M})$ as in Section 12.2 by integrating over the string modes. The fourth order differential operator acting on the wave packets $\psi_{k;y}$ leads again to

$$V_{4,\mathcal{M}}[\psi_{k;y}] \approx V_4(T)^{\mu\nu\rho\sigma}k_\mu k_\nu k_\rho k_\sigma \tag{12.3.34}$$

as in (12.2.8). All KK modes on $\mathcal{K}$ contribute in this way, and their mass enters in the propagator. Therefore, the local contribution to the effective action takes the form

$$i\text{Tr}\left(\frac{V_{4,\mathcal{M}}}{(\Box - i\varepsilon)^4}\right) = \frac{i}{(2\pi)^4}\int_\mathcal{M} d^4x \int d^4k \sum_\Lambda \frac{V_4(T)^{\mu\nu\rho\sigma}k_\mu k_\nu k_\rho k_\sigma}{(k \cdot k + m_\Lambda^2 - i\varepsilon)^4} \tag{12.3.35}$$

summing over the KK modes on $\mathcal{K}$. The integral over k is again log-divergent as in Section 12.2, but regulated by the fully nonlocal form (12.2.5) of the induced action. This is a subleading higher-derivative contribution to four-dimensional gravity, which will not be considered any further.

12.4 Gravity on covariant quantum spacetime

Now we want to compute the one-loop effective action on a covariant $\mathcal{M}^{3,1}$ background given by some deformation of $T^\alpha \sim t^\alpha$ (10.1.1) in the weak gravity regime. Recall that the underlying quantum space is a six-dimensional bundle

$$\mathbb{C}P^{1,2} \cong S_n^2 \,\bar{\times}\, \mathcal{M}^{3,1}, \tag{12.4.1}$$

where $\bar{\times}$ indicates the twisted bundle structure. The space of functions is then spanned by the following modes:

$$\phi_{sm} = \phi_{sm}(x)Y_{sm}(t). \tag{12.4.2}$$

Here, x are coordinates on $\mathcal{M}^{3,1}$, and $Y_{sm}(t)$ are (normalized) fuzzy spherical harmonics on S_n^2 corresponding to the expansion into $\mathfrak{hs}$ modes in Section 5.4. According to (5.4.73), these are eigenvectors of $\Box$ with eigenvalue

$$\Box = k \cdot k + m_s^2. \tag{12.4.3}$$

Here, $k \cdot k = k_\mu k_\nu \gamma^{\mu\nu}$ where

$$\gamma^{\mu\nu} = \eta_{\alpha\beta} E^{\alpha\mu} E^{\beta\nu}, \tag{12.4.4}$$

and the "higher-spin mass" on S^2 is given by (5.4.73)

$$m_s^2 = \frac{3s}{R^2}. \tag{12.4.5}$$

This is an IR mass in the late-time regime. On more general (deformed) backgrounds, this holds in local normal coordinates, where the possible $\mathfrak{h}s$ components of the frame vanish, cf. Section 10.2.5.

The desired local contribution to the one-loop effective action can now be computed in the semi-classical setting. It is crucial here that the internal space S_n^2 supports only finitely many modes[4]. We will therefore evaluate the trace over $\mathrm{End}(\mathcal{H})$ in a hybrid way, expanding into the finite tower of $\mathfrak{h}s$ harmonics on the internal S_n^2, and subsequently using the semi-classical trace formula for localized wavepackets on $\mathcal{M}^{3,1}$. Since the Poisson structure on the bundle space does not respect the local factorization, there will be mixed $S^2 - \mathcal{M}$ contributions with a nonstandard different behavior. These are suppressed in the presence of fuzzy extra dimensions, which span the transversal six dimensions in target space. In that case, a physically acceptable gravity action can arise, as elaborated in Section 12.4.2.

12.4.1 Gravitational action on $\mathcal{M}^{3,1}$ without extra dimensions

We start again with the general one-loop effective action (11.1.28), where now

$$\mathcal{F}^{\alpha\beta} = -\{t^\alpha, t^\beta\} = \frac{1}{r^2 R^2} \theta^{\alpha\beta}. \tag{12.4.6}$$

Expanding the modes into normalized harmonics on S_n^2 (12.4.2), each derivative $\delta \mathcal{F}^{\alpha\beta} = [\mathcal{F}^{\alpha\beta}, .]$ in

$$V_4 = \left(\delta\mathcal{F}^{\alpha\beta} \delta\mathcal{F}_{\alpha\beta}\right)\left(\delta\mathcal{F}^{\gamma\delta} \delta\mathcal{F}_{\gamma\delta}\right) - 4\left(\delta\mathcal{F}^{\alpha\beta} \delta\mathcal{F}_{\beta\gamma} \delta\mathcal{F}^{\gamma\delta} \delta\mathcal{F}_{\delta\alpha}\right) \tag{12.4.7}$$

acts either on $\phi_{sm}(x)$ or on $Y_{sm}(t)$ via

$$[\mathcal{F}^{\alpha\beta}, \phi_{sm}(x) Y_{sm}(t)] = [\mathcal{F}^{\alpha\beta}, \phi_{sm}(x)] Y_{sm}(t) + \phi_{sm}(x)[\mathcal{F}^{\alpha\beta}, Y_{sm}(t)]. \tag{12.4.8}$$

This leads to contributions from $\mathcal{M}$, from S_n^2, and mixed $S_n^2 - \mathcal{M}$ contributions, where S_n^2 plays a role somewhat analogous to $\mathcal{K}$.

Pure $\mathcal{M}$ contribution

The pure $\mathcal{M}$ contribution is obtained by keeping only the first term in (12.4.8). This is completely analogous as in Section 12.2 summing over the internal harmonics; however, $\Box$ now acts on the nontrivial modes (12.4.2) with eigenvalue

$$\Box = k \cdot k + m_s^2. \tag{12.4.9}$$

---

[4] This internal S_n^2 is not captured by standard IIB supergravity, because it is not embedded in target space. It might be possible to describe it via nonabelian Dirac–Born–Infeld theory, but this is not straightforward.

This gives a similar nonlocal contribution as in (12.2.5), which we will not consider any further.

Mixed $S^2 - \mathcal{M}$ contributions

Now we take into account contributions from the second term in (12.4.8). Since the mixed $S^2 - \mathcal{M}$ contribution will turn out to be subleading compared with those from $\mathcal{K}$, we simplify the computation by assuming that the local structure is the same as on the undeformed cosmic background. Then $[\mathcal{F}^{\alpha\beta}, t^i] = -\frac{i}{R^2}(\eta^{\alpha i} t^\beta - \eta^{\beta i} t^\alpha)$, and $\Box$ respects the local $SO(3)$. One then finds locally

$$\sum_m \langle Y_{sm}(t), [\mathcal{F}^{\alpha\beta}, Y_{sm}(t)] \rangle = 0$$

$$\sum_m \langle Y_{sm}(t), [\mathcal{F}^{\alpha\beta}, [\mathcal{F}^{\gamma\delta}, Y_{sm}(t)]] \rangle = \frac{1}{4R^4} \kappa_s^{\alpha\beta;\gamma\delta}$$

$$\sum_m \langle Y_{sm}(t), [\mathcal{F}^{\alpha\beta}, [\mathcal{F}^{\gamma\delta}, [\mathcal{F}^{\alpha'\beta'}, [\mathcal{F}^{\gamma'\delta'}, Y_{sm}(t)]]]] \rangle = \frac{1}{4R^8} \kappa_s^{\alpha\beta;\gamma\delta;\alpha'\beta';\gamma'\delta'}, \quad (12.4.10)$$

where the tensors $\kappa_s^{\alpha\beta;\gamma\delta}$ etc. are invariant under the local $SO(3)$ as well as under the spacelike $SO(3,1)$ isometry. The remaining derivations $\delta\mathcal{F}^{\alpha\beta}$ act on $\mathcal{C}(\mathcal{M}^{3,1})$ as in (12.3.22). Taking into account the form of V_4, one obtains the following mixed $S^2 - \mathcal{M}$ contributions from V_4:

$$\sum_m \langle Y_{sm}(t), V_{4,\text{mix}} Y_{sm}(t) \rangle = \frac{1}{2R^4} \kappa_s^{\alpha\beta;\alpha\beta} \left(\delta\mathcal{F}^{\gamma\delta} \delta\mathcal{F}^{\gamma\delta}\right) + \frac{1}{R^4} \kappa_s^{\alpha\beta;\gamma\delta} \delta\mathcal{F}^{\alpha\beta} \delta\mathcal{F}^{\gamma\delta}$$

$$- \frac{4}{R^4} \kappa_s^{\alpha\beta;\beta\gamma} \delta\mathcal{F}^{\gamma\delta} \delta\mathcal{F}^{\delta\alpha} - \frac{2}{R^4} \kappa_s^{\alpha\beta;\gamma\delta} \delta\mathcal{F}^{\beta\gamma} \delta\mathcal{F}^{\delta\alpha}$$

$$=: \frac{2}{R^4} \tilde{k}_{\alpha\beta\gamma\delta}^s \delta\mathcal{F}^{\alpha\beta} \delta\mathcal{F}^{\gamma\delta}, \quad (12.4.11)$$

where $\delta\mathcal{F}^{\alpha\beta} \delta\mathcal{F}^{\gamma\delta}$ are second order derivations acting on $\mathcal{C}(\mathcal{M}^{3,1})$ which needs to be traced over. This should be compared with (12.3.20) for fuzzy extra dimensions. The trace decomposes into a finite sum of higher spin sectors and the trace over the scalar functions $\phi_{sm}(x)$. We can evaluate the latter using the classical trace formula similarly as in (12.3.24), which leads to

$$\Gamma_{1\,\text{loop}}^{S^2 - \mathcal{M}} = \frac{3i}{4} \text{Tr}\left(\frac{V_{4,\text{mix}}}{(\Box - i\varepsilon)^4}\right)$$

$$= \frac{3i}{2(2\pi)^4 R^4} \int_{\mathcal{M}} d^4x \sqrt{G}\, T^{\alpha\beta\mu} T^{\gamma\delta\nu} \sum_s \tilde{k}_{\alpha\beta\gamma\delta}^s \int \frac{d^4k}{\sqrt{G}} \frac{k_\mu k_\nu}{(k \cdot k + m_s^2 - i\varepsilon)^4}$$

$$= \frac{3i}{2(2\pi)^4} \frac{1}{R^4} \int_{\mathcal{M}} d^4x \sqrt{G} \left(\sum_s \tilde{\kappa}_{(4)}(m_s^2) \tilde{k}_{\alpha\beta\gamma\delta}^s\right) T^{\alpha\beta\mu} T^{\gamma\delta\nu} G_{\mu\nu}$$

$$= -\frac{1}{(2\pi)^4} \frac{\pi^2}{8R^2} \int_{\mathcal{M}} d^4x \sqrt{G} \rho^{-6} \left(\sum_{s>0} \frac{\tilde{k}_{\alpha\beta\gamma\delta}^s}{3s}\right) T^{\alpha\beta\mu} T^{\gamma\delta\nu} G_{\mu\nu}, \quad (12.4.12)$$

recalling that (12.3.27)

$$\tilde{\kappa}_{(4)}(m_s^2) = \frac{i\pi^2}{12m_s^2}\rho^{-6} = \frac{i\pi^2 R^2}{36s}\rho^{-6}. \tag{12.4.13}$$

Note that the contractions of the torsion via $\tilde{k}_{\alpha\beta\gamma\delta}$ is not Lorentz invariant a priori, due to the timelike vector field τ (5.4.15) on the cosmic background. However, we will argue that this is strongly suppressed compared with the contribution from $\mathcal{K}$ considered in the next section provided

$$c_{\mathcal{K}}^2 m_{\mathcal{K}}^2 \gg \frac{1}{R^2}. \tag{12.4.14}$$

Pure S_n^2 contribution

Finally, we consider the pure contribution from S_n^2. The trace then reduces to the form

$$\Gamma_{1\,\text{loop}}^{S^2} = \frac{3i}{4}\text{Tr}\left(\frac{V_{4,S^2}}{(\Box - i\varepsilon)^4}\right) = \frac{3i}{4(2\pi)^4}\int_{\mathcal{M}} d^4x \sum_s \frac{\tilde{V}_{4,s}}{R^8} \int d^4k \frac{1}{(k\cdot k + m_s^2 - i\varepsilon)^4}$$

$$= -\frac{3}{4(2\pi)^4}\int_{\mathcal{M}} d^4x\sqrt{G} \sum_s \frac{\tilde{V}_{4,s}}{R^8}\kappa_{(4)}(m_s^2)$$

$$= -\frac{1}{8(4\pi)^2}\int_{\mathcal{M}} \Omega\,\rho^{-2}\frac{1}{R^4}\sum_{s\geq 1} \frac{\tilde{V}_{4,s}}{(3s)^2} \tag{12.4.15}$$

for some dimensionless $\tilde{V}_{4,s}$. This amounts to a contribution to the effective potential, which a priori could have either sign.

12.4.2 Einstein–Hilbert action from fuzzy extra dimensions $\mathcal{K}$

Finally, we combine these considerations and consider the case of fuzzy extra dimensions $\mathcal{K}$ on covariant quantum spaces, so that the full background has the structure

$$S_n^2 \,\bar{\times}\, \mathcal{M}^{3,1} \times \mathcal{K}. \tag{12.4.16}$$

Such a background will typically involve all $9 + 1$ matrices for suitable $\mathcal{K}$, thus exploiting the full IKKT matrix model. The space of functions is then spanned by the following nontrivial modes on both $\mathcal{K}$ and S_n^2,

$$\phi_{\Lambda,sn}(x)Y_{sn}(t)\Upsilon_\Lambda, \tag{12.4.17}$$

where Υ_Λ are the eigenmodes on $\mathcal{K}$. According to (5.4.73), these are eigenvectors of $\Box$ with eigenvalue

$$\Box = k\cdot k + m_s^2 + m_\Lambda^2. \tag{12.4.18}$$

We should distinguish the contributions from the trivial and nontrivial modes on $\mathcal{K}$. The trivial contributions coincide with those of Section 12.4.1 in the absence of $\mathcal{K}$. To simplify the discussion, we will therefore focus on the contribution from the nontrivial modes in the following.

Mixed $\mathcal{M}^{3,1} - \mathcal{K}$ contribution and induced gravity

Consider the contributions of the mixed $4D - 6D$ terms (12.3.19)

$$V_{4,\text{mix}} = 2\big(\delta \mathcal{F}^{\alpha\beta} \delta \mathcal{F}_{\alpha\beta}\big)\big(\delta \mathcal{F}^{ij} \delta \mathcal{F}_{ij}\big) \tag{12.4.19}$$

acting on nontrivial modes on $\mathcal{K}$ and $\mathcal{M}^{3,1}$,

$$\phi_\Lambda(x) \Upsilon_\Lambda. \tag{12.4.20}$$

The mixed term from V_4 acting on $\mathcal{K}$ and functions on $\mathcal{M}^{3,1}$ have the same structure as in Section (12.3), and lead to the same effective action as in (12.3.24) and (12.3.29),

$$
\begin{aligned}
\Gamma^{\mathcal{K}-\mathcal{M}}_{1\,\text{loop}} &= \frac{3i}{4}\text{Tr}\bigg(\frac{V_{4,\text{mix}}}{(\square - i\varepsilon)^4}\bigg) \\
&= i\frac{3}{2(2\pi)^4}\int_{\mathcal{M}} d^4x\, T^{\alpha\beta\mu} T^{\alpha\beta\nu} \sum_{\Lambda,s}(2s+1)C_\Lambda^2 m_{\mathcal{K}}^4 \int d^4k\, \frac{k_\mu k_\nu}{(k\cdot k + m_\Lambda^2 + m_s^2 - i\varepsilon)^4} \\
&= -\frac{1}{(2\pi)^4}\int_{\mathcal{M}} d^4x \sqrt{G}\, \rho^{-2} m_{\mathcal{K}}^2 c_{\mathcal{K}}^2\, T^\rho{}_{\sigma\mu} T_\rho{}^\sigma{}_\nu\, G^{\mu\nu},
\end{aligned}
\tag{12.4.21}
$$

where $c_{\mathcal{K}}^2$ (12.3.30) is replaced by

$$c_{\mathcal{K}}^2 = \frac{\pi^2}{8}\sum_{\Lambda,s}\frac{(2s+1)C_\Lambda^2}{\mu_\Lambda^2 + \frac{m_s^2}{m_{\mathcal{K}}^2}} > 0. \tag{12.4.22}$$

Again, we can rewrite the effective action in terms of an Einstein–Hilbert term (12.3.32)

$$\boxed{\Gamma^{\mathcal{K}-\mathcal{M}}_{1\,\text{loop}} = \int d^4x\, \frac{\sqrt{|G|}}{16\pi G_N}\bigg(\mathcal{R} + \frac{1}{2}\tilde{T}_\nu \tilde{T}_\mu G^{\mu\nu} - 2\rho^{-2}\partial_\mu\rho\,\partial^\mu\rho + 2\rho^{-1}\partial_\mu\rho\, G_N^{-1}\partial^\mu G_N\bigg)}$$

$$\tag{12.4.23}$$

with effective Newton constant (12.3.31)

$$\frac{1}{G_N} = \frac{2c_{\mathcal{K}}^2}{\pi^3}\rho^{-2} m_{\mathcal{K}}^2. \tag{12.4.24}$$

Note that this action has a similar structure as the $S^2 - \mathcal{M}$ contribution (12.4.12). However, the latter is not Lorentz invariant, and should thus be suppressed in order to obtain a reasonable gravity theory. Since (12.4.21) is governed by the scale factor $m_{\mathcal{K}}^2$ rather than $\frac{1}{R^2}$, this is the case provided

$$c_{\mathcal{K}}^2 m_{\mathcal{K}}^2 \gg \frac{1}{R^2} \sim m_s^2 \tag{12.4.25}$$

so that the contribution from $\mathcal{K}$ dominates. This is reasonable because the Kaluza–Klein modes arising from $\mathcal{K}$ should be very heavy, while m_s^2 should be a very small IR mass scale. The scale of $m_\mathcal{K}^2$ is presumably set by quantum effects, which will be discussed later.

Mixed $S^2 - \mathcal{K}$ contribution

Now consider contributions of the mixed $4D - 6D$ terms (12.3.19)

$$V_{4,\mathrm{mix}} = 2\big(\delta\mathcal{F}^{\alpha\beta}\delta\mathcal{F}_{\alpha\beta}\big)\big(\delta\mathcal{F}^{ij}\delta\mathcal{F}_{ij}\big), \tag{12.4.26}$$

where $\delta\mathcal{F}^{\alpha\beta}\delta\mathcal{F}_{\alpha\beta}$ acts on $Y_{sn}(t)$ rather than $\phi_\Lambda(x)$, and the transversal derivations act on Υ_Λ as in (12.3.20), with

$$\big(\delta\mathcal{F}^{ij}\delta\mathcal{F}_{ij}\big)\Upsilon_\Lambda = [\mathcal{F}^{ij}, [\mathcal{F}_{ij}, \Upsilon_\Lambda]] = m_\mathcal{K}^4 C_\Lambda^2 \Upsilon_\Lambda, \qquad C_\Lambda^2 > 0 \tag{12.4.27}$$

under the same assumptions. Now

$$C_\mathcal{M}^2 = \delta\mathcal{F}^{\alpha\beta}\delta\mathcal{F}_{\alpha\beta} \tag{12.4.28}$$

is the quadratic Casimir of $\mathfrak{so}(3,1)$, which is diagonal for the spherical harmonics on S_n^2 with positive eigenvalue

$$C_\mathcal{M}^2 Y_m^s = \frac{2}{R^4} s(s+2) Y_m^s. \tag{12.4.29}$$

As a check, we compute

$$C_\mathcal{M}^2 t^i = [\mathcal{F}_{\alpha\beta}, [\mathcal{F}^{\alpha\beta}, t^i]] = -\frac{i}{R^2}[\mathcal{F}_{\alpha\beta}, \eta^{\alpha i}t^\beta - \eta^{\beta i}t^\alpha] = \frac{6}{R^4}t^i. \tag{12.4.30}$$

Summing over the modes (12.4.17), $V_{4,\mathrm{mix}}$ reduces to

$$V_{4,\mathrm{mix}}[\Upsilon_\Lambda Y_{sn}(t)] = \frac{4m_\mathcal{K}^4}{R^4} s(s+2)C_\Lambda^2 \,\Upsilon_\Lambda Y_{sn}(t) \tag{12.4.31}$$

and the trace can be evaluated as

$$\frac{3i}{4}\mathrm{Tr}\Big(\frac{V_{4,\mathcal{K}}}{(\Box - i\varepsilon)^4}\Big) = i\frac{3m_\mathcal{K}^4}{(2\pi)^4 R^4}\int_\mathcal{M} d^4x \sum_{\Lambda;sn} s(s+2)C_\Lambda^2 \int d^4k \frac{1}{(k\cdot k + m_s^2 + m_\Lambda^2 - i\varepsilon)^4}$$

$$= i\frac{3m_\mathcal{K}^4}{(2\pi)^4 R^4}\int_\mathcal{M} d^4x\sqrt{G}\sum_{\Lambda;sn} s(s+2)C_\Lambda^2 \kappa_{(4)}(m_s^2 + m_\Lambda^2)$$

$$= -\frac{\pi^2}{2(2\pi)^4}\int_\mathcal{M}\Omega\,\rho^{-2}\frac{m_\mathcal{K}^4}{R^4}\sum_{\Lambda;s}\frac{s(s+2)(2s+1)C_\Lambda^2}{(m_s^2 + m_\Lambda^2)^2}. \tag{12.4.32}$$

This gives a positive contribution to the vacuum energy since $S = T - V$, which for $\frac{1}{R^2} \ll m_\Lambda^2$ simplifies as

$$\Gamma_{1\,\mathrm{loop}}^{\mathcal{K}-S^2} = \frac{3i}{4}\mathrm{Tr}\Big(\frac{V_{4,\mathcal{K}}}{(\Box - i\varepsilon)^4}\Big) \approx -\frac{1}{32\pi^2}\int_\mathcal{M}\Omega\,\rho^{-2}\frac{1}{R^4}\sum_{\Lambda;s}\frac{s(s+2)(2s+1)C_\Lambda^2}{\mu_\Lambda^4} \tag{12.4.33}$$

similar to (12.3.17), which is independent of $m_\mathcal{K}^2$.

Pure $\mathcal{K}$ contribution

Now consider the contributions arising from V_4 acting solely on the $6D$ terms on $\mathcal{K}$. This has a similar structure as in Section 12.3, However, $\Box$ acts on the nontrivial modes (12.4.17) with eigenvalue

$$\Box = k \cdot k + m_s^2 + m_\Lambda^2. \tag{12.4.34}$$

Thus, the trace reduces to

$$
\begin{aligned}
\frac{3i}{4}\mathrm{Tr}\left(\frac{V_{4,\mathcal{K}}}{(\Box - i\varepsilon)^4}\right) &= \frac{3i}{4(2\pi)^4} \int_{\mathcal{M}} d^4x\, m_{\mathcal{K}}^8 \sum_{\Lambda;sn} V_{4,\Lambda} \int d^4k\, \frac{1}{(k \cdot k + m_s^2 + m_\Lambda^2 - i\varepsilon)^4} \\
&= \frac{3i}{4(2\pi)^4} \int_{\mathcal{M}} d^4x\sqrt{G}\, m_{\mathcal{K}}^8 \sum_{\Lambda;sn} V_{4,\Lambda} \kappa_{(4)}\left(m_s^2 + m_\Lambda^2\right) \\
&= -\frac{\pi^2}{8(2\pi)^4} \int_{\mathcal{M}} \Omega\, \rho^{-2} m_{\mathcal{K}}^8 \sum_{\Lambda;s} \frac{(2s+1)V_{4,\Lambda}}{\left(m_s^2 + m_\Lambda^2\right)^2}. \tag{12.4.35}
\end{aligned}
$$

Again assuming $m_\Lambda^2 \gg \frac{1}{R^2}$, this simplifies as

$$
\Gamma^{\mathcal{K}}_{1\,\text{loop}} = \frac{3i}{4}\mathrm{Tr}\left(\frac{V_{4,\mathcal{K}}}{(\Box - i\varepsilon)^4}\right) \approx -\frac{\pi^2}{8(2\pi)^4} \int_{\mathcal{M}} \Omega\, \rho^{-2} m_{\mathcal{K}}^4 \sum_{\Lambda;s} \frac{(2s+1)V_{4,\Lambda}}{\mu_\Lambda^4}. \tag{12.4.36}
$$

This is a contribution to the effective potential for $m_{\mathcal{K}}$, with the same structure as (12.3.17).

Pure S_n^2 contribution

Finally, we consider the pure contribution from S_n^2, which leads along the lines of (12.4.15) to

$$
\begin{aligned}
\Gamma^{S^2}_{1\,\text{loop}} = \frac{3i}{4}\mathrm{Tr}\left(\frac{V_{4,S^2}}{(\Box - i\varepsilon)^4}\right) &= -\frac{3}{4(2\pi)^4} \int_{\mathcal{M}} d^4x\sqrt{G} \sum_{s,\Lambda} \frac{\tilde{V}_{4,s}}{R^8}\kappa_{(4)}\left(m_s^2 + m_\Lambda^2\right) \\
&= -\frac{\pi^2}{8(2\pi)^4} \int_{\mathcal{M}} \Omega\, \rho^{-2} \sum_{s,\Lambda} \frac{\tilde{V}_{4,s}}{R^8} \frac{1}{\left(m_s^2 + m_\Lambda^2\right)^2} \tag{12.4.37}
\end{aligned}
$$

analogous to (12.4.15). Assuming that $m_\Lambda^2 \gg \frac{1}{R^2}$, this reduces to

$$
\Gamma^{S^2}_{1\,\text{loop}} \approx -\frac{\pi^2}{8(2\pi)^4} \int_{\mathcal{M}} \Omega\, \rho^{-2} \frac{1}{m_{\mathcal{K}}^4 R^8} \sum_{s,\Lambda} \frac{\tilde{V}_{4,s}}{\mu_\Lambda^4} \tag{12.4.38}
$$

which has a similar structure as (12.4.15), and could a priori have either sign.

12.4.3 Induced gravity in Euclidean signature

For completeness, we also compute the induced gravity term in the case of Euclidean signature. In the presence of fuzzy extra dimensions $\mathcal{K}$, (12.3.24) is replaced by

$$\text{Tr}\left(\frac{V_{4,\text{mix}}}{\Box^4}\right) = \frac{2m_\mathcal{K}^4}{(2\pi)^4} \int_\mathcal{M} d^4x\sqrt{G}\left(\sum_\Lambda C_\Lambda^2 \tilde{\kappa}_{(4)}(m_\Lambda)\right) T^{\alpha\beta\mu} T_{\alpha\beta}{}^\nu G_{\mu\nu}. \tag{12.4.39}$$

Here,

$$\tilde{\kappa}_{(4)}G_{\mu\nu} = \int d^4k \frac{k_\mu k_\nu}{(k\cdot k + m^2)^4} = \frac{\pi^2}{12m^2}\rho^{-6}G_{\mu\nu}. \tag{12.4.40}$$

Recalling that the mass scale of the KK modes is set by the radius $m_\mathcal{K}$ through (8.3.7), we obtain

$$\Gamma_{1\text{ loop}} = -\frac{3}{4}\text{Tr}\left(\frac{V_{4,\text{mix}}}{\Box^4}\right) = -\frac{c_\mathcal{K}^2}{(2\pi)^4}\int_\mathcal{M} d^4x\sqrt{G}\, m_\mathcal{K}^2\, T_\rho{}^\sigma{}_\mu T^\rho{}_{\sigma\nu}\, G^{\mu\nu}, \tag{12.4.41}$$

which has the same form as (12.3.32), with $c_\mathcal{K}^2$ given by (12.3.30). This can again be written in the Einstein–Hilbert form $\Gamma_{1\text{ loop}} \sim \int G_N^{-1}\mathcal{R} + c\tilde{T}^\mu\tilde{T}_\mu + \dots$. Recalling our non-standard sign convention for the action, we recover the appropriate form $\int G_N^{-1}\mathcal{R} + \mathcal{L}_{\text{matter}}$ of a Euclidean gravitational action.

Discussion

Before collecting the contributions, we should discuss some technical points. In the computations of this section, the spacetime brane $\mathcal{M}$ and the compact space $\mathcal{K}$ were treated on a different footing, using string modes on $\mathcal{M}$ and harmonics on $\mathcal{K}$. It was crucial that $\mathcal{K}$ is a compact fuzzy space, which admits only a finite number of internal KK modes; otherwise, the sums over these modes would diverge.

On the other hand, it may seem more natural to treat $\mathcal{M}$ and $\mathcal{K}$ on the same footing. This is tricky, because the product space $\mathcal{M} \times \mathcal{K}$ is higher-dimensional, so that the integral over the string states on $\mathcal{M} \times \mathcal{K}$ would no longer be almost-local. An appropriate setup for such a computation would be to use factorized quasi-coherent states $|x\rangle|\xi\rangle$ to define the string modes as in (6.2.96), where $|\xi\rangle$ are coherent states on $\mathcal{K}$. Then the sum over the KK modes is replaced by an integral over the string modes on $\mathcal{K}$. One can then first carry out the integral over $\mathcal{M}$, which for fixed $\xi, \xi' \in \mathcal{K}$ would have precisely the same form as e.g. in (12.3.24), replacing m_Λ^2 by $|\xi - \xi'|^2$. The (compact) integral over $\mathcal{K}$ would then lead to an equivalent result as the sum over KK modes, and there is no contradiction between the two approaches. This setup should be particularly useful if $\mathcal{K}$ has a self-intersecting geometry as discussed in Section 8.4; then the lowest KK modes are typically string states linking different sheets at their intersections, and should provide the leading contribution to the effective potential and the low-energy gauge theory on $\mathcal{M}$. Finally, it is important to keep in mind that the present one-loop computations can be trusted only in the weak coupling regime.

12.4.4 One-loop gravitational action and vacuum stability

Combining the bare matrix model action with the one-loop contributions, we obtain the following one-loop effective action on the background brane $\mathcal{M}^{3,1} \times \mathcal{K}$

$$S_{\text{1loop}} = S_{\text{YM}} + S_{\text{grav}} + S_{\text{vac}}. \tag{12.4.42}$$

Here,

$$S_{\text{YM}} = -\frac{1}{g^2} \int_{\mathcal{M}} \Omega \, \text{tr}_{\mathcal{K}} \mathcal{F}_{ab} \mathcal{F}^{ab} \tag{12.4.43}$$

is the bare action of the IKKT model; note that the coupling to matter including fermions[5] as well as nonabelian bosons arising from $\mathcal{K}$ are part of S_{YM}. The induced gravitational action at one loop has the form (12.4.21)

$$S_{\text{grav}} = -\int_{\mathcal{M}} d^4x \, \frac{\sqrt{G}}{16\pi G_N} \frac{1}{2} T_\rho{}^\sigma{}_\mu T^\rho{}_{\sigma\nu} G^{\mu\nu} + \Gamma_{\text{h.o.}}(\mathcal{M}, \mathcal{K}). \tag{12.4.44}$$

The first term can be written in terms of the Einstein–Hilbert action as in (12.4.23), and the second term indicates higher-derivative contributions from the one loop effective action. Finally, S_{vac} subsumes the remaining contributions of the induced vacuum energy that are independent of the geometry of $\mathcal{M}^{3,1}$, including the contributions (12.4.36), (12.4.38), and (12.4.33) from $\mathcal{K}$, from S_n^2, and from $S_n^2 - \mathcal{K}$:

$$S_{\text{vac}} = -\int_{\mathcal{M}} \Omega \rho^{-2} \left(C_1 m_{\mathcal{K}}^4 + \frac{C_2}{R^8 m_{\mathcal{K}}^4} + \frac{C_3}{R^4} \right). \tag{12.4.45}$$

Note that the last two terms arise only on covariant quantum spacetime.

The most important point is the presence of an induced Einstein–Hilbert term implicit in S_{grav}, with Newton constant set by the KK scale $m_{\mathcal{K}}^2$. This contribution has two more derivatives than the bare action S_{YM}, and is therefore expected to dominate at shorter length scales. Moreover, both the bare and the one-loop induced action admit Ricci-flat linearized metric fluctuations. Therefore, the deviations from general relativity should be small at shorter length scales. On the other hand, the bare action S_{YM} is expected to govern the extreme IR regime, so that the pre-gravity as discussed in Section 10.4 should dominate at cosmic scales, where matter no longer acts as a source of gravity. The cross-over behavior between these regimes should be studied in detail elsewhere.

A detailed analysis of the resulting gravity theory and its physical viability is beyond the scope of this book. However, we can provide some justification for the stability of the $\mathcal{M} \times \mathcal{K}$ background at one loop.

---

[5] Fermions do indeed couple appropriately to the background geometry, as discussed in [25].

Effective potential and stabilization for basic $\mathcal{M} \times \mathcal{K}$ branes

Consider first the effective potential for basic (non-covariant, i.e. without internal S_n^2 fiber) spacetime branes $\mathcal{M}$ in the presence of $\mathcal{K}$. The semi-classical MM action contributes

$$S_{\text{YM}} = - \int_{\mathcal{M}} \Omega \, \text{tr}_{\mathcal{K}}(\mathcal{F}_{\dot{\alpha}\dot{\beta}} \mathcal{F}^{\dot{\alpha}\dot{\beta}} + \mathcal{F}_{ij}\mathcal{F}^{ij} + 2\{T^{\dot{\alpha}}, T^i\}\{T_{\dot{\alpha}}, T_i\}) \tag{12.4.46}$$

(dropping the fermions), where the second term depends on the scale $m_{\mathcal{K}}^2$ of $\mathcal{K}$ as in (8.3.1)

$$\text{tr}_{\mathcal{K}}(\mathcal{F}_{ij}\mathcal{F}^{ij}) = m_{\mathcal{K}}^4 F_{\mathcal{K}}^2, \tag{12.4.47}$$

and $F_{\mathcal{K}}^2$ is some discrete number depending on the structure of $\mathcal{K}$. Note that the coupling constant g is absorbed in the matrices. We assume for now that $m_{\mathcal{K}}$ is constant, so that $\{T^{\dot{\alpha}}, T^i\} = 0$. Together with the one-loop contributions (12.3.17) and (12.3.32) while dropping (12.3.35) which is expected to be suppressed, we can combine the $m_{\mathcal{K}}$-dependent terms of the one-loop effective action for the background in terms of an effective potential for $m_{\mathcal{K}}$

$$S_{\text{1loop}} \ni - \int_{\mathcal{M}} \Omega \, V(m_{\mathcal{K}}) \tag{12.4.48}$$

using $S = \int T - V$ in Minkowski signature. This gives

$$V(m_{\mathcal{K}}) = \frac{c_{\mathcal{K}}^2}{(2\pi)^4} m_{\mathcal{K}}^2 T^{\rho}{}_{\sigma\mu} T_{\rho}{}^{\sigma}{}_{\nu} G^{\mu\nu} + m_{\mathcal{K}}^4 (F_{\mathcal{K}}^2 + C_1 \rho^{-2}). \tag{12.4.49}$$

Then the potential $V(m_{\mathcal{K}})$ has a global minimum determined by

$$m_{\mathcal{K}}^2 = -\frac{1}{2(2\pi)^4} \frac{c_{\mathcal{K}}^2}{(F_{\mathcal{K}}^2 + C_1 \rho^{-2})} T^{\rho}{}_{\sigma\mu} T_{\rho}{}^{\sigma}{}_{\nu} G^{\mu\nu} \tag{12.4.50}$$

provided the torsion of the background spacetime is negative in the sense that

$$T^{\rho}{}_{\sigma\mu} T_{\rho}{}^{\sigma}{}_{\nu} G^{\mu\nu} < 0. \tag{12.4.51}$$

This is indeed possible in Minkowski signature; e.g. on the covariant FLRW spacetime $\mathcal{M}^{3,1}$, we can compute this curvature contribution explicitly. This gives

$$T^{\rho}{}_{\sigma\mu} T_{\rho}{}^{\sigma}{}_{\nu} G^{\mu\nu} = \rho^{-2} T^{\dot{\alpha}\dot{\beta}\mu} T_{\dot{\alpha}\dot{\beta}}{}^{\nu} \gamma_{\mu\nu} = -\frac{6}{R^2 \sinh^3(\eta)} \sim -\frac{6}{a(t)^2} \tag{12.4.52}$$

using (10.1.9), (12.3.28) and (5.4.44), which is set by the cosmic curvature scale. Then the effective potential for $m_{\mathcal{K}}$ takes the form

$$V(m_{\mathcal{K}}) = -\frac{6 c_{\mathcal{K}}^2}{(2\pi)^4} \frac{m_{\mathcal{K}}^2}{a(t)^2} + m_{\mathcal{K}}^4 (F_{\mathcal{K}}^2 + C_1 \rho^{-2}), \tag{12.4.53}$$

which has a stable minimum with

$$\boxed{V(m_{\mathcal{K}}) < 0.} \tag{12.4.54}$$

Hence, a background $\mathcal{M} \times \mathcal{K}$ with fuzzy extra dimensions is energetically favorable compared with a background without $\mathcal{K}$; note that local variations of $m_{\mathcal{K}}$ due to local curvature variations will be suppressed by the kinetic term for $m_{\mathcal{K}}^2$, as discussed shortly. This provides some justification for the presence of $\mathcal{K}$, although the mechanism is delicate. A more robust mechanism independent of the curvature arises on covariant spacetime.

Effective potential and stabilization for covariant $\mathcal{M}^{3,1} \times \mathcal{K}$

Now consider the covariant spacetime $\mathcal{M}^{3,1}$ discussed in Section 5.4. Then the vacuum energy is modified by the $\mathfrak{h}\mathfrak{s}$ modes contributing to the trace, as computed in (12.4.36) and (12.4.38). The $S^2 - \mathcal{K}$ contribution (12.4.33) is (almost) independent of $m_{\mathcal{K}}^2$ and will therefore be dropped here. Then the effective potential for $m_{\mathcal{K}}$ takes the form

$$V(m_{\mathcal{K}}) = -\frac{6c_{\mathcal{K}}^2}{(2\pi)^4} \frac{m_{\mathcal{K}}^2}{a(t)^2} + m_{\mathcal{K}}^4 \left(F_{\mathcal{K}}^2 + C_1 \rho^{-2} \right) + \frac{C_2}{R^8 m_{\mathcal{K}}^4} \rho^{-2}, \tag{12.4.55}$$

where $c_{\mathcal{K}}^2$ is given by (12.4.22). Assuming that $C_{1,2} \gg 0$ such that the curvature term is subleading, this potential has a stable minimum for

$$m_{\mathcal{K}}^8 = \frac{C_2}{C_1 + \rho^2 F_{\mathcal{K}}^2} \frac{1}{R^8}. \tag{12.4.56}$$

This provides a robust mechanism to stabilize $\mathcal{M}^{3,1} \times \mathcal{K}$ independent of the local geometry, and a large hierarchy of scales arises naturally for large N.

It is interesting to note that the Newton constant $G_N \sim \rho^2 m_{\mathcal{K}}^{-2}$ is found to increase with ρ^2. Similarly, the gauge coupling g_{YM} (9.1.72) also depends on ρ^2. Of course, the computations in this section are incomplete and only preliminary; for example, the dilaton and the background will also be modified at one loop. Nevertheless, these results strongly suggest that backgrounds of the structure $\mathcal{M}^{3,1} \times \mathcal{K}_N$ can be stabilized through quantum effects, and may lead to a qualitatively reasonable emergent gravity, with a large hierarchy due to the discrete nature of $\mathcal{K}_N$. This should suffice to motivate more detailed investigations.

Kinetic term and rigidity for $m_{\mathcal{K}}$

Since the compactification scale $m_{\mathcal{K}} = m_{\mathcal{K}}(x)$ is dynamical, we should also consider its response to local deformations of the spacetime geometry. The kinetic term for $m_{\mathcal{K}}$ arises from the matrix model as

$$-2 \int_{\mathcal{M}} \Omega \, \mathrm{tr}_{\mathcal{K}} \{T^\alpha, T^i\}\{T_\alpha, T_i\} = -2 \int_{\mathcal{M}} \Omega \, \{T^\alpha, m_{\mathcal{K}}\}\{T_\alpha, m_{\mathcal{K}}\} \mathrm{tr}_{\mathcal{K}}(\mathcal{K}_i \mathcal{K}^i)$$

$$= -2 \int_{\mathcal{M}} d^4 x \sqrt{G} \, \mathrm{tr}_{\mathcal{K}}(\mathcal{K}_i \mathcal{K}^i) \partial^\mu m_{\mathcal{K}} \partial_\mu m_{\mathcal{K}} \tag{12.4.57}$$

along the lines of Section 9.1.1. Then $m_{\mathcal{K}}(x)$ should be determined such that the overall energy is minimal. Assuming that $m_{\mathcal{K}}^2$ is a very high energy scale, local variations $\bar{m}_{\mathcal{K}} + \delta m_{\mathcal{K}}(x)$ would imply a huge kinetic energy, so that $\frac{\delta m_{\mathcal{K}}}{m_{\mathcal{K}}} \ll 1$ at low energies. Therefore, local variations $\delta m_{\mathcal{K}}$ are strongly suppressed, thus shielding the Newton constant G_N from local variations of the geometry such as in (12.4.50).

Vacuum energy versus cosmological constant

Consider again the quantum contribution to the vacuum energy (12.4.45)

$$S_{\mathrm{vac}} = -\int_{\mathcal{M}} \Omega \rho^{-2}\left(C_1 m_{\mathcal{K}}^4 + \frac{C_2}{R^8 m_{\mathcal{K}}^4} + \frac{C_3}{R^4}\right), \tag{12.4.58}$$

where

$$\Omega = \rho_M d^4 x = d^4 x \sqrt{|G|}\rho^{-2} \tag{12.4.59}$$

is the symplectic volume form. In orthodox quantum field theory coupled to general relativity, the vacuum energy has the form

$$S_{\mathrm{c.c.}} = \int d^4 x \sqrt{|G|}\Lambda^4, \tag{12.4.60}$$

which contributes to the cosmological constant Λ. This contribution is typically huge in any type of path-integral approach to QFT coupled to gravity, at least of the order of *TeV* (which is the lowest conceivable scale of SUSY breaking), unless some ad-hoc fine-tuning is assumed. In contrast, astronomical observations interpreted in terms of GR imply a bound on the cosmological constant of order *meV*. This clash constitutes the cosmological constant problem, which is one of the most profound open problems in theoretical physics.

The present framework offers a possible solution of this puzzle, since the quantum contributions (12.4.58) to the vacuum energy are not proportional to the Riemannian volume form but to the symplectic volume Ω multiplied with the dilaton ρ^{-2}. Recall that the symplectic volume form is rigid and basically measures the number of states per volume, cf. (4.2.50)

$$\int \Omega = (2\pi)^n \dim \mathcal{H}. \tag{12.4.61}$$

Of course, the vacuum energy will influence the cosmic background to some extent via the dilaton, which remains to be clarified in detail. However, recall that $\rho \sim a(t)$ at late times for the cosmic background under consideration. Therefore, the vacuum energy[6] is not constant as in GR, but decreases with the cosmic expansion; naively comparing (12.4.58) with (12.4.60) leads to

$$\Lambda \sim \rho^{-4} \sim a(t)^{-4} \tag{12.4.62}$$

for the background under consideration. This suggests that vacuum energy does not lead to a cosmological constant problem in the present framework, but its physical role is to stabilize the $\mathcal{M} \times \mathcal{K}$ background and the Kaluza–Klein scale $m_{\mathcal{K}}^2$, as discussed before. Even though this is not in line with the present standard lore for cosmology and gravity, the resulting picture seems consistent and compelling enough to seriously study its physical viability.

---

[6] The vacuum energy is nonvanishing even for the maximally supersymmetric IKKT model, since SUSY is spontaneously broken.

13 Matrix quantum mechanics and the BFSS model

The main focus of this book are matrix models of Yang–Mills type, whose prime representative is the IKKT model. There exists a related type of matrix models, where the matrices are considered as explicitly time-dependent $X^a = X^a(t)$, with $t \in \mathbb{R}$ interpreted as time variable. This leads to *Matrix Quantum Mechanics*. Among all these models, there is again one special model, which is singled out by maximal supersymmetry. That model was introduced by de Wit, Hoppe, and Nicolai [57] in the context of membrane models, and it came to fame 10 years later under the name BFSS model, due to a seminal paper [21] revealing its relation to string or M-theory. We briefly discuss some features of this model and its relation with the IKKT model.

13.1 Matrix quantum mechanics

We start with the familiar description of point particles. As a classical system, a single particle in three dimensions is described in terms of three functions $x^i(t)$ and a Hamiltonian function $H(x^i, p_i)$. As a quantum mechanical system, this is described by hermitian operators $\hat{x}^i$ and $\hat{p}_j$ satisfying the Heisenberg equations of motion. These operators can be realized in the Schrödinger representation as multiplication and derivative operators acting on $\mathcal{H} = L^2(\mathbb{R}^3)$.

Now consider a system of $D \times N^2$ degrees of freedom, which are arranged as D time-dependent complex matrices $X^a = (X^a)^i_j$ subject to the hermiticity constraint

$$X^{a\dagger} = X^a, \qquad \text{i.e.} \quad (X^a)^{i*}_j = (X^a)^j_i. \tag{13.1.1}$$

To specify their classical dynamics, we choose the following quartic Lagrangian

$$L = \text{Tr}\left(\frac{1}{2}\dot{X}^a \dot{X}_a + \frac{1}{4}[X^a, X^b][X_a, X_b]\right) \tag{13.1.2}$$

where $\dot{X}^a = \frac{d}{dt}X^a$. Here, indices are contracted with δ_{ab}, and Tr denotes the trace is over the indices i, j. Then the equations of motion are

$$\ddot{X}^a = -\Box_X X^a, \qquad \Box_X = [X^b, [X_b, .]]. \tag{13.1.3}$$

One may also add a mass term $\frac{m^2}{2}X^a X_a$ if desired. The corresponding canonical momenta are

$$P^a = \dot{X}^a = (P^a)^i_j \tag{13.1.4}$$

subject to the same hermiticity constraint. Hence, classical configurations in such a system are given by hermitian matrices that evolve in time according to the Hamiltonian equation of motion

$$\dot{X}^a(t) = \frac{\partial H}{\partial P_a}, \qquad \dot{P}_a(t) = -\frac{\partial H}{\partial X^a} \qquad (13.1.5)$$

suppressing the matrix indices, with the Hamilton function

$$H = \mathrm{Tr}\Big(\frac{1}{2}P^a P_a - \frac{1}{4}[X^a, X^b][X_a, X_b]\Big). \qquad (13.1.6)$$

For example, any solution of the matrix equations

$$\Box_T T^a = \omega^2 T^a \qquad (13.1.7)$$

(cf. (7.0.10)) provides a time-dependent rotating solution of (13.1.3) via

$$X^a(t) = \cos(\omega t)T^a, \qquad X^{a+3}(t) = \sin(\omega t)T^a. \qquad (13.1.8)$$

One can thus obtain classical solutions given e.g. by rotating fuzzy spheres (3.3.13) in a six-dimensional matrix model, which can be viewed as quantized two-dimensional membranes.

Now consider the corresponding quantum theory. Then the system is described by D hermitian matrix-valued operators

$$(\hat{X}^a)^\dagger = \hat{X}^a, \qquad \text{i.e.} \qquad (\hat{X}^a)_j^{i\,+} = (\hat{X}^a)_i^j \qquad (13.1.9)$$

acting on $\mathcal{H} = L^2(\mathbb{R}^{D \times N^2})$. The dagger † here implies transposition followed by hermitian conjugation $^+$ of the operators. Together with the matrix-valued momentum operators $\hat{P}_b$, they satisfy the canonical commutation relations

$$[(\hat{X}^a)_j^i, (\hat{P}_b)_l^k] = i\delta_b^a \delta_l^i \delta_j^k, \qquad [\hat{X}^a, \hat{X}^b] = 0 = [\hat{P}_a, \hat{P}_b], \qquad (13.1.10)$$

which are realized in the Schrödinger representation via

$$(\hat{X}^a)_i^j |\psi(X)\rangle = (X^a)_i^j |\psi(X)\rangle$$
$$(\hat{P}^a)_j^i |\psi(X)\rangle = -i\frac{\partial}{\partial(X^a)_i^j} |\psi(X)\rangle. \qquad (13.1.11)$$

Note that the matrix indices i, j have nothing to do with the operator nature of $\hat{X}^a$ in matrix quantum mechanics, and each $(\hat{X}^a)_j^i$ is an operator acting on the infinite-dimensional Hilbert space $\mathcal{H}$; this is in contrast to the pure Yang–Mills matrix models considered before. We are interested in the properties of the quantum mechanical system governed by the Hamiltonian

$$\hat{H} = \mathrm{Tr}\Big(\frac{1}{2}\hat{P}^a \hat{P}_a - \frac{1}{4}[\hat{X}^a, \hat{X}^b][\hat{X}_a, \hat{X}_b]\Big). \qquad (13.1.12)$$

In the "deep quantum" regime (typically characterized by small quantum numbers), the properties are expected to be quite distinct from the classical case. However, there should also be a semi-classical regime (typically characterized by large quantum numbers), where the quantum system is close to the classical case. This is realized for states $|\psi(t)\rangle \in \mathcal{H}$ whose expectation values are close to a classical matrix trajectory

$$(X^a)^i_j(t) \approx \langle \psi(t)|(\hat{X}^a)^i_j|\psi(t)\rangle = \langle \psi|(\hat{X}^a)^i_j(t)|\psi\rangle, \tag{13.1.13}$$

using the Heisenberg picture in the second form. Such a semi-classical state could have the form

$$|\psi\rangle \approx \otimes_{i,j,a}|(X^a)^i_j\rangle, \tag{13.1.14}$$

where $|(X^a)^i_j\rangle$ are coherent states localized at $(X^a)^i_j$.

Matrix quantum mechanics can be described alternatively in terms of a Feynman path integral. Since the Hamiltonian is quadratic in the momenta, the time evolution in the Schrödinger or position space representation can then be obtained as

$$\begin{aligned}
|\psi(t_2)\rangle &= U(t_2 - t_1)|\psi(t_1)\rangle \\
&= \frac{1}{Z}\int_{t_1}^{t_2} DX^a(t)\, \exp\left(i\int dt\, \mathrm{Tr}\big(\tfrac{1}{2}\dot{X}^a\dot{X}_a + \tfrac{1}{4}[X^a,X^b][X_a,X_b]\big)\right)\psi(X,t_1)
\end{aligned} \tag{13.1.15}$$

as familiar from quantum mechanics. Correlation functions of $\hat{X}^a(t)$ can be computed similarly.

Gauged matrix model

So far, we have not assigned any physical interpretation to the matrix degrees of freedom. One possible interpretation of $(X^a)^i_j$ in the context of string theory is as a description of (the lightest dof of) open strings stretching between $D0$-branes labeled by i and j. The $D0$-branes should be thought of as point-like objects in target space $\mathbb{R}^D$, which do not carry any intrinsic degrees of freedom. Being identical objects, the description should be invariant under $U(N)$ transformations acting on them. More precisely, we require that the action governing this system should be invariant under $U(N)$ gauge transformations at any time t, with physical Hilbert space $\mathcal{H}_{\mathrm{phys}}$ defined as $U(N)$-invariant sector of $\mathcal{H}$. As familiar from gauge theory, this is achieved by replacing

$$\partial_t \rightarrow D_t = \partial_t - i[A(t),.]. \tag{13.1.16}$$

This leads to the action

$$S = \int dt\, \mathrm{Tr}\Big(\tfrac{1}{2}(D_t X^a)(D_t X_a) + \tfrac{1}{4}[X^a,X^b][X_a,X_b]\Big), \tag{13.1.17}$$

which is invariant under the following time-dependent gauge transformations:

$$\begin{aligned}
X^a(t) &\rightarrow U^{-1}(t)X^a(t)U(t), \\
A(t) &\rightarrow U^{-1}(t)A(t)U(t) + iU^{-1}(t)\partial_t U(t).
\end{aligned} \tag{13.1.18}$$

Moreover, the variation w.r.t. $A(t)$ imposes a "Gauss law" constraint

$$[X^a, D_t X_a] = 0, \tag{13.1.19}$$

which in the quantum theory forces states to be invariant under $SU(N)$.

13.2 The BFSS model and its relation with M-theory

So far, we considered only bosonic matrices. As in the context of time-independent matrix models, there are particular supersymmetric models that are distinguished by mild quantum effects. In purely bosonic models and other models without SUSY, quantum effects lead to a significant attractive interaction between bosonic objects or branes, which arises from the zero point energies of the strings linking the branes. These interactions induced by bosonic and fermionic modes largely cancel in supersymmetric models. This is crucial for the existence of extended and decoupled solutions, which can be interpreted as (stacks of) branes moving in target space.

The BFSS model (named after Banks–Fischer–Shenker–Susskind [21], and known previously as de Wit–Hoppe–Nicolai model [57]) is defined through the following action for $9 + 1$ time-dependent hermitian matrices $X^a = X^a(t)$ and $A = A(t)$ and matrix-valued Majorana spinors $\Psi = \Psi(t)$ of $SO(9)$:

$$
S_{\text{BFSS}} = \frac{1}{g^2} \int dt \text{Tr}\Big(\frac{1}{2}(D_t X^a)(D_t X_a) + \frac{1}{4}[X^a, X^b][X_a, X_b]
$$
$$
- \frac{i}{2}\Psi^T D_t \Psi + \frac{1}{2}\Psi^T \Gamma_a [X^a, \Psi]\Big). \tag{13.2.1}
$$

Here, indices $a, b = 1, \ldots, 9$ are contracted with δ_{ab}, and Γ_a are $SO(9)$ Gamma matrices. This model is obtained by reducing $\mathcal{N} = 1$ super-Yang–Mills theory from $9 + 1$ dimensions to $0 + 1$ dimensions (removing all dependence on space), and $A(t)$ originates from the timelike gauge field. The action is invariant under the time-dependent gauge transformations (13.1.18) extended to fermions as $\Psi(t) \to U^{-1}(t)\Psi(t)U(t)$, and a global nine-dimensional rotational and translational invariance amended by the supersymmetry transformations inherited from the $9 + 1$ dimensional SYM theory. Dropping the fermionic contributions, this action leads to the following equation of motion for the matrices X^a:

$$
D_t^2 X^a + \Box_X X^a = 0. \tag{13.2.2}
$$

These equations admit a rich variety of solutions, and there are different ways of looking at them. In [57], this model was considered as a description for quantized relativistic two-dimensional membranes such as (13.1.8) moving in target space, viewed as a generalization of strings. However, such membranes admit massless deformations called "spikes," much like those on a coronavirus. It is easy to see that such spikes cost negligible energy, even in the quantum theory due to maximal SUSY. This gives rise to a continuous spectrum of excitations of such a membrane, in stark contrast with the case of strings. Therefore, such membranes should not be viewed as analogs of fundamental strings with excitations interpreted as elementary particles; instead, they should be viewed as $2n + 1$-dimensional branes, analogous to the $2n$-dimensional branes in the IKKT model.

A more succinct interpretation was put forward in [21], who proposed that this model describes a system of N point-like time-dependent "Dirichlet particles" (or $D0$ branes) in type IIA string theory [21], which in turn are constituents of the higher-dimensional brane solutions of the model. Moreover, BFSS conjectured that the model should give a complete description of M-theory compactified on a lightlike circle S^1 in the light-front or infinite momentum frame, with N quanta of momentum $p^+ = N/R$ along S^1. M-theory is a hypothetical 11-dimensional theory proposed to understand the dualities among the known superstring theories in 10 dimensions. The $D0$ branes in type IIA string theory can then be interpreted as massive KK graviton modes in 11-dimensional M-theory compactified on S^1; a single graviton with minimal momentum $p^+ = 1/R$ is modeled by a diagonal 1×1 block of the matrices. M-theory is believed to underlie superstring theory, and reduce to IIA string theory in 10 dimensions upon compactification on spacelike S^1. The above-mentioned issue of the continuous spectrum of the BFSS model is then understood naturally in terms of many-particle states of M-theory.

There is a considerable body of evidence supporting this conjecture. In particular, the interaction between two objects in this model can be computed at one loop, in complete analogy of the calculations in the IKKT model in Sections 11.1 and in particular 11.1.7. For backgrounds given by block-diagonal matrices with two blocks, one finds an effective interaction which is the IIA analog of the result in (11.1.65), consistent with linearized supergravity in 11 dimensions compactified on S^1. Again, the cancellations due to maximal supersymmetry are crucial. These computations have been pushed to higher loop levels for simple backgrounds, where the agreement with supergravity is less clear; a detailed review of these results can be found in [198], for some more recent developments see e.g. [202].

Notwithstanding these remarkable insights, the relation with M-theory is not essential for the focus of the present book, which is the weakly coupled regime of the matrix gauge theory on nontrivial classical vacua, i.e. branes playing the role of spacetime. Such branes can also be studied in the BFSS model, but their dimension is $2n + 1$ with an explicit time, which does not include $3 + 1$ dimensional spacetime. We will therefore restrict the discussion of the BFSS model to a few selected aspects in the following, which are of relevance or interest in this context.

Compactification on circles and tori

In the string theory literature, compactifications of the BFFS model on S^1 are usually defined by imposing the following constraints analogous to (11.1.81)

$$X^1 + R_1 \mathbb{1} = U^{-1} X^1 U,$$
$$[U, X^a] = 0, \qquad a \neq 1. \tag{13.2.3}$$

To be specific, we choose to compactify the first matrix X^1. If this is imposed as a constraint including fluctuations, then the BFSS model becomes a $1 + 1$ dimensional matrix-valued field theory on $S^1 \times \mathbb{R}$ following the lines of Section 11.1.8

$$S_{\text{BFSS}} = \frac{1}{g^2} \int dt d\varphi \text{Tr}\Big(\frac{1}{2}(D_t A^1)(D_t A^1) + \sum_{a \neq 1} \frac{1}{2}(D_t X^a)(D_t X_a) - \frac{i}{2}\Psi^T D_t \Psi$$

$$- \frac{1}{2} D_\varphi X^a D_\varphi X_a - \frac{i}{2}\Psi^T \Gamma_1 D_\varphi \Psi + \sum_{a,b \neq 1} \frac{1}{4}[X^a, X^b][X_a, X_b] + \frac{1}{2}\Psi^T \Gamma_a[X^a, \Psi]\Big)$$

$$\tag{13.2.4}$$

where $X^1 = -iR_1\frac{d}{d\varphi} + A^1(\varphi)$ and $D_\varphi = R\partial_\varphi + i[A^1, .] = i[X^1, .]$. In doing so, we have introduced infinitely many dof per volume, and the trace was renormalized by an infinite factor. This can be generalized to compactifications on (non)commutative tori, and stringy features such as T-duality can be recovered, cf. [38, 76, 197].

On the other hand, if we consider (13.2.3) as a particular background without imposing the constraint also for the fluctuations, then the most general matrix-valued fluctuations would contain extra higher-order sectors $(\frac{\partial}{\partial\varphi})^n$. This includes even more degrees of freedom[1] than the constrained version leading to a matrix-valued field theory. Either way, the compactified matrix model no longer provides a nonperturbative framework for a quantum theory, since it involves infinitely many dof per volume. This is in contrast to the fuzzy compact extra dimensions $\mathcal{K}_N$ discussed in Chapters 8 and 12, where the effective dimensions are reduced through spontaneous symmetry breaking on suitable vacua with finitely many dof per volume, thus preserving the nonperturbative nature of matrix models.

13.3 Relation between the BFFS and IKKT models

The BFSS model is clearly related to the IKKT model, as both are obtained by dimensional reduction from the same $9 + 1$ dimensional $\mathcal{N} = 1$ super-Yang–Mills theory to 1 or 0 dimensions, respectively. In particular, reducing the BFSS model to static matrices leads precisely to the IKKT model. At the level of solutions, this can be seen as follows: given a classical solution X^a of the IKKT model, i.e. $\Box_X X^a = 0$, a corresponding static solution of the BFSS model is obtained by

$$A(t) = X^0$$
$$X^a(t) = X^a, \qquad a = 1, \ldots, 9 \tag{13.3.1}$$

where both A and X^a are independent of t. Using a suitable gauge transformation (13.1.18), we can then eliminate $A(t)$ by introducing an explicit time-dependence as follows:

$$A \to U^{-1}(t)AU(t) + iU^{-1}(t)\partial_t U(t) \overset{!}{=} 0. \tag{13.3.2}$$

However, then $X^a \to U^{-1}(t)X^a U(t)$ is time-dependent. This gauge transformation is determined by

$$\partial_t U(t) = iAU(t) \tag{13.3.3}$$

which is solved by

$$U(t) = e^{iAt}. \tag{13.3.4}$$

Therefore, the gauge-transformed matrices

$$X^a(t) = U^{-1}(t)X^a U(t) = e^{-i[A, .]t}X^a \tag{13.3.5}$$

---

[1] The relations (13.2.3) can be interpreted in terms of a fuzzy cylinder, which amounts to a two-dimensional non-compact space for each compactified direction.

are indeed solutions of the BFSS model

$$\frac{d^2}{dt^2}X^a(t) = -[A,[A,X^a(t)]] = -\Box_X X^a(t) \tag{13.3.6}$$

using the IKKT equation of motion. For example, the $1+1$ dimensional quantum plane is a solution of the IKKT model given by $[X^0, X^1] = i\theta\mathbb{1}$. The corresponding solution of the BFFS model is given by $A = X^0$, $X^1(t) = X^1$, which is equivalent to the time-dependent solution

$$X^1(t) = e^{-i[X^0,.]t}X^1 = e^{\theta t}X^1 \tag{13.3.7}$$

with $A = 0$. This is a hermitian matrix evolving exponentially in time. Hence, the time evolution in the BFSS model is generated here by the timelike matrix of the IKKT model acting in the adjoint.

Conversely, one can formally obtain the BFSS model by "compactifying" the IKKT model on a timelike circle, as discussed in Section 11.1.8. However, this does not mean that the models are equivalent, since that compactification involves a nontrivial constraint.

Another interesting relation of these models arises in the high-temperature limit of the thermal BFSS model, which reduces to the bosonic part of the Euclidean IKKT model [117]. Indeed, the (bosonic part of the) thermal BFSS model is defined by

$$S_{\text{BFSS}} = \frac{1}{g^2}\int_0^\beta d\tau \text{Tr}\Big(\frac{1}{2}(D_\tau X^a)(D_\tau X^a) - \frac{1}{4}[X^a, X^b][X^a, X^b]\Big) \tag{13.3.8}$$

imposing periodic and antiperiodic boundary condition in τ for the bosonic and fermionic matrices, respectively. Here $\beta = \frac{1}{T}$ is the inverse temperature. All fields are then Fourier expanded in periodic Euclidean time τ in terms of Matsubara frequencies,

$$X^a(\tau) = \sum_{n\in\mathbb{Z}} X_n^a e^{i\omega_n\tau}, \qquad \omega_n = \frac{2\pi n}{\beta} \tag{13.3.9}$$

and similarly for fermions. In the high-temperature limit $\beta \to 0$, all nontrivial modes are then suppressed and decouple, so that the model reduces to the (bosonic part of the) Euclidean IKKT model; in particular, the fermions decouple due to the antiperiodic boundary conditions, which imply nonvanishing Matsubara frequencies. Based on this relation, it was proposed [39] to use the high-temperature BFFS model as a means to introduce thermal fluctuations in a cosmology based on the IKKT model. In fact, (13.3.8) is obtained by formally compactifying the Euclidean IKKT model on a circle [128]. Pushing this idea further, one might use the real-time thermal BFSS model as a definition of a thermal IKKT model in Minkowski signature, which could then be applied to the cosmological solutions discussed in Section 5.4.

13.4 The BMN model

Besides the BFSS matrix model, suitably modified matrix models have been proposed as description of M-theory on nontrivial target space backgrounds. The most notable example is the BMN model, which is supposed to describe M-theory on a maximally supersymmetric plane-wave background in the large N limit with fixed $p^+ = N/R$ [30]. This model is defined via the following Hamiltonian:

$$H_{\text{BMN}} = R \, \text{Tr} \left(\frac{1}{2} \hat{P}_a \hat{P}^a - \frac{1}{4} [X_a, X_b]^2 - \frac{1}{2} \Psi^{\text{T}} \Gamma^a [X_a, \Psi] \right)$$
$$+ \frac{R}{2} \text{Tr} \left(\sum_{i=1}^{3} \left(\frac{\mu}{3R} \right)^2 X_i^2 + \sum_{a=4}^{9} \left(\frac{\mu}{6R} \right)^2 X_a^2 + i \frac{\mu}{4R} \Psi^{\text{T}} \Gamma^{123} \Psi + i \frac{2\mu}{3R} \epsilon^{ijk} X_i X_j X_k \right),$$

$$(13.4.1)$$

where the first line coincides with the Hamiltonian of the BFSS model. The model is remarkable because it admits the same number of supersymmetries as the BFSS model, despite the nontrivial quadratic and cubic potential in the second line, and therefore enjoys the same benign behavior upon quantization. The special feature is that the model admits stacks of fuzzy two-spheres $S_{N_i}^2$ as classical solutions, much like in the ARS model (7.3.2). These solutions can be interpreted in terms of membrane solutions of M-theory, and they turn out to be supersymmetric, which means that they are protected from quantum corrections. This in turn allows to extrapolate some of their physical properties to the strong coupling regime, and make contact with further features of M-theory such as five-branes. However, these aspects are beyond the scope of this book, and for a detailed discussion we refer to the literature [136, 199].

13.5 Holographic aspects and deep quantum regime

There is a large body of literature on a seemingly similar but in fact very different approach to quantum gravity using matrix models, based on holographic ideas. This should be seen in the context of the AdS/CFT correspondence, where a (conformal) field theory on the boundary of anti-de Sitter space is viewed as dual to an effective gravity theory in the bulk. We have encountered the basic mechanism in Section 11.1.7, where a background given by noncommutative branes in the IKKT matrix model was shown to give rise to interactions consistent with IIB supergravity in the bulk, via quantum effects. More specifically, a stack of coincident $3 + 1$ dimensional branes was seen to generate an effective metric in target space which describes a "black brane," and reduces to the $AdS^5 \times S^5$ metric in the near horizon limit. Analogous statements clearly apply to lower-dimensional brane backgrounds, while higher-dimensional branes are typically unstable at weak coupling.

As pointed out before, that holographic $9+1$ dimensional bulk geometry is most relevant in the strong coupling regime of the underlying matrix model for large t'Hooft coupling $\lambda = g^2 N$, where it can be viewed as a useful dual description of the gauge theory defined by the matrix model. This is the deep quantum regime of the matrix model, where quantum fluctuations of the matrices play the dominant role, much like in random matrix theory as discussed in Section 7.2.1.

However, that dual, holographic description should not be taken as physical in the weakly coupled regime of the matrix model around some background brane, where the fluctuations are governed by a weakly coupled gauge theory. Then there are no propagating modes in the bulk, as emphasized before. Moreover, the $9 + 1$ dimensional holographic bulk geometry would need to be compactified to $3 + 1$ dimensions to be related to real physics, which would entail the issues discussed in Sections 13.2 and 11.1.8, including the arbitrariness issue in string theory known as the landscape. Nevertheless, the holographic point of view is very rich and interesting. Since this is not the subject of the present book, we shall only mention some pertinent results in the next paragraph, and refer to the literature [10, 44, 71, 93, 117] for more details.

For example, it was found in numerical studies that some observables of the BFSS model at low temperature such as the internal energy agree with predictions based on a dual holographic geometry describing a non-extremal $9 + 1$ dimensional black hole in target space. In particular, the area of the horizons of these black holes was observed to correspond to the entropy of the corresponding thermal state in the BFSS model. This suggests in particular that the Bekenstein–Hawking entropy of the black hole has a microscopic origin in terms of the open strings attached to the constituent N $D0$-branes. At high temperatures, the observed results appear to be consistent with the classical equipartition of N^2 deconfined gluonic states, recalling that the fermions disappear in this limit. Matrix quantum mechanics provides a powerful framework to understand and test such statements in the nonperturbative regime.

Appendix A Gaussian integrals over matrix spaces

We recall the following well-known Gaussian integral formulas:

$$\int_{\mathbb{R}^N} dx \, e^{-x^T A x} = \frac{\pi^{N/2}}{\sqrt{\det A}} = \pi^{N/2} \exp\left(-\frac{1}{2}\text{Tr}(\ln A)\right)$$

$$\int_{\mathbb{R}^N} dx \, e^{-x^T A x + Jx} = \frac{\pi^{N/2}}{\sqrt{\det A}} e^{\frac{1}{4}J^T A^{-1} J}$$

$$\int_{\mathbb{C}^N} dz d\bar{z} \, e^{-z^\dagger B z} = \frac{\pi^N}{\det B} = \pi^N \exp\left(-\text{Tr}(\ln B)\right)$$

$$\int_{\mathbb{C}^N} dz d\bar{z} \, e^{-z^\dagger B z + J^\dagger \phi + \phi^\dagger J} = \frac{\pi^N}{\det B} e^{J^T B^{-1} J}, \tag{A.0.1}$$

where A is an orthogonal and B a hermitian $N \times N$ matrix, and $dz d\bar{z}$ indicates the standard measure on $\mathbb{C}^N \cong \mathbb{R}^{2N}$. In particular, the integral over hermitian matrices $\phi = \phi^\dagger$ is obtained as

$$\int d\phi \, e^{-\text{tr}(\phi^\dagger \mathcal{K} \phi)} = c \, \exp\left(-\frac{1}{2}\text{Tr}(\ln \mathcal{K})\right) \tag{A.0.2}$$

where

$$\mathcal{K}: \quad \text{End}(\mathbb{C}^N) \to \text{End}(\mathbb{C}^N) \tag{A.0.3}$$

is a positive definite hermitian operator acting on the space $\text{End}(\mathbb{C}^N) \cong \mathbb{R}^{2N^2}$ of complex $N \times N$ matrices equipped with the standard inner product

$$\langle \phi, \psi \rangle = \text{tr}(\phi^\dagger \psi). \tag{A.0.4}$$

Then $\mathcal{K}$ can be diagonalized on $\text{End}(\mathbb{C}^N)$ with positive real eigenvalues. A typical example for $\mathcal{K}$ is the Matrix Laplacian $\square$. The factor $\frac{1}{2}$ in (A.0.2) arises because the integral is over hermitian rather than all complex matrices. This generalizes straightforwardly to general complex matrices ϕ and source terms J.

Appendix B Some $SO(D)$ group theory

B.1 Some properties of $\mathfrak{so}(6)$ representations

The Lie algebra $\mathfrak{so}(6)$ is defined by

$$[M_{ab}, M_{cd}] = i\left(\delta_{ac}M_{bd} - \delta_{ad}M_{bc} + \delta_{bd}M_{ac} - \delta_{bc}M_{ad}\right), \tag{B.1.1}$$

for $a, b, \ldots = 1, \ldots, 6$. We label the irreducible representations ("irreps") V_Λ of $\mathfrak{so}(6)$ by their highest weights with Dynkin labels $\Lambda = (m_1, m_2, m_3)$. Weyl's dimension formula then states that their dimension is given by

$$\dim(V_\Lambda) = \frac{1}{12}(m_1 + 1)(m_2 + 1)(m_3 + 1)(m_1 + m_2 + 2)(m_2 + m_3 + 2)$$
$$\cdot (m_1 + m_2 + m_3 + 3). \tag{B.1.2}$$

For example,

$$\dim(V_{(m,0,m)}) = \frac{1}{12}(m + 1)^2(m + 2)^2(2m + 3)$$
$$\dim(V_{(0,m,0)}) = \frac{1}{12}(m + 1)(m + 2)^2(m + 3)$$
$$\dim(V_{(m,0,0)}) = \frac{1}{6}(m + 1)(m + 2)(m + 3). \tag{B.1.3}$$

In particular, $\dim(V_{(1,0,0)}) = 4$, $\dim(V_{(0,1,0)}) = 6$, and $\dim(V_{(1,0,1)}) = 15$. We can thus identify $(1, 0, 0) \equiv (4)$ as fundamental (spinorial) representation, $(0, 1, 0) \equiv (6)$ as vector representation, and $(1, 0, 1) \equiv (15)$ as adjoint representation.

The quadratic Casimir of $\mathfrak{so}(6)$ on V_Λ can be computed as follows:

$$C^2[\mathfrak{so}(6)] = \sum_{a<b\leq 6} M_{ab}M_{ab} = \langle \Lambda, \Lambda + 2\rho \rangle = m_i \langle \Lambda_i, \Lambda_j \rangle (m_j + 2), \tag{B.1.4}$$

where Λ_i are the fundamental weights, $\rho = \sum_i \Lambda_i$, and

$$\langle \Lambda_i, \Lambda_j \rangle = \frac{1}{4}\begin{pmatrix} 3 & 2 & 1 \\ 2 & 4 & 2 \\ 1 & 2 & 3 \end{pmatrix} \tag{B.1.5}$$

is the inner product matrix of $\mathfrak{so}(6)$. For the $(0, 1, 0)$ vector representation, this gives

$$C^2[\mathfrak{so}(6)]\big|_{(0,1,0)} = \frac{1}{4}(0, 1, 0) \begin{pmatrix} 3 & 2 & 1 \\ 2 & 4 & 2 \\ 1 & 2 & 3 \end{pmatrix} \begin{pmatrix} 2 \\ 3 \\ 2 \end{pmatrix} = 5 \tag{B.1.6}$$

and for $(n, 0, n)$ we obtain

$$C^2[\mathfrak{so}(6)]\big|_{(n,0,n)} = \frac{1}{4}(n, 0, n) \begin{pmatrix} 3 & 2 & 1 \\ 2 & 4 & 2 \\ 1 & 2 & 3 \end{pmatrix} \begin{pmatrix} n+2 \\ 2 \\ n+2 \end{pmatrix} = 2n(n+3). \tag{B.1.7}$$

B.2 Basic relations for $\mathfrak{so}(5) \subset \mathfrak{so}(6)$

Consider the subalgebra $\mathfrak{so}(5) \subset \mathfrak{so}(6)$ defined by restricting the indices in (B.1.1) to $a, b = 1, \ldots, 5$. Weyl's dimension formula states that the dimensions of the highest weight irrep V_Λ with Dynkin labels $\Lambda = (m_1, m_2)$ is given by

$$\dim(V_\Lambda) = \frac{1}{6}(m_1 + 1)(m_2 + 1)(m_1 + m_2 + 2)(2m_1 + m_2 + 3), \tag{B.2.1}$$

where α_1 is the long root and α_2 is the short root of $\mathfrak{so}(5)$. The quadratic Casimir of V_Λ with $\Lambda = (n - s, 2s)$ is found to be

$$C^2[\mathfrak{so}(5)]\big|_{(n-s,2s)} = \frac{1}{2}(n - s, 2s) \begin{pmatrix} 2 & 1 \\ 1 & 1 \end{pmatrix} \begin{pmatrix} n-s+2 \\ 2s+2 \end{pmatrix} = n(n+3) + s(s+1). \tag{B.2.2}$$

Gamma matrices

The Clifford algebra associated to this $\mathfrak{so}(5)$ is generated by gamma matrices γ_a, $a = 1, \ldots, 5$, which satisfy

$$\{\gamma_a, \gamma_b\}_+ = 2\delta_{ab}. \tag{B.2.3}$$

A useful explicit realization is given by

$$\gamma_m = i \begin{pmatrix} 0 & -\sigma_m \\ \sigma_m & 0 \end{pmatrix} \qquad \gamma_4 = \begin{pmatrix} 0 & \mathbb{1}_2 \\ \mathbb{1}_2 & 0 \end{pmatrix}, \qquad \gamma_5 = \begin{pmatrix} \mathbb{1}_2 & 0 \\ 0 & -\mathbb{1}_2 \end{pmatrix} \tag{B.2.4}$$

for $m = 1, 2, 3$. Then the

$$\Sigma^{ab} = \frac{1}{4i}[\gamma^a, \gamma^b], \qquad a, b = 1, \ldots, 5 \tag{B.2.5}$$

provide the spinorial representation (4) of $\mathfrak{so}(5)$ on $\mathbb{C}^4$. This extends to the representation $(4) \cong (1, 0, 0)$ of $\mathfrak{so}(6)$ in terms of the generators Σ^{ab}, $a, b = 1, \ldots, 6$ given by

$$\Sigma^{ab} = \frac{1}{4i}[\gamma^a, \gamma^b], \qquad \Sigma^{a6} = \frac{1}{2}\gamma^a \tag{B.2.6}$$

for $a, b = 1, \ldots, 5$. We also note the following useful identity

$$\sum_{a \leq 5} \left(\gamma_a \otimes \gamma^a \right)_S = \mathbb{1} \tag{B.2.7}$$

or more explicitly

$$(\gamma_a)^\alpha{}_\beta (\gamma^a)^\delta{}_\kappa + (\beta \leftrightarrow \kappa) = \delta^\alpha{}_\beta \delta^\delta{}_\kappa + (\beta \leftrightarrow \kappa). \tag{B.2.8}$$

This provides the projector onto the symmetric part of the $\mathfrak{so}(5)$ representation

$$(4) \otimes (4) = (10)_S \oplus (5)_{AS} \oplus (1)_{AS}. \tag{B.2.9}$$

In the more informative highest weight notation using Dynkin labels, this reads as

$$(0, 1) \otimes (0, 1) = (0, 2)_S \oplus (1, 0)_{AS} \oplus (0, 0)_{AS}. \tag{B.2.10}$$

Note that as $\mathfrak{so}(6)$ representation, the antisymmetric part of the tensor product

$$\left((4) \times (4) \right)_{AS} = (6) \tag{B.2.11}$$

is irreducible, but it decomposes into $(5) \oplus (1)$ under $\mathfrak{so}(5)$. The projector on the trivial mode in (B.2.9) encodes an invariant antisymmetric tensor of $\mathfrak{so}(5)$ with spinorial indices:

$$\kappa^{\alpha\beta} = -\kappa^{\beta\alpha} \tag{B.2.12}$$

which can be used to raise and lower spinorial indices α. Decomposed under $\mathfrak{su}(2)_L \times \mathfrak{su}(2)_R \subset \mathfrak{so}(5)$, $\kappa^{\alpha\beta}$ is simply the sum of the two invariant ε tensors.[1] This implies in particular that

$$(\gamma_a)^{\alpha\beta} = -(\gamma_a)^{\beta\alpha} \tag{B.2.13}$$

is antisymmetric. As a consequence, the spinorial representation of a vector V^a is antisymmetric,

$$V^{\alpha\beta} := V^a (\gamma_a)^{\alpha\beta} = -V^{\beta\alpha}. \tag{B.2.14}$$

Similarly, the spinorial representation of an antisymmetric rank 2 tensor $\theta^{ab} = -\theta^{ba}$ has the form

$$\theta^{\alpha\alpha';\beta\beta'} := \theta^{ab} (\gamma_a)^{\alpha\beta} (\gamma_b)^{\alpha'\beta'} = \kappa^{\alpha\alpha'} \theta^{\beta\beta'} + \theta^{\alpha\alpha'} \kappa^{\beta\beta'} = -\theta^{\beta\beta';\alpha\alpha'} \tag{B.2.15}$$

Here $\theta^{\alpha\alpha'} = \theta^{\alpha'\alpha}$ is symmetric and can be identified with $(0, 2)$.

B.3 Basic relations for $\mathfrak{so}(4, 2)$ and $\mathfrak{su}(2, 2)$

Non-compact versions of $\mathfrak{so}(D)$ are obtained by multiplying some generators with i. In particular, the Lie algebra $\mathfrak{so}(4, 2)$ is defined by

$$[M_{ab}, M_{cd}] = i \left(\eta_{ac} M_{bd} - \eta_{ad} M_{bc} + \eta_{bd} M_{ac} - \eta_{bc} M_{ad} \right), \tag{B.3.1}$$

---

[1] There is no such invariant tensor for $\mathfrak{so}(6)$, because then (4) is not equivalent to its dual $(\bar{4})$.

where $\eta_{ab} = \mathrm{diag}(-1,1,1,1,1,-1)$ with $a,b,\ldots = 0,\ldots,5$. Unitary representations of $SO(4,2)$ arise for Hermitian generators $M_{ab} = (M_{ab})^\dagger$ acting on some Hilbert space $\mathcal{H}$. The maximal compact subgroup of $SO(4,2)$ is $SU(2)_L \times SU(2)_R \times U(1)_E$, generated by the following generators:

$$L_m = \frac{1}{2}\left(\frac{1}{2}\varepsilon_{mnl}M_{nl} + M_{m4}\right) \qquad \text{generating} \quad SU(2)_L$$

$$R_m = \frac{1}{2}\left(\frac{1}{2}\varepsilon_{mnl}M_{nl} - M_{m4}\right) \qquad \text{generating} \quad SU(2)_R$$

with $m,n,l = 1,2,3$, and the $U(1)_E$ generator $E = M_{05}$. They satisfy

$$\begin{aligned}
[L_m, L_n] &= i\varepsilon_{mnl}L_l, \\
[R_m, R_n] &= i\varepsilon_{mnl}R_l, \\
[L_m, R_n] &= [E, L_n] = [E, R_n] = 0.
\end{aligned} \tag{B.3.2}$$

E can be identified as conformal Hamiltonian, whose spectrum is positive in a positive energy representation. Denoting the corresponding maximal compact Lie subalgebra as $\mathcal{L}^0 = \mathfrak{su}(2)_L \times \mathfrak{su}(2)_R \times \mathfrak{u}(1)_E$, the conformal algebra $\mathfrak{g}$ has the graded decomposition

$$\mathfrak{g} = \mathcal{L}^+ \oplus \mathcal{L}^0 \oplus \mathcal{L}^- \tag{B.3.3}$$

with respect to E where $\mathcal{L}^\pm$ are the non-compact generators, such that

$$[\mathcal{L}^0, \mathcal{L}^\pm] = \mathcal{L}^\pm, \qquad [E, \mathcal{L}^\pm] = \pm\mathcal{L}^\pm. \tag{B.3.4}$$

The six roots of $\mathfrak{so}(6)_\mathbb{C}$ decompose accordingly into two compact roots $X_{\beta_i}^\pm$ and four non-compact roots $X_{\pm\alpha_{ij}}$. The latter transform as $(2)_L \otimes (2)_R$ i.e. as complex vectors of $SO(4)$, and satisfy $(X_{\pm\alpha_{ij}})^\dagger = -X_{\mp\alpha_{ij}}$ in a unitary representation, whereas $(X_{\pm\beta_i})^\dagger = X_{\mp\beta_i}$.

Gamma matrices

We now adopt the following conventions for the Gamma matrices of $\mathfrak{so}(4,1) \subset \mathfrak{so}(4,2)$

$$\{\gamma_a, \gamma_b\} = -2\eta_{ab}, \qquad a,b = 0,\ldots,4 \tag{B.3.5}$$

where $\eta_{ab} = \mathrm{diag}(-1,1,1,1,1)$, with

$$\gamma_a\gamma^a = -5\mathbb{1} \tag{B.3.6}$$

and

$$\gamma_0 = \begin{pmatrix} \mathbb{1}_2 & 0 \\ 0 & -\mathbb{1}_2 \end{pmatrix} \qquad \gamma_m = \begin{pmatrix} 0 & -\sigma_m \\ \sigma_m & 0 \end{pmatrix} \qquad \gamma_4 = i\begin{pmatrix} 0 & \mathbb{1}_2 \\ \mathbb{1}_2 & 0 \end{pmatrix}. \tag{B.3.7}$$

Here σ_m, $m = 1,2,3$ are the Pauli matrices, and $\gamma_4 := \gamma_0\gamma_1\gamma_2\gamma_3$. Even though this seems at odds with (B.2.3), this convention leads to a transparent form for γ_0. They satisfy the reality conditions

$$\begin{aligned}
\gamma_a^\dagger &= -\gamma_b\eta^{ba} = \gamma_0\gamma_a\gamma_0^{-1}, & a,b &= 0,1,2,3,4 \\
\Sigma_{ab}^\dagger &= \Sigma_{a'b'}\eta^{aa'}\eta^{bb'} = \gamma_0\Sigma_{ab}\gamma_0^{-1}, & a,b &= 0,1,2,3,4,5.
\end{aligned} \tag{B.3.8}$$

Then generators of $\mathfrak{so}(4,2)$ are given explicitly by

$$\Sigma_{\mu\nu} := \frac{1}{4i}\left[\gamma_\mu, \gamma_\nu\right] \qquad \Sigma_{\mu 4} := -\frac{i}{2}\gamma_\mu\gamma_4 \qquad \Sigma_{\mu 5} := \frac{1}{2}\gamma_\mu \qquad \Sigma_{45} := \frac{1}{2}\gamma_4 \qquad \text{(B.3.9)}$$

for $\mu, \nu = 0, 1, 2, 3$. They generate the universal covering group $SU(2,2)$ of $SO(4,2)$, which is the group of 4×4 complex matrices satisfying

$$U^{-1} = \gamma_0 U^\dagger \gamma_0^{-1} \qquad \text{(B.3.10)}$$

and respecting the following indefinite sesquilinear form on $\mathbb{C}^4$

$$\bar{\psi}_1\psi_2 = \psi_1^\dagger \gamma^0 \psi_2. \qquad \text{(B.3.11)}$$

The 15-dimensional Lie algebra $\mathfrak{su}(2,2) = \mathfrak{so}(4,2)$ can thus be identified with the space of traceless complex 4×4 matrices $Z = Z^\alpha_\beta$ with real structure

$$Z^\dagger = \gamma_0 Z \gamma_0^{-1}. \qquad \text{(B.3.12)}$$

Some useful identities

We can evaluate the $SO(4,1)$ intertwiner

$$\sum_{a,b\leq 4} \Sigma_{ab} \otimes \Sigma^{ab} = \mathcal{C}^2[\mathfrak{so}(4,1)]_{(4)\otimes(4)} - 2\mathcal{C}^2[\mathfrak{so}(4,1)]_{(4)} \qquad \text{(B.3.13)}$$

acting on

$$(4) \otimes (4) = \left((10)_S \oplus (6)_{AS}\right)_{\mathfrak{so}(4,2)} = \left((10)_S \oplus (5)_{AS} \oplus (1)_{AS}\right)_{\mathfrak{so}(4,1)}. \qquad \text{(B.3.14)}$$

Using the well-known eigenvalues of the quadratic Casimirs (which coincide with those of the compact group), it follows that

$$\left(\sum_{a,b\leq 4} \Sigma_{ab} \otimes \Sigma^{ab}\right)_S = \mathbb{1}, \qquad \text{(B.3.15)}$$

$$\left(\sum_{a,b\leq 5} \Sigma_{ab} \otimes \Sigma^{ab}\right)_S = \frac{3}{2}\mathbb{1}. \qquad \text{(B.3.16)}$$

This implies

$$\sum_{a\leq 5} \left(\gamma_a \otimes \gamma^a\right)_S = -\mathbb{1} \qquad \text{and} \qquad \sum_{a,b\leq 5} \Sigma_{ab}\Sigma^{ab} = 5. \qquad \text{(B.3.17)}$$

Similarly, there is an $\mathfrak{so}(4,2)$ identity

$$\eta_{cc'}\left(\Sigma^{ac} \otimes \Sigma^{bc'} + \Sigma^{bc} \otimes \Sigma^{ac'}\right)_S = \frac{1}{2}\eta_{ab}. \qquad \text{(B.3.18)}$$

To see this, note that both sides are symmetric, and therefore act on $(0,0,1)^{\otimes s2} = (0,0,2)$; there is no non-trivial generator $\mathcal{S}^{ab}$ on the rhs, since the resulting symmetric tensor operator would have to be in $(0,1,0)^{\otimes s2} = (0,2,0) + (0,0,0)$, but $(0,2,0) \notin \mathrm{End}(0,0,2)$, thus only η_{ab} can occur. We also note the following $\mathfrak{so}(4,2)$ identities:

$$\left(\epsilon_{abcdef}\Sigma^{ab} \otimes \Sigma^{cd}\right)_S = 2(\Sigma_{ef} \otimes \mathbb{1} + \mathbb{1} \otimes \Sigma_{ef}) \qquad \text{(B.3.19)}$$

and

$$\{\Sigma^{ab}, \Sigma^{cd}\}_+ = (\eta^{ac}\eta^{bd} - \eta^{ad}\eta^{bc})\mathbb{1} + \frac{1}{2}\epsilon^{abcdef}\Sigma^{ef}. \tag{B.3.20}$$

In particular,

$$\epsilon^{abcdef}\Sigma^{ab}\Sigma^{cd} = 12\Sigma^{ef}. \tag{B.3.21}$$

B.4 Trace relations for $SO(D)$

The one-loop effective action of the IKKT model and similar matrix models involves traces of $SO(D)$ generators of the form $T_{A...C} := \mathrm{Tr}\Sigma_A \ldots \Sigma_C$, where $A \equiv (ab)$ for $a < b$ labels the generators $\Sigma_A^{(Y)}$ or $\Sigma_A^{(\psi)}$ of $SO(D)$ in the vector or spinor representation, respectively. The resulting $T_{A...C}$ are invariant tensors of $SO(D)$, and we assume that D is even for simplicity. The generators are clearly traceless,

$$\mathrm{Tr}\Sigma_A = \mathrm{Tr}\Sigma_{ab} = 0. \tag{B.4.1}$$

For the higher invariants some more work is needed:

Quadratic invariants

Consider first the quadratic invariant

$$\mathrm{Tr}\Sigma_A\Sigma_B = \frac{g_{AB}}{(g_{CD}g^{CD})}\mathrm{Tr}\,C^2 = \frac{2g_{AB}}{D(D-1)}\mathrm{Tr}\,C^2, \tag{B.4.2}$$

where $C^2 = g^{AB}\Sigma_A\Sigma_B$ is the quadratic Casimir and $g^{AB} = \delta^{AB}$ is the Killing metric. This can be obtained using standard results from $SO(D)$ group theory (cf. (B.1.4))

$$C^2\big|_{\mathrm{spinor}} = g^{AB}\Sigma_A^{(\psi)}\Sigma_B^{(\psi)} = \frac{1}{2}\Sigma_{ab}^{(\psi)}\Sigma_{ab}^{(\psi)} = \frac{1}{8}(D^2 - D)\mathbb{1},$$

$$C^2\big|_{\mathrm{vector}} = g^{AB}\Sigma_A^{(Y)}\Sigma_B^{(Y)} = \frac{1}{2}\Sigma_{ab}^{(Y)}\Sigma_{ab}^{(Y)} = (D-1)\mathbb{1}, \tag{B.4.3}$$

since $A = (a, b)$; $a < b$ is a basis of $SO(D)$; summation over ab is understood. Hence,

$$\frac{1}{2}\mathrm{Tr}\Sigma_{ab}^{(\psi)}\Sigma_{ab}^{(\psi)} = \frac{1}{8}(D^2 - D)\mathrm{Tr}_\psi\mathbb{1} = 2^{D/2-4}D(D-1),$$

$$\frac{1}{2}\mathrm{Tr}_Y\Sigma_{ab}^{(Y)}\Sigma_{ab}^{(Y)} = (D-1)\mathrm{Tr}_Y\mathbb{1} = D(D-1) \tag{B.4.4}$$

using the fact that chiral spinors of $SO(D)$ have dimension $2^{D/2-1}$. In particular, for $SO(10)$ we get

$$\frac{1}{2}\mathrm{Tr}\Sigma_{ab}^{(\psi)}\Sigma_{ab}^{(\psi)} = 180, \qquad\qquad \frac{1}{2}\mathrm{Tr}_Y\Sigma_{ab}^{(Y)}\Sigma_{ab}^{(Y)} = 90. \tag{B.4.5}$$

This implies that the $\mathcal{O}(\mathcal{F}^2)$ contributions to the one-loop effective action in (11.1.28) cancel, which reflects the special structure due to supersymmetry. For $SO(6)$, we would get

$$\frac{1}{2}\mathrm{Tr}\Sigma_{ab}^{(\psi)}\Sigma_{ab}^{(\psi)} = 15, \qquad\qquad \frac{1}{2}\mathrm{Tr}_Y\Sigma_{ab}^{(Y)}\Sigma_{ab}^{(Y)} = 30. \qquad (B.4.6)$$

Moreover, in a $D = 6$ Yang–Mills matrix model there would be no factor $\frac{1}{2}$ in front of the fermionic contribution in (11.1.28), because there is no Majorana condition. Therefore, the $\mathcal{O}(\mathcal{F}^2)$ contributions would not cancel, so that the model would not be one-loop finite on four-dimensional branes and display UV/IR divergences. For $SO(8)$, we get

$$\frac{1}{2}\mathrm{Tr}\Sigma_{ab}^{(\psi)}\Sigma_{ab}^{(\psi)} = 56, \qquad\qquad \frac{1}{2}\mathrm{Tr}_Y\Sigma_{ab}^{(Y)}\Sigma_{ab}^{(Y)} = 56. \qquad (B.4.7)$$

There is again no Majorana condition so that the $\mathcal{O}(\mathcal{F}^2)$ contributions would cancel; however, that model isn't even supersymmetric and already the $\mathrm{Tr}\log\square$ term would diverge.

Cubic invariants

The only cubic invariant tensor of $SO(D)$ is the structure constant[2] f_{ABC}. Thus,

$$\mathrm{Tr}\Sigma_A\Sigma_B\Sigma_C = \frac{i}{2}\frac{f_{ABC}}{g^{DE}g_{DE}}\mathrm{Tr}\,C^2 = i\frac{f_{ABC}}{D(D-1)}\mathrm{Tr}\,C^2 \qquad (B.4.8)$$

since $\mathrm{Tr}[\Sigma_A, \Sigma_B]\Sigma_C = i\frac{f_{ABC}}{g^{A'B'}g_{A'B'}}\mathrm{Tr}\,C^2$. This gives

$$\mathrm{Tr}\Sigma_A\Sigma_B\Sigma_C = if_{ABC}\frac{D}{D(D-1)}(D-1) = if_{ABC}, \qquad\qquad \text{vector}$$

$$\mathrm{Tr}\Sigma_A\Sigma_B\Sigma_C = if_{ABC}\frac{2^{D/2-1}}{D(D-1)}\frac{1}{8}D(D-1) = i2^{D/2-4}f_{ABC}, \qquad \text{spinor} \qquad (B.4.9)$$

which differ by a factor 2 in the $SO(10)$ case,

$$\mathrm{Tr}\Sigma_A\Sigma_B\Sigma_C = if_{ABC}, \qquad \text{vector}$$

$$\mathrm{Tr}\Sigma_A\Sigma_B\Sigma_C = 2if_{ABC}, \qquad \text{spinor}. \qquad (B.4.10)$$

This implies that these cubic terms cancel in the IKKT model (11.1.28).

Quartic invariants

The quartic terms no longer cancel for $SO(10)$ or $SO(9, 1)$, and we need to work them out explicitly. For the vector representation, one obtains

$$\Theta^{a_1b_1}\ldots\Theta^{a_4b_4}\mathrm{Tr}\left(\Sigma_{a_1b_1}^{(Y)}\ldots\Sigma_{a_4b_4}^{(Y)}\right) = 16\Theta^{a_1b_1}\ldots\Theta^{a_4b_4}g_{b_1a_2}g_{b_2a_3}g_{b_3a_4}g_{b_4a_1}$$

$$= 16\Theta^{ab}\Theta^{bc}\Theta^{cd}\Theta^{da}, \qquad (B.4.11)$$

---

[2] For $SO(6)$ there exists also a symmetric invariant tensor d_{ABC} that would arise in the spinorial representation; we shall ignore this case here.

where $g_{ab} = \delta_{ab}$ or η_{ab}, according to the signature. To compute the spinorial traces, we can use the standard contraction formulas for Dirac spinors

$$\text{Tr}\left(\gamma_{a_1} \cdots \gamma_{a_{2n}}\right) = \sum_{\text{contractions}\,\mathcal{C}} (-1)^{\mathcal{C}} g_{a_i a_j} \cdots g_{a_k a_l} \text{Tr}\mathbb{1},$$

$$\text{Tr}\left(\gamma_{a_1} \cdots \gamma_{a_{2n}} \gamma\right) = 0, \qquad 2n < 10, \tag{B.4.12}$$

where $\gamma = i\gamma_1 \cdots \gamma_{10}$ in the Euclidean case, so that[3]

$$\text{Tr}\left(\gamma_{a_1} \cdots \gamma_{a_{2n}} \left(\frac{1+\gamma}{2}\right)\right) = 16 \sum_{\text{contractions}\,\mathcal{C}} (-1)^{\mathcal{C}} g_{a_i a_j} \cdots g_{a_k a_l}, \quad 2n < 10.$$

For the 16-dimensional chiral spinor representations of $SO(10)$ or $SO(9,1)$, this gives

$$\Theta^{a_1 b_1} \cdots \Theta^{a_4 b_4} \text{Tr}\left(\Sigma^{(\psi)}_{a_1 b_1} \cdots \Sigma^{(\psi)}_{a_4 b_4}\right)$$

$$= 4\Theta^{a_1 b_1} \cdots \Theta^{a_4 b_4} \Big(4 g_{b_1 a_2} g_{b_2 a_3} g_{b_3 a_4} g_{b_4 a_1} - 4 g_{b_1 a_2} g_{b_2 a_4} g_{b_4 a_3} g_{b_3 a_1} - 4 g_{b_1 a_3} g_{b_3 a_2} g_{b_2 a_4} g_{b_4 a_1}$$

$$+ g_{b_1 a_2} g_{b_2 a_1} g_{b_3 a_4} g_{b_4 a_3} + g_{b_1 a_3} g_{b_3 a_1} g_{b_2 a_4} g_{b_4 a_2} + g_{b_1 a_4} g_{b_4 a_1} g_{b_2 a_3} g_{b_3 a_2}\Big). \tag{B.4.13}$$

The first three terms are connected contractions while the last three are disconnected contractions. Hence

$$\Theta^{a_1 b_1} \cdots \Theta^{a_4 b_4} \text{Tr}\left(\frac{1}{2}\Sigma^{(\psi)}_{a_1 b_1} \cdots \Sigma^{(\psi)}_{a_4 b_4} - \Sigma^{(Y)}_{a_1 b_1} \cdots \Sigma^{(Y)}_{a_4 b_4}\right)$$

$$= 2\Theta^{a_1 b_1} \cdots \Theta^{a_4 b_4}\Big(-4 g_{b_1 a_2} g_{b_2 a_3} g_{b_3 a_4} g_{b_4 a_1} - 4 g_{b_1 a_2} g_{b_2 a_4} g_{b_4 a_3} g_{b_3 a_1} - 4 g_{b_1 a_3} g_{b_3 a_2} g_{b_2 a_4} g_{b_4 a_1}$$

$$+ g_{b_1 a_2} g_{b_2 a_1} g_{b_3 a_4} g_{b_4 a_3} + g_{b_1 a_3} g_{b_3 a_1} g_{b_2 a_4} g_{b_4 a_2} + g_{b_1 a_4} g_{b_4 a_1} g_{b_2 a_3} g_{b_3 a_2}\Big)$$

$$= 6\left(-4(\Theta g \Theta g \Theta g \Theta g) + (\Theta g \Theta g)^2\right), \tag{B.4.14}$$

for the IKKT model (11.1.28).

---

[3] For 10 insertions, the γ gives an imaginary contribution to the effective action, which corresponds to a Wess–Zumino term.

Appendix C **Torsion identities**

We need the following contraction formulas for the contorsion (cf. (7.54) and (7.55) in [187])

$$K^{\sigma\rho\mu}K_{\rho\sigma\nu} = \frac{1}{4}\left(2T^{\mu\sigma\rho}(T_{\nu\rho\sigma} + T_{\nu\sigma\rho}) - T^{\rho\sigma\mu}T_{\rho\sigma\nu}\right) \tag{C.0.1}$$

$$K_{\sigma}{}^{\rho\mu}K_{\rho}{}^{\sigma}{}_{\mu} = \frac{1}{4}T^{\mu\sigma\rho}(2T_{\mu\rho\sigma} + T_{\mu\sigma\rho}). \tag{C.0.2}$$

The totally antisymmetric torsion is defined by (9.3.14)

$$T^{(AS)\nu}{}_{\rho\mu} = T^{\nu}{}_{\rho\mu} + T_{\mu}{}^{\nu}{}_{\rho} + T_{\rho\mu}{}^{\nu}. \tag{C.0.3}$$

Then $T^{(AS)}{}_{\nu\rho\mu}$ can naturally be interpreted as a three-form $T^{(AS)}$, which is related by the Hodge star $\star$ corresponding to $G_{\mu\nu}$ to a one-form $\tilde{T}_\sigma dx^\sigma$ via

$$T^{(AS)} := \frac{1}{6}G_{\nu\nu'}T^{(AS)\nu'}{}_{\rho\mu}dx^\nu \wedge dx^\rho \wedge dx^\mu = \star(\tilde{T}_\sigma dx^\sigma). \tag{C.0.4}$$

In coordinates, this amounts to

$$T^{(AS)\nu}{}_{\rho\mu} = -\sqrt{|G|}G^{\nu\nu'}\varepsilon_{\nu'\rho\mu\sigma}G^{\sigma\sigma'}\tilde{T}_{\sigma'},$$
$$\varepsilon^{\nu'\rho\mu\kappa}G_{\nu'\nu}T^{(AS)\nu}{}_{\rho\mu} = 6\sqrt{|G|}G^{\kappa\sigma'}\tilde{T}_{\sigma'} \tag{C.0.5}$$

using the conventions $\varepsilon_{0123} = 1 = -\varepsilon^{0123}$ so that

$$\varepsilon_{\nu\rho\sigma\mu}G^{\nu\nu'}G^{\sigma\sigma'}G^{\mu\mu'}G^{\rho\rho'} = |\det G^{\mu\nu}|\varepsilon^{\nu\rho'\sigma\mu'}. \tag{C.0.6}$$

This implies

$$\begin{aligned}
T^{(AS)\sigma}{}_{\rho\mu}T^{(AS)}{}_{\sigma}{}^{\rho}{}_{\nu} &= |G|G^{\sigma\sigma'}\varepsilon_{\sigma'\rho\mu\kappa}G^{\rho\rho'}\varepsilon_{\sigma\rho'\nu\eta}G^{\kappa\kappa'}\tilde{T}_{\kappa'}G^{\eta\eta'}\tilde{T}_{\eta'} \\
&= \varepsilon^{\sigma\rho\mu'\kappa}\varepsilon_{\sigma\rho\nu\eta}G_{\mu'\mu}\tilde{T}_\kappa G^{\eta\eta'}\tilde{T}_{\eta'} \\
&= 2(\tilde{T}_\nu\tilde{T}_\mu - G_{\nu\mu}G^{\kappa\eta}\tilde{T}_\kappa\tilde{T}_\eta)
\end{aligned} \tag{C.0.7}$$

where $|G| = |\det G_{\mu\nu}|$ and the contraction gives

$$T^{(AS)\sigma}{}_{\rho\mu}T^{(AS)}{}_{\sigma}{}^{\rho}{}_{\nu}G^{\mu\nu} = -6\tilde{T}_\nu\tilde{T}_\mu G^{\mu\nu}. \tag{C.0.8}$$

We also note the following identity

$$T^{(AS)\sigma}{}_{\rho\mu} T^{(AS)}{}_{\sigma}{}^{\rho}{}_{\nu} = -2(T_{\mu}{}^{\sigma}{}_{\rho} T^{\rho}{}_{\sigma\nu} + T_{\nu}{}^{\sigma}{}_{\rho} T^{\rho}{}_{\sigma\mu}) + 2(T_{\mu}{}^{\sigma}{}_{\rho} T_{\nu\sigma}{}^{\rho} - T_{\mu}{}^{\sigma}{}_{\rho} T_{\nu}{}^{\rho}{}_{\sigma})$$
$$+ T^{\sigma}{}_{\rho\mu} T_{\sigma}{}^{\rho}{}_{\nu}. \tag{C.0.9}$$

Contracting this with $G^{\mu\nu}$ gives

$$T^{(AS)\sigma}{}_{\rho\mu} T^{(AS)}{}_{\sigma}{}^{\rho}{}_{\nu} G^{\mu\nu} = 3\left(T_{\mu}{}^{\sigma}{}_{\rho} T_{\nu\sigma}{}^{\rho} G^{\mu\nu} - 2 T^{\sigma}{}_{\mu\rho} T_{\sigma}{}^{\rho}{}_{\nu} G^{\mu\nu} \right)$$
$$T^{\sigma}{}_{\mu\rho} T_{\sigma}{}^{\rho}{}_{\nu} G^{\mu\nu} = \frac{1}{2} T_{\mu}{}^{\sigma}{}_{\rho} T_{\nu\sigma}{}^{\rho} G^{\mu\nu} + \tilde{T}_{\nu} \tilde{T}_{\mu} G^{\mu\nu}. \tag{C.0.10}$$

Consider the four-dimensional integral

$$\kappa_{(n)} := \int \frac{d^4 k_\mu}{\sqrt{|G_{\mu\nu}|}} \frac{1}{(k \cdot k + m^2 - i\varepsilon)^n}$$

$$= \rho^{-4} \int \frac{d^4 k_\mu}{\sqrt{|G_{\mu\nu}|}} \frac{1}{(k_\mu k_\nu G^{\mu\nu} + m^2 - i\varepsilon)^n}$$

$$= \rho^{-4} \int d^4 k \frac{1}{(k_\mu k_\nu \eta^{\mu\nu} + m^2 - i\varepsilon)^n}, \tag{D.0.1}$$

where $k \cdot k = \rho^2 G^{\mu\nu} k_\mu k_\nu$ has signature $(-+++)$, using an appropriate redefinition of k in the last step. We can compute this by integrating first over k^0 via a contour rotation $\int dk^0 = i \int dk_E^4$. Then the integral coincides with the Euclidean one up to a factor i, which is

$$\kappa_{(n)} = i\rho^{-4} \int d^4 k_E \frac{1}{(k_\mu k_\nu \eta^{\mu\nu} + m^2)^n} = i 2\pi^2 \rho^{-4} \int_0^\infty dk k^3 \frac{1}{(k^2 + m^2)^n}. \tag{D.0.2}$$

This is well-defined for $n \geq 3$ and gives

$$\kappa_{(4)} = \frac{i\pi^2}{6m^4} \rho^{-4}$$

$$\kappa_{(3)} = \frac{i\pi^2}{2m^2} \rho^{-4}. \tag{D.0.3}$$

Note that the region that contributes significantly to the integral is given by $|k| \lesssim m$. This means that in the trace computations in Section 12, k has effectively a cutoff at m, which should ensure that the integrals are safely within the semi-classical regime. The size L of the semi-classical wave-packets $\psi_{x,k}^{(L)}$ can then be sufficiently large, so that the approximation $k > \frac{1}{L}$ (6.2.69) easily holds for all k that contribute significantly to these integrals.

Appendix E **Functions on coadjoint orbits**

Quantized coadjoint orbits of semi-simple Lie groups provide the prime examples of quantized symplectic spaces. Due to the rigidity of symplectic spaces exhibited in Darboux theorem, they can be considered as universal models underlying all symplectic spaces with the same topology. This justifies their use as models for physical spacetime, and their group-theoretic origin allows an explicit description of the space of functions or harmonics on them, both in the classical and the quantum case.

The mode decomposition on the classical orbit $\mathcal{O}[\Lambda]$

Consider the coadjoint orbit of a semi-simple Lie group G

$$\mathcal{O}[\Lambda] = \{g\Lambda g^{-1}; \ g \in G\} \tag{E.0.1}$$

through some weight $\Lambda \in \mathfrak{g}_0^*$ in the (dual of the) Cartan subalgebra; all (co)adjoint orbits of G are of this structure. They are homogeneous spaces of G, and can be identified with

$$\mathcal{O}[\Lambda] \cong G/K, \tag{E.0.2}$$

where $K \subset G$ is the stabilizer group of Λ. This identification is given by the map

$$G/K \to \mathcal{O}[\Lambda],$$
$$gK \mapsto g\Lambda g^{-1}$$

which is clearly well-defined and bijective. This map is equivariant, in the sense that the adjoint action of G on $\mathcal{O}[\Lambda]$ translates into the *left* action on G/K. Hence, we wish to decompose functions on G/K under the left action of G.

Functions on G/K can be considered as functions on G which are invariant under the *right* action of K. Now the Peter–Weyl theorem states that the space of functions on G is isomorphic as a bimodule to

$$\mathcal{C}(G) \cong \bigoplus_{\lambda \in P^+} \mathcal{H}_\lambda \otimes \mathcal{H}_\lambda^*. \tag{E.0.3}$$

Here λ runs over all dominant integral weights P^+, and $\mathcal{H}_\lambda$ is the corresponding highest-weight module. Let $\mathrm{mult}_{\lambda^+}^{(K)}$ be the dimension of the subspace of $\mathcal{H}_\lambda^* \equiv \mathcal{H}_{\lambda^+}$ which is invariant under the action of K. We thus obtain the following explicit mode decomposition (3.3.85)

$$\mathcal{C}(\mathcal{O}[\Lambda]) \cong \bigoplus_{\lambda \in P^+} \mathrm{mult}_{\lambda^+}^{(K)} \ \mathcal{H}_\lambda. \tag{E.0.4}$$

In the example of $S^2 \cong SU(2)/U(1)_z$ viewed as a coadjoint orbit of $SU(2)$ through (the dual of) L_z, the $\mathcal{H}_\lambda \equiv \mathcal{H}_j$ are irreps with half-integer spin j, and the invariant subspace under the stabilizer $K = U(1)_z$ is given by the states $|j, 0\rangle$ with zero weight $m = 0$, which implies $j = l \in \mathbb{N}$. This leads to (3.3.11)

$$\mathcal{C}(S^2) = \bigoplus_{l\in\mathbb{N}} \mathcal{H}_l. \tag{E.0.5}$$

As a less trivial example, consider $\mathbb{C}P^n \cong SU(n+1)/_{SU(n)\times U(1)} \cong \mathcal{O}[\Lambda_1]$, where Λ_i are the fundamental weights of the Lie algebra $\mathfrak{su}(n+1) \cong A_n$. Then the stabilizer $K \cong SU(n) \times U(1)$ is generated by the root generators $X_j^\pm$ with $j = 2, \ldots, n$ and H_{Λ_1}, noting that $\langle \Lambda_1, \alpha_j \rangle = 0$ for $j = 2, \ldots, n$ and therefore $[H_{\Lambda_1}, X_j^\pm] = 0$. Here α_j are the simple roots. In particular, all Cartan generators of $\mathfrak{su}(n+1)$ are contained in K, so that only states with vanishing weight can be invariant under K. It is not hard to see that the only irreps $\mathcal{H}_\lambda$ that contain states at weight zero that are invariant under K are the self-dual ones with highest weights given by multiples of the highest root $\alpha_1 + \ldots + \alpha_n = \Lambda_1 + \Lambda_n = (1, 0, \ldots, 0, 1)$ in Dynkin notation, so that $\lambda = j(\Lambda_1 + \Lambda_n)$ for $j \in \mathbb{N}$. Then $\mathrm{mult}_{\lambda^+}^{(K)} = 1$, and we obtain (3.3.86)

$$\mathcal{C}(\mathbb{C}P^n) = \bigoplus_{j\in\mathbb{N}} \mathcal{H}_{(j,0,\ldots,0,j)}. \tag{E.0.6}$$

Quantized coadjoint orbits

Consider any given coadjoint orbit $\mathcal{O}[\Lambda]$ of G as above, where $\Lambda \in P^+$ is an integral weight in the fundamental Weyl chamber (this does not restrict the geometry of the orbit). We claim that there is a series of quantized algebras of function (3.3.87)

$$\mathrm{End}(\mathcal{H}_{n\Lambda}) = \mathcal{H}_{n\Lambda} \otimes \mathcal{H}_{n\Lambda}^* \cong \oplus_\mu N_{n\Lambda n\Lambda^+}^\mu \mathcal{H}_\mu \tag{E.0.7}$$

for $n \in \mathbb{N}$, which can be viewed as quantizations of $\mathcal{O}[\Lambda]$. More precisely, there exists an equivariant quantization map

$$\mathcal{Q}: \quad \mathcal{C}(\mathcal{O}(\Lambda)) \to \mathrm{End}(\mathcal{H}_{n\Lambda}), \tag{E.0.8}$$

which is a norm-preserving isomorphism of the modes with sufficiently low $\mu \leq O(n)$, and thus reproduces (E.0.4) for large n. This follows from

$$N_{\lambda\lambda^+}^\mu = \mathrm{mult}_{\mu^+}^{(K_\lambda)} \tag{E.0.9}$$

which holds for sufficiently small μ, i.e. smaller than all *nonvanishing* Dynkin labels of λ. That relation can be shown directly with some effort using Weyl's character formula, see e.g. [153]. However, a much more transparent way to understand (E.0.8) is through coherent states, as explained in Section 4.6.3.

Appendix F **Glossary and notations**

Abbreviations

- ARS: A Yang–Mills-type matrix model in terms of three matrices with cubic term, introduced by Alekseev, Recknagel, and Schomerus [3].
- BFSS: The maximally supersymmetric model of matrix quantum mechanics named after Banks, Fischler, Shenker, and Susskind [21], and introduced by de Wit, Hoppe, and Nicolai [57].
- CCR: Canonical commutation relations.
- CFT: Conformal field theory: a quantum field theory that respects conformal transformations, often involving central extensions of the conformal algebra with nonvanishing central charges.
- GR: General relativity.
- IR: Infrared (=long wavelength).
- Irrep: Irreducible representation.
- IKKT: The maximally supersymmetric matrix model introduced by Ishibashi, Kitazawa, Kawai, and Tsuchiya [102].
- KK: Kaluza–Klein modes, i.e. harmonics on compact extra dimensions.
- MW: Majorana–Weyl.
- NC: Noncommutative.
- NCFT: Noncommutative field theory. This is some analog of field theory living on a noncommutative space, rather than an ordinary manifold.
- NCQFT: Noncommutative quantum field theory. This is some analog of quantum field theory living on a noncommutative space.
- ON: Orthonormal.
- QFT: Quantum field theory.
- SSB: Spontaneous symmetry breaking.
- SYM: Super-Yang–Mills.
- UV: Ultraviolet (=short wavelength).
- VEV: Vacuum expectation value.
- YM: Yang–Mills.

Mathematical symbols

- $\mathcal{C}(\mathcal{M})$: The space of smooth functions on a manifold $\mathcal{M}$.
- $\mathcal{C}_{\mathrm{IR}}(\mathcal{M})$: "Infrared" regime of functions on $\mathcal{M}$. Section 4.3.1.

- $\mathcal{C}$: The space of modes on fuzzy H_n^4 or covariant quantum spacetime $\mathcal{M}^{3,1}$.
- $\mathcal{C}^s$: The space of spin s modes on H_n^4 or covariant quantum spacetime $\mathcal{M}^{3,1}$, defined in (5.2.30).
- $\mathbb{C}^*$: The set of complex numbers without 0, which forms a multiplicative group.
- End(V): Algebra of endomorphisms i.e. linear maps $V \to V$ on a vector space V.
- HS($\mathcal{H}$): The space of Hilbert–Schmidt operators on a Hilbert space $\mathcal{H}$. This is the subspace of End($\mathcal{H}$) consisting of operators with finite Hilbert–Schmidt norm $\|A\|_{\mathrm{HS}} := \sqrt{\mathrm{Tr}(A^\dagger A)}$. It forms a Hilbert space with inner product $\langle A, B \rangle_{\mathrm{HS}} = \mathrm{Tr}(A^\dagger B)$.
- $\mathfrak{hs}$: the vector space generated by the θ^{ab} that generate the higher-spin modes on covariant quantum space(time) or their semi-classical limit, such as H_n^4, S_N^4, or $\mathcal{M}^{3,1}$.
- $\mathcal{L}_V$: Lie derivative along the vector field V.
- Loc($\mathcal{H}$): The space of almost-local functions on an almost-local quantum space, which can be (isometrically) identified with classical functions on the underlying symplectic space. Sections 4.3.1 and 4.4.
- L_{NC}: length scale of noncommutativity, defined by $L_{\mathrm{NC}}^2 = |\det \theta^{\mu\nu}|^{1/2n}$.
- L_{coh}: coherence length on a noncommutative space, defined in terms of the quantum metric by $L_{\mathrm{coh}}^2 = |\det g_{\mu\nu}|^{1/2n}$.
- $\mathbb{PT}$: twistor space, essentially defined as complex projective space $\mathbb{C}P^3$ or $\mathbb{C}P^{1,2}$, possibly with some points removed. Section 5.3.
- $\{f, g\}$: Poisson bracket of functions $f, g \in \mathcal{C}(\mathcal{M})$.
- $[F, G]$: Commutator $[F, G] = FG - GF$ of operators $F, G \in$ End($\mathcal{H}$).
- $|x\rangle$: (quasi-)coherent state localized at x.
- $\|z\rangle$: local holomorphic section of the coherent states. Sections 4.5 and 4.6.3.
- $\eth_a$: derivative operator on fuzzy H_n^4 defined in (5.2.51).

Mathematical and physical concepts

- Brane: In the context of matrix models, these are irreducible matrix configurations, which are solutions of the Yang–Mills matrix models. Reducible solutions are denoted as stacks of branes. Section 7.4.5.

  In string theory, branes are typically submanifolds in target space that are described from the worldsheet point of view in terms of boundary conditions of the CFT, or as solutions of the supergravity theory in target space.
- Equivariant bundle: A bundle $\pi : \mathcal{B} \to \mathcal{M}$ equipped with group actions $G \times \mathcal{B} \to \mathcal{B}$ and $G \times \mathcal{M} \to \mathcal{M}$ compatible with the bundle projection π. This is particularly relevant for covariant quantum spaces. Sections 2.1, 2.2 and Chapter 5.
- Contorsion: A combination of the torsion tensor, which provides the relation between the Weitzenböck connection and the Levi–Civita connection. Section 9.2.
- Fubini–Study metric: The $SU(n)$-invariant metric on $\mathbb{C}P^{n-1}$ that – together with the appropriate symplectic form – makes $\mathbb{C}P^{n-1}$ a Kähler manifold.
- Fuzzy space: A quantized symplectic space defined in terms of matrix configurations X^a, and supporting only finitely many degrees of freedom per volume. Sections 3.3, 3.4, 8.

- Kähler manifold: A symplectic manifold $\mathcal{M}$ with a compatible complex structure, such that $g(u, v) = \omega(u, Jv)$ is a (positive definite) Riemannian metric.
- Kirillov–Kostant symplectic form: The unique G-invariant symplectic form on coadjoint orbits of a Lie group G. Section 2.2.
- Moyal–Weyl quantum plane $\mathbb{R}_\theta^{2n}$: Noncommutative space defined by the commutation relations $[X^a, X^b] = i\theta^{ab}\mathbb{1}$ in terms of an antisymmetric (numerical) matrix θ^{ab}. Section 3.1.
- Quasi-coherent states: Optimally localized states $|x\rangle$ defined for any given matrix configurations as ground states of the displacement Hamiltonian H_x. Section 4.1.
- Quantum metric: Metric on $\mathcal{M}$ which characterizes the absolute value of $\langle x|y\rangle$ between quasi-coherent states, given by the pullback of the Fubini–Study metric on $\mathbb{C}P^N$ via the quasi-coherent states on $\mathcal{M}$. Section 4.2.1.
- Singular set: The set of points in target space of a matrix configuration where the lowest eigenspace of the displacement Hamiltonian is degenerate.
- String mode: Elements $|x\rangle\langle y| \in \text{End}(\mathcal{H})$ given in terms of quasi-coherent states.
- Symplectic form: A closed nondegenerate two-form $\omega \in \Omega^2(\mathcal{M})$ on a manifold $\mathcal{M}$. Section 1.1.
- Symplectomorphism: A smooth map $\Phi : \mathcal{M} \to \mathcal{M}$ on a symplectic manifold $(\mathcal{M}, \omega)$ which preserves the symplectic form. Section 1.1.
- Target space: In the context of matrix models with matrices X^a, $a = 1, \ldots, D$, this is the "virtual" space $\mathbb{R}^D$ where the semi-classical maps $x^a \sim X^a : \mathcal{M} \to \mathbb{R}^D$ take values in. Heuristically, the branes $\mathcal{M} \subset \mathbb{R}^D$ are submanifolds embedded in target space $\mathbb{R}^D$.

  In string theory, the target space $\mathcal{T}$ is the space wherein the string $X : \mathcal{M}^{(2)} \to \mathcal{T}$ propagates, which in the simplest case is $\mathcal{T} = \mathbb{R}^{9,1}$.

  In quantum field theory, one considers more general sigma models in terms of fields $\Phi : \mathcal{M} \to \mathcal{T}$ taking values in some manifold $\mathcal{T}$ considered as target space.
- Torsion: The antisymmetric part of the covariant derivative of the frame. Section 9.2.
- Twistor space $\mathbb{PT}$: loosely defined as complex projective space $\mathbb{C}P^3$ or $\mathbb{C}P^{1,2}$, possibly with some points removed. This is a S^2 bundle over space or spacetime.
- Weitzenböck connection: The unique connection defined in terms of some given frame, with the property that the frame is covariantly constant. Its curvature vanishes, but it has torsion. Section 9.2.
- Weyl invariance: Invariance of a Lagrangian $\mathcal{L}(\phi, g_{\mu\nu})$ for some fields ϕ coupled to the metric $g_{\mu\nu}$ under Weyl rescalings $g_{\mu\nu} \to \alpha g_{\mu\nu}$. This is the case e.g. for two-dimensional scalar field theory with $\mathcal{L} = \int d^2x\sqrt{|g|}g^{\mu\nu}\partial_\mu\phi\partial_\nu\phi$.

References and Further Reading

[1] Adamo, Tim. 2018. Lectures on twistor theory. *PoS*, **Modave2017**, 1–57.

[2] Aldrovandi, Ruben, and Pereira, José Geraldo. 2013. *Teleparallel Gravity: An Introduction*. Springer.

[3] Alekseev, Anton Yu., Recknagel, Andreas, and Schomerus, Volker. 2000. Brane dynamics in background fluxes and noncommutative geometry. *JHEP*, **05**, 010.

[4] Alishahiha, Mohsen, Oz, Yaron, and Sheikh-Jabbari, Mohammmad M. 1999. Supergravity and large N noncommutative field theories. *JHEP*, **11**, 007.

[5] Anagnostopoulos, Konstantinos N., Azuma, Takehiro, and Nishimura, Jun. 2013. Monte Carlo studies of the spontaneous rotational symmetry breaking in dimensionally reduced super Yang–Mills models. *JHEP*, **11**, 009.

[6] Anagnostopoulos, Konstantinos N., Azuma, Takehiro, and Nishimura, Jun. 2016. Monte Carlo studies of dynamical compactification of extra dimensions in a model of nonperturbative string theory. *PoS*, **LATTICE2015**, 307.

[7] Anagnostopoulos, Konstantinos N., Azuma, Takehiro, Ito, Yuta, Nishimura, Jun, and Papadoudis, Stratos Kovalkov. 2018. Complex Langevin analysis of the spontaneous symmetry breaking in dimensionally reduced super Yang–Mills models. *JHEP*, **02**, 151.

[8] Anagnostopoulos, Konstantinos N., Azuma, Takehiro, Ito, Yuta. et al. 2020. Complex Langevin analysis of the spontaneous breaking of 10D rotational symmetry in the Euclidean IKKT matrix model. *JHEP*, **06**, 069.

[9] Anagnostopoulos, Konstantinos N., Azuma, Takehiro, Hatakeyama, Kohta. et al. 2023. Progress in the numerical studies of the type IIB matrix model. *Eur. Phys. J. Spec. Top.*, **10**.

[10] Anous, Tarek, Karczmarek, Joanna L., Mintun, Eric, Van Raamsdonk, Mark, and Way, Benson. 2020. Areas and entropies in BFSS/gravity duality. *SciPost Phys.*, **8**(4), 1–26.

[11] Aoki, Hajime, Ishibashi, Nobuyuki, Iso, Satoshi. et al. 2000. Noncommutative Yang–Mills in IIB matrix model. *Nucl. Phys.*, **B565**, 176–192.

[12] Aoki, Hajime, Iso, Satoshi, Kawai, Hikaru. et al. 1999. IIB matrix model. *Prog. Theor. Phys. Suppl.*, **134**, 47–83.

[13] Asano, Yuhma, and Steinacker, Harold C. 2022. Spherically symmetric solutions of higher-spin gravity in the IKKT matrix model. *Nucl. Phys. B*, **980**, 115843.

[14] Asano, Yuhma, Kawai, Hikaru, and Tsuchiya, Asato. 2012. Factorization of the effective action in the IIB matrix model. *Int. J. Mod. Phys. A*, **27**, 1250089.

[15] Austing, Peter, and Wheater, John F. 2001a. Convergent Yang–Mills matrix theories. *JHEP*, **04**, 019.

[16] Austing, Peter, and Wheater, John F. 2001b. The convergence of Yang–Mills integrals. *JHEP*, **02**, 028.

[17] Azuma, Takehiro, Bal, Subrata, Nagao, Keiichi, and Nishimura, Jun. 2004. Nonperturbative studies of fuzzy spheres in a matrix model with the Chern–Simons term. *JHEP*, **05**, 005.

[18] Azuma, Takehiro, Bal, Subrata, Nagao, Keiichi, and Nishimura, Jun. 2005. Perturbative versus nonperturbative dynamics of the fuzzy $S^2 \times S^2$. *JHEP*, **09**, 047.

[19] Azuma, Takehiro, Bal, Subrata, Nagao, Keiichi, and Nishimura, Jun. 2006. Dynamical aspects of the fuzzy CP^2 in the large N reduced model with a cubic term. *JHEP*, **05**, 061.

[20] Balasin, Herbert, Blaschke, Daniel N., Gieres, Francois, and Schweda, Manfred. 2015. On the energy-momentum tensor in Moyal space. *Eur. Phys. J. C*, **75**(6), 284.

[21] Banks, Tom, Fischler, W., Shenker, Stephen H., and Susskind, Leonard. 1997. M theory as a matrix model: A conjecture. *Phys. Rev.*, **D55**, 5112–5128.

[22] Barrett, John W. 2015. Matrix geometries and fuzzy spaces as finite spectral triples. *J. Math. Phys.*, **56**(8), 082301.

[23] Barrett, John W., and Glaser, Lisa. 2016. Monte Carlo simulations of random noncommutative geometries. *J. Phys. A*, **49**(24), 245001.

[24] Battista, Emmanuele, and Steinacker, Harold C. 2022. On the propagation across the big bounce in an open quantum FLRW cosmology. *Eur. Phys. J. C*, **82**(10), 909.

[25] Battista, Emmanuele, and Steinacker, Harold C. 2023. Fermions on curved backgrounds of matrix models. *Phys. Rev. D*, **107**(4), 046021.

[26] Bayen, F., Flato, Moshe, Fronsdal, Christian, Lichnerowicz, Andre, and Sternheimer, Daniel. 1978. Deformation theory and quantization. 2. Physical applications. *Annals Phys.*, **111**, 111–151.

[27] Bellissard, Jean, van Elst, Andreas, and Schulz-Baldes, Hermann. 1994. The noncommutative geometry of the quantum Hall effect. *J. Math. Phys.*, **35**(10), 5373–5451.

[28] Belyaev, Dmitry V., and Samsonov, Igor B. 2011. Wess-Zumino term in the N=4 SYM effective action revisited. *JHEP*, **04**, 112.

[29] Berenstein, David, and Dzienkowski, Eric. 2012. Matrix embeddings on flat R^3 and the geometry of membranes. *Phys. Rev.*, **D86**, 086001.

[30] Berenstein, David Eliecer, Maldacena, Juan Martin, and Nastase, Horatiu Stefan. 2002. Strings in flat space and pp waves from N=4 super-Yang–Mills. *JHEP*, **04**, 013.

[31] Berman, David S., Campos, Vanicson L., Cederwall, Martin. et al. 2001. Holographic noncommutativity. *JHEP*, **05**, 002.

[32] Berry, Michael V. 1984. Quantal phase factors accompanying adiabatic changes. *Proc. Roy. Soc. Lond. A*, **392**, 45–57.

[33] Bietenholz, Wolfgang, Hofheinz, Frank, and Nishimura, Jun. 2003. Simulating noncommutative field theory. *Nucl. Phys. B Proc. Suppl.*, **119**, 941–946.

[34] Bigatti, Daniela, and Susskind, Leonard. 2000. Magnetic fields, branes and noncommutative geometry. *Phys. Rev. D*, **62**, 066004.

[35] Blaschke, Daniel N., and Steinacker, Harold. 2011. On the 1-loop effective action for the IKKT model and non-commutative branes. *JHEP*, **10**, 120.

[36] Blaschke, Daniel N., Steinacker, Harold, and Wohlgenannt, Michael. 2011. Heat kernel expansion and induced action for the matrix model Dirac operator. *JHEP*, **03**, 002.

[37] Bordemann, Martin, Meinrenken, Eckhard, and Schlichenmaier, Martin. 1994. Toeplitz quantization of Kahler manifolds and $gl(N), N \rightarrow \infty$ limits. *Commun. Math. Phys.*, **165**, 281–296.

[38] Brace, Daniel, Morariu, Bogdan, and Zumino, Bruno. 1999. Dualities of the matrix model from T duality of the Type II string. *Nucl. Phys. B*, **545**, 192–216.

[39] Brahma, Suddhasattwa, Brandenberger, Robert, and Laliberte, Samuel. 2022. Emergent metric space-time from matrix theory. *JHEP*, **09**, 031.

[40] Brezin, Edouard, Itzykson, Claude, Parisi, Giorgio, and Zuber, Jean-Bernard. 1978. Planar diagrams. *Commun. Math. Phys.*, **59**, 35–51.

[41] Carow-Watamura, Ursula, Steinacker, Harold, and Watamura, Satoshi. 2005. Monopole bundles over fuzzy complex projective spaces. *J. Geom. Phys.*, **54**, 373–399.

[42] Castelino, Judith, Lee, Sangmin, and Taylor, Washington. 1998. Longitudinal five-branes as four spheres in matrix theory. *Nucl. Phys.*, **B526**, 334–350.

[43] Cattaneo, Alberto S., and Felder, Giovanni. 2001. Poisson sigma models and deformation quantization. *Mod. Phys. Lett. A*, **16**, 179–190.

[44] Catterall, Simon, and Wiseman, Toby. 2008. Black hole thermodynamics from simulations of lattice Yang–Mills theory. *Phys. Rev. D*, **78**, 041502.

[45] Chamseddine, Ali H., and Connes, Alain. 1997. The spectral action principle. *Commun. Math. Phys.*, **186**, 731–750.

[46] Chatzistavrakidis, Athanasios, Steinacker, Harold, and Zoupanos, George. 2011. Intersecting branes and a standard model realization in matrix models. *JHEP*, **09**, 1–36.

[47] Chepelev, Iouri, and Tseytlin, Arkady A. 1998a. Interactions of type IIB D-branes from D instanton matrix model. *Nucl. Phys.*, **B511**, 629–646.

[48] Chepelev, Iouri, and Tseytlin, Arkady A. 1998b. On membrane interaction in matrix theory. *Nucl. Phys. B*, **524**, 69–85.

[49] Chepelev, Iouri, and Roiban, Radu. 2000. Renormalization of quantum field theories on noncommutative $\mathbb{R}^d$. 1. Scalars. *JHEP*, **05**, 037.

[50] Chu, Chong-Sun, and Ho, Pei-Ming. 1999. Noncommutative open string and D-brane. *Nucl. Phys. B*, **550**, 151–168.

[51] Chu, Chong-Sun, Madore, John, and Steinacker, Harold. 2001. Scaling limits of the fuzzy sphere at one loop. *JHEP*, **08**, 038.

[52] Connes, Alain, and Rieffel, Marc A. 1987. Yang–Mills for noncommutative two-tori. *Contemp. Math.*, **62**, 237–266.

[53] Connes, Alain. 1994. *Noncommutative Geometry*. Academic Press.

[54] Connes, Alain, Douglas, Michael R., and Schwarz, Albert S. 1998. Noncommutative geometry and matrix theory: Compactification on tori. *JHEP*, **02**, 003.

[55] de Badyn, Mathias Hudoba, Karczmarek, Joanna L., Sabella-Garnier, Philippe, and Yeh, Ken Huai-Che. 2015. Emergent geometry of membranes. *JHEP*, **11**, 089.

[56] De Wilde, Marc, and Lecomte, Pierre. 1983. Existence of star-products and of formal deformations of the Poisson Lie algebra of arbitrary symplectic manifolds. *Lett. Math. Phys.*, **7**(6), 487–496.

[57] de Wit, Bernard, Hoppe, Jens, and Nicolai, Hermann. 1988. On the quantum mechanics of supermembranes. *Nucl. Phys.*, **B305**, 545–581.

[58] Di Francesco, Philippe, Ginsparg, Paul H., and Zinn-Justin, Jean. 1995. 2-D gravity and random matrices. *Phys. Rept.*, **254**, 1–133.

[59] Didenko, Vyacheslav E., and Skvortsov, Evgeny D. 2014. Elements of Vasiliev theory. 1–61. arXiv:1401.2975.

[60] Dixmier, Jacques. 1961. Représentations intégrables du groupe de De Sitter. *Bull. Soc. Math. Fr.*, **89**, 9–41.

[61] Dong, Zhihuan, and Senthil, T. 2020. Noncommutative field theory and composite Fermi liquids in some quantum Hall systems. *Phys. Rev. B*, **102**(20), 205126.

[62] Doplicher, Sergio, Fredenhagen, Klaus, and Roberts, John E. 1995. The quantum structure of space-time at the Planck scale and quantum fields. *Commun. Math. Phys.*, **172**, 187–220.

[63] Douglas, Michael R., and Nekrasov, Nikita A. 2001. Noncommutative field theory. *Rev. Mod. Phys.*, **73**, 977–1029.

[64] Douglas, Michael R., and Taylor, Washington. 1998. Branes in the bulk of Anti-de Sitter space. arXiv:hep-th/9807225.

[65] Drinfeld, Vladimir G. 1986. Quantum groups. *Zap. Nauchn. Semin.*, **155**, 18–49.

[66] Dubois-Violette, Michel, Kerner, Richard, and Madore, John. 1990. Noncommutative differential geometry of matrix algebras. *J. Math. Phys.*, **31**, 316–322.

[67] Dunne, Gerald, and Jackiw, Roman. 1993. "Peierls substitution" and Chern–Simons quantum mechanics. *Nucl. Phys. B Proc. Suppl.*, **33**(3), 114–118.

[68] Eguchi, Tohru, and Kawai, Hikaru. 1982. Reduction of dynamical degrees of freedom in the large N Gauge theory. *Phys. Rev. Lett.*, **48**, 1063–1066.

[69] Faddeev, Ludvig D., Reshetikhin, Nicolai Yu., and Takhtajan, Leon A. 1989. Quantization of Lie groups and Lie algebras. *Alg. Anal.*, **1**(1), 178–206.

[70] Felder, Laurin J., and Steinacker, Harold C. 2023. Oxidation, reduction and semi-classical limit for quantum matrix geometries. 1–25. e-Print: 2306.10771.

[71] Filev, Veselin G., and O'Connor, Denjoe. 2016. The BFSS model on the lattice. *JHEP*, **05**, 167.

[72] Filk, Thomas. 1996. Divergencies in a field theory on quantum space. *Phys. Lett. B*, **376**, 53–58.

[73] Fredenhagen, Stefan, and Steinacker, Harold C. 2021. Exploring the gravity sector of emergent higher-spin gravity: Effective action and a solution. *JHEP*, **05**, 183.

[74] Fronsdal, Christian. 1978. Massless fields with integer spin. *Phys. Rev.*, **D18**, 3624.

[75] Fukuma, Masafumi, Kawai, Hikaru, Kitazawa, Yoshihisa, and Tsuchiya, Asato. 1998. String field theory from IIB matrix model. *Nucl. Phys. B*, **510**, 158–174.

[76] Ganor, Ori J., Ramgoolam, Sanjaye, and Taylor, Washington. 1997. Branes, fluxes and duality in M(atrix) theory. *Nucl. Phys. B*, **492**, 191–204.

[77] Gilkey, Peter B. 1995. *Invariance Theory: The Heat Equation and the Atiyah–Singer Index Theorem*. Online access: EMIS: Electronic Library of Mathematics. CRC-Press.

[78] Gilmore, Robert. 2008. *Lie Groups, Physics, and Geometry: An Introduction for Physicists, Engineers and Chemists*. Cambridge University Press.

[79] Giulini, Domenico. 2003. That strange procedure called quantization. *Lect. Notes Phys.*, **631**, 17–40.

[80] Gonzalez-Arroyo, Antonio, and Okawa, Masanori. 1983. The twisted Eguchi–Kawai model: A reduced model for large N lattice gauge theory. *Phys. Rev. D*, **27**, 2397.

[81] Grosse, Harald, and Madore, John. 1992. A noncommutative version of the Schwinger model. *Phys. Lett. B*, **283**, 218–222.

[82] Grosse, Harald, and Presnajder, Peter. 1993. The construction on noncommutative manifolds using coherent states. *Lett. Math. Phys.*, **28**, 239–250.

[83] Grosse, Harald, Klimcik, Ctirad, and Presnajder, Peter. 1996a. On finite 4-D quantum field theory in noncommutative geometry. *Commun. Math. Phys.*, **180**, 429–438.

[84] Grosse, Harald, Klimcik, Ctirad, and Presnajder, Peter. 1996b. Towards finite quantum field theory in noncommutative geometry. *Int. J. Theor. Phys.*, **35**, 231–244.

[85] Grosse, Harald, and Wulkenhaar, Raimar. 2005. Renormalization of ϕ^4 theory on noncommutative R^4 in the matrix base. *Commun. Math. Phys.*, **256**, 305–374.

[86] Grosse, Harald, and Wulkenhaar, Raimar. 2014. Self-dual noncommutative ϕ^4 -theory in four dimensions is a non-perturbatively solvable and non-trivial quantum field theory. *Commun. Math. Phys.*, **329**, 1069–1130.

[87] Gubser, Steven S., and Sondhi, Shivaji L. 2001. Phase structure of noncommutative scalar field theories. *Nucl. Phys. B*, **605**, 395–424.

[88] Guhr, Thomas, Muller-Groeling, Axel, and Weidenmuller, Hans A. 1998. Random matrix theories in quantum physics: Common concepts. *Phys. Rept.*, **299**, 189–425.

[89] Gutt, Simone. 1979. Equivalence of deformations and associated*-products. *Lett. Math. Phys.*, **3**(4), 297–309.

[90] Gutt, Simone, and Rawnsley, John. 1999. Equivalence of star products on a symplectic manifold: An introduction to Deligne's Čech cohomology classes. *J. Geom. Phys.*, **29**(4), 347–392.

[91] Hanada, Masanori, and Shimada, Hidehiko. 2015. On the continuity of the commutative limit of the 4d N=4 non-commutative super Yang–Mills theory. *Nucl. Phys.*, **B892**, 449–474.

[92] Hanada, Masanori, Kawai, Hikaru, and Kimura, Yusuke. 2006. Describing curved spaces by matrices. *Prog. Theor. Phys.*, **114**, 1295–1316.

[93] Hanada, Masanori, Hyakutake, Yoshifumi, Ishiki, Goro, and Nishimura, Jun. 2014. Holographic description of quantum black hole on a computer. *Science*, **344**, 882–885.

[94] Hasebe, Kazuki. 2012. Non-compact Hopf maps and fuzzy ultra-hyperboloids. *Nucl. Phys.*, **B865**, 148–199.

[95] Hawkins, Eli. 1999. Quantization of equivariant vector bundles. *Commun. Math. Phys.*, **202**, 517–546.

[96] Heckman, Jonathan, and Verlinde, Herman. 2015. Covariant non-commutative space–time. *Nucl. Phys.*, **B894**, 58–74.

[97] Heckman, Jonathan J., and Verlinde, Herman. 2011. Instantons, twistors, and emergent gravity. 1–64. e-Print: 1112.5210.

[98] Heidenreich, W. 1981. Tensor products of positive energy representations of SO(3,2) and SO(4,2). *J. Math. Phys.*, **22**, 1566–1574.

[99] Hellerman, Simeon, and Van Raamsdonk, Mark. 2001. Quantum Hall physics equals noncommutative field theory. *JHEP*, **10**, 039.

[100] Ho, Pei-Ming, and Li, Miao. 2001. Fuzzy spheres in AdS / CFT correspondence and holography from noncommutativity. *Nucl. Phys.*, **B596**, 259–272.

[101] Hoppe, Jens R. 1982. *Quantum Theory of a Massless Relativistic Surface and a Two-Dimensional Bound State Problem*. Ph.D. thesis, Massachusetts Institute of Technology.

[102] Ishibashi, Nobuyuki, Kawai, Hikaru, Kitazawa, Yoshihisa, and Tsuchiya, Asato. 1997. A Large N reduced model as superstring. *Nucl. Phys.*, **B498**, 467–491.

[103] Ishiki, Goro. 2015. Matrix geometry and coherent states. *Phys. Rev.*, **D92**(4), 046009.

[104] Ishiki, Goro, Matsumoto, Takaki, and Muraki, Hisayoshi. 2016. Kähler structure in the commutative limit of matrix geometry. *JHEP*, **08**, 042.

[105] Iso, Satoshi, Kawai, Hikaru, and Kitazawa, Yoshihisa. 2000. Bilocal fields in noncommutative field theory. *Nucl. Phys.*, **B576**, 375–398.

[106] Iso, Satoshi, Kimura, Yusuke, Tanaka, Kanji, and Wakatsuki, Kazunori. 2001. Noncommutative gauge theory on fuzzy sphere from matrix model. *Nucl. Phys.*, **B604**, 121–147.

[107] Jack, Ian, and Jones, D. R. Timothy. 2001. Ultraviolet finiteness in noncommutative supersymmetric theories. *New J. Phys.*, **3**, 1–9.

[108] Jimbo, Michio. 1985. A q-difference analogue of U(g) and the Yang–Baxter equation. *Lett. Math. Phys.*, **10**(1), 63–69.

[109] Jurco, Branislav, and Schupp, Peter. 2000. Noncommutative Yang–Mills from equivalence of star products. *Eur. Phys. J. C*, **14**, 367–370.

[110] Jurco, Branislav, Moller, Lutz, Schraml, Stefan, Schupp, Peter, and Wess, Julius. 2001. Construction of non-Abelian gauge theories on noncommutative spaces. *Eur. Phys. J. C*, **21**, 383–388.

[111] Karabali, Dimitra, and Nair, Velayudhan Parameswaran. 2002. Quantum Hall effect in higher dimensions. *Nucl. Phys.*, **B641**, 533–546.

[112] Karabali, Dimitra, and Nair, Velayudhan Parameswaran. 2004. The effective action for edge states in higher dimensional quantum Hall systems. *Nucl. Phys.*, **B679**, 427–446.

[113] Karabali, Dimitra, and Nair, Velayudhan Parameswaran. 2006. Quantum Hall effect in higher dimensions, matrix models and fuzzy geometry. *J. Phys.*, **A39**, 12735–12764.

[114] Karczmarek, Joanna L., and Sabella-Garnier, Philippe. 2014. Entanglement entropy on the fuzzy sphere. *JHEP*, **03**, 129.

[115] Karczmarek, Joanna L., and Steinacker, Harold C. 2023. Cosmic time evolution and propagator from a Yang–Mills matrix model. *J. Phys. A*, **56**(17), 175401.

[116] Karczmarek, Joanna L., and Yeh, Ken Huai-Che. 2015. Noncommutative spaces and matrix embeddings on flat $\mathbb{R}^{2n+1}$. *JHEP*, **11**, 146.

[117] Kawahara, Naoyuki, Nishimura, Jun, and Takeuchi, Shingo. 2007. High temperature expansion in supersymmetric matrix quantum mechanics. *JHEP*, **12**, 103.

[118] Kim, Sang-Woo, Nishimura, Jun, and Tsuchiya, Asato. 2012. Expanding (3+1)-dimensional universe from a Lorentzian matrix model for superstring theory in (9+1)-dimensions. *Phys. Rev. Lett.*, **108**, 011601.

[119] Kirillov, Aleksandr Aleksandrovich. 2004. *Lectures on the Orbit Method*. Vol. 64. American Mathematical Society.

[120] Kitazawa, Yoshihisa. 2002. Vertex operators in IIB matrix model. *JHEP*, **04**, 004.

[121] Kitazawa, Yoshihisa, and Nagaoka, Satoshi. 2007. Graviton propagators in supergravity and noncommutative gauge theory. *Phys. Rev. D*, **75**, 046007.

[122] Klinkhamer, Frans R. 2021. IIB matrix model: Emergent spacetime from the master field. *PTEP*, **2021**(1), 013B04.

[123] Kolb, Edward W. 1989. A coasting cosmology. *Astrophys. J.*, **344**, 543–550.

[124] Kontsevich, Maxim. 2003. Deformation quantization of Poisson manifolds. 1. *Lett. Math. Phys.*, **66**, 157–216.

[125] Kováčik, Samuel, and O'Connor, Denjoe. 2018. Triple point of a scalar field theory on a fuzzy sphere. *JHEP*, **10**, 010.

[126] Krasnov, Kirill. 2020. *Formulations of General Relativity*. Cambridge Monographs on Mathematical Physics. Cambridge University Press.

[127] Krauth, Werner, and Staudacher, Matthias. 1998. Finite Yang-Mills integrals. *Phys. Lett.*, **B435**, 350–355.

[128] Laliberte, Samuel, and Brahma, Suddhasattwa. 2023. IKKT thermodynamics and early universe cosmology. *JHEP*, **11**, 161.

[129] Langmann, Edwin, and Szabo, Richard J. 2002. Duality in scalar field theory on noncommutative phase spaces. *Phys. Lett. B*, **533**, 168–177.

[130] Lukierski, Jerzy, Ruegg, Henri, and Zakrzewski, Wojciech J. 1995. Classical quantum mechanics of free kappa relativistic systems. *Annals Phys.*, **243**, 90–116.

[131] Madore, John. 1992. The fuzzy sphere. *Class. Quant. Grav.*, **9**, 69–88.

[132] Madore, John. 1999. *An Introduction to Noncommutative Differential Geometry and Its Physical Applications*. Vol. 257. Cambridge University Press.

[133] Madore, John, Schraml, Stefan, Schupp, Peter, and Wess, Julius. 2000. Gauge theory on noncommutative spaces. *Eur. Phys. J.*, **C16**, 161–167.

[134] Maldacena, Juan Martin. 1999. The large N limit of superconformal field theories and supergravity. *Int. J. Theor. Phys.*, **38**, 1113–1133.

[135] Maldacena, Juan Martin, and Russo, Jorge G. 1999. Large N limit of noncommutative gauge theories. *JHEP*, **09**, 025.

[136] Maldacena, Juan Martin, Sheikh-Jabbari, Mohammad M., and Van Raamsdonk, Mark. 2003. Transverse five-branes in matrix theory. *JHEP*, **01**, 038.

[137] Marsden, Jerrold E., and Ratiu, Tudor S. 2013. *Introduction to Mechanics and Symmetry: A Basic Exposition of Classical Mechanical Systems*. Vol. 17. Springer Science & Business Media.

[138] Matusis, Alec, Susskind, Leonard, and Toumbas, Nicolaos. 2000. The IR/UV connection in the noncommutative gauge theories. *JHEP*, **12**, 002.

[139] Mehta, Madan Lal. 2004. *Random Matrices*. Elsevier.

[140] Melia, Fulvio, and Shevchuk, Andrew. 2012. The $R_h = ct$ universe. *Mon. Not. Roy. Astron. Soc.*, **419**, 2579–2586.

[141] Minwalla, Shiraz, Van Raamsdonk, Mark, and Seiberg, Nathan. 2000. Noncommutative perturbative dynamics. *JHEP*, **02**, 020.

[142] Nair, Velayudhan Parameswaran. 2022. Remarks on entanglement for fuzzy geometry and gravity. arXiv:2207.04776.

[143] Nair, Velayudhan Parameswaran, Polychronakos, Alexios P., and Tekel, Jurai. 2012. Fuzzy spaces and new random matrix ensembles. *Phys. Rev. D*, **85**, 045021.

[144] Nekrasov, Nikita, and Schwarz, Albert S. 1998. Instantons on noncommutative $\mathbb{R}^4$ and $(2, 0)$ superconformal six-dimensional theory. *Commun. Math. Phys.*, **198**, 689–703.

[145] Nishimura, Jun. 2022. Signature change of the emergent space-time in the IKKT matrix model. *PoS*, **CORFU2021**, 255.

[146] Nishimura, Jun, and Sugino, Fumihiko. 2002. Dynamical generation of four-dimensional space-time in the IIB matrix model. *JHEP*, **05**, 001.

[147] Nishimura, Jun, and Tsuchiya, Asato. 2019. Complex Langevin analysis of the space-time structure in the Lorentzian type IIB matrix model. *JHEP*, **06**, 077.

[148] Nishimura, Jun, Okubo, Toshiyuki, and Sugino, Fumihiko. 2011. Systematic study of the SO(10) symmetry breaking vacua in the matrix model for type IIB superstrings. *JHEP*, **10**, 135.

[149] O'Connor, Denjoe, and Ydri, Badis. 2006. Monte Carlo simulation of a NC gauge theory on the fuzzy sphere. *JHEP*, **11**, 016.

[150] Okawa, Yuji, and Yoneya, Tamiaki. 1999. Multibody interactions of D particles in supergravity and matrix theory. *Nucl. Phys. B*, **538**, 67–99.

[151] Okuno, Shizuka, Suzuki, Mariko, and Tsuchiya, Asato. 2016. Entanglement entropy in scalar field theory on the fuzzy sphere. *PTEP*, **2016**(2), 023B03.

[152] Panero, Marco. 2016. The numerical approach to quantum field theory in a non-commutative space. *PoS*, **CORFU2015**, 099.

[153] Pawelczyk, Jacek, and Steinacker, Harold. 2002. A quantum algebraic description of D branes on group manifolds. *Nucl. Phys.*, **B638**, 433–458.

[154] Pawelczyk, Jacek, and Steinacker, Harold. 2003. Algebraic brane dynamics on SU(2): Excitation spectra. *JHEP*, **12**, 010.

[155] Perelomov, Askold M. 1986. *Generalized Coherent States and Their Applications.* Springer.

[156] Peskin, Michael E., and Schroeder, Daniel V. 1995. *An Introduction to Quantum Field Theory.* Addison-Wesley.

[157] Polychronakos, Alexios P. 2000. Flux tube solutions in noncommutative gauge theories. *Phys. Lett. B*, **495**, 407–412.

[158] Polychronakos, Alexios P. 2001. Quantum Hall states as matrix Chern–Simons theory. *JHEP*, **04**, 011.

[159] Polychronakos, Alexios P. 2013. Effective action and phase transitions of scalar field on the fuzzy sphere. *Phys. Rev. D*, **88**, 065010.

[160] Polychronakos, Alexios P., Steinacker, Harold, and Zahn, Jochen. 2013. Brane compactifications and 4-dimensional geometry in the IKKT model. *Nucl. Phys. B*, **875**, 566–598.

[161] Ramgoolam, Sanjaye. 2001. On spherical harmonics for fuzzy spheres in diverse dimensions. *Nucl. Phys.*, **B610**, 461–488.

[162] Rivelles, Victor O. 2003. Noncommutative field theories and gravity. *Phys. Lett.*, **B558**, 191–196.

[163] Sakharov, Andrej D. 1968. Vacuum quantum fluctuations in curved space and the theory of gravitation. *Sov. Phys. Dokl.*, **12**, 1040–1041.

[164] Schleich, Wolfgang P. 2011. *Quantum Optics in Phase Space.* John Wiley.

[165] Schlichenmaier, Martin. 2018. Berezin–Toeplitz quantization and naturally defined star products for Kähler manifolds. *Anal. Math. Phys.*, **8**(4), 691–710.

[166] Schneiderbauer, Lukas. 2016. *BProbe: A Wolfram Mathematica Package.* http://dx.doi.org/10.5281/zenodo.45045

[167] Schneiderbauer, Lukas, and Steinacker, Harold C. 2016. Measuring finite quantum geometries via quasi–coherent states. *J. Phys.*, **A49**(28), 285301.

[168] Schomerus, Volker. 1999. D-branes and deformation quantization. *JHEP*, **06**, 030.

[169] Schreivogl, Paul, and Steinacker, Harold. 2013. Generalized fuzzy torus and its modular properties. *SIGMA*, **9**, 060.

[170] Seiberg, Nathan, and Witten, Edward. 1999. String theory and noncommutative geometry. *JHEP*, **09**, 032.

[171] Sinova, Jairo, Meden, V., and Girvin, S. M. 2000. Liouvillian approach to the integer quantum Hall effect transition. *Phys. Rev. B*, **62**(3), 2008–2015.

[172] Snyder, Hartland S. 1947. Quantized space-time. *Phys. Rev.*, **71**, 38–41.

[173] Sperling, Marcus, and Steinacker, Harold C. 2018. Intersecting branes, Higgs sector, and chirality from $\mathcal{N} = 4$ SYM with soft SUSY breaking. *JHEP*, **04**, 116.

[174] Sperling, Marcus, and Steinacker, Harold C. 2019a. Covariant cosmological quantum space-time, higher-spin and gravity in the IKKT matrix model. *JHEP*, **07**, 010.

[175] Sperling, Marcus, and Steinacker, Harold C. 2019b. The fuzzy 4-hyperboloid H_n^4 and higher-spin in Yang–Mills matrix models. *Nucl. Phys.*, **B941**, 680–743.

[176] Steinacker, Harold. 2004. Quantized gauge theory on the fuzzy sphere as random matrix model. *Nucl. Phys.*, **B679**, 66–98.

[177] Steinacker, Harold. 2005. A non-perturbative approach to non-commutative scalar field theory. *JHEP*, **03**, 075.

[178] Steinacker, Harold. 2007. Emergent gravity from noncommutative gauge theory. *JHEP*, **12**, 049.

[179] Steinacker, Harold. 2009a. Covariant field equations, gauge fields and conservation laws from Yang-Mills matrix models. *JHEP*, **02**, 044.

[180] Steinacker, Harold. 2009b. Emergent gravity and noncommutative branes from Yang–Mills matrix models. *Nucl. Phys.*, **B810**, 1–39.

[181] Steinacker, Harold. 2010. Emergent geometry and gravity from matrix models: An introduction. *Class. Quant. Grav.*, **27**, 133001.

[182] Steinacker, Harold, and Tran, Tung. 2023. Spinorial higher-spin gauge theory from IKKT in Euclidean and Minkowski signatures. *JHEP*, **12**, 010.

[183] Steinacker, Harold C. 2015. One-loop stabilization of the fuzzy four-sphere via softly broken SUSY. *JHEP*, **12**, 115.

[184] Steinacker, Harold C. 2016. String states, loops and effective actions in noncommutative field theory and matrix models. *Nucl. Phys.*, **B910**, 346–373.

[185] Steinacker, Harold C. 2018. Cosmological space-times with resolved Big Bang in Yang–Mills matrix models. *JHEP*, **02**, 033.

[186] Steinacker, Harold C. 2019. Scalar modes and the linearized Schwarzschild solution on a quantized FLRW space-time in Yang–Mills matrix models. *Class. Quant. Grav.*, **36**(20), 205005.

[187] Steinacker, Harold C. 2020. Higher-spin gravity and torsion on quantized space-time in matrix models. *JHEP*, **04**, 111.

[188] Steinacker, Harold C. 2021a. Higher-spin kinematics & no ghosts on quantum space-time in Yang–Mills matrix models. *Adv. Theor. Math. Phys.*, **25**(4), 1025–1093.

[189] Steinacker, Harold C. 2021b. Quantum (matrix) geometry and quasi-coherent states. *J. Phys. A*, **54**(5), 055401.

[190] Steinacker, Harold C., and Tekel, Juraj. 2022. String modes, propagators and loops on fuzzy spaces. *JHEP*, **06**, 136.

[191] Steinacker, Harold C., and Tran, Tung. 2022. A twistorial description of the IKKT-matrix model. *JHEP*, **11**, 146.

[192] Steinacker, Harold C., and Zahn, Jochen. 2015. Self-intersecting fuzzy extra dimensions from squashed coadjoint orbits in $\mathcal{N} = 4$ SYM and matrix models. *JHEP*, **02**, 027.

[193] Susskind, Leonard. 2001. The quantum Hall fluid and noncommutative Chern–Simons theory. arXiv:hep-th/0101029.

[194] Szabo, Richard J. 2003. Quantum field theory on noncommutative spaces. *Phys. Rept.*, **378**, 207–299.

[195] Szabo, Richard J., and Tirelli, Michelangelo. 2022. Noncommutative instantons in diverse dimensions. arXiv:2207.12862.

[196] 't Hooft, Gerard. 1974. A planar diagram theory for strong interactions. *Nucl. Phys. B*, **72**, 461–473.

[197] Taylor, Washington. 1997. D-brane field theory on compact spaces. *Phys. Lett. B*, **394**, 283–287.

[198] Taylor, Washington. 2001. M(atrix) theory: Matrix quantum mechanics as a fundamental theory. *Rev. Mod. Phys.*, **73**, 419–462.

[199] Taylor, Washington, and Van Raamsdonk, Mark. 1999. Multiple D0-branes in weakly curved backgrounds. *Nucl. Phys. B*, **558**, 63–95.

[200] Taylor, Washington, and Van Raamsdonk, Mark. 2000. Multiple Dp-branes in weak background fields. *Nucl. Phys. B*, **573**, 703–734.

[201] Tekel, Juraj. 2015. Matrix model approximations of fuzzy scalar field theories and their phase diagrams. *JHEP*, **12**, 176.

[202] Tropper, Adam, and Wang, Tianli. 2023. Lorentz symmetry and IR structure of the BFSS matrix model. *JHEP*, **07**, 150.

[203] Tseytlin, Arkady A. 2000. Born–Infeld action, supersymmetry and string theory. In: *The Many Faces of the Superworld*. Vol. 2. World Scientific (Singapore), pp. 417–452.

[204] Tseytlin, Arkady A., and Zarembo, K. 2000. Magnetic interactions of D-branes and Wess–Zumino terms in super Yang–Mills effective actions. *Phys. Lett. B*, **474**, 95–102.

[205] Vaidya, Sachindeo. 2001. Perturbative dynamics on the fuzzy S^2 and $\mathbb{R}P^2$. *Phys. Lett. B*, **512**, 403–411.

[206] Valenzuela, Mauricio. 2015. From phase space to multivector matrix models. *J. Math. Phys.*, **59**, 062302.

[207] Van Suijlekom, Walter D. 2015. *Noncommutative Geometry and Particle Physics*. Springer.

[208] Vasiliev, Mikhail A. 1990. Consistent equation for interacting gauge fields of all spins in (3+1)-dimensions. *Phys. Lett.*, **B243**, 378–382.

[209] Vassilevich, Dmitri V. 2003. Heat kernel expansion: User's manual. *Phys. Rept.*, **388**, 279–360.

[210] Voisin, Claire. 2003. *Hodge Theory and Complex Algebraic Geometry II*. Vol. 2. Cambridge University Press.

[211] Waldmann, Stefan. 2007. *Poisson–Geometrie*. Springer.

[212] Wess, Julius, and Zumino, Bruno. 1991. Covariant differential calculus on the quantum hyperplane. *Nucl. Phys. B Proc. Suppl.*, **18**, 302–312.

[213] Woronowicz, Stanislaw L. 1987. Compact matrix pseudogroups. *Commun. Math. Phys.*, **111**, 613–665.

[214] Yang, Chen N. 1947. On quantized space-time. *Phys. Rev.*, **72**, 874.

[215] Yang, Hyun Seok. 2009a. Emergent gravity from noncommutative spacetime. *Int. J. Mod. Phys.*, **A24**, 4473–4517.

[216] Yang, Hyun Seok. 2009b. Emergent spacetime and the origin of gravity. *JHEP*, **05**, 012.

[217] Zhang, Shou-Cheng, and Hu, Jiang-ping. 2001. A four-dimensional generalization of the quantum Hall effect. *Science*, **294**, 823–828.

Index